Springer Proceedings in Physics

Volume 239

Indexed by Scopus

The series Springer Proceedings in Physics, founded in 1984, is devoted to timely reports of state-of-the-art developments in physics and related sciences. Typically based on material presented at conferences, workshops and similar scientific meetings, volumes published in this series will constitute a comprehensive up-to-date source of reference on a field or subfield of relevance in contemporary physics. Proposals must include the following:

- name, place and date of the scientific meeting
- a link to the committees (local organization, international advisors etc.)
- scientific description of the meeting
- list of invited/plenary speakers
- an estimate of the planned proceedings book parameters (number of pages/ articles, requested number of bulk copies, submission deadline).

More information about this series at http://www.springer.com/series/361

Alvaro Ferraz · Kumar S. Gupta ·
Gordon Walter Semenoff ·
Pasquale Sodano
Editors

Strongly Coupled Field Theories for Condensed Matter and Quantum Information Theory

Proceedings, International Institute of Physics, Natal, Rn, Brazil, 2–21 August 2015

Springer

Editors
Alvaro Ferraz
International Institute of Physics-UFRN
Natal-Rn, Rio Grande do Norte, Brazil

Kumar S. Gupta
Theory Division
Saha Institute of Nuclear Physics
Kolkata, India

Gordon Walter Semenoff
Department of Physics and Astronomy
University of British Columbia
Vancouver, BC, Canada

Pasquale Sodano
International Institute of Physics-UFRN
Natal-Rn, Rio Grande do Norte, Brazil

ISSN 0930-8989 ISSN 1867-4941 (electronic)
Springer Proceedings in Physics
ISBN 978-3-030-35472-5 ISBN 978-3-030-35473-2 (eBook)
https://doi.org/10.1007/978-3-030-35473-2

This Springer imprint is published by the registered company Springer Nature Switzerland AG
The registered company address is: Gewerbestrasse 11, 6330 Cham, Switzerland

Why with the time do I not glance aside
to new-found methods and to compounds
strange?
William Shakespeare,
Sonnet 76.

Preface

The discovery and the control of new quantum behaviors of strongly interacting systems is by now a crucial issue for the advancement of the quantum technologies needed for quantum information processing and quantum devices. The aim of engineering and manipulating new classes of quantum devices poses, at the same time, many fundamental science problems since one has to describe the collective behavior of strongly interacting quantum particles when they are confined to pertinent geometries and topologies at the nanometer scale.

It is well known that, as one proceeds to smaller and smaller length scales, quantum properties can be substantially modified by the presence of interactions and correlations leading to the emergence of non-perturbative regimes which cannot be described by conventional mean-field theory approaches. In their most spectacular form, interactions at the nanoscale lead to strongly correlated states of matter already evidenced in fractional quantum Hall fluids, high critical temperature superconductors, some topological insulators and superconductors, graphene bilayers, and in "synthetic materials" such as Josephson superconducting networks and arrays, or junctions of quantum wires and one-dimensional ultracold atomic systems. In many instances, non-perturbative, as well as counterintuitive, properties of these materials may be used to engineer new platforms for the solid-state implementation of quantum devices.

Impurities are ubiquitous in condensed matter systems: the discovery of the Kondo effect and of Anderson localization have brought to the forefront their relevance for stabilizing new and unexpected behaviors of a quantum system. When realized in a condensed matter system, such features are expected as emergent phenomena. There is, indeed, growing evidence that by judiciously engineering the coupling of impurities to their environment, i.e., to the other modes of the condensed matter system containing the impurity, one can:

(A) Set new stable phases, frustrate decoherence, and facilitate enhanced responses to external perturbations in quantum devices;
(B) Engineer devices useful for topological quantum computation;

(C) Generate long-distance correlations (entanglement) and improve the efficiency of quantum communication in spin systems;
(D) Apply to the engineering of solid-state devices analytical methods recently developed in string and boundary conformal field theories.

Paradigmatic examples of experimentally well controllable systems, where Kondo-like impurities lead to the emergence of new behaviors relevant for their applications to quantum devices, are provided by spin chains, ultracold atoms trapped in optical lattices, anti-dots in fractional quantum Hall (FQH) systems, and junctions of either quantum wires or superconducting Josephson junction arrays. For these systems, boundary field theories together with the Tomonaga–Luttinger liquid description have been providing useful insights for the existence of new fixed points escaping conventional mean-field theory analyses as well as for establishing remarkable connections with Kondo physics.

It is by now well known that progresses in condensed matter physics and other disciplines may influence each other. Long ago, the study of superconductivity has been closely related to structures underlying the standard model of elementary particle physics such as spontaneous symmetry breaking and the Higgs mechanism. A generation later, the understanding of renormalization in field theories describing elementary particles was beautifully mirrored in an almost complete solution of the theory of critical phenomena and second-order phase transition.

In more recent years, in addition to quantum field theory, some new approaches have entered the fray. In fact, condensed matter theory has been more and more informed by string theory, through the holographic study of strongly coupled systems. This approach maps certain quantum dynamical systems in their strong coupling limit onto purely classical gravitational systems in higher dimensions; questions about the highly correlated deep quantum limit of the dynamical system are then answered by solving the Einstein equations which govern classical gravity. This has the promise of a quantitative approach to certain quantum systems in the limit where the interactions among their constituents are very strong, to the point where their analysis is inaccessible to any other known analytic technique short of brute force numerical simulations. Not only would this strongly correlated limit be analyzable in a quantitative way, but it would be systematically correctable. Whether it can be used to study realistic models of condensed matter systems is still an open question which is currently under investigation. At the very least, this approach will provide models which give a qualitative description of entirely new states of matter and, more importantly, new paradigms which should aid physicists in understanding and classifying the behavior of such systems. This idea is relevant since some unsolved problems in condensed matter physics—such as high-temperature superconductivity or heavy fermion systems—almost invariably involve strong coupling physics and perhaps new states of matter.

Condensed matter physics shows also an interesting interface with quantum information theory. In fact, condensed matter physicists are gradually learning that, in a study of the states of a quantum mechanical system with a large number of degrees of freedom, there is invariably an advantage to be gained in taking into account the quantum information that is stored in the states, and in describing the states themselves in an information-theoretic framework. The storage, transport, and manipulation of quantum information have become relevant issues in modern condensed matter physics; for instance, systems with minimal environmental decoherence will no doubt lead to many interesting developments for condensed matter. Moreover, one of the most utile diagnostics of the storage of quantum information, the entanglement entropy, in strongly coupled systems is, these days, in many examples, computed using a string theory holographic formula.

Another line of development is a new interface with mathematics and considers the classification of the topological phases of matter such as topological insulators and topological superconductors. The K-theoretical classification of topological insulators is the prototypical example. This subject is only at its beginning, as the most complete developments so far apply only to non-interacting and to weakly interacting systems. A full understanding of the topological phases of strongly correlated matter is still an open problem. Approaches to this problem could well make use of holographic duality and already have made use of ideas from quantum information theory. Properties of the reduced density matrix and the information about quantum entanglement that they contain, for example, are thought to carry the information as to whether a system is in a "topological" or a "non-topological" phase. It is likely that string theory holography will have something to say about these systems at strong coupling.

Another important interface of condensed matter is with atomic physics and cold atoms. Cold atoms provide an experimental technique where important statistical mechanical systems can be engineered and studied in a wide array of circumstances. Analogs of systems which are strongly correlated in the real world can be simulated at various values of the coupling and studies in a variety of regimes. This is combined with the fact that the experimental measurements that can be performed on a cold atom system are often complementary to those available in the condensed matter system. This has the potential of yielding unprecedented information about models, like the Hubbard model, in regimes which are important for their physical applications, for example, in the regime where they model high-temperature superconductivity.

Recently, new low-energy neutral fermionic excitations (Majorana edge modes) have been claimed to be relevant in a variety of condensed matter systems, also providing new insights for the investigation of non-Fermi liquid states in correlated systems. Majorana fermions were first proposed in 1937 by Ettore Majorana (19) who considered a modification to the relativistic Dirac equation for conventional spin ½ particles (Dirac fermions) that gave purely real (as opposed to complex) solutions. These Majorana fermions are particles that are their own antiparticles—their creation operator in quantum theory is equal to their annihilation operator, unlike the case for conventional (Dirac) fermions such as electrons and holes which are different due to their opposite charge. Being real they are neutral excitations.

Low-energy Majorana modes have been recently the object of many theoretical and experimental investigations. Located at the edges of one-dimensional (1D) devices are responsible for the emergence of stretched non-local electron states allowing for distance-independent tunneling, crossed Andreev reflection, teleportation-like coherent transfer of a fermion, and fractional Josephson effects. Their effects emerge in a variety of condensed matter platforms.

The huge interest in Majorana fermions goes beyond fundamental curiosity since there is enormous potential for future quantum information technology if devices can be made based on manipulation of Majorana fermions, which are effectively fractionalized particles (anyons) obeying non-Abelian rather than Fermi–Dirac statistics. This would allow a qubit to be stored non-locally in a pair of widely separated Majorana bound states. These could be more insensitive to the effect of local sources of decoherence, which is currently the major obstacle to realizing a scalable quantum computer.

One can easily convince oneself that the breadth of the technical skills needed to synthesize these research directions is truly staggering. Condensed matter systems such as high Tc superconductors, topological insulators, and superconductors, and Kondo systems realized with pertinently engineered quantum devices all exhibit a variety of new behaviors, which may be accounted by pertinent strongly coupled field theories. The book will provide an insightful panorama on some of these topics emphasizing interesting connections between them. The contents are based on the lectures given at the program (conference +workshop) on "Strongly Coupled Field Theories for Condensed Matter and Quantum Information Theory" held at the International Institute of Physics in Natal during August 2015. The book suffered an undue delay caused by a serious illness of the corresponding editor who is very grateful to all the authors for the extra effort done in updating their lectures.

Acknowledgements: We thank all the IIP staff for the qualified help given during the Conference and Workshop held in Natal. We thank Margareth Barqueiro and Bia Pessoa for their kind and efficient help in assisting us during all the stages of the conference and for taking good care of all the organizational tasks. In addition, we are very grateful to Margareth Barqueiro for helping in the preparation of this book; without her qualified help, we could not have prepared the final version submitted to the editor. P. S. is very grateful to IIP, Natal, for a stimulating and pleasant hospitality at the Institute from Aug 2011 to October 2016; unfortunately, his stay suddenly terminated due to a very serious illness (He partially recovered at last!). Two of us (P. S. and A. F.) are grateful to the Ministry of Science, Technology, Innovation and Communication (MCTIC) and to the CNPq for financial support for granting them a "Bolsa de Produtividade em Pesquisa". Finally, we also want to acknowledge the agencies CAPES and CNPq for the support given for our Workshop/Conference as well as the Minister of Education for generously supporting the IIP throughout the past years.

Natal-Rn, Brazil — Alvaro Ferraz
Kolkata, India — Kumar S. Gupta
Vancouver, Canada — Gordon Walter Semenoff
Natal-Rn, Brazil — Pasquale Sodano

Contents

Contributors

Tony J. G. Apollaro Department of Physics, University of Malta, Msida, Malta

Abolfazl Bayat Department of Physics and Astronomy, University College London, London, UK;
Institute of Fundamental and Frontier Sciences, University of Electronic Science and Technology of China, Chengdu, China

Sougato Bose Department of Physics and Astronomy, University College London, London, UK

Francesco Buccheri Institute for Theoretical Physics, Heinrich-Heine-University Düsseldorf, Düsseldorf, Germany

Michele Burrello Center for Quantum Devices, Niels Bohr Institute, Niels Bohr International Academy, University of Copenhagen, Copenhagen, Denmark

David K. Campbell Department of Physics, Boston University, Boston, MA, USA

Mucio A. Continentino Centro Brasileiro de Pesquisa Físicas, Rio de Janeiro, RJ, Brazil

Gabriele De Chiara School of Mathematics and Physics, Centre for Theoretical Atomic, Molecular, and Optical Physics, Queen's University Belfast, London, UK

Reinhold Egger Institute for Theoretical Physics, Heinrich-Heine-University Düsseldorf, Düsseldorf, Germany

Johanna Erdmenger Institut für Theoretische Physik und Astrophysik, Julius-Maximilians-Universität Würzburg, Würzburg, Germany

Domenico Giuliano Dipartimento di Fisica, Università della Calabria, Cosenza, Italy;
I.N.F.N., Gruppo collegato di Cosenza, Cosenza, Italy

Thors Hans Hansson Department of Physics, Stockholm University, Stockholm, Sweden;
Nordita, KTH Royal Institute of Technology and Stockholm University, Stockholm, Sweden

Henrik Johannesson Department of Physics, University of Gothenburg, Gothenburg, Sweden;
Beijing Computational Science Research Center, Beijing, China

Thomas Klein Kvorning Department of Physics, KTH Royal Institute of Technology, Stockholm, Sweden;
Department of Physics, University of California, Berkeley, CA, USA

Vladimir Korepin C.N.Yang Institute for Theoretical Physics, Stony Brook University, Stony Brook, NY, USA

Tommaso Macrì Departamento de Física Teórica e Experimental and International Institute of Physics, Universidade Federal do Rio Grande do Norte, Natal, RN, Brazil

Javier Martinez Magan Instituto Balseiro, San Carlos de Bariloche, Argentina

Mariana Malard Faculdade UnB Planaltina, University of Brasilia, Planaltina, DF, Brazil

Eduardo C. Marino Instituto de Física, Universidade Federal do Rio de Janeiro, Rio de Janeiro, RJ, Brazil

Dmitry Melnikov International Institute of Physics, Federal University of Rio Grande do Norte, Natal, Brazil

Simone Paganelli Dipartimento di Scienze Fisiche e Chimiche, Università dell'Aquila, Coppito-L'Aquila, Italy

Rodrigo G. Pereira International Institute of Physics and Departamento de Física Teórica e Experimental, Universidade Federal do Rio Grande do Norte, Natal, RN, Brazil

João C. Pinto Barros Albert Einstein Center for Fundamental Physics, Institute for Theoretical Physics, University of Bern, Bern, Switzerland

Giovanni Ramírez Instituto de Investigación en Ciencias Físicas y Matemáticas (ICFM-USAC), Universidad de San Carlos de Guatemala, Guatemala City, Guatemala

Javier Rodríguez-Laguna Departamento de Física Fundamental, Universidad Nacional de Educacion a Distancia (UNED), Madrid, Spain

Griffith M. Rufo Centro Brasileiro de Pesquisa Físicas, Rio de Janeiro, RJ, Brazil

Sabrina Rufo Centro Brasileiro de Pesquisa Físicas, Rio de Janeiro, RJ, Brazil

Germán Sierra Instituto de Física Teórica (UAM-CSIC), Madrid, Spain

Pasquale Sodano I.N.F.N., Sezione di Perugia, Perugia, Italy;
International Institute of Physics, Universidade Federal do Rio Grande do Norte, Natal-RN, Brazil

Fumihiko Sugino Center for Theoretical Physics of the Universe, Institute for Basic Science (IBS), Yuseong-gu, Daejeon, Republic of Korea

Erik Tonni SISSA and INFN, Trieste, Italy

Andrea Trombettoni CNR-IOM DEMOCRITOS Simulation Center and SISSA, Trieste, Italy

Jacopo Viti International Institute of Physics, UFRN, Campos Universitário, Natal, Brazil

Chapter 1
Effective Field Theories for Topological States of Matter

Thors Hans Hansson and Thomas Klein Kvorning

Abstract Since the discovery of the quantum Hall effect in the 1980s it has been clear that there exists states of matter characterized by subtle quantum mechanical effects that renders certain properties surprisingly stable against dirt and noise. The theoretical understanding of these topological quantum phases have continued to develop during the last few decades and it has really surged after the discovery of the time-reversal invariant topological insulators. There are many examples of topological phases that have been important for the theoretical understanding of topological states of matter as well as being of great physical relevance. In this chapter we will focus on some examples that we find particularly enlightening and relevant, but we will not make a complete classification. Some of the most important tools for the understanding of topological quantum matter are based on effective field theory methods. We shall employ two different types of effective field theories. The first, which is valid at intermediate length and time-scales, will not capture the physics at microscopic scales. Such theories are the analogs, for topological phases, of the Ginzburg–Landau theories used to describe the usual symmetry breaking non-topological phases. The second type of theories describe the physics on scales where non-topological gapped states would be very boring, namely at distances and times much larger than the correlation length and the time set by the inverse gap. On these scales everything is independent of any distance and the theories will be topological field theories, which do not describe any dynamics in the bulk, but do carry informa-

T. H. Hansson (✉)
Department of Physics, Stockholm University, 10691 Stockholm, Sweden
e-mail: hansson@fysik.su.se

Nordita, KTH Royal Institute of Technology and Stockholm University, Stockholm, Sweden

T. Klein Kvorning
Department of Physics, KTH Royal Institute of Technology, 10691 Stockholm, Sweden
e-mail: kvorning@kth.se

Department of Physics, University of California, Berkeley, CA 94720, USA

A. Ferraz et al. (eds.), *Strongly Coupled Field Theories for Condensed Matter and Quantum Information Theory*, Springer Proceedings in Physics 239,
https://doi.org/10.1007/978-3-030-35473-2_1

tion about topological properties of the excitations, and also about excitations at the boundaries of the system. Finally, we will also study effective response actions. In a strict sense these are not effective theories, since they do not have any dynamical content, but encode the response of the system to external perturbations, typically an electromagnetic field. As we shall see, however, the effective response action for topological states can be used to extract parts of the dynamic theory through a method called functional bosonization.

1.1 Introduction

In school we learn about three states of matter: solids, liquids and gases. Typically these occur at different temperatures so that heating a solid first melts it into a liquid, then evaporates the liquid into a gas. The transition between two such phases of matter does not happen gradually as the temperature changes, but is a drastic event, a *phase transition*, that occur at a precisely defined transition temperature. The classification of matter according to the three states just mentioned is not at all exhaustive. For example, solids can be insulating or conducting and depending on the temperatures they can, or cannot, be magnetic.

Over the years, this picture has become more and more refined, but most of the fundamental ideas governing phases of matter was for a long time based on the work of Lev Landau in the 1930s. However, a few decades ago this completely changed with the advent of *topological phases of matter*. As you will see, these phases are of a fundamentally different nature, and they are the subject of these notes. But before we get there some words on terminology:

Matter denote systems with many degrees of freedom, typically collections of "particles" that can be atoms or electrons, but also quantum mechanical spins or quasiparticles such as phonons or magnons. In these notes we shall however only deal with systems composed of fermions, and typically these fermions should be thought of as electrons in a solid.

With a *phase of matter* we mean matter with certain characteristic properties such as rigidity, superfluid density, or magnetization. A phase is however not defined by any specific values for these quantities, but by whether they are at all present—for example, the transition between a solid and a liquid happens when the rigidity vanishes. Thus, by definition, one cannot gradually interpolate between two phases—they are only connected via the drastic event of a phase transition.

We shall here study properties of matter which is kept at such low temperatures that the thermal fluctuations can be neglected; the temperature is effectively zero. In a quantum system where the ground-state is separated from the first excited state by a finite energy gap, ΔE, this amounts to having $k_B T \ll \Delta E$, where k_B is Bolzmann's constant. Although there is a lot of current interest in gapless phases, we shall only consider those that are gapped.

Contrary to what you might think, not all matter solidifies even at the lowest temperatures. Many interesting phenomena such as superconductivity and superfluidity persist even down to zero temperature and this is due to quantum rather than thermal fluctuations. Phase transitions occur also at zero temperature, but are then driven by quantum rather than by thermal fluctuations, and are induced by changes in some external parameter. The transition from a superconductor to a normal metal by changing the magnetic field is a well know example of such a *quantum phase transition*. While the usual finite temperature phase transitions show up as abrupt changes in various thermodynamical quantities, quantum phase transitions are signaled by qualitative changes in the quantum mechanical ground-state wave function.

Since these quantum phases occur at zero temperature their properties are encoded in qualitative properties of the ground-state of the system. Thus, classifying and characterizing zero temperature states of matter is equivalent to doing this for ground-state wave functions with a very large number of degrees of freedom. More precisely, two systems at zero temperature are said to be in different phases when, in the thermodynamic, i.e., large volume, limit, there is no continuous way to transform one state (i.e., the ground-state wave function) into the other while remaining at zero temperature and not closing the energy gap. So when we say that one state cannot continuously be transformed into another we mean that this cannot be done in the thermodynamic limit while keeping the energy gap finite.

There are two complementary ways to precisely characterize (quantum) phases of matter. The first, which goes back to the work of Lev Landau in the 1930s, is based on symmetries [1]. The basic idea is that of *spontaneous symmetry breaking* which means that the ground-state has less symmetry than the microscopic Hamiltonian. An important concept is the order parameter, ψ_a, which transforms according to some representation of a symmetry group. The archetypal example is a ferromagnet where the order parameter is the magnetization, $\mathbf{M}$, which transforms as a vector under spatial rotations, and is non-zero below the Curie temperature. While the order parameter is essentially a classical concept, and the phase transitions studied by Landau are driven by thermal fluctuations, the Landau approach equally well applies to quantum systems. These can also be classified according to their pattern of spontaneous symmetry breaking, and the order parameter appears as the ground-state expectation value, $\psi_a = \left\langle \hat{\psi}_a \right\rangle$, where $\hat{\psi}_a(\mathbf{x}, t)$ is an operator with appropriate symmetry properties that can be measured locally. Low-lying excitations around the symmetry broken ground-state corresponds to long wave length oscillations in $\psi_a(\mathbf{x}, t)$ which are gapless if the broken symmetry is continuous—these are the famous Goldstone modes.

The second way phases can be distinguished, which is special to quantum systems, was developed in the last decades, and is drastically different from the Landau paradigm. Such phases differ by properties of the quantum entanglement of the ground-state wave functions, and they cannot directly be distinguished by local measurements. They are referred to as *topological phases of matter* or *topological quantum matter* and are the main topic of these notes.

Topology is the mathematical study of properties that are preserved under continuous transformations, and in this context it refers to properties of the ground-state wave function preserved under continuously changing external parameters. The distinctions between different topological phases of matter are somewhat intricate and require a systematic study using tools developed in the field of mathematical topology, motivating "topological" in the expression topological quantum matter.

There are two classes of topological phases with quite different properties: the *symmetry protected topological states* (SPT-states) and the *topologically ordered states* (TO-states). As already mentioned, two states are in different phases if one cannot continuously transform the ground-state wave function of one into that of the other. In many situations this turns out to be too restrictive and would not allow us to identify important topological phases. The reason is that there can be symmetries that *all* physically realizable perturbations respect. The natural question to ask is then what the possible phases are if one is restricted to systems with a certain symmetry—*symmetry protected phases*. That is, we widen the definition and say that two states which respect a specific symmetry S are in different SPT phases as long as any continuous transformations between them violate S.

A most striking property of the SPT states is that they support boundary states which can be used to classify them. The boundary state of a d-dimensional SPT state is described by a $d-1$ dimensional field theory which cannot describe a *bona fide* $d-1$ system which preserves the symmetry (we will use lower-case d to denote the number of spatial dimension, while upper-case D will denote the number of space-time dimensions, i.e., $D = d + 1$). Typically, what happens is that the application of an external field makes the conserved charge, corresponding to the symmetry, flow from bulk to edge, thus preventing them from being individually consistent. As you will see, for some of these topological states this goes hand in hand with a quantization of certain transport coefficients, most notably the Hall conductance.

Symmetry protected phases have been known, at least as a theoretical possibility, since the work of Haldane on topological effects in $1d$ spin chains [2]. However, the field got a renaissance after the fairly recent both theoretical and experimental discovery of the time-reversal invariant topological insulators, see e.g., [3]. These states can be realized by non-interacting fermions and their discovery led to a systematic study of those states that continuously can be transformed into each other, restricting one self to *only* non-interacting systems. Soon, there was a complete classification of non-interacting fermionic systems with $U(1)$ symmetry and/or non-unitary symmetries in terms of non-interacting topological invariants [4, 5].

Since, in the real world, there are generally some interactions present, this classification would seem to apply only to very fine-tuned situations and be, at best, of marginal interest. Fortunately, however, for many (but not all) of the systems with non-interacting topological invariants there are characterizing properties, such as boundary states and quantized transport coefficients that do not dependent on interactions being absent. At a more theoretical level, it has also been shown that many of the non-interacting topological classes are characterized by various quantum anomalies, which are known to be insensitive to (at least weak) interactions [6].

The non-interacting classification has been essential for pinpointing these characteristics and is of great importance for the understanding of topological quantum matter in general. All SPT states that we will discuss in these notes can be realized with non-interacting fermions, but note that SPT phases appear in more general settings. For instance, all non trivial bosonic SPT states are interacting.

The meaning of TO has varied over time and there still no complete consensus concerning the nomenclature. However, some states are considered topologically ordered regardless of convention, namely the ones that support localized fractionalized bulk excitations with non-trivial topological interactions. The most famous examples are in 2d where these excitations are *anyonic* quasi-particles, *anyons*, which have a remarkable type of interaction. At first sight they do not seem to interact at all, at least not at long distances, but a closer look will reveal that there is a subtle form of interaction, referred to as anyonic, or fractional, statistics[1]: the state of the system will depend not only on the positions of the individual particles, but also on their history. Or more precisely, on how their world lines have *braided*.

In fact, one can classify the TO ground-states in terms of the properties of the quasi-particles they support without specifying the Hamiltonian. At first this may seem strange: If you only know the ground-state, any states could be made the low lying excited ones by judiciously picking the Hamiltonian. However, if you assume that the Hamiltonian is local, meaning a sum of local terms, $\hat{H} = \sum_{\{\mathbf{x}\}} \hat{h}_{\mathbf{x}}$ where each $\hat{h}_{\mathbf{x}}$ only has exponentially small support outside of a region close to $\mathbf{x}$, then there is a close connection between the properties of the ground-state, and those of the excitations, see e.g., [7, 8].

We discuss 2d states that support anyons but also touch upon analogous states in other dimensions. In 1d, there are none, and in dimensions higher than two the states will support higher dimensional excitations (such as strings and membranes) that can realize higher dimensional analogs of anyons.

There are also phases that are topologically ordered in the sense that they are well defined without requiring the presence of a symmetry, but they have no excitations which interact topologically and their physical characteristics resemble that of SPTs. It is for these states that the conventions on whether they are topologically ordered or not vary. With an abuse of the concept they can be considered SPTs but the "symmetries" that protect them are not really symmetries, in the sense that they cannot be broken. An important example (that we will discuss) is the Majorana nano-wire [9] which can be considered protected by fermion parity conservation [10]. Fermion parity conservation is not really a symmetry since it cannot be broken, it is a property that always is present.

There are many examples of topological quantum matter that have been important for the development of the theoretical understanding of the field as well as being of great physical relevance. However, in these notes you will neither find an exhaustive list of such examples nor attempts to a complete classification. But rather focus on a few examples that we find particularly enlightening and relevant.

[1]The topological interaction between anyons is a generalization of the Berry phase of -1 acquired by the exchange of two identical fermions. This minus sign is directly related to Fermi statistics and in the same way the topological interaction is related to a specific "exclusion statistics". This is the reason for the term "statistics" in anyonic statistics.

Since we only consider gapped local Hamiltonians we have characteristic time and length scales $\tau = \hbar/E_{\text{gap}}$ and the correlation length ξ. We will employ two different types of effective field theories. The first, which is valid at length and time scales at the order of ξ, τ and longer, will not capture the physics at microscopic scales (like the lattice spacing $a \sim \mathring{A}$ or the bandwidth $\Delta\varepsilon \sim eV$). Such theories are the analogs, for topological phases, of the Ginzburg–Landau theories used to describe the usual symmetry breaking non-topological phases. Examples are the Chern–Simons–Ginzburg–Landau theories for QH liquids, see e.g., [11], and the Ginzburg–Landau–Maxwell theories for superconductors.[2] These theories have information both about topological quantities, such as charges and statistics of quasi-particles, and of collective bosonic excitations such as plasmons or magnetorotons.

The second type of theories describe the physics on scales where non-topological gapped states would be very boring, namely at distances and times much larger than ξ and τ. On these scales everything is independent of any distance and the theories will be *topological field theories*, which do not describe any dynamics in the bulk, but do carry information about topological properties of the excitations, and also about excitations at the boundaries of the system. Typical examples we shall study are the Wen-Zee Chern–Simons theories for QH liquids [16, 17] and the BF-theories for superconductors and topological insulators [18, 19].

Finally, we will also study *effective response actions*. In a strict sense these are not effective theories, since they do not have any dynamical content, but encode the response of the system to external perturbations, typically an electromagnetic field. As you will see the effective response action for topological states can however be used to extract parts of the dynamic theory through a method called functional bosonization.

We have tried to make these notes reasonably self-contained, but of course we will often refer to other texts. The list of references is by no mean exhaustive; when there are good reviews we often cite these rather than the original papers.

1.2 The Quantized Hall Conductance

Figure 1.1 shows a schematic view of the experiment by von Klitzing et al. [20], where the quantum Hall effect was discovered, and also the original data. The quantum Hall effect will be discussed again later, but for now it suffices to know that the quantum Hall effect is observed in a two-dimensional electron gas, which is gated so that the chemical potential increases monotonically with the gate voltage V_g shown in Fig. 1.1. von Klitzing's experiment was performed at a temperature of 1.5 K and a magnetic field of 18 T. A constant current $I = 1\,\mu\text{A}$ was driven through the system

[2] Most textbooks in condensed matter theory will cover the Ginzburg–Landau–Maxwell theory. For a modern text see e.g., [12]. Reference [13], by S. Weinberg, one of the founders of effective field theory, gives a good presentation from the field theoretic point of view. There are also several excellent recent textbooks, as for instance [14, 15], on the general subject of these notes.

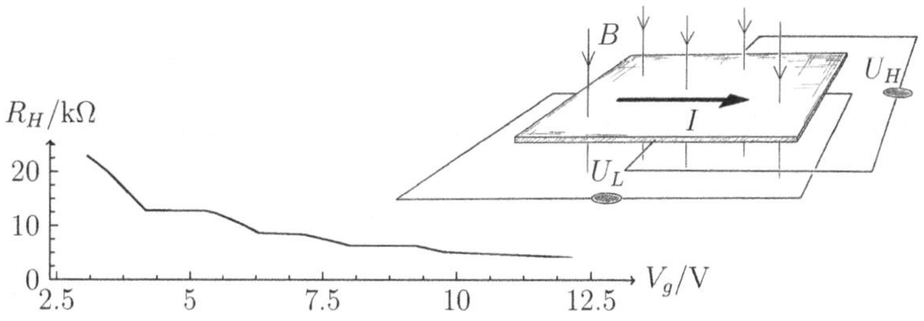

Fig. 1.1 Data from the original paper [20] with a set-up as is schematically depicted above: a two dimensional electron gas is subject to a magnetic field, B, and a constant current, I, is driven through the system. The chemical potential of the electron gas increases monotonically with V_g and there are clear plateaux with constant Hall resistance R_H. *Figure by S. Holst*

and the perpendicular voltage U_H was measured. In Fig. 1.1 you can see clear plateaux where the resistance is constant, and on a closer inspection these plateaux are located at integer multiples of the von Klitzing constant, R_K up to relative errors of the order $\lesssim 10^{-8}$.

Since then, similar experiments been performed a large number of times on different systems and in different parameter ranges including at room temperature [21]. The result is always the same, the Hall resistance is quantized at multiples or rational fractions of R_K. This is remarkable! Remember that we are dealing with macroscopic systems which depend on a practically infinite number of parameters. Even so, if you keep a constant current the voltage will be exactly the same in samples that can vary extensively. How can this be? In this section you will not only find the answer to this question and its related consequences, you will also become familiar with response actions, and topological field theories. These are important tools for analyzing topological states of matter, and they will be used extensively throughout these notes.

1.2.1 The Hall Conductance as a Chern Number

We now explain why the Hall conductance is quantized in gapped $2d$ system at temperatures $k_B T \ll \Delta E$. This is a most important fact: a quantized value cannot change continuously, so there has to be a phase transition between states with different Hall conductance—the quantized value is one of the phase-distinguishing characteristics mentioned in the introduction.

The argument given here is based on the work by Niu, Thouless, and Wu [22]. There are however many important earlier contributions that lay the ground work for the understanding, most notably [23–26].

By definition, the conductivity tensor gives the linear current density response to an electric field and in two spatial dimension it can be parametrized as,

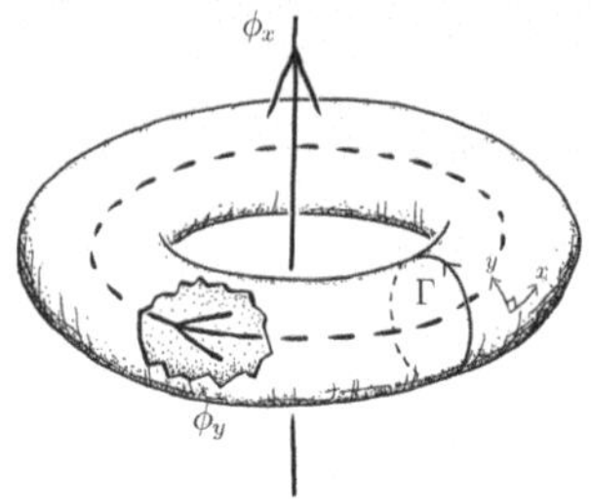

Fig. 1.2 A torus with flux ϕ_x/ϕ_x encircled by the two independent non-contractible loops and an example of a curve Γ which encircle the flux ϕ_y. *Figure by S. Holst*

$$j^i = \sigma^{ij} E_j = \sigma_H \varepsilon^{ij} E_j + \sigma_L E^i \ , \tag{1.1}$$

which defines the usual longitudinal conductivity σ_L, and the transverse, or Hall, conductivity σ_H. What is measured in an experiment, is however, not the conductivity, but the conductance for some macroscopic (or mesoscopic) sample. The conductance gives the current response to a voltage, or an electromotive force, $\mathcal{E}$, which is defined by,

$$\mathcal{E} = \int_\gamma d\mathbf{l} \cdot \mathbf{E} \ , \tag{1.2}$$

where the integral is along some curve γ. For an open curve, $\mathcal{E}$ is just the voltage difference between the endpoints, as in the Hall bar shown in Fig. 1.1, while in toroidal, cylindrical or Corbino geometry, the curve is closed, and the electromotive force should be understood as generated by a time dependent magnetic field through a hole, as in Fig. 1.2. Multiplying the definition of longitudinal and Hall conductances (1.1) with ε^{ij} and integrating along a curve γ gives,

$$I_\gamma = \int_\gamma d\mathbf{l} \cdot \mathbf{E} \sigma_H + \int_\gamma d\mathbf{l} \times \mathbf{E} \sigma_L \ , \tag{1.3}$$

where conductances I_γ is the current passing through the curve γ. Assuming a homogeneous sample, so that the Hall conductivity is constant, this become

$$I_\gamma = \sigma_H \, \mathcal{E}_\gamma + \sigma_L \int_\gamma d\mathbf{l} \times \mathbf{E} \, , \tag{1.4}$$

which shows that in a pure $2d$ sample the Hall conductance actually equals the Hall conductivity which is a material property. In particular notice that no geometrical factors, that would be hard to measure with high precision, enter the relation.[3]

[3]This is not true in higher dimensions, and it is also not true for the longitudinal $2d$ conductance. Even for a pure sample does not equal the conductivity. For a rectangular Hall bar as in Fig. 1.1 the longitudinal conductance is $(W/L)\sigma$ where W and L are the widths and length of the bar respectively.

Now consider a $2d$ Hall bar which is large enough that the transport properties do not depend on the boundary conditions. We can then pick them to be periodic, which is the same as assuming that the space is a torus. Furthermore assume that the spectrum on the torus has a gap $\Delta E \gg k_B T$ above an N-dimensional subspace of degenerate states (up to splittings that vanish in the thermodynamic limit). We will also imagine having magnetic fluxes, $\phi_{x/y}$, through the torus as illustrated in Fig. 1.2. Our assertion is then that for a big system the Hall conductance will not depend on these fluxes, and will also equal that for a physical Hall bar.

Inserting an arbitrary flux, $\phi_{x/y}$, through any of the non-contractible loops on the torus (see Fig. 1.2) does not change the conductance, but it does change the Hamiltonian.[4] However, for the special case of inserting a flux quantum, $\phi_0 = hc/e$, the resulting Hamiltonian is identical to the one where there is no flux. Thus, the Hamiltonian depends on the parameters $\phi_{x/y}$ which are defined on a space where the points (ϕ_x, ϕ_y) and $(\phi_x + n_x\phi_0, \phi_y + n_y\phi_0)$ are identified. Put differently, the parameter space $T^2_\phi = \{(\phi_x, \phi_y)\}$ is a torus, which we will refer to as the flux-torus to distinguish it from the physical space, which, because of the periodic boundary conditions, also is a torus.

The idea is now to consider maps from the parameter space into the space of ground-state wave functions. Such maps are characterized by an integer ch_1, called the first Chern number. The proper mathematical setting for this concept is the theory of fiber bundles, and more precisely, the degenerate ground-state wave functions form a complex vector bundle over the parameter space T^2_ϕ. For a brief introduction to Berry phases and Chern numbers we refer to Sect. 1.8.1.

The basic result is that the Hall conductance σ_H is given by the formula,

$$\sigma_H = \sigma_0 \frac{ch_1}{N} , \tag{1.5}$$

where $\sigma_0 = e^2/h$ and N is the number of degenerate ground-states. We are now ready for the actual calculation.

Let us first pick the gauge potential as

$$\mathbf{A} = \tilde{\mathbf{A}} + \frac{\phi_x}{L_x}\hat{x} + \frac{\phi_y}{L_y}\hat{y} , \tag{1.6}$$

where the integral of $\tilde{A}_i$ along any of the non-contractible torus-loops is zero. Changing $\phi_{x/y} \to \phi_{x/y} + \phi_0$ would give back the same physical Hamiltonian but in a different gauge; so, $\phi_{x/y}$ is used not only label the fluxes through the holes, but also the gauge choice. Assume now that we have a monotonically increasing $\phi_y(t)$ such that in time τ a full unit of flux is inserted in the hole, i.e., $\phi_y(\tau) = \phi_y(0) + \phi_0$. According to Faraday's law, this generates an electromotive force,

[4] In some references you will come across the notion of "twisted boundary conditions" this is equivalent to inserting flux through the holes of the torus.

$$\mathscr{E}_{\gamma_y} = \int_{\gamma_y} d\mathbf{l}\cdot\mathbf{E} = \frac{1}{c}\frac{\partial \phi_y}{\partial t}\,, \tag{1.7}$$

where γ_y is any of the non-contractible loops encircling the flux ϕ_y once. The conductance is defined in the limit of vanishing electric field so we should assume $\tau \to \infty$ and, since there is an energy gap to all excited states, the time dependence is thus given by the adiabatic theorem. We choose an orthonormal basis of the ground-state manifold, for each ϕ_x, at $\phi_y = \phi_y(0) = 0$,

$$\left\{\left|(\phi_x,\phi_y);\alpha\right\rangle\right\}_{\alpha=1,\dots N}\Big|_{\phi_y=0}\,, \tag{1.8}$$

which is taken as a smooth function of ϕ_x. Under the adiabatic time evolution $U(t) \equiv U(\phi_y(t))$ (recall that $\phi_y(t)$ is monotonic) this evolves into,

$$\left|(\phi_x,\phi_y);\alpha\right\rangle = U(\phi_y)\left|(\phi_x,0);\alpha\right\rangle\,. \tag{1.9}$$

Now we are ready to calculate the current. With no loss of generality we shall take the curve γ to be a straight line in the y-direction, and get,

$$\begin{aligned} I_{\gamma_y} &= \frac{1}{L_x}\int \mathrm{d}^2x\,\hat{x}\cdot\mathbf{j}(\mathbf{x}) = \frac{1}{L_x}\int \mathrm{d}^2x\,\psi^*(\mathbf{x},t)\frac{c\,\partial H(\mathbf{A})}{\partial A_x}\psi(\mathbf{x},t) \\ &= \int \mathrm{d}^2x\,\psi^*(\mathbf{x},t)\frac{c\,\partial H(\mathbf{A})}{\partial \phi_x}\psi(\mathbf{x},t) = c\left\langle\psi\right|\left(\partial_{\phi_x}H\right)\left|\psi\right\rangle\,, \end{aligned} \tag{1.10}$$

where we used the definition of current density operator $\mathbf{j}(\mathbf{x}) = c\,\partial H(\mathbf{A})/\partial\mathbf{A}$, and our decomposition of the gauge potential, (1.6). We can, with out loss of generality, assume that we start out in the state $\left|(\phi_x,\phi_y);1\right\rangle$, and we then have the expression,

$$I_{\gamma_y}(\phi_y) = c\left\langle(\phi_x,\phi_y);1\right|\left(\partial_{\phi_x}H\right)\left|(\phi_x,\phi_y);1\right\rangle \tag{1.11}$$

which by, repeated use of Leibniz rule and using

$$H\left|(\phi_x,\phi_y);1\right\rangle = i\hbar\frac{\partial\phi_y}{\partial t}\partial_{\phi_y}\left|(\phi_x,\phi_y);1\right\rangle\,, \tag{1.12}$$

and $E_0 = \left\langle(\phi_x,\phi_y);1\right|H\left|(\phi_x,\phi_y);1\right\rangle$, can be rewritten as

$$I_{\gamma_y}(\phi_y) = c\partial_{\phi_x}E_0 + i\hbar c\frac{\partial\phi_y}{\partial t}\,. \tag{1.13}$$

When averaging over ϕ_y and τ, the first term vanishes since $E_0(\phi_x,\phi_y) = E_0(\phi_x + 2\pi,\phi_y)$ and we get

$$\bar{I}_{\gamma_y} = \frac{hc}{\phi_0}\frac{1}{\tau}\int_0^{\phi_0}\int_0^{\phi_0} \mathrm{d}\phi_x \mathrm{d}\phi_y \varepsilon^{ij}\partial_{\phi_i}\left\langle(\phi_x, \phi_y); 1\right| \partial_{\phi_j}\left|(\phi_x, \phi_y); 1\right\rangle . \tag{1.14}$$

The integrand equals a term in the trace of the Berry field strength corresponding to the state $\left|(\phi_x, \phi_y); 1\right\rangle$, see (1.186) in Sect. 1.8.1; so, if we average over the different states in the degenerate ground-state manifold we end up with

$$\bar{I}_{\gamma_y} = \frac{hc}{N\phi_0}\frac{1}{\tau}\int_0^{\phi_0}\int_0^{\phi_0} \mathrm{d}\phi_x \mathrm{d}\phi_y \frac{\mathrm{Tr}\,(\mathscr{F})}{2\pi} = \frac{1}{N}\frac{e}{\tau}ch_1 , \tag{1.15}$$

where the first Chern number, ch_1, is defined in Sect. 1.8.1. We now argue that without loss of generality we can do just that: Let us for simplicity assume that there are two ground-states, and compare their conductivity in some bounded region. (This can in principle can be measured by a local probe.) One possibility is that the conductivity is in fact the same in the two states, so that the conductance trivially equals the average of the conductances in the two states. The other possibility is that there is a region where the conductivities do differ, which means that there are *local* operators with different expectation values in the two states. Now think of weakly perturbing the Hamiltonian in some region with such terms. This will break the degeneracy, and result in a unique ground-state an the question of averaging is gone.

To get the conductance we are left with calculating the electromotive force. From Faraday's law, (1.7) we get,

$$\bar{\mathscr{E}}_y = \frac{1}{\phi_0 \tau c}\int_0^{\phi_0}\int_0^{\tau} \mathrm{d}\phi_y \mathrm{d}t \frac{\partial \phi_x}{\partial t} = \frac{\phi_0}{\tau c} , \tag{1.16}$$

and then finally,

$$\sigma_H = \frac{\bar{I}_x}{\bar{\mathscr{E}}_y} = \sigma_0 \frac{ch_1}{N} , \tag{1.17}$$

which concludes the proof of (1.5)—the quantization of the Hall conductance.

This formula allows for a conductance which is a fraction of the quantum of conductance σ_0, but only if the ground-state degeneracy cannot be broken by any local operator. This is in fact a characteristic property of topologically ordered states—the ground-state degeneracy is a topologically protected number. We will return to this point briefly in Sect. 1.6, and you should note that this type of degeneracy is very different from what you are used to. Normally a degeneracy is either "accidental", that is dependent some fine tuning of parameters, or due to some symmetry. In both cases the degeneracy can be broken by adding local terms, which in the second case has to violate the symmetry.

As we just mentioned the states with a conductance of a fraction of σ_0, are topologically ordered. But what about the states which have a conductance of an integer times σ_0. Are they topologically ordered or symmetry protected?

The Hall conductance is a charge response and one could at least in principle imagine breaking the $U(1)$-charge conservation symmetry by proximity to a superconductor. The Hall-conductance would then no longer be well-defined and could no longer be used as a characteristic for a phase of matter. You might therefore think that the states with integer quantized Hall conductance are SPTs protected by $U(1)$-symmetry. And you would be right; One could, in principle, imagine an SPT with quantized Hall conductance. However, in all realistic situations the quantized charge Hall conductance goes hand in hand with a quantized *thermal* conductance: the heat current is proportional to a quantized constant, the temperature, and the temperature gradient. This quantized constant is harder to access experimentally than the charge Hall conductance, but it is by no means impossible, see e.g., [27]. However, it is the theoretical importance that is of main interest here; energy conservation is part of the definition of quantum matter (without it you could not define zero-temperature), so one cannot get rid of the thermal conductance in the same way as with the charge conductance. The states with integer quantized Hall conductance are therefore topologically ordered but of the kind, mentioned in the introduction, that are similar to SPTs.

1.2.2 The Chern–Simons Response Action

In this section we shall first encode the quantum Hall response, derived above, in an *effective response action*. From now on we put $c = 1$, and often $\hbar = 1$. We assume there is a conserved $U(1)$ current (typically the electric current) which we can couple to a gauge field A_μ. The effective action, $\Gamma[A_\mu]$, is the generating functional for connected n-point functions,

$$\Gamma[A_\mu] \equiv -i \log \mathscr{Z}[A_\mu] \equiv -i \log \left\langle GS|\mathscr{T} e^{i\int \mathrm{d}t\, H(A_\mu(t))}|GS\right\rangle , \tag{1.18}$$

where $|GS\rangle$ is the ground-state and $\mathscr{T}$ denotes time-ordering. Taking derivatives of $\Gamma[A_\mu]$ gives time-ordered connected current n-point functions. Most importantly the current expectation value,

$$\left.\frac{\delta}{c\,\delta A'_\mu(\mathbf{x},t)}\Gamma[A'_\mu]\right|_{A'_\mu=A_\mu} \equiv j^\mu(\mathbf{x},t) = \left\langle GS|\hat{j}^\mu(\mathbf{x},t)|GS\right\rangle ; \tag{1.19}$$

$$\hat{j}^\mu(\mathbf{x},t) \equiv \mathscr{T} e^{i\int_t^\infty \mathrm{d}t'\, H(A_\mu(t'))}\,\hat{j}^\mu(\mathbf{x})\,\mathscr{T} e^{i\int_{-\infty}^{t} \mathrm{d}t'\, H(A_\mu(t'))} , \tag{1.20}$$

where A_μ is a back-ground gauge potential.

Since we consider gapped systems, which by definition have have no mobile charge carriers, and thus do not conduct, we first remind ourselves of the effective action for usual insulators. In these materials, external electric and magnetic fields gives rise to dielectric and diamagnetic effects, such as a polarization charge, $\rho_{pol} = -\chi_e \nabla \cdot \mathbf{E}$. This *linear response* is captured by the effective action,

$$\Gamma_{\text{med}} = \int dt d^3x \left(\frac{\chi_e}{2} \mathbf{E}^2 - \frac{\chi_m}{2} \mathbf{B}^2 \right) , \tag{1.21}$$

where χ_e and χ_m are the electric and magnetic susceptibilities respectively. This is the expression with the lowest number of derivatives, which is quadratic in the fields (and thus gives linear response), and is invariant under rotations, reflections (parity), time reversal and gauge transformations. Thus it describes the response of a large class of isotropic materials in weak and slowly varying electromagnetic fields.

We now turn to the Hall response. The relation $j^i = \sigma_H \varepsilon^{ij} E_j$ can be written as

$$j^i = \sigma_H \varepsilon^{ij} \left(\partial_0 A_j - \partial_j A_0 \right) , \tag{1.22}$$

where 0 is the time index, the roman letters $i, j, \ldots$ are the space indices. Equation (1.22) together with current conservation, $\partial_i j^i = \partial_0 j^0$ gives the *Streda formula*,

$$\partial_0 j^0 = \partial_i j^i = \sigma_H \varepsilon^{ij} \partial_i \left(\partial_0 A_j - \partial_j A_0 \right) = \sigma_H \partial_0 B ,$$

where B is the component of the magnetic field perpendicular to the $2d$ system. Assuming $B = 0$ at $t = -\infty$, this can be combined with (1.22) to give

$$j^\mu = \sigma_H \, \varepsilon^{\mu\nu\sigma} \partial_\nu A_\sigma , \tag{1.23}$$

and, by integration, the corresponding term,

$$\Gamma_H[A] = \frac{\sigma_H}{2} \int d^3x \, \varepsilon^{\mu\nu\sigma} A_\mu(x) \partial_\nu A_\sigma(x) , \tag{1.24}$$

in the effective response action.

As opposed to $\Gamma_{\text{med.}}$, the Hall term, Γ_H, violates both time-reversal and parity symmetry. Thus we can conclude that in a system where these symmetries are present we have zero Hall conductance. In the quantum Hall systems the symmetry is broken by a background magnetic field, but as you will see, a magnetic field is not necessary; other physical systems which violates the symmetry in other ways can also produce a non-zero Hall conductance.

Another point to notice is that the Hall term is not written only in terms of field strengths, so one might worry that it is not gauge invariant. Under the gauge transformation

$$A_\mu \to A_\mu + \partial_\mu \lambda \tag{1.25}$$

one gets the variation

$$\delta \int_V d^3x \, \varepsilon^{\mu\nu\sigma} A_\mu(x) \partial_\nu A_\sigma(x) = \int_{\partial V} dx^i E_i(x) \lambda(x) , \tag{1.26}$$

where ∂V is the boundary of the space-time volume V. The Hall term is thus gauge invariant only up to a boundary term. Since gauge invariance is a consequence of current conservation, this means that we do not have current conservation if the system of consideration has a boundary. The resolution to this quandary is that there is an extra piece in the effective action that only resides on the boundary and describes an edge current in the quantum Hall sample. Such edge currents are known to be present and it is important to find a formulation of the effective low energy theory that incorporates them in a natural way.

1.2.3 The Topological Field Theory

The basic tool to find a formulation of the effective low energy theory will be that of *topological field theory*. We shall return to this concept several times later, but for now just look at the simplest example and see that it has the desired properties. We take the Lagrangian

$$\mathscr{L}(b; A, j) = -\frac{1}{4\pi}\varepsilon^{\mu\nu\sigma} b_\mu \partial_\nu b_\sigma - \frac{e}{2\pi}\varepsilon^{\mu\nu\sigma} b_\mu \partial_\nu A_\sigma - j_q^\mu b_\mu \,, \tag{1.27}$$

where b is an auxiliary gauge field,[5] and j_q a quasi-particle current. The first term in (1.27) is called the Chern–Simons (CS) term, and this particular topological field theory is thus called CS theory. To understand the meaning of the field, b, we calculate the electric current j,

$$j^\mu = \frac{\delta \mathscr{L}}{\delta A_\mu} = -\frac{e}{2\pi}\varepsilon^{\mu\nu\sigma}\partial_\nu b_\sigma \,. \tag{1.28}$$

So, b is just a way to parametrize j. Note that j, which by definition is conserved, is invariant under the gauge transformation

$$b_\mu \to b_\mu + \partial_\mu \chi \,, \tag{1.29}$$

where χ is a scalar, since it is the field strength corresponding to the vector potential b. Since b is related to the conserved current, we shall refer to it as "hydrodynamic".

Why is this theory is referred to as topological? First you notice that it does not depend on the metric tensor. A normal kinetic term has the general covariant form $\sim g^{\mu\nu} D_\mu\phi D_\nu\phi$, and thus depends on the geometry of the space on which it is defined. If the action does not depend on the metric, correlation functions of operators cannot depend on the metric either, specifically they cannot depend on any distance or time. Furthermore, the equation of motion for the b field is,

---

[5]Note that the conventions for this field differ. We use the notation from [11], while in the work by Wen [17] the field here denoted by b is denoted by a.

$$\varepsilon^{\mu\nu\sigma}\partial_\nu b_\sigma = -e\varepsilon^{\mu\nu\sigma}\partial_\nu A_\sigma - 2\pi j_q^\mu , \tag{1.30}$$

that is *the field strength is completely determined by the external sources*. This means that, as opposed to usual Maxwell electrodynamics, there are no propagating photons—the equations of motion are just constraints. For instance, the zeroth component of (1.30) is $2\pi\rho = \varepsilon^{ij}\partial_i b_j \equiv B^{(b)}$, which relates $B^{(b)}$, to the charge density, $\rho = j^0$, of the external sources. The analysis just given is, only for a system on an infinite plane, the case of boundaries will be discussed below, in Sect. 1.2.4.

Since the Lagrangian, (1.27), is quadratic in b we can integrate it out to get an effective action for A only. We can use the following path integral formula,

$$e^{i\Gamma[A,j]} = \int \mathscr{D}[\mathbf{b}]e^{i\int d^3r\,\mathscr{L}(b;A,j)} , \tag{1.31}$$

to get the response action. Performing the integral we get,

$$\Gamma[A,j] = \int d^3x \left[\frac{\sigma_H}{2}\varepsilon^{\mu\nu\sigma}A_\mu(x)\partial_\nu A_\sigma(x) + ej_q^\mu(x)A_\mu(x)\right] + \int d^3x d^3y\, j_q^\mu(x)\left(\frac{\pi}{d}\right)_{\mu\nu}(x-y)j_q^\nu(y) , \tag{1.32}$$

where you should recall that $\sigma_H = e^2/2\pi$, and where $(1/d)_{\mu\nu}(x-y)$ is the inverse of the Chern–Simons operator kernel $\varepsilon^{\mu\nu\sigma}\partial_\sigma$. The first term in this expression is just the Chern–Simons response term, (1.24), derived earlier, while the last term is a *topological interaction* between the particles described by the source j_q. The last term provides the minus sign that the wave function acquires when two identical fermions are exchanged. A simple way to understand this phase is to recall that the equation of motion (1.30) associates charge with flux and that the resulting charge-flux composites will pick up an Aharanov–Bohm like phase when encircling each other. (The story is a little more subtle and we will come back to in the last section on the fractional quantum Hall effect.) We again stress that the above result is only correct on an infinite plane, since it does not conserve current at the edge.

1.2.3.1 Functional Bosonization

You might find the above discussion somewhat unsatisfactory since the topological field theory in the previous section was merely postulated. Here we amend this by describing a rather general method to actually *derive* a topological field theory given an effective response action [28]. After a general exposition of the method we then specialize to Hall response.

The starting point is the path-integral formula for the partition function,

$$\mathscr{Z}[A_\mu] = \int \mathscr{D}[\bar{\psi}, \psi] e^{iS[\bar{\psi},\psi,A]}, \tag{1.33}$$

where the action S describes the system of interest. We will not do the integral explicitly, but make use of the fact that for a gapped system at zero temperature the functional $\mathscr{Z}[A_\mu]$ will be local.

Because of current conservation, the effective response action, and therefore also $\mathscr{Z}$, needs to be gauge invariant meaning that

$$\mathscr{Z}[A_\mu + a_\mu] = Z[A_\mu], \tag{1.34}$$

for any a being a pure gauge, i.e., satisfying,

$$f_{\mu\nu} \equiv \partial_\mu a_\nu - \partial_\mu a_\nu = 0. \tag{1.35}$$

Thus one can express $\mathscr{Z}$ as

$$\mathscr{Z}[A] = \int \mathscr{D}[a]\, \mathscr{Z}[A + a] \prod_{\mu\nu\cdots} \varepsilon^{\mu\nu\lambda\ldots\alpha\beta} \delta[f_{\alpha\beta}(a(x))], \tag{1.36}$$

where the delta functionals under the product sign enforce the zero field-strength constraint. Here, x is a point in $D = d + 1$ dimensional space-time, and $\varepsilon^{\mu\nu\lambda\ldots\alpha\beta}$ is the D-dimensional totally anti-symmetric Levi-Civita symbol. Introducing an auxiliary tensor field $b_{\mu_1\mu_1\ldots\mu_{D-2}}$ to express the delta functional as a functional Fourier integral, we get,

$$\mathscr{Z}[A] = \int \mathscr{D}[a]\mathscr{D}[b]\, \mathscr{Z}[A + a] e^{i\frac{1}{2}\int d^D x\, \varepsilon^{\mu\nu\lambda\ldots\alpha\beta} b_{\mu\nu\lambda\ldots} f_{\alpha\beta}(a)}, \tag{1.37}$$

and by the shift $a \to a - A$, finally,

$$\mathscr{Z}[A] = \int \mathscr{D}[a]\mathscr{D}[b]\mathscr{Z}[a] e^{i\frac{1}{2}\int d^D x\, \varepsilon^{\mu\nu\lambda\ldots\alpha\beta} b_{\mu\nu\lambda\ldots}[f_{\alpha\beta}(a) - F_{\alpha\beta}(A)]}$$

$$\equiv \int \mathscr{D}[a]\mathscr{D}[b] e^{i\int d^D x \mathscr{L}}, \tag{1.38}$$

where the last equality defines the Lagrangian $\mathscr{L}$. Note that this action, by construction, is invariant under gauge transformations of the electromagnetic potential A_μ, since the electrical current is conserved in the model of consideration. Below you will see that this implies the existence of edge modes.

Given this one can calculate the expectation value of the $U(1)$ current as

$$\langle j^\mu(x)\rangle = i\frac{\delta \ln \mathscr{Z}[A]}{\delta A_\mu(x)} = \left\langle \varepsilon^{\mu\nu\lambda\rho\ldots} \partial_\nu b_{\lambda\rho\ldots}(x)\right\rangle, \tag{1.39}$$

and similarly for higher order correlation functions. Note that by construction the current is conserved. To appreciate the meaning of the field $b_{\mu\nu\lambda\ldots}$, let us look at the simplest special cases. For $D = 2$, b is a scalar, and the above relation reads,

$$\langle j^{\mu}(x)\rangle = \langle \varepsilon^{\mu\nu}\partial_{\nu}b(x)\rangle \ , \tag{1.40}$$

which you might recognize if you are familiar with the method of bozonization in $1 + 1$ dimension. This case is special, in the sense that it holds even if the average is removed, that is it holds as an *operator* identity. A concise account of the fascinating physics and mathematics of $(1 + 1)D$ systems can be found in the books [14, 29].

For $D = 3$, b is vector field and

$$\langle j^{\mu}(x)\rangle = \langle \varepsilon^{\mu\nu\sigma}\partial_{\nu}b_{\sigma}(x)\rangle \ . \tag{1.41}$$

Up to a normalization, which we will discuss below this is the same as the previously derived relation for the electric current, (1.28).

Clearly the expression for $\mathscr{Z}[A_{\mu}]$ derived above, (1.38), is useful only if we can, at least approximately, evaluate the fermionic functional integral to get $Z[a]$. In $1 + 1$ dimensions this can sometimes be done exactly. In higher dimensions this is not possible. However one can find an approximation by assuming there is a gap and making a derivative expansion.

In the case of $2d$ systems with quantum Hall response, we already know one piece in $Z[a]$ that will for sure be present namely the Hall term, (1.24). Combining this with the just derived expression for $\mathscr{Z}[A_{\mu}]$ (1.38), gives the effective Lagrangian,

$$\mathscr{L} = -\frac{1}{2\pi}\varepsilon^{\mu\nu\sigma}b_{\mu}\partial_{\nu}(a_{\sigma} - A_{\sigma}) + \frac{\sigma_H}{2}\varepsilon^{\mu\nu\sigma}a_{\mu}\partial_{\nu}a_{\sigma} \ , \tag{1.42}$$

where we renormalized the field b so that it, up to the factor $(-e)$, is identical to the previous expression for the electromagnetic current, (1.28).

Above we used a seemingly arbitrary argument to fix the normalization of the field b, and you should worry about this since a different convention would give a different value for σ_H when b is integrated over to obtain the effective response action. To understand this point, we must look closer at the first term in the above Lagrangian, (1.42) which is the 2d incarnation of the topological *BF theory* which is defined in any dimension as,

$$\mathscr{L}_{BF} = -\frac{1}{2}\varepsilon^{\mu\nu\lambda\ldots\alpha\beta}b_{\mu\nu\lambda\ldots}f_{\alpha\beta}(a)\,. \tag{1.43}$$

In Sect. 1.6.2 we shall briefly discuss the $3d$ case in the connection with fluctuating superconductors. There is a rich mathematical literature on BF theory, [30, 31], but here we shall only cover material of direct relevance for physics. One such point is the question of normalization brought up above. With the chosen normalization the ground-state is unique, as is required for a number of filled Landau levels. A derivation of this result is given in Sect. 1.8.2.

1.2.4 The Bulk-Boundary Correspondence

We now show how the Chern–Simons topological field theory (1.27) in a natural way incorporates the presence of edge excitations. For a more thorough discussion you should consult the paper by Stone [32] and the review by Wen [33]. We specify the action by integrating the CS Lagrangian, (1.27), over a bounded and simply connected region D, and for simplicity we will put $\sigma_H = \sigma_0$ throughout this section,

$$S[b; A] = \int_D \mathrm{d}^3x \, \mathscr{L}(b, A) \, . \tag{1.44}$$

This action is not gauge invariant since we get a non-zero variation at the boundary ∂D. What this means is that the pure gauge mode, $\partial_\mu \chi$ which in the bulk has no physical meaning, (and would be removed by gauge fixing) will at the edge manifest itself as a physical degree of freedom. To see this explicitly we substitute $b_\mu = \partial_\mu \chi$ into the Lagrangian to get,

$$S[b, \chi; A] = -\frac{1}{4\pi} \int_D \mathrm{d}^3x \, \varepsilon^{\mu\nu\sigma} b_\mu \partial_\nu b_\sigma + \int_{\partial D} \mathrm{d}t \mathrm{d}x \, \chi \, \partial_x (\partial_t - v\partial_x) \chi(x, t) \, , \tag{1.45}$$

where for simplicity we neglected the external field A (which is easily reintroduced), and where the field $\chi(x, t)$ has support only on the boundary ∂D parametrized by the coordinate x. We also added an extra term $\sim \chi \, \partial_x^2 \, \chi$ that does not follow from the Chern–Simons action (1.44), but which will be present if there is an electrostatic confining potential [33], which is needed e.g., in the quantum Hall effect to define the quantum Hall droplet. The meaning of this term is clear from the equation of motion for the χ-field,

$$(\partial_t - v\partial_x)\chi(x, t) = 0 \, , \tag{1.46}$$

which shows that v is the velocity of a gapless edge-mode propagating in one direction. The physical origin of this velocity is obvious: it is the $\mathbf{E} \times \mathbf{B}$ drift velocity of the electrons in the external magnetic field and the confining electric field at the boundary. If we reintroduce the electromagnetic field and study the current conservation at the boundary, we will see that the non-conservation of the bulk current, which follows from the non gauge-invariant part of the Chern–Simons action, (1.26), is compensated by a corresponding non-conservation of the boundary current; so the total charge is conserved [32].

The mathematics related to this result is quite interesting. The boundary theory should after all just be a model for electrons moving in one direction along a line, and as such we would expect the theory just to be that of a Fermi gas, or if interactions are present, a Luttinger liquid. In both cases we would expect the boundary charge to be conserved. What is special here is that the mode is *chiral*, i.e., it only propagates in one direction. From the theory of Luttinger liquids, we learn that in the presence of an electric field, the right and left moving currents are not separately conserved, but only their sum, which is the electromagnetic current. The difference, which defines the *axial current*, which in Dirac notation reads $j^A_\mu = \bar{\psi}\gamma_3\gamma_\mu\psi$, is not conserved because of the *axial anomaly*. The subject of anomalies in quantum field theory is fascinating, but will not be discussed in these lectures.

1.3 Physical Systems with Quantized Hall Conductance

In this section we move from the general discussion to real physical systems that have a quantized Hall response. We begin with the integer quantum Hall effect (IQHE), which started the whole field of topological states of matter. Given the previous general analysis, we can use a simple symmetry argument to explain it.

We then turn to the systems that have been game changers for the last decade—the various kinds of topological band insulators. Recall that one of the early successes of quantum mechanics was the division of crystalline materials into conductors, semiconductors and insulators depending on whether or not the Fermi level is inside a band gap (there is no sharp distinction between insulators and semi-conductors; only a conventional classification depending on the size of the gap). The insulators seem to be the most boring states, and it was an important discovery that they can belong to different topological classes which differ in their quantum Hall response.

In an intermediate step, we study a system with both a magnetic field, and a weak periodic potential. This is instructive not only for providing a more realistic model for the quantum Hall effect, but also for showing how to calculate a topological invariant for a clean (non-interacting) band insulator. We then turn to the first example a system with a quantized Hall effect without any magnetic field, the Chern Hall insulator, and stress the importance of breaking time-reversal invariance.

1.3.1 The Integer Quantum Hall Effect

The integer quantum Hall effect is observed when a very clean two dimensional electron gas is cooled and subjected to a strong perpendicular magnetic field.

If we neglect electron–electron interactions, this is the famous Landau problem and we know that the energy is quantized as $E_n = n\hbar\omega_c$ with the cyclotron frequency $\omega_c = eB/m$. Each of these Landau levels (LLs) has a macroscopic degeneracy such that there is one quantum state in the area $2\pi\ell_B^2 \equiv 2\pi\hbar/eB$ which corresponds to a unit flux, $\phi_0 = h/e = 2\pi/e$.

In Sect. 1.2.1 you learnt that for a gapped state the Hall conductance is always quantized in integer multiples of the quantum of conductance divided by the degeneracy of the ground-state. For a number of filled Landau levels, the ground-state is non-degenerate, and one can obtain the Hall conductance by a simple symmetry argument: Let us start from the idealized case of non-interacting electrons moving in the xy-plane, and no impurities. Having p completely filled LLs corresponds to a charge density $\rho = -ne$, where $n = p/(2\pi\ell_B^2)$ is the electron number density. From this we get

$$B = \frac{\rho}{p}\frac{2\pi\hbar}{e^2} = \frac{1}{p\,\sigma_0}\rho\,, \tag{1.47}$$

where $\mathbf{B} = B\hat{z}$.

Now assume that in our frame we have an electric field $\mathbf{E} = 0$ and vanishing current density $\mathbf{j} = 0$. Then consider a frame moving with velocity $-\mathbf{v}$, $v \ll c$, relative to us. In this frame $\mathbf{B}$ is unchanged, but $\mathbf{E} = -\mathbf{v} \times \mathbf{B}$, so

$$\mathbf{j} = \rho\,\mathbf{v} = p\sigma_0\,\hat{z} \times \mathbf{E}\,, \tag{1.48}$$

from which follows that the Hall conductivity is $\sigma_H = p\sigma_0$, and since our system is invariant under Galilean transformation, this result holds in any inertial frame. We now return to a realistic system with electron–electron interactions, and impurities. Since we have already shown that the conductance is quantized as long as the gap remains we know that the conductance must stay the same as these potentials are turned on under the assumption that the gap does not close.

1.3.2 The Hall Conductance in a Periodic Potential

We shall now present a special case of a problem originally treated in a very influential paper by Thouless et al. [24]. Recall that in a constant magnetic field, the (magnetic) translation operators, T_1 and T_2, which commute with the Hamiltonian, do not commute among themselves. This follows because $T_1^{-1}T_2^{-1}T_1T_1$ does not equal identity, since it amounts to a closed path that encloses flux and thus by the Aharonov-Bohm argument gives a phase to the wave function. From this we learn that if one picks a *flux lattice*, which is defined such that there is exactly one flux-quantum through a unit cell, the lattice translation operators will all commute and can be simultaneously diagonalized.

We shall consider the case where the Bravais lattice of the periodic potential is such that the unit cell of the flux lattice contains an integer number of unit cells of the potential.[6] In this case we can still diagonalize the magnetic translations, and then invoke Bloch's theorem to express the wave functions as,

$$\psi_{\mathbf{k}\,n}(\mathbf{x}) = e^{i\mathbf{k}\cdot\mathbf{x}} u_{\mathbf{k}\,n}(\mathbf{x}) \,, \tag{1.49}$$

where n is a band index (here the LL index) and $\mathbf{k}$ the crystal, or quasi, momentum that lives in the "magnetic Brillouin zone", $|k_i| \leq \pi/\ell_B = \pi\sqrt{eB}$.[7] The Bloch functions $u_{\mathbf{k}\,n}(\mathbf{x})$ are eigenfunctions of the Bloch Hamiltonian,

$$H_{Bl} = \frac{\hbar^2}{2m}(-i\nabla + e\mathbf{A} + \mathbf{k})^2 + V_{lat} \,. \tag{1.50}$$

We will assume that the periodic potential is weak enough that the cyclotron gap persist, and that the lowest N bands are completely filled. For each $\mathbf{k}$ in the Brillouin-zone (BZ) there are N wave functions

$$\left\{e^{i\mathbf{k}\cdot\mathbf{x}} u_{\mathbf{k}\,n}(\mathbf{x})\right\}_{n=1,\ldots,N} \,, \tag{1.51}$$

so associated to each $\mathbf{k} \in BZ$ there is an N dimensional sub-Hilbert-space, $h(\mathbf{k})$, of the full single particle Hilbert space. The set $\{\mathbf{k}, h(\mathbf{k})\}_{\mathbf{k}}$ where $\mathbf{k} \in BZ$, is thus a fiber bundle over the Brillouin-zone, see Sect. 1.8.1. By definition the Berry connection on this fiber bundle is

$$\mathscr{A}^{nm}_{k_i}(\mathbf{k}) = -i\left\langle \psi_{\mathbf{k}\,n}(\mathbf{x}) | \partial_{k_i} | \psi_{\mathbf{k}\,m}(\mathbf{x})\right\rangle \equiv -i \int d^2x\, \psi^*_{\mathbf{k}\,n}(\mathbf{x}) \partial_{k_i} \psi_{\mathbf{k}\,m}(\mathbf{x}) \,.$$

This can also be written as the anti-commutator of the creation and annihilation operators,

$$\mathscr{A}^{nm}_{i} = -i\left\{a_{\mathbf{k}\,n}, \partial_{k_i} a^\dagger_{\mathbf{k}\,m}\right\} \tag{1.52}$$

where $a^\dagger_{\mathbf{k}\,n}$ and $a_{\mathbf{k}\,n}$ are the Fourier components of the electron creation and annihilation operators $\psi(\mathbf{x})$ and $\psi^\dagger(\mathbf{x})$ satisfying $\left\{\psi^\dagger(\mathbf{x}), \psi(\mathbf{x}')\right\} = \delta^2(\mathbf{x} - \mathbf{x}')$. The corresponding Berry field-strength, which we denote by $\mathscr{B}$ to distinguish it from the previously defined flux-torus field strength, becomes

$$\mathscr{B}^{nm}_{k_i,k_j} = \partial_{k_i}\mathscr{A}^{nm}_{k_j} - \partial_{k_i}\mathscr{A}^{nm}_{k_j} + i\mathscr{A}^{np}_{k_i}\mathscr{A}^{pm}_{k_j} - i\mathscr{A}^{np}_{k_i}\mathscr{A}^{pm}_{k_j} \,, \tag{1.53}$$

where the repeated index p should be summed over. Since the Brillouin-zone is two-dimensional, the field strength has only one independent component

---

[6] Reference [24] treated the case where the ratio between the areas of the unit cells in the flux lattice and the Bravais lattis of the potential is a rational number q/p. In case one has to consider a larger unit cell, and each filled band will in general have a larger Hall conductance.

[7] In a translationally invariant system, the shape of this zone is arbitrary, but the area is fixed to support n units of magnetic flux. In our case the shape has to be taken as to be commensurate with the Bravais lattice of the potential.

$$\mathscr{B} \equiv \varepsilon^{ij} \mathscr{B}^{nm}_{k_i,k_j} \ . \tag{1.54}$$

From now on we will suppress the upper indices m, n etc. and multiplications of $\mathscr{B}$ or $\mathscr{A}$ has the meaning of matrix multiplication. In this short hand notation we have

$$\mathscr{B} = \varepsilon^{ij} \left(\partial_{k_i} \mathscr{A}_{k_j} + i \mathscr{A}_{k_i} \mathscr{A}_{k_j} \right) \ , \tag{1.55}$$

for the the Berry "magnetic" field in the Brillouin-zone.

Although this *Brillouin zone bundle* is conceptually quite different from the flux bundle, with Berry curvature $\mathscr{F}$, introduced in Sect. 1.2.1, they turn out to be closely related in the present case where electron–electron interactions and random impurities are neglected. To see this, first recall that the non-interacting many body groundstate is given by

$$|GS\rangle = \prod_{n=1}^{N} \prod_{\mathbf{k} \in BZ} a^{\dagger}_{\mathbf{k}\,n} |0\rangle \ . \tag{1.56}$$

Secondly, from the Bloch Hamiltonian, H_{Bl}, we can, by taking a vector potential describing fluxes through the holes in the torus, infer that the Bloch functions at finite flux are related to those at zero flux by,

$$u^{\phi_x,\phi_y}_{\mathbf{k}\,n}(\mathbf{x}) = u^{0,0}_{\mathbf{k}'\,n}(\mathbf{x}); \qquad \mathbf{k}' = \left(k_x + \frac{2\pi}{L_x} \frac{\phi_x}{\phi_0}, k_y + \frac{2\pi}{L_y} \frac{\phi_y}{\phi_0} \right) \ , \tag{1.57}$$

where ϕ_x and ϕ_y denote the fluxes encircled by the two independent non-contractible loops on the torus, see Fig. 1.2. This means that a derivative with respect to a the flux ϕ_i can be turned into a derivative with respect to the crystal momentum k_i. We can also define the Chern number for the flux-torus fiber bundle, defined by the N first LLs,

$$|GS, \boldsymbol{\phi}\rangle = \prod_{n=1}^{N} \prod_{\mathbf{k} \in BZ} a^{\dagger}_{\mathbf{k}\,n\,\boldsymbol{\phi}} |0\rangle \ ; \qquad a^{\dagger}_{\mathbf{k}\,n} = \int \mathrm{d}^2 x \, \psi^{\dagger}(\mathbf{x}) \psi^{\phi_x,\phi_y}_{\mathbf{k}\,n}(\mathbf{x}) \ , \tag{1.58}$$

i.e., with Berry connection

$$\mathscr{A}_{\phi_i} = \left\langle GS, \boldsymbol{\phi} | \partial_{\phi_i} | GS, \boldsymbol{\phi} \right\rangle \ . \tag{1.59}$$

In Sect. 1.2.1 we showed that the Chern number of this connection is proportional to the Hall conductance, and in Sect. 1.8.1.4 we show that the first Chern number for the flux-torus fiber bundle, and the Brillouin zone bundle are equal. Combining these results yields the formula

$$\sigma_H = \frac{\sigma_0}{2\pi} \int d^2k \, \mathrm{Tr}\, \mathscr{B} \,, \tag{1.60}$$

for the Hall conductance. Thus, in this case we can express the Hall conductance in terms of the first Chern number of the BZ bundle. This result was originally obtained in [24] by using linear response and properties of the Bloch wave functions.

The above calculation using the Brillouin bundle relies on translational invariance, and the absence of interactions and is thus much less general than the result derived in Sect. 1.2.1. When applicable the formula derived here is however much simpler to handle, and it has been essential in developing the classification of non-interacting topological matter—its theoretical importance should not be underestimated.

1.3.3 The Chern Insulator

The formula (1.60) for the Hall conductance opens the possibility of having a topological phase in a crystalline system even in the absence of a magnetic field. Importantly, it demonstrates that *topological band theory* can be used to determine the actual values of the topological invariants.

In an important paper from 1988, Haldane showed that one can have a quantum Hall effect without any net magnetic field [34]. He constructed an effective model of electrons hopping on a hexagonal lattice penetrated by a staggered magnetic field that is on average zero. The model did, however, break time-reversal invariance and is today referred to as a *Chern insulator* that exhibits a *quantized anomalous Hall effect*. A model that is slightly simpler than the one used by Haldane is free electrons hopping on a square lattice, with a π-flux on each elementary plaquette [35]. Since the main theme of these notes are continuum field theory descriptions, we will not give the position space lattice Hamiltonian, which you can find in the original work [35]. For the present purpose, it suffices to say that *the Chern insulator* can be modelled by the following two-band momentum space Hamiltonian,

$$H_C = \sum_{\mathbf{k}} c^\dagger_{\mathbf{k}\,\alpha} h^{\alpha\beta}(\mathbf{k}) c_{\mathbf{k}\,\beta} \,, \tag{1.61}$$

with

$$h^{\alpha\beta}(\mathbf{k}) = d_a(\mathbf{k})\sigma^a_{\alpha\beta}; \qquad \mathbf{d} = (\sin k_x, \sin k_y, M + \cos k_x + \cos k_y) \,, \tag{1.62}$$

where both energy and M are measured in units of some hopping strength, t.

To calculate the Berry flux, we note that $h^{\alpha\beta}(\mathbf{k})$ is nothing but the Hamiltonian for a spin-half particle moving in a magnetic field $\mathbf{d}$. The spectrum, given by the Zeeman energy, is thus $\pm|\mathbf{d}|$, and the eigenfunctions satisfy,

$$\hat{d} \cdot \boldsymbol{\sigma} \, |E_k; \pm\rangle = \pm \, |E_k; \pm\rangle \,. \tag{1.63}$$

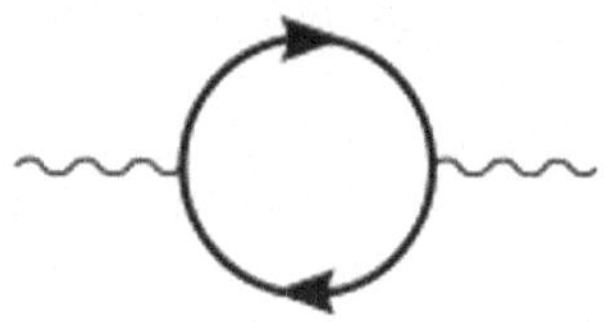

Fig. 1.3 Diagram giving rise to the CS action as the lowest term in a derivative expansion

Assuming that there is no gap closing, i.e., $|\mathbf{d}| > 0$, and taking the Fermi energy to be zero, the Berry field strength can be shown to be

$$\mathscr{B}^{\pm}(\mathbf{k}) = \varepsilon_{ij}\partial_i \mathscr{A}^{\pm}_j = \mp\frac{1}{2}\varepsilon_{ij}\hat{d}\cdot\partial_i\hat{d}\times\partial_j\hat{d}\,, \tag{1.64}$$

where the lower sign corresponds to the filled band and we used the short hand notation $\partial_{i/j} \equiv \partial_{k_{i/j}}$. The most direct way to show the above formula, although a bit tedious, is to first find $|E_k; -\rangle$ and then just calculate. An alternative derivation that does not require the explicit wave functions is given in Sect. 1.8.3.

The expression on the right hand side of the above expression is the Jacobian of the function $\hat{d}(\mathbf{k})$, and we define the integer valued Pontryagin index by

$$\mathscr{Q} = \frac{1}{8\pi}\int \mathrm{d}^2k\, \varepsilon_{ij}\hat{d}\cdot\partial_i\hat{d}\times\partial_j\hat{d}\,, \tag{1.65}$$

which measures how many times the surface of the unit sphere on which $\hat{d}$ is defined, is covered by the map from the compact manifold where $\mathbf{k}$ is defined.

It remains to determine the value of $n = \mathscr{Q}$. For large $|M|$, where the hopping can be neglected, the eigenfunctions $|E_k\rangle$ become $\mathbf{k}$-independent and the Pontryagin density is identically zero. ($M \gg 1$ is the atomic limit where the wave functions are sharply localized at the lattice sites.) $\mathscr{Q}$ is a topological quantity, so it can only change when the gap in the Fermi spectrum closes and $\hat{d}$ no longer is a smooth function of $\mathbf{k}$. From the expression $E_k = -|\mathbf{d}(\mathbf{k})|$, one realizes that the gap closes for $M = -2$ (at $\mathbf{k} = 0$), for $M = 2$ (at $\mathbf{k} = (\pi, \pi)$) and for $M = 0$ (at $\mathbf{k} = (0, \pi)$ and $\mathbf{k} = (\pi, 0)$).

Let us now analyze what happens when M increases from a large negative value towards -2. Putting $M = -2 + m$, and linearizing the Hamiltonian one gets,

$$H_{lin} = k_x\sigma_1 + k_y\sigma_2 + m\sigma_3\,, \tag{1.66}$$

which we recognize as the Hamiltonian for a $D = 2 + 1$ Dirac particle. Since the topological nature of a phase is a low-energy property, one should be able to capture the change in phase by analyzing the continuum theory in the vicinity of $m = 0$. We now show how to do this.

The continuum D = 2 + 1 Dirac theory is defined by the Lagrangian,

$$\mathscr{L}_D = \bar{\psi}\left(\gamma^\mu(i\partial_\mu - eA_\mu) - m\right)\psi \tag{1.67}$$

with $\gamma^\mu = (\sigma_3, i\sigma_2, -i\sigma_1)$. We calculate the electromagnetic response by integrating out the fermions, which, to lowest order in A_μ, amounts to calculating the loop diagram in Fig. 1.3. This was originally done by Redlich [36] with the result,

$$\Gamma_D[A] = \frac{m}{|m|}\frac{e^2}{8\pi}\int \mathrm{d}^3x\, \varepsilon^{\mu\nu\sigma} A_\mu \partial_\nu A_\sigma \,. \tag{1.68}$$

Note that a Dirac mass term in $2d$ breaks the parity symmetry, so it is not surprising that a Chern–Simons term appears in the effective action. The term persists even in the limit $m \to 0$, where the classical Lagrangian respects parity, and is therefore often referred to a parity anomaly.[8] In Sect. 1.8.4 we give an alternative derivation of this result by calculating the response to a constant magnetic field. Just as the Schrödinger case, the eigenvalues of the Dirac equation falls into Landau levels, $E_n = \pm\sqrt{neB + m^2}$, for $n > 0$, and the contribution from these states to σ_H cancel. Only the lowest Landau level, with $n = 0$ and energy $E_0 = m$, contributes. This derivation stresses that even though anomalies seem to be related to the short distance behaviour of the theory, they should be considered as a low energy effect.

Note that the coefficient in the response action derived from the Dirac Lagrangian, (1.68), translates into a Hall conductivity $\sigma_H = \pm\sigma_0/2$ which is half of the one calculated above. This is surprising since we have argued that the Hall conductivity, for topological reasons, must be an integer times σ_0. The solution to this apparent contradiction, is that it is not possible to consistently formulate the Dirac equation on a lattice without "doubling" the number of low-energy fermions. This result, first obtained in the context of high energy physics by [37], basically says that the low-energy physics of fermions in a band of finite width, cannot be faithfully represented by a single Dirac field.[9] We now return to the two-band model (1.61). Recall that we put $M = -2 + m$, and we want to know what happens as m is tuned from a small negative value to a small positive value. Doing this changes the spectrum only in the close vicinity of $\mathbf{k} = 0$, so the *change* in σ_H should be faithfully represented by the linearized Dirac theory. From the response action derived from the Dirac Lagrangian (1.68) we get the change $\Delta\sigma_H = \frac{1}{2}(1 - (-1))\sigma_0 = \sigma_0$. A similar analysis can be made for the other gap closing points. The result is that σ_H, in units of σ_0 changes as $0 \to 1 \to -1 \to 0$ as M is tuned from $-\infty$ to ∞. It should now also be clear, that the effective topological theory for the Chern insulator is identical to that for the IQHE given by (1.42). Note that at $M = 0$ the gap closes at two points, so to model this transition we need two Dirac fields, and thus the change of two units in σ_H.

In 2013, a quantized Hall effect was observed in thin films of Cr-doped $(\mathrm{Bi,Sb})_2\mathrm{Te}_3$ at zero magnetic field, thus providing the first experimental detection of a Chern insulator [38].

---

[8] As explained in [36] a gauge invariant regularization of the ultraviolet divergence (e.g., using the Pauli-Villars method) gives rise to the anomaly term, but does not fix the sign.

[9] See Sect. 16.3.3 in [14] for more details.

1.4 Generalizing to Other Dimensions

1.4.1 The 1d Case

Until now we discussed the $2d$ systems with $U(1)$ symmetry and showed that in the topological scaling limit, their response action is the Hall term

$$\Gamma[A_\mu] = \frac{\sigma_H}{2} \int \mathrm{d}^3x \, \varepsilon^{\mu\nu\sigma} A_\mu F_{\nu\sigma} \,, \tag{1.69}$$

which encodes the Hall response. We again emphasize that this expression is independent of the metric.

In $2d$ we knew from the start that we where looking for, a Hall response. Now assume that we had not known that, but would anyway have asked the question: Is there a possible topological response? A strategy in this hypothetical case would have been to write down all possible response actions that are gauge invariant and independent of the metric (and thus of any length or time scale). On a technical level the absence of a metric implies that the only way to contract indices is by the anti-symmetric epsilon tensor. In $(2+1)D$ this leaves only one option namely the Chern–Simons action (1.69). In $(1+1)D$, there is also only one choice namely

$$\Gamma[A_\mu] = \frac{e}{4\pi} \int \mathrm{d}^2x \, \theta \, \varepsilon^{\mu\nu} F_{\mu\nu} \,, \tag{1.70}$$

which we will refer to as the $1d$ θ-term. This is the integral of the electric field strength, and is thus, as opposed to the Chern Simons term, fully gauge invariant, and it does not contribute to the equations of motion. The choice of the symbol θ, which indicates an angle, is not accidental as will be clear below.

The θ-term is different from the Chern–Simons term in $(2+1)D$ in that uniform fields do not induce any currents but only polarization. In the static case, polarization amounts to creating a dipole density that partially screens the external electric field, $\mathbf{D} = (1 + \chi_e)\,\mathbf{E}$, or equivalently it creates edge charges. To see how the topological term alters this, consider a line segment with endpoints at $x = x_\pm$. Choosing the gauge with $A_x = 0$, and assuming θ in the $1d$ θ-term, (1.70), to be constant gives,

$$\Gamma = -e \frac{\theta}{2\pi} \int \mathrm{d}t \, [A_0(x_+, t) - A_0(x_-, t)] \,. \tag{1.71}$$

Varying Γ with respect to $A_0(x_\pm, t)$ gives the charge at $x_\pm$,

$$Q^\pm = \mp e \frac{\theta}{2\pi} \,, \tag{1.72}$$

where Q^+ and Q^- are the charges on the right and left ends of the line segment respectively. Thus the topological term adds a constant to the edge charge. Including the usual non-topological action $\sim \int d^2x\, \chi_e \mathbf{E}^2$ (cf. the $2d$ linear response action (1.21)) for a dielectric gives

$$Q^{\pm} = \pm \chi_e e V \mp e \frac{\theta}{2\pi} , \tag{1.73}$$

where $V = A_0(x_+, t) - A_0(x_-, t)$ is the voltage difference between the left and right end.

We can now see why θ should be regarded as an angular variable. Making the shift $\theta \to \theta + k\, 2\pi$, amounts simply to adding k unit charges at the ends of the wire which is a local effect that for instance can be due to impurities. A non integer value of $\theta/2\pi$, on the other hand, is a bulk polarization effect, and we now show that it differs from the usual polarization $\sim\chi_e$ in being quantized as long as certain symmetries are respected.

For this, we again look at a system with periodic boundary conditions, i.e., a circle. Now imagine that we adiabatically transform a system from a trivial insulator to an insulator with a non-trivial $1d$ θ-term (1.70). During this process, a charge, Q, will be transported around the circle and this charge is, as we will see, given purely by θ. This is directly related to the bulk polarization, since to create a polarization charge Q, it has to flow past every point except close to the edges where it accumulates.

Varying the $1d$ θ-term, with respect to A_μ, the current

$$j_x = e \frac{\partial_t \theta(t)}{2\pi} , \tag{1.74}$$

and the total charge that has been transported around when $\theta(t)$ is changed from θ_1 to the final value θ_2, is

$$Q = \int dt\; j_x = \int dt\, e \frac{\partial_t \theta(t)}{2\pi} = e \frac{\theta_2 - \theta_1}{2\pi} .$$

Now lets calculate this in a microscopic picture using a many body state $|\psi(t, \phi)\rangle$ that start out in the atomic limit, and then evolves adiabatically to some state $|\psi(t_2, \phi)\rangle$ (ϕ denotes the magnetic flux passing through the circle). A similar set of manipulations as those used to show that the Hall conductance is the first Chern number gives

$$j_x = \langle \partial_t \psi | \partial_\phi \psi \rangle - \langle \partial_\phi \psi | \partial_t \psi \rangle \stackrel{def.}{=} \mathscr{F}_{t\phi} . \tag{1.75}$$

Here $\mathscr{F}_{t\phi}$ is the Berry field strength of the fiber bundle with the cartesian product of the time-interval $[t_1, t_2]$ and the space of flux-values through the circle as base space and on-dimensional fibers spanned by $|\psi(t, \phi)\rangle$. The total charge passing through a point on the circle during the process is now obtained by integrating over t. Also, as in the case of the Hall conductance, we invoke locality to average over the flux (here meaning that the polarizability cannot depend on the flux through the circle). We get,

$$Q = \frac{1}{2\pi}\int_0^{2\pi} d\phi \int_{t_1}^{t_2} dt\, \mathscr{F} = \int_0^{2\pi} d\phi\, \mathscr{A}_\phi(t_2) - \int_0^{2\pi} d\phi\, \mathscr{A}_\phi(t_1)\,, \tag{1.76}$$

where

$$\mathscr{A}_\phi(t) = \langle\psi|\partial_\phi\psi\rangle - \langle\partial_\phi\psi|\psi\rangle \tag{1.77}$$

is the ϕ component of the Berry connection of the just mentioned fiber bundle. We now refer to Sect. 1.8.1, where the $1d$ Chern–Simons invariant is defined as,

$$CS_1[\mathscr{A}(t)] = \frac{1}{2\pi}\int_0^{2\pi} \mathscr{A}_\phi(t) d\phi\,, \tag{1.78}$$

and where it is shown that 2π times the exponent of this is a well defined basis independent property. At t_1 the state is just a product state of localized electrons, so we can choose a gauge where $\mathscr{A}_\phi(t_1) = 0$ and it follows that

$$e^{2\pi CS_1[\mathscr{A}(t_1)]} = 1; \qquad e^{i2\pi Q/e} = e^{2\pi CS_1[\mathscr{A}]}\,, \tag{1.79}$$

where t_2 is suppressed since the state $|\psi(t_2)\rangle$ is the many-body state of interest i.e., we define $\mathscr{A}_\phi \equiv \langle\psi(t_2)|\partial_\phi\psi(t_2)\rangle - \langle\partial_\phi\psi(t_2)|\psi(t_2)\rangle$.

The above phase (1.79) has an alternative interpretation as the Berry phase accumulated when a unit flux is adiabatically inserted through the circle. Since both time-reversal and chiral symmetry (that is charge conjugation composed with time reversal) maps inserting an upward flux to inserting a downward flux, one can conclude that if any of these symmetries are present during the adiabatic process, one has

$$e^{\int_0^{2\pi} \mathscr{A}_\phi d\phi} = e^{-\int_0^{2\pi} \mathscr{A}_\phi d\phi}\,, \tag{1.80}$$

which leaves only two possibilities,

$$e^{iQ/e} = e^{2\pi CS_1[\mathscr{A}]} = \begin{cases} e^{i\pi} & \text{non-trivial} \\ 0 & \text{trivial} \end{cases}\,, \tag{1.81}$$

corresponding to having $\theta = 0$ or $\theta = \pi \bmod 2\pi$.

If we have a non-interacting system with lattice translation invariance we will as in $2+1$ dimensions have a fiber bundle defined by the Bloch states over the Brillouin zone circle. Again, in the same way as the Chern number, the exponent of the Chern–Simons invariant of the flux-circle bundle will be the same as the exponent of the Chern–Simons invariant of the Brillouin zone circle. We will make use of this in the next section where we study a model which has a topological polarization response.

1.4.2 Realization with Dirac Fermions

Rather than studying a lattice model, we shall pick a continuum model with a global chiral symmetry, and would therefore be expected to be characterized by the Chern–Simons invariant and thus fall into one of the two classes we found above. Since topological response is a long distance effect, we expect that such a continuum theory will also describe chiral symmetric lattice models.

With this motivation, we shall investigate the $1d$ Dirac fermion ψ, with mass m, coupled to a gauge field a using path integral methods. Our starting point is the partition function,

$$Z[a] = \int \mathscr{D}[\bar{\psi}, \psi] e^{i\int d^2x\, \bar{\psi}(\gamma^\mu(i\partial_\mu - a_\mu) - m)\psi} . \tag{1.82}$$

We parametrize the gauge field as $a_\mu = \varepsilon_{\mu\nu}\partial_\nu\xi + \partial_\mu\lambda$, so that $F = \varepsilon^{\mu\nu}\partial_\mu a_\nu = -\partial^2\xi$; the term containing λ is just a gauge transformation which does not contribute to the action (provided we are on a simply connected manifold). One can now verify that the *chiral transformation*

$$\psi \rightarrow e^{-i\gamma_3\xi}\psi , \tag{1.83}$$

where $\gamma_3 = i\gamma_0\gamma_1$, eliminates the transverse gauge field $\varepsilon^{\mu\nu}\partial_\nu\xi$ from the action, while the mass term is changed,

$$\bar{\psi}(\gamma^\mu(i\partial_\mu - a_\mu) - m)\psi \rightarrow \bar{\psi}(\gamma^\mu i\partial_\mu - me^{-2i\gamma_3\xi})\psi . \tag{1.84}$$

This looks very strange, since for the massless case it seems like we by this transformation can get rid of a non-trivial field. The resolution of the apparent contradiction is that the path integral measure is not invariant under the transformation. Using techniques pioneered by Fujikawa [39], one can show that under the chiral transformation (1.83),

$$\mathscr{D}[\bar{\psi}, \psi] \rightarrow \mathscr{D}[\bar{\psi}, \psi] e^{-\frac{i}{2\pi}\int d^2x\, \xi\partial^2\xi} , \tag{1.85}$$

which is the path integral incarnation of the axial anomaly referred to at the end of Sect. 1.2.4. In particular we shall be interested in a (space-time) constant chiral transformation $\xi(x) = -\theta/2$, which does not change the coupling to a_μ but only affects the mass term, and introduces a $1d$ θ-term, (1.70), in the response action.

For the case of the continuum Dirac equation we shall follow the same logic as in the $2+1$ dimensional case, and only calculate how the value of θ *differs* between different phases. From (1.84) one realizes that taking $2\xi = \theta = \pi$ amounts to changing the sign of the fermion mass. Taking the gamma matrices,

$$\gamma^0 = \sigma^1; \qquad \gamma^1 = i\sigma^3; \qquad \gamma^3 = i\sigma^2 , \tag{1.86}$$

we have

$$H = \begin{pmatrix} 0 & m - ik \\ m + ik & 0 \end{pmatrix} = \begin{pmatrix} 0 & Q^\dagger \\ Q & 0 \end{pmatrix} . \quad (1.87)$$

It is straightforward to obtain the wave functions, and calculate the Brillouin-zone Berry potential,

$$\mathscr{A} = \frac{1}{2} \frac{m}{k^2 + m^2} . \quad (1.88)$$

We can now form the corresponding Chern–Simons invariant by integrating over the filled states labeled by k,

$$CS_1[\mathscr{A}] = \frac{1}{2\pi} \int_{-\infty}^{\infty} \mathrm{d}k\, \mathscr{A} . \quad (1.89)$$

Using the expression for the Berry connection of the Brillouin-zone fiber bundle (1.88) and the definition of the Chern–Simons invariant (1.89) we get

$$CS_1 = \frac{1}{2\pi} \int_{-\infty}^{\infty} \mathrm{d}k \, \frac{1}{2} \frac{m}{k^2 + m^2} = \frac{1}{4} \frac{m}{|m|} \quad (1.90)$$

for the filled Dirac sea. Previously, in the discussion of the Chern insulator, we saw that the Chern number of the Brillouin-zone fiber bundle and the flux-torus fiber bundle were equal. Analogously the just derived Brillouin zone Chern–Simons invariant equals the flux-circle Chern–Simons invariant, which was introduced in the previous section and was shown to be proportional to the topological polarisation.

There is an alternative way to characterize the topology of Hamiltonian of the form on the right hand side of (1.87). (It can be shown that a general Hamiltonian with chiral symmetry can be written in this form [5], so it applies to more than the Dirac equation.) This alternative characteristic is by the *winding number* defined by,

$$w = \frac{i}{2\pi} \int \mathrm{d}k \, Q^{-1} \partial_k Q = \frac{i}{2\pi} \int \mathrm{d}k \, \frac{-im}{k^2 + m^2} = \frac{1}{2} \frac{m}{|m|} , \quad (1.91)$$

i.e., it equals twice the invariant CS_1. Note that the winding number changes by one unit when the sign of the mass changes, consistent with θ in (1.70) changing by π.

Just as in the discussion of the Chern number for the Dirac sea, you might wonder how something that is called a winding number can be non-integer. The resolution is again related to the regularization of the continuum Dirac theory. If we instead consider the lattice version,

$$H_{lat} = \sin(k)\sigma^2 + (m - 1 + \cos(k))\sigma^1 \quad (1.92)$$

so that $Q = -i \sin(k) - (m - 1 + \cos(k))$ we get

$$w = \frac{i}{2\pi} \int_{\pi}^{\pi} \mathrm{d}k \, \partial_k \ln(m - 1 + e^{-ik}) \,. \tag{1.93}$$

For $m < 0$ the curve $m - 1 + e^{-ik}$ does not wind around the origin, so the logarithm can be picked to be single valued and thus $w = 0$. For $0 < m < 2$ it winds one turn in the negative direction and $w = 1$. The change of the winding at $m = 0$ is the same as in the continuum model.

1.4.3 Higher Dimensions

Both in the $2 + 1$ and the $1 + 1$ dimensional cases, there is only one possible topological $U(1)$ symmetric response term. This is the case in any dimension, and it turns out that all odd space time dimension cases are similar to the $(2 + 1)D$ case while all even ones are similar to the $(1 + 1)D$ case.

To find topological response terms in D space-time dimension we need a D-form to contract with the anti-symmetric epsilon tensor with a result that is gauge invariant in the bulk. These conditions are very restrictive, and leave us with,

$$\Gamma[A_\mu] \propto \int \mathrm{d}^2x \, \varepsilon^{\mu\nu} F_{\mu\nu} \qquad D = 2 \tag{1.94}$$

$$\Gamma[A_\mu] \propto \int \mathrm{d}^3x \, \varepsilon^{\mu\nu\sigma} A_\mu F_{\nu\sigma} \qquad D = 3 \tag{1.95}$$

$$\Gamma[A_\mu] \propto \int \mathrm{d}^4x \, \varepsilon^{\mu\nu\sigma\lambda} F_{\mu\nu} F_{\sigma\lambda} \qquad D = 4 \tag{1.96}$$

$$\Gamma[A_\mu] \propto \int \mathrm{d}^5x \, \varepsilon^{\mu\nu\sigma\lambda\kappa} A_\mu F_{\nu\sigma} F_{\lambda\kappa} \qquad D = 5 \tag{1.97}$$

$$\vdots \qquad\qquad \vdots$$

You notice a difference between even and odd space time dimensions. In the even case the actions are Chern–Simons terms that only are gauge invariant up to edge terms. The most important example is the already discussed $2d$ case, and the higher dimensional analogs are very similar:

Using similar arguments to the ones for the $(2 + 1)D$ case one can show that the response in $D = 2k + 1$ following from the response above (1.97) is the kth Chern number of the flux-torus bundle and for a non-interacting system in a periodic potential this equals the kth Chern number of the bundle of filled states over Brillouin zone (there are $2k$ independent fluxes that one can thread in a $d = 2k$ dimensional torus). Thus, just as in $2d$, there is a quantized current response.

Using similar arguments as in the $(1 + 1)D$ case one can show that the response in $D = 2k$ is given by the exponent of 2π times the kth Chern–Simons invariant of the flux torus bundle. This is as in $(1+1)D$, not quantized unless one can assume either time-reversal or chiral symmetry.

Let us briefly mention the most important example of these topological responses, namely the one in $(3+1)D$ that describes a $3d$ time-reversal invariant topological insulator:

$$\Gamma[A_\mu] = \frac{\theta\sigma_0}{16\pi}\int d^4x\, \varepsilon^{\mu\nu\sigma\lambda} F_{\mu\nu}F_{\sigma\lambda} = \frac{\theta\sigma_0}{2\pi}\int d^4x\, \mathbf{E}\cdot\mathbf{B}\,.$$

This term has many important implications that we will list without any derivations:

- A time-reversal invariant system with $\theta = \pi \mod 2\pi$, i.e., a $3d$ topological insulator, has a gapless surface mode.
- If the time-reversal invariance is weakly broken at the surface, the system will have a quantized Hall response with $\sigma_H = \frac{1}{2}\sigma_0 \mod \sigma_0$ [40].
- The *Witten effect* [41]: A magnetic monopole will be accompanied by a localized electric charge of $\frac{1}{2}e \mod e$, and conversely, putting an electric charge near the surface of a topological insulator, will induce a "magnetic monopole" in the bulk! [42].

1.5 Systems Characterized Only by Edge Modes

The topological phases we have considered so far were all characterized by their $U(1)$ response. All of them also supported gapless edge modes; so, you might think that these characteristics go hand in hand. As we shall now discuss, this is not true: there are topological phases of matter which have no topological response, but still exhibit topologically protected gapless edge modes. We shall exemplify with two of the most important cases: the $2d$ time-reversal invariant topological insulator, and the Kitaev chain with and without time-reversal invariance, which is also referred to as a $1d$ topological superconductor.

1.5.1 The Time-Reversal Invariant Topological Insulator

Although you learned that a magnetic field is not necessary for having a quantized Hall response, clearly time-reversal invariance has to be broken since the current is flowing in a particular direction. This invites the question of whether it is possible to have topologically non-trivial states which are time-reversal invariant. An obvious way to get such a system is to add two copies of a Chern insulator, and since the particles are electrons, the natural candidates for the two "species" are the two spin directions: up and down. Such a system will exhibit a quantized *spin Hall effect* that can be described very similarly to the Chern insulator. Just as in case of the integer QH effect, we can construct a topological field theory to describe the quantum spin

Hall effect. Since there are now two conserved currents, corresponding to spin up and spin down, we expect a topological action with two gauge fields $b^{\uparrow}$ and $b^{\downarrow}$.

$$\mathscr{L}_{QSH} = -\frac{1}{4\pi}\left(\varepsilon^{\mu\nu\sigma} b^{\uparrow}_{\mu}\partial_{\nu} b^{\uparrow}_{\sigma} + \varepsilon^{\mu\nu\sigma} b^{\downarrow}_{\mu}\partial_{\nu} b^{\downarrow}_{\sigma}\right). \tag{1.98}$$

This is an example of a *doubled Chern–Simons theory*. There is no QH effect, since the contributions to σ_H obtained by coupling to an external electromagnetic field come with different signs and cancel each other. Similarly, there is no chiral electric edge current, but instead a chiral spin current. This is however only relevant in the cases where one component of the spin is conserved, which is normally not the case in real materials. Surprisingly, however, there are topologically distinct states even when the spin current is not conserved. It is these time-reversal invariant states that are commonly referred to as topological insulators. To arrive at this conclusion using field theory would require that we break the symmetry as $U(1) \times U(1) \to U(1)$, and show that the resulting theory still has protected gapless edge modes. As of now, we do not know of any way to do this. However, since the seminal work by Kane and Mele [43, 44], it has been known that the gapless edge modes persists even when the symmetry is broken. Furthermore they showed that there is new topological "Z_2 invariant" that tells whether the insulator is trivial with no edge modes, or topological with gapless edge modes. The analysis by Kane and Mele was for non-interacting systems, and, at least in this limit, the topological insulator is a time-reversal protected SPT state. The edge modes are seen in many experiments with topological insulators, and are believed to be a generic feature of a large class of materials, see e.g., [3].

1.5.2 The Kitaev Chain

All phases we have discussed so far have a $U(1)$ symmGennestry corresponding to conservation of electric charge. There are, however, important phases of matter which are most easily described using a formalism where this symmetry is broken. This is when there is a condensate, either of fundamental bosons (like in ^{4}He), or Cooper pairs (as in a superconductor) that act as a reservoir of charged particles. Mathematically this amounts to having terms of the type $\sim\Delta\psi\psi$ in the Hamiltonian, and in this section we will consider the "reservoir" Δ as static, and the electromagnetic fields as a background with no dynamics. An important physical situation where this happens realized is when Δ is generated by proximity to a superconductor.

We will refer to quadratic Hamiltonians with terms breaking $U(1)$ as Bogolliobov–de Gennes (BdG) Hamiltonians. It has been realized for quite some time [4, 5] that BdG systems are also part of the comprehensive classification schemes of free fermion theories that was alluded to in the introduction.

Of particular interest are the chiral systems of p-wave type, since their boundary theories typically support *Majorana* modes which can be thought of as a "half fermion". This is in fact one of the most striking examples of edges of topological phases that cannot be realized as *bona fide* boundary systems not connected to a bulk. The simplest example of this is the Kitaev chain, which is of large current interest since it can be experimentally realized in quantum wires with strong spin-orbit coupling [45] or in chains of magnetic atoms on top of a superconductor [46].

The following presentation is a shortened, and somewhat simplified, adaption of [9]. The *Kitaev chain* is a model for spinless fermions hopping on a $1d$ lattice, and is given by the Hamiltonian

$$H_K = \sum_j \left[-t(a_j^\dagger a_{j+1} + a_{j+1}^\dagger a_j) - \mu(a_j^\dagger a_j - \tfrac{1}{2}) + \Delta^\star a_j a_{j+1} + \Delta a_{j+1}^\dagger a_j^\dagger \right] . \tag{1.99}$$

Here t is a hopping amplitude, μ a chemical potential, and Δ an induced superconducting gap. In terms of the Majorana fields,

$$c_{2j-1} = a_j + a_j^\dagger ; \qquad c_{2j} = \frac{1}{i}(a_j - a_j^\dagger) , \tag{1.100}$$

the Hamiltonian becomes

$$H_K = \frac{i}{2} \sum_j \left[-\mu c_{2j-1} c_{2j} + (t + \Delta) c_{2j} c_{2j+1} + (-t + \Delta) c_{2j-1} c_{2j+2} \right] . \tag{1.101}$$

Let us now consider two special cases.

1. The trivial case: $\Delta = t = 0, \ \mu < 0$.

$$H_1 = -\mu \sum_j \left(a_j^\dagger a_j - \frac{1}{2} \right) = \frac{i}{2}(-\mu) \sum_j c_{2j-1} c_{2j} \tag{1.102}$$

   The Majorana operators c_{2j-1}, c_{2j} related to the fermion ψ_j on the site j are paired together to form a ground-state with the occupation number 0.

2. $\Delta = t > 0, \ \mu = 0$. In this case

$$H_2 = it \sum_j c_{2j} c_{2j+1} , \tag{1.103}$$

   and here the Majorana operators c_{2j}, c_{2j+1} from different sites are paired. We can define new annihilation and creation operators

$$\tilde{a}_j = \frac{1}{2}(c_{2j} + i\, c_{2j+1}) ; \qquad \tilde{a}_j^\dagger = \frac{1}{2}(c_{2j} - i\, c_{2j+1}) , \tag{1.104}$$

   which are shared between sites j and $j + 1$. The Hamiltonian becomes

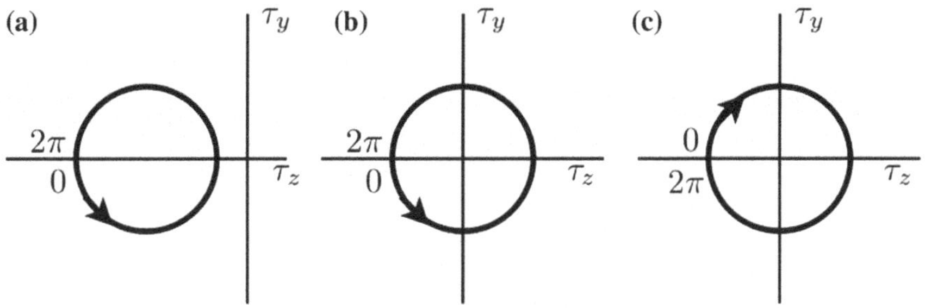

Fig. 1.4 Winding numbers ν of $\mathbf{d}(k)$ for the full Kitaev chain, in **a** trivial phase with $w = 0$, for $0 < t < \mu/2, \Delta > 0$, **b** topological phase with $w = 1$ for $\mu = 0, 0 < t = \Delta$ and **c** topological phase with $w = -1$ for $\mu = 0, 0 < t = -\Delta$. The arrows denote the direction in which k increases

$$H_2 = 2t \sum_{j=1}^{L-1} \left(\tilde{a}_j^\dagger \tilde{a}_j - \frac{1}{2} \right) . \tag{1.105}$$

Note that neither c_1 nor c_{2L} is part of the Hamiltonian, and as a consequence the ground-state is degenerate, since these two Majorana operators can be combined to a fermion,

$$\Psi = \frac{i}{2}(c_1 + i\, c_{2L}) , \tag{1.106}$$

which can be occupied or unoccupied, corresponding to a two-fold degeneracy. The fermion number corresponding to this field, is, however, delocalized at the two ends of the chain. Thus, no local perturbation can change the occupation of this state. It is this property that has made fractionalized fermions an interesting object for quantum information technology. *If a qubit could be stored in a pair of spatially separated Majorana fermions, it would be very robust against noise* [47].

Although the above analysis was only for two very special points in the parameter space, Kitaev established that the whole parameter region $\Delta \neq 0$, and $|\mu| < 2|t|$ is topological.

It is illustrative to see how a topological index appears if we assume $\Delta \in \mathbb{R}$, which corresponds to time-reversal invariance. To this end, we introduce the Nambu spinor $\Psi_k^\dagger = (a_k^\dagger, a_{-k})$ in terms of which the Hamiltonian (1.99) in momentum space becomes,

$$H_K = \sum_k \Psi_k^\dagger \mathscr{H}_K(k) \Psi_k , \tag{1.107}$$

with $\mathscr{H}_K(k)$ given by

$$\mathscr{H}_K(k) = \big(-\mu/2 - t\cos(k)\big)\tau_z - \Delta \sin(k)\tau_y = -\mathbf{d}(k) \cdot \boldsymbol{\tau} \tag{1.108}$$

where the Pauli-matrices τ_i act in the particle-hole spinor space. Just as in the case of the $1d$ insulator discussed, the topological invariant takes the form of a *winding number*. To show this, consider the curve traced out by the vector $\mathbf{d}(k) = (0, \Delta \sin(k), \mu/2 + t \cos(k))$ in the (τ_y, τ_z)-plane (i.e., the space of Hamiltonians), as k sweeps through the full Brillouin zone. This is illustrated in Fig. 1.4, where we schematically show the curve swept by $\mathbf{d}(k)$ in the trivial phase, with winding $w = 0$, and the two different topological phases, with winding $w = \pm 1$.

One can construct Hamiltonians that realize any integer number of windings (a simple way is just to take many copies of the Kitaev wire), and one would think that there is one topologically distinct state for every integer, a so called $\mathbb{Z}$ classification. However, it turns out that this is true only in the absence of interaction. Fidkowski and Kitaev [48] showed that including carefully chosen four-fermion interactions, there are only eight distinct phases which amounts to a $\mathbb{Z}_8$ classification.

Note that the result so far is based on time-reversal invariance symmetry that ensures that Δ is real. So, the eight states are time-reversal protected SPT states. If the time-reversal symmetry is broken (e.g., by a current in the s-wave superconductor inducing Δ) the phase of Δ will vary, and the winding number will no longer be well defined. However, the Majorana modes are still present for general Δ, but there is only a $\mathbb{Z}_2$ index—either there is a Majorana or not.[10] The Majorana modes are present even in the presence of interactions, and no symmetry was assumed; So, this is an example of TO, but of the kind, mentioned in the introduction, that resembles SPT states. One can thus regard the Majorana chain as being protected by fermion parity [10], which is not a symmetry but a property common to all fermionic systems.

1.6 Superconductors Are Topologically Ordered

In this section we shall discuss the perhaps simplest example of a topologically ordered state, namely an usual s-wave paired BCS superconductor coupled to electromagnetism described by Maxwell theory. We start by some general remarks and then turn to examples.

It is known that even a weak attractive interaction will turn a Fermi liquid into a superconductor at sufficiently low temperature. The mechanism is the formation of Cooper pairs composed of two electrons with equal but opposite momenta. Such pairs form, not because of the strength of the interaction, but because of the large available phase space, given by the Fermi surface. This means that even though the theory is weakly coupled (in conventional superconductors by an electron-phonon interaction) the ground-state is non-perturbative.

---

[10] This has a direct consequence for the spectrum of Josephson junctions. In the first (real) case, the junction between two topological states with winding number ± 1, will host two Majorana zero-modes, which amounts to a single Dirac zero-mode, while in the second (complex) case such a junction will have no zero-mode, see e.g., [49].

Most common superconductors have Cooper pairs where the spins form a singlet, which forces the orbital wave function to be symmetric. The simplest possibility is an s-wave, and this is in fact the symmetry of the order parameter in most conventional, or low T_c, superconductors. From the quantum field theory point of view, the BCS approach to superconductivity amounts to a self-consistent mean field approximation, based on the pairing field $\Delta(x) = \psi_\uparrow(x)\psi_\downarrow(x)$.

The fermions are now gapped and can be integrated out. The resulting field theory is to a good approximation the Ginzburg–Landau theory for superconductors.

The Ginzburg–Landau theory supports vortex solutions. The elementary vortex has a core that captures half a unit of magnetic flux, since the Cooper pair has charge $2e$. This means that a quasi-particle will pick up a phase $e^{i\pi} = -1$ when encircling a vortex at a distance large enough for it not to penetrate the vortex core. This is an example of a *topological braiding phase*, which can, as we shall show, readily be captured by a topological field theory. We will now clarify the distinction between a real, "fluctuating", superconductor coupled to electromagnetism, and a model superconductor described by a BdG theory without a dynamical electromagnetic field. The difference, which was first clearly pointed out by [50], is that the former totally screens the electromagnetic current while the second does not. Thus, in the fluctuating superconductor, the *only* low energy degrees of freedom are the vortices, and electrically neutral fermionic quasi-particles. That this is a topologically ordered phase was first pointed out by [51]. The topological field theory that describes the low-energy properties of the s-wave superconductor is the very same BF theory that we already discussed in connection with the IQHE and the Chern insulator. For simplicity we will focus on the $2d$ case, but with a short comment on the $3d$ case in Sect. 1.6.2. For pedagogical reasons we shall first give a heuristic derivation of the BF theory, using an analogy with the Chern–Simons theory discussed in Sect. 1.2.3. Later, in Sect. 1.6.3, we outline a derivation of the BF theory from a microscopic model.

1.6.1 BF Theory of s-Wave Superconductors—Heuristic Approach

We first consider the $(2+1)D$ case where both quasi-particles and vortices are particles, and we can proceed in close analogy to the bosonic Chern–Simons theory for the quantum Hall effect. Recall that in that case the equations of motion relating charge and flux, and the statistics of the quasi-particles (which in that case is simply holes in the filled Landau level) followed from the coupling to the gauge field. The present case differs from the above in that we have two distinct excitations, quasi-particles and vortices, and we will describe them with two conserved currents, j_q^μ and j_v^μ, which we couple to two different gauge fields, a and b, by the Lagrangian,

$$\mathscr{L}_{\text{curr}} = -a_\mu j_q^\mu - b_\mu j_v^\mu \,. \tag{1.109}$$

A simple calculation shows that in order to get a phase π when moving a j_q quantum around a j_v quantum we should take

$$\mathscr{L}_{BF} = \frac{1}{2\pi}\varepsilon^{\mu\nu\sigma} b_\mu f^{(a)}_{\nu\sigma} , \tag{1.110}$$

where $f^{(a)}_{\mu\nu} = \partial_\mu a_\nu - \partial_\nu a_\mu$. This we recognize as the BF action, but with a coefficient that differs from (1.42) derived in the QH context. Putting the parts together we have the topological action,[11]

$$\mathscr{L}_{\text{top}} = \frac{1}{\pi}\varepsilon^{\mu\nu\sigma} b_\mu \partial_\nu a_\sigma - a_\mu j^\mu_q - b_\mu j^\mu_v . \tag{1.111}$$

The topological nature of $\mathscr{L}_{\text{top}}$ is clear from the equations of motion,

$$j^\mu_v = \frac{1}{\pi}\varepsilon^{\mu\nu\sigma}\partial_\nu a_\sigma = \frac{1}{2\pi}\varepsilon^{\mu\nu\sigma} f^{(a)}_{\nu\sigma}; \qquad j^\mu_q = \frac{1}{\pi}\varepsilon^{\mu\nu\sigma}\partial_\nu b_\sigma = \frac{1}{2\pi}\varepsilon^{\mu\nu\sigma} f^{(b)}_{\nu\sigma} , \tag{1.112}$$

which show that the gauge invariant field strengths are fully determined by the currents, just as in the Chern–Simons theory. These equations both have a very direct physical interpretation. For instance, if we write (1.112) as $j^\mu_q + (1/\pi)\varepsilon^{\mu\nu\sigma}\partial_\nu b_\sigma = 0$ this expresses that the quasi-particle current is totally screened by the superconducting condensate shows that $(-1/\pi)\varepsilon^{\mu\nu\sigma}\partial_\nu b_\sigma$ should be interpreted as the screening current. This observation can indeed be used to give an alternative derivation, or rather motivation, for our topological field theory; the potential a_μ is nothing but a Lagrange multiplier that enforces the constraint of total screening of the current j_q. For a more detailed discussion, see [18].

It is interesting to consider the quantization and conservation of charge in our topological field theory. Quantisation of charge implies that the current couples to a $U(1)$ gauge field A_μ that is not only invariant under

$$A_\mu \to A_\mu + \partial_\mu \Lambda , \tag{1.113}$$

where Λ is a continues well-defined (i.e., single valued) function, but the more general transformation

$$A_\mu \to A_\mu + \frac{1}{q} e^{i\Lambda} i\partial_\mu e^{-i\Lambda} , \tag{1.114}$$

[11]The symmetry properties of the Lagrangian (1.111) are worth a comment. Under the parity transformation $(x, y) \to (-x, y)$ the two potentials transform as $(a_0, a_x, a_y) \to (a_0, -a_x, a_y)$ and $(b_0, b_x, b_y) \to (-b_0, b_x, -b_y)$, while under time reversal the transformations are, $(a_0, a_x, a_y) \to (a_0, -a_x, -a_y)$ and $(b_0, b_x, b_y) \to (-b_0, b_x, b_y)$, respectively. The unusual transformation properties of the potential b_μ follow from those of the vortex current. It is easy to check that the BF action is invariant under both P and T.

where q is the minimal charge, and $e^{i\Lambda}$ is a well defined function, although Λ is not a continues well-defined function. We call a gauge field with this property *compact*. A simple example will clarify the distinction between compact and non-compact gauge fields. Consider a $2d$ gauge field and use polar coordinates (ρ, φ). An usual gauge transformation (1.113) with $\Lambda = n\varphi$ is not allowed since Λ is not a continues well-defined function. However, $e^{in\varphi}$ is, and the gauge transformation (1.114) reads $A_\varphi \to A_\varphi + n/q\rho$. Although this transformation introduces a singularity in the gauge field, it leaves the Wilson loop $\exp(iq\oint A_\varphi)$, for any closed curve (including one that encircles the singularity at the origin) invariant, and you can think of the gauge transformation as inserting an invisible flux in the system.

If we want the two currents j_v^μ and j_q^μ to be integer valued, the above argument leads us to require the gauge fields a_μ and b_μ to be compact and transform as

$$a_\mu \to a_\mu + e^{i\Lambda_a} i\partial_\mu e^{-i\Lambda_a}; \qquad b_\mu \to b_\mu + e^{i\Lambda_b} i\partial_\mu e^{-i\Lambda_b} . \tag{1.115}$$

The question of current conservation is more subtle. Since the world line of a point-like vortex can be thought of as a vortex line in space-time, non-conservation of the vortex charge would amount to having such world lines ending at a point. This could happen if there were unit charge magnetic monopoles in space-time, on which two such world lines could terminate. Such monopoles in space-time are called *instantons* and are known to exist in many field theories. Since we do not have any magnetic monopoles, this is however not a realistic option, and the vortex charge is conserved. The situation is quite different when it comes to the electric charge. Here the Cooper-pair condensate acts as a source of pairs of electrons, and in our topological theory such processes, corresponding to formation or breaking of pairs, this could be incorporated by having instantons in the b field.

1.6.2 The 3 + 1 Dimensional BF Theory

Turning to the case of $3+1$ dimensions, we have essentially the same construction, but with the difference that the vortices are now strings, and the field b is an antisymmetric tensor, $b_{\mu\nu}$. The action is still again of the BF type and reads,

$$\mathscr{L}_{BF} = \frac{1}{\pi}\varepsilon^{\mu\nu\sigma\lambda} b_{\mu\nu}\partial_\sigma a_\lambda . \tag{1.116}$$

The gauge transformations of the b field are given by

$$b_{\mu\nu} \to b_{\mu\nu} + \partial_\mu \xi_\nu - \partial_\nu \xi_\mu , \tag{1.117}$$

where ξ_μ is a vector-valued gauge parameter. The minimal coupling of the b potential to the world sheet, Σ, of the string is given by the action,

$$S_{\text{vort}} = -\int_{\Sigma} \mathrm{d}\tau \mathrm{d}\sigma^{\mu\nu} b_{\mu\nu} = -\int_{\Sigma} \mathrm{d}\tau \mathrm{d}\sigma \left| \frac{d(x^{\mu}, x^{\nu})}{d(\tau, \sigma)} \right| b_{\mu\nu} , \tag{1.118}$$

where (τ, σ) are time and space like coordinates on the worldsheet. This is a direct generalization of the coupling of a to the world line, Γ, of a spinon,

$$S_{sp} = -\int_{\Gamma} \mathrm{d}x^{\mu} a_{\mu} = -\int_{\Gamma} \mathrm{d}\tau \frac{\mathrm{d}x^{\mu}}{\mathrm{d}\tau} a_{\mu} . \tag{1.119}$$

Combining these elements we get the topological action for the $3+1$ dimensional superconductor,

$$S_{\text{top}} = \int \mathrm{d}^4 x \, \mathscr{L}_{BF} + S_{sp} + S_{\text{vort}} . \tag{1.120}$$

The proof that this action indeed gives the correct braiding phases can be found e.g., in [31], and a discussion of this action in the context of superconductivity has appeared before in [52].

1.6.3 Microscopic Derivation of the BF Theory

So far, we did not derive the BF theory, but rather constructed it, or guessed it, from the braiding properties of quasi-particles and vortices. It would obviously be more reassuring if the theory could be *derived* from a microscopic model. Here we outline such a derivation starting not from the original fermionic theory, but from an effective Ginzburg–Landau model coupled to a quasi-particle source. (The derivation of this theory from a model with paired electrons is a standard exercise that can be found e.g., in the book by [12].) For simplicity, we shall follow [18] and consider a toy version of the Ginzburg–Landau theory, namely the $(2+1)D$ relativistic Abelian Higgs models defined by the Lagrangian,[12]

$$\mathscr{L}_{AH} = \frac{1}{2M} |i D_{\mu}\phi|^2 - \frac{\lambda}{4}(\phi^{\dagger}\phi)^2 - \frac{m^2}{2}\phi^{\dagger}\phi - \frac{1}{4}F_{\mu\nu}^2 - eA_{\mu} j_q^{\mu} , \tag{1.121}$$

where we used a standard particle physics notation where ϕ is the charge $-2e$ scalar field representing the Cooper pair condensate, $iD_{\mu} = i\partial_{\mu} + 2eA_{\mu}$ is the covariant derivative, $F_{\mu\nu}$ is the electromagnetic field strength and the conserved current j_q^{μ},

[12] This is a toy model not only because we use a relativistic form for the kinetic energy, but also because we use $2+1D$ Maxwell theory, which amounts to a logarithmic Coulomb interaction. The generalization to the more realistic case is straightforward, and the result is qualitatively the same. The derivation, however, becomes less transparent. For the interested reader [53] is a good reference to see how to include fermions and also see how the chiral d-wave case works.

with charge e, is introduced to describe the gapped quasi-particles discussed above.[13] Just as in the usual Ginzburg–Landau theory the Abelian Higgs model supports vortex solutions, which are characterized by a singularity in the phase of the Cooper-pair field. Defining $\phi = \sqrt{\rho}\, e^{i\varphi}$, and writing $\varphi = \tilde{\varphi} + \eta$, where η is regular, and can be removed by a proper gauge transformation, the vorticity is encoded in the gauge field $a_\mu = \frac{1}{2}\partial_\mu \varphi$, which depends on the vortex positions. The corresponding conserved vortex current is parametrized as $j_\nu = (1/\pi)\varepsilon^{\mu\nu\sigma}\partial_\mu a_\sigma$. It is now a matter of algebraic manipulations, involving integrating out gapped degrees of freedom, to derive the topological theory (1.111), see [18]. For this to work, it is crucial that we include a dynamical electromagnetic field, it is only then that the external currents are completely screened and all bulk modes are gapped and can be integrated over. Making a derivative expansion, and keeping terms to second order results in the effective Lagrangian,

$$\mathscr{L}_{eff} = \frac{1}{\pi}\varepsilon^{\mu\nu\sigma} b_\mu \partial_\nu a_\sigma - \frac{1}{4e^2}(f^{(a)}_{\mu\nu})^2 - \frac{1}{4}\left(\frac{e}{m_s \pi}\right)^2 (f^{(b)}_{\mu\nu})^2 - a_\mu j^\mu_q - b_\mu j^\mu_v \,. \tag{1.122}$$

where $m_s^2 = 4e^2\bar{\rho}/M$, and $\bar{\rho}$ is the average density. Note that the topological theory emerges as the leading term in this expansion! The higher derivative terms, which are of Maxwell form, are not topological, and have the effect of introducing bulk degrees of freedom. These are however gapped, and can be identified as the plasmon mode. At low energies the plasmons can be neglected and we can retain only the topological part. Another physical effect captured by the Maxwell terms is the London penetration length λ_L which is the size of the magnetic flux tube associated to a vortex. In the purely topological theory, the vortices are strictly point-like. In this section we strictly dealt with s-wave superconductors. The extension to the d-wave case is relatively straightforward, but since there are gapless quasi-particles associated to the nodal lines, the effective theory must include these, and becomes quite a bit more complicated [54]. The p-wave case is considerably more difficult because of the Majorana modes associated to vortices.

1.6.4 The Two-Dimensional p-Wave Superconductor

In the previous section we assumed that the fermionic part of the Lagrangian was gapped and could be integrated out. In the p-wave case this cannot be done since there are Majorana modes associated with vortices. These modes form a finite degenerate subspace of the Hilbert space, and the system has a most amazing feature: Quasi-adiabatic braidings of the vortices correspond to non-commuting unitary operators acting on the finite-dimensional Hilbert space. Since the vortices should are identical

[13]Note that in spite of the relativistic form we normalize the kinetic term such that $|\phi|^2$ has the dimension of density.

particles, this amounts to having non-Abelian fractional statistics. There is a very close resemblance between this state and the strongly correlated Moore-Read pfaffian quantum Hall state [55], that is a candidate wave function for the plateau observed at filling fraction $\nu = 5/2$. (This state will be discussed briefly in the last section.)

The simplest realization of a $2d$ chiral p-wave superconductor is by spin-less electrons with a parabolic dispersion relation, i.e., taking the single-particle Hamiltonian to be $H_1 = (-i\nabla - e\mathbf{A})^2/2m$ and adding an attractive short-range interaction,

$$\hat{V} = \frac{\lambda}{2}\int d^2x d^2y \left[\psi^\dagger(\mathbf{r}')\psi^\dagger(\mathbf{r})\nabla^2\delta^2(\mathbf{r}-\mathbf{r}')\psi(\mathbf{r})\psi(\mathbf{r}')\right]_{\text{reg}} , \tag{1.123}$$

where reg denotes that the potential can be approximated by $\lambda\nabla^2\delta^2(x-y)$ only for small momentum transfers close to the fermi surface. (The precise form of the potential is not needed for the coming discussion.)

Since we expect a condensation of Cooper pairs, it is practical to introduce a Hubbard-Stratonovich boson, $\mathbf{\Delta}$, which mediates the interaction. More precisely, we add an action for $\mathbf{\Delta}$ so that the classical solution is $\mathbf{\Delta}_{cl} = \lambda(\psi\nabla\psi)_{\text{reg}}$, meaning that we can, at the mean-field level, replace the quartic interaction by a term $\mathbf{\Delta}\cdot\psi^\dagger\nabla\psi^\dagger + h.c.$ ($h.c.$ means Hermitean conjugate).

To proceed we assume a chiral condensate and make the mean-field ansatz $\Delta \equiv \Delta_{\bar{z}} = \lambda\langle\psi\partial_{\bar{z}}\psi\rangle_{\text{reg}}$ and $\Delta_z = \lambda\langle\psi\partial_z\psi\rangle_{\text{reg}} = 0$. For simplicity we will from now on drop the subscript *reg*, but all momentum sums involving Δ should be understood to have a cutoff. With these assumptions we get the Hamiltonian,

$$H = \int d^2x \; \psi^\dagger\,(H_1 - \mu)\,\psi + \psi^\dagger\Delta\partial_z\psi^\dagger + h.c., \tag{1.124}$$

and proceed to find a homogeneous solution $\Delta = const$. Fourier transforming the field operator, $\psi(\mathbf{r}) = \sum_{\mathbf{k}} e^{-i\mathbf{k}\cdot\mathbf{r}}c_{\mathbf{k}}/\sqrt{V}$, we get

$$H = \frac{1}{2}\sum_{\mathbf{k}} \left(c^\dagger_{\mathbf{k}} \; c_{-\mathbf{k}}\right) \underbrace{\begin{pmatrix} \xi_{\mathbf{k}} & \Delta(k_y + ik_x) \\ \Delta(k_y - ik_x) & -\xi_{\mathbf{k}} \end{pmatrix}}_{H_{\mathbf{k}}} \begin{pmatrix} c_{\mathbf{k}} \\ c^\dagger_{-\mathbf{k}} \end{pmatrix} ,$$

where $\xi_{\mathbf{k}} = \mathbf{k}^2/2m - \mu$. From the property

$$\sigma_x H^*_{\mathbf{k}}\sigma_x = -H_{-\mathbf{k}} , \tag{1.125}$$

it follows that all negative-energy solutions at $-\mathbf{k}$ can be written in terms of positive-energy solutions $H_{\mathbf{k}}(U_{\mathbf{k}}, V_{\mathbf{k}})^T = E_{\mathbf{k}}(U_{\mathbf{k}}, V_{\mathbf{k}})^T$, at $\mathbf{k}$. We can thus diagonalize the Hamiltonian with the positive-energy solutions $\Gamma^\dagger_{\mathbf{k}} = U_{\mathbf{k}}c^\dagger_{\mathbf{k}} + V_{\mathbf{k}}c_{-\mathbf{k}}\; that$,

$$H = E_0 + \sum_{\mathbf{k}} E_{\mathbf{k}}\Gamma^\dagger_{\mathbf{k}}\Gamma_{\mathbf{k}} ,$$

and the positive eigenvalues are obtained by directly diagonalizing H_k,

$$E_{\mathbf{k}} = \sqrt{\xi_{\mathbf{k}}^2 + |\Delta \mathbf{k}|^2} \, .$$

Note that the normalization $|U_{\mathbf{k}}|^2 + |V_{\mathbf{k}}|^2 = 1$ follows from requiring $\left\{ \Gamma_{\mathbf{k}}^{\dagger}, \Gamma_{\mathbf{k}'} \right\} = \delta_{\mathbf{k}\mathbf{k}'}$. The ground-state ($|BCS\rangle$) is thus the state annihilated by all $\Gamma_{\mathbf{k}}$, i.e., $\propto \prod_{\mathbf{k}} \Gamma_{\mathbf{k}} |0\rangle$. To get the normalization right, we note that

$$\Gamma_{\mathbf{k}} \Gamma_{-\mathbf{k}} |0\rangle = V_{-\mathbf{k}}^* \left(U_{\mathbf{k}}^* + V_{\mathbf{k}}^* c_{-\mathbf{k}}^{\dagger} c_{\mathbf{k}}^{\dagger} \right) |0\rangle \, ,$$

and from the requirement $|U_{\mathbf{k}}|^2 + |V_{\mathbf{k}}|^2 = 1$, we realize that the correctly normalized ground-state is

$$|BCS\rangle = \prod_{\mathbf{k}}{}' \left(U_{\mathbf{k}}^* + V_{\mathbf{k}}^* c_{-\mathbf{k}}^{\dagger} c_{\mathbf{k}}^{\dagger} \right) |0\rangle \, ,$$

where $\prod'$ indicates that the product is over half of the $\mathbf{k}$ values, as for instance,

$$\mathbf{k} \in \left\{ k_x > 0 \text{ or } k_x = 0, \; k_y \geq 0 \right\} \, . \tag{1.126}$$

To show that the ansatz $\Delta_z = 0$ and $\Delta_{\bar{z}} = constant$ is self-consistent, we calculate the mean field $\Delta_{x \pm iy}$,

$$\Delta_{x \pm iy} = \lambda \langle \psi \partial_{x \pm iy} \psi \rangle = \sum_{\mathbf{k}} (k_y \pm i k_x) U_{\mathbf{k}} V_{\mathbf{k}}^* = \sum_{\mathbf{k}} f(|\mathbf{k}|, \Delta)(k_x^2 \pm k_y^2) \, , \tag{1.127}$$

where f is a function determined by $U_{\mathbf{k}} V_{\mathbf{k}}^*$. In the last step we used $(U_{\mathbf{k}}, V_{\mathbf{k}}) \propto \left(f(|\mathbf{k}|, \Delta)(k_y + i k_x), 1 \right)$ which is obtained by explicitly diagonalizing $H_{\mathbf{k}}$. Since the above sum is symmetric in interchange of k_x and k_y we infer that $\Delta_z = 0$. For Δ we instead get a gap-equation which determines its ground-state value.

We are now ready to see how the Majorana modes appear in the vortices. In a vortex solution the phase of Δ winds around some points (vortex cores). Close to a vortex core $\mathbf{r}'$ we have

$$\Delta_z(\mathbf{r}') = 0; \qquad \Delta(\mathbf{r}) = |\Delta| \, e^{in \arg(\mathbf{r}-\mathbf{r}') + i\lambda} \qquad n \in \mathbb{Z} \, ,$$

where λ is a regular function. There are no analytical self-consistent solutions to the vortex problem, but we do not need the precise form, just the fact that a solution exists. Since Δ carries charge $2e$ we infer that for a solution to have finite energy there must be a magnetic flux with strength $nhc/2e$ associated to the vortex. We will now show that in the presence of vortices with odd strength there are fermionic zero-modes, and that hese zero-modes imply non-Abelian statistics see [56, 57].

The fermionic mean field Hamiltonian (1.124) is quadratic so it can be diagonalized by single-particle operators

$$\phi_{u,v}^{\dagger} = \int \mathrm{d}^2 x \left(\psi^{\dagger}(\mathbf{r})\ \psi(\mathbf{r})\right) \begin{pmatrix} u(\mathbf{r}) \\ v(\mathbf{r}) \end{pmatrix} , \tag{1.128}$$

and the Heisenberg equation of motion $[\phi_{u,v}^{\dagger}, H] = E\phi_{u,v}^{\dagger}$ becomes

$$\underbrace{\begin{pmatrix} H_1 - \mu & \frac{1}{2}\{\Delta, \partial_z\} \\ -\frac{1}{2}\{\Delta^*, \partial_{\bar{z}}\} & -H_1^* + \mu \end{pmatrix}}_{\mathscr{H}} \begin{pmatrix} u(\mathbf{r}) \\ v(\mathbf{r}) \end{pmatrix} = E \begin{pmatrix} u(\mathbf{r}) \\ v(\mathbf{r}) \end{pmatrix} . \tag{1.129}$$

The single particle Hamiltonian $\mathscr{H}$ has a number of discrete modes and a continuum. Using the property

$$\sigma_x \mathscr{H}^* \sigma_x = -\mathscr{H} \tag{1.130}$$

we notice that an odd number of discrete modes implies an odd number of zero-modes. Changing the parameters, modes can come down from the continuum and become discrete and vice versa, but because of the property (1.130) they must always come in pairs. Therefore we can with certainty say that one cannot change the parity of the number of zero-modes. If we have one zero-mode we will continue to have (at least) one. If, by chance, there is an extra pair of zero-modes, interactions would most likely gap out two of the three, so for generic parameters we expect exactly one zero mode. Note that (1.130) is not a physical symmetry that can be broken by changing the Hamiltonian. Instead it follows from the structure of the second-quantized Hilbert-space.

The above argument is only valid for a single vortex on the infinite plane. For any finite system there will necessarily be an even number of discrete modes and it seems like the argument fails. However, as long as the vortex modes are exponentially localized, and the vortices are well separated compared to this length scale, the vortices can still be treated as isolated up to small perturbations.

These perturbations will allow tunneling of one vortex zero-mode into another, which turns them into two non-localized non-zero energy modes. But the splitting will be proportional to the tunneling rate, and therefore exponentially small in λ_Z/L, where L is the distance between the vortices, and λ_Z is the size of the vortex zero-modes. We are thus left with proving that the zero-modes are (exponentially) localized, and that there is only one (or an odd number) of them at each vortex. Although we cannot solve the (1.129) in general, we can find analytical expressions for the zero-modes if we make some approximations. We begin by considering an isolated vortex and no disorder potential. With this assumption, and taking (r, θ) as polar coordinates centered around the vortex core, the only θ dependence in (1.129) is in the phase of Δ, and we can assume $\Delta = f(r)e^{i\Omega}$ with $\Omega = in\theta + i\lambda$ and λ a constant. Next we notice that the ansatz,

$$\begin{pmatrix} u \\ v \end{pmatrix} = e^{il\theta} \begin{pmatrix} e^{i(n-1)\theta/2+i\lambda/2} u_l(r) \\ e^{-i(n-1)\theta/2-i\lambda/2} v_{-l}(r) \end{pmatrix} , \tag{1.131}$$

with $(u_l(r), v_{-l}(r))$ real, diagonalizes $\mathscr{H}$ in the θ variable so we are left with a one-dimensional problem $\mathscr{H}_l(u_l(r), v_{-l}(r))^T = E(u_l(r), v_{-l}(r))^T$. Note that if n is odd l must be an integer, and if n is even, l must be a half integer. Analogous to (1.125) the effective 1d Hamiltonian $\mathscr{H}_l$ has the property $\sigma_x H_l^* \sigma_x = -H_{-l}$. So, for each zero energy solution $(u_l(r), v_{-l}(r))^T$ there is necessarily a dual solution, $(v_{-l}(r), u_l(r))^T$. It therefore seems that all zero-modes come in pairs, and we could not possibly get an odd number. However, when the vortex strength n is odd, we have a self-dual solution,

$$(u_0(r), v_0(r))^T = (v_0(r), u_0(r))^T ,$$

and thus an odd number of zero-modes.

Let us examine the equation for the self-dual mode in more detail. Since $u_0(r) = v_0(r)$ we are left with a single one-dimensional second-order equation. It can be brought to a more familiar form by the gauge choice $\mathbf{A} = A_\theta(r)\hat{\theta}$ and the transformation,

$$u_0(r) = \chi(r) \exp\left(-\frac{m}{2}\int^r f(r')dr'\right) .$$

Substituting this into (1.129) gives

$$-\frac{\chi''(r)}{2m} - \frac{\chi'(r)}{2mr} + \frac{((n-1)/2)^2}{2mr^2}\chi(r) + \frac{(n-1)eA_\theta(r)}{2m}\chi(r) + \frac{mf^2}{8}\chi(r) = \mu\chi(r),$$

which we identify as a Schrödinger equation for a particle of mass m, with angular momentum $k = (n-1)/2$, moving in a potential $mf^2/8$, and a radially symmetric magnetic field $B = \frac{1}{r}A_\theta + \partial_r A_\theta$. Further assuming that the coherence length ξ (i.e., the radius of the region where $|\Delta|$ differs substantially from its asymptotic value $|\Delta_0|$) is much smaller than λ_Z, and the London length λ_L (i.e., the radius of region with non-zero magnetic field) is much larger than the same length scale, we obtain the solution,

$$u_0(r) = \begin{cases} e^{-m|\Delta_0|r/2} J_k\left(r\sqrt{2\mu m - (m|\Delta_0|)^2/4}\right) & \text{for } \mu > m|\Delta_0|^2/8 \\ e^{-m|\Delta_0|r/2} I_k\left(r\sqrt{(m|\Delta_0|)^2/4 - 2\mu m}\right) & \text{for } 0 < \mu < m|\Delta_0|^2/8 , \end{cases}$$

which is exponentially localized. The approximations we made were to assume that we have an extreme type II superconductor, but the result holds independent of this approximation as long as $\mu > 0$.

We here found the vortex Majorana zero modes from a detailed calculation. However, we want to stress that they are directly related to anyonic nature of the vortices and therefore also the topological order. So, the existence of the modes does not rely on any of the detailed assumptions. As long as the energy gap remains open the zero modes cannot disappear even if the system is changed considerably.

1.7 Fractional Quantum Hall Liquids

Our second example of states with topological interactions are the archetypical ones, namely the quantum Hall liquids. We start with the most celebrated ones.

In 1982, not long after the discovery of the IQHE, Tsui, Stormer and Gossard observed a plateau in the conductance at $\sigma_H = 1/3\sigma_0$ [58]. Later, many more states were discovered with $\sigma_H = p/q\sigma_0$, the vast majority with an odd denominator q. This is the celebrated fractional quantum Hall effect (FQHE). The FQHE poses a much more difficult theoretical challenge than the integer one. The basic difficulty is the massive degeneracy of the free electron states in an partially filled Landau level. Neglecting the lattice potential, the *only* energy scale is that of the Coulomb interaction $E_C \sim e^2\rho^{-1/2}$, where $\rho^{1/2}$ is the mean distance between the particles. (We assume that the cyclotron gap, $E_B \sim eB/m$ is large, so that for all practical purposes $E_C/E_B = 0$.) As a consequence there is no small parameter, and thus no hope to understand the FQHE by using perturbation theory.

The first, and in a sense most successful, approach to the FQH problem was due to [59], who, by an ingenious line of arguments, managed to guess a many-electron wave function, that gives an essentially correct description of the states with conductance $\sigma_H = \sigma_0/m$.

Another approach, which which we will take, is to try to find an effective low energy theory, and we will outline how this can be done. This resulting low-energy theory is called the Chern–Simons–Ginzburg–Landau theory.

1.7.1 The Chern–Simons–Ginzburg–Landau Theory

The starting point is the microscopic Hamiltonian for N electrons in a constant transverse magnetic field B, interacting via a two-body potential $V(r)$, which should be thought of as a (suitably screened) Coulomb potential,

$$H = \frac{1}{2m}\sum_{i=1}^{N}(\mathbf{p}_i - eA(\mathbf{x}_i))^2 + \sum_{i<j}^{N} V(|\mathbf{x}_i - \mathbf{x}_j|)\,, \tag{1.132}$$

where $\mathbf{A}(\mathbf{x}) = \frac{B}{2}(-y, x)$. Our aim is to find an equivalent *bosonic* formulation of this theory, which should be amenable to a mean-field description. A direct application of the method of functional bosonization described in Sect. 1.2.3.1, will not work since we would not be able to compute the partition function $Z[a]$, for a partially filled Landau level. Another approach that might come to mind is to invoke pairing and introduce a Cooper pair field. This will however also not work since it would describe a superconductor, not an insulating QH state. Instead we proceed by first performing a *statistics changing transformation* on the electrons.

The idea is to relate the fermionic wave functions to their bosonic counterparts by the unitary transformation

$$\Psi_F(\mathbf{x}_1, \ldots \mathbf{x}_N) = \phi_k(\mathbf{x}_1 \ldots \mathbf{x}_N)\Psi_B(\mathbf{x}_1, \ldots \mathbf{x}_N) \tag{1.133}$$

where the phase factor ϕ_k is given by

$$\phi_k(\mathbf{x}_1 \ldots \mathbf{x}_N) = e^{ik\sum_{a<b}\alpha_{ab}}\,, \tag{1.134}$$

with k an odd integer, and α_{ab} the polar angle between the vectors $\mathbf{x}_a$ and $\mathbf{x}_b$. The corresponding bosonic Hamiltonian is identical to the fermionic one, except that it includes a coupling to a *statistical, or Chern–Simons, gauge potential,*

$$\mathbf{a}(\mathbf{x}_a) = k\,\nabla \sum_{b\neq a} \alpha_{ab}\,. \tag{1.135}$$

Thus, instead of studying the original fermionic Hamiltonian (1.132) we can study the bosonic Hamiltonian

$$H_B = \frac{1}{2m}\sum_{a=1}^{N}\left[\mathbf{p}_a - e\mathbf{A}(\mathbf{x}_a) + \mathbf{a}(\mathbf{x}_a)\right]^2 - \sum_{a<b} V(|\mathbf{x}_a - \mathbf{x}_b|)\,. \tag{1.136}$$

To proceed we first notice that although $\mathbf{a}$ looks like a pure gauge, it is singular at the positions of the particles, so that the statistical magnetic field is given by,

$$\varepsilon^{ij}\partial_i a_j \equiv b^{(a)} = 2\pi k \sum_{b\neq a}\delta^2(\mathbf{x}_a - \mathbf{x}_b)\,, \tag{1.137}$$

which amounts to *attaching a singular flux tube of strength k to each particle.* The statistical exchange phase, $\theta = k\pi$ can thus be seen as an Aharanov–Bohm effect.[14]

We are now ready to construct a quantum field theory describing our bosonized electron in a path integral formulation. The variables will be a non-relativistic boson field, ϕ, describing the electrons, and the statistical gauge field $\mathbf{a}$. The expression for the statistical magnetic field (1.137) implies that $\mathbf{a}$ is such that there is a flux tube tied to each particle, i.e.,

$$2\pi k\rho = 2\theta\phi^{\star}\phi = \varepsilon^{ij}\partial_i a_j\,. \tag{1.138}$$

[14]The naive picture of the "composite bosons" flux-charge composites, is however slightly misleading since that would implies that you get a phase $2\times k2\pi$ when taking one particle a full turn around another; there are equal contributions from the charge circling the flux and the flux circling the charge. This is not what happens, the correct phase is $k2\pi$ corresponding to the exchange phase $k\pi$ [60].

This *local* constraint is implemented by a Lagrange multiplier field a_0, and the result is the non-relativistic Chern–Simons–Ginzburg–Landau (CSGL) field theory,

$$\mathscr{L}_B = \mathscr{L}_\phi + \frac{1}{2\pi k} a_0 \varepsilon^{ij} \partial_i a_j \,, \tag{1.139}$$

where

$$\mathscr{L}_\phi = \phi^\star (i\partial_0 - a_0 + eA_0)\phi - \frac{1}{2m} |(\mathbf{p} + e\mathbf{A} - \mathbf{a})\phi|^2 - V(|\phi|) \,, \tag{1.140}$$

and where A^μ is an external electromagnetic field, that includes the constant background magnetic field B, and $\rho = \phi^\star \phi$ is the density. The term $\sim a_0 \varepsilon^{ij} \partial_i a_j$ is nothing but the Coulomb gauge version (i.e., $\nabla \cdot \mathbf{a} = 0$) of the full CS action $\sim \varepsilon^{\mu\nu\sigma} a_\mu \partial_\nu a_\sigma$; so, we can finally write the partition function as

$$Z[A_\mu] = \int \mathscr{D}[\phi^\star] \mathscr{D}[\phi] \mathscr{D}[a_\mu] \, e^{i \int d^3x \, \mathscr{L}_{CSGL}(\phi, a; A)} \tag{1.141}$$

with

$$\mathscr{L}_{CSGL} = \mathscr{L}_\phi + \frac{1}{2\pi k} \varepsilon^{\mu\nu\sigma} a_\mu \partial_\nu a_\sigma \,. \tag{1.142}$$

To proceed we employ a mean field approach. To do so we first assume V in $\mathscr{L}_\phi$ to be a contact potential,

$$V(|\phi|) = -\frac{\mu}{2} |\phi|^2 + \frac{\lambda}{4} |\phi|^4 \,,$$

where both μ and λ are positive, so the minimum occurs at finite density. In the mean field approach we assume that the thin flux tubes that make up the statistical gauge field $b = \varepsilon^{ij} \partial_i a_j$, can be replaced by a smeared out field that can be cancelled against the constant external field B, i.e., $\varepsilon^{ij} \partial_i a_j = e \varepsilon^{ij} \partial_i A_j$. With this, we can choose a gauge where the combination $eA - \mathbf{a} = 0$, and the vector potential terms in $\mathscr{L}_\phi$ then disappear. The equation of motion for the a_0 field gives the constraint,

$$\varepsilon^{ij} \partial_i a_j = 2\pi k \phi^* \phi = 2\pi k \rho \,, \tag{1.143}$$

which combined with the mean field assumption results in

$$\rho = \phi^* \phi = \frac{eB}{2\pi k} \,.$$

To get the full mean field solution, the mean density $\bar{\rho}$ is picked as to minimise $V(\rho)$, and in summary we have,

$$\phi = \phi_0 = \bar{\rho} = \sqrt{\frac{\mu}{\lambda}}; \qquad \mathbf{a} = e\mathbf{A}\,. \tag{1.144}$$

This is only possible if

$$\bar{\rho} = \frac{1}{2\theta} b = \frac{1}{2\pi k} eB = \frac{\rho_0}{k} \tag{1.145}$$

(where ρ_0 is the density of a filled Landau level). Or in other terms, the filling fraction is $\nu = 1/k$, where k can be any odd integer.

Just as in the usual Ginzburg–Landau theory, the GLCS theory supports mean field vortex solutions, which in Coulomb gauge, for a unit strength vortex is,

$$\phi \underset{r\to\infty}{\sim} \sqrt{\bar{\rho}}\, e^{i\varphi}; \qquad a_\varphi \underset{r\to\infty}{\sim} \frac{1}{r}; \qquad a_r = 0\,. \tag{1.146}$$

Since the statistical magnetic field is tied to the charge density (1.138), we can calculate the excess charge related to the vortex by integrating the expression for the statistical magnetic field (1.145),

$$Q_\nu = \int \mathrm{d}^2 r\, \rho(\mathbf{x}) = \frac{\nu e}{2\pi} \int \mathrm{d}^2 x \varepsilon^{ij} \partial_i a_j = \frac{\nu}{2\pi} e \int \mathrm{d}\mathbf{x} \cdot \mathbf{a} = \nu e\,. \tag{1.147}$$

So, the vortex describes a quasi-particle with fractional charge νe. In [11, 61] you find a more detailed analysis of the CSGL theory including the demonstration that the quasi-particles are Abelian anyons with a statistics angle $\nu\pi$. Here we shall show how it can be used to extract an effective topological theory for the Laughlin states.

1.7.2 From the CSGL Theory to the Effective Topological Theory

In the presence of a collection of vortices, we parametrize the field ϕ as

$$\phi = \sqrt{\rho(\mathbf{x})} e^{i\theta(\mathbf{x})} \xi_v(\mathbf{x})\,, \tag{1.148}$$

where we have extracted the singularities in ξ_v and θ is a smooth fluctuating phase. Next we substitute this parametrization (1.148) into the CSLG Lagrangian (1.142) and expand the Lagrangian around the mean-field solution,

$$\phi = \left(\sqrt{\bar{\rho}} + \frac{\delta\rho}{2\sqrt{\bar{\rho}}}\right) \xi_v e^{i\theta}\,, \tag{1.149}$$

and the temporal term then become

$$\mathcal{L}_{\text{temp}} = \delta\rho\left(-\partial_t\theta + i\xi_v^\star\partial_t\xi_v - a_0 + eA_0\right) + \mathcal{O}\left(\delta\rho^2\right) + \mathcal{O}\left(\delta\rho(eA_0 - a_0)\right) . \tag{1.150}$$

The term $\bar{\rho}\partial_t\delta\rho$ vanishes when the derivative is integrated over and does not contribute. The leading contribution from the kinetic energy is

$$\mathcal{L}_{\text{kin}} = -\frac{\bar{\rho}^2 + \mathcal{O}(\delta\rho)}{2m}\left(\nabla\theta - i\xi_v^*\nabla\xi_v - \mathbf{a} + e\mathbf{A}\right)^2 + \mathcal{O}\left(\partial_i\delta\rho\right) , \tag{1.151}$$

and we linearize this quadratic term by introducing a Hubbard Stratonovich field $\mathbf{X}$, to get,

$$\mathcal{L}_{\text{kin}} = \frac{m}{2\bar{\rho}}\mathbf{X}^2 + \mathbf{X}\cdot\left(\nabla\theta - i\xi_v^*\nabla\xi_v - \mathbf{a} + e\mathbf{A}\right) + h.o. , \tag{1.152}$$

where $h.o.$ denote terms of higher order in the expansion. We now recognise θ as a Lagrange multiplier field, imposing conservation of the three-current $(\delta\rho, \mathbf{X})$, which therefore can be parametrized in terms of a field b_μ as

$$(\delta\rho, \mathbf{X})^\mu = \frac{1}{2\pi}\varepsilon^{\mu\nu\sigma}\partial_\nu b_\sigma . \tag{1.153}$$

The terms containing ξ_v can now be taken together and written as

$$\frac{1}{2\pi i}\varepsilon^{\mu\nu\sigma}\left(\xi_v^*\partial_\mu\xi_v\right)\partial_\nu b_\sigma . \tag{1.154}$$

To interpret this term, let's look closer at ξ_v. The integral

$$\int_\gamma d\mathbf{l}\cdot\xi_v^*\nabla\xi_v \tag{1.155}$$

gives the change of the phase of ξ_v along the curve γ and thus, by definition, if γ is a closed curve

$$\frac{1}{2\pi i}\int_\gamma d\mathbf{l}\cdot\xi_v^*\nabla\xi_v = q_v , \tag{1.156}$$

where q_v denotes the number of right handed minus the number of left handed vortices encircled by γ, i.e., the vortex charge encircled by γ. Using the $2d$ version of Stokes theorem we can thus conclude

$$\frac{1}{2\pi i}\int_S \varepsilon^{ij}\partial_i\xi_v^*\partial_j\xi_v = q_v , \tag{1.157}$$

where q_v is the vortex charge in the region S. It follows that the vortex charge density is

$$\rho_v = \frac{1}{2\pi i}\varepsilon^{ij}\partial_i\xi_v^*\partial_j\xi_v . \tag{1.158}$$

Generalizing this argument by letting γ be a general space-time curve, we can write the full vortex current as

$$j_v^\mu = \frac{1}{2\pi i}\varepsilon^{\mu\nu\sigma}\partial_\nu\xi_v^*\partial_\sigma\xi_v \,. \tag{1.159}$$

By integration by parts, of the derivative ∂_ν, and using the above identity for j_v^μ, (1.159), the term that contain the ξ_v of the Lagrangian (1.154) can be written as $b_\mu j_v^\mu$. Collecting everything, we get the full Lagrangian,

$$\mathscr{L} = \frac{1}{2\pi}\varepsilon^{\mu\nu\sigma}(eA_\mu - a_\mu)\partial_\nu b_\sigma + b_\mu j_v^\mu + \frac{1}{4\pi k}\varepsilon^{\mu\nu\sigma} a_\mu\partial_\nu a_\sigma + \, h.o \,, \tag{1.160}$$

where *h.o.* denote all higher derivative terms in b_μ. Finally, integrating out a_μ we end up with

$$\mathscr{L}_{\text{top}} = -\frac{k}{4\pi}\varepsilon^{\mu\nu\sigma} b_\mu\partial_\nu b_\sigma - \frac{e}{2\pi}\varepsilon^{\mu\nu\sigma} A_\mu\partial_\nu b_\sigma + b_\mu j_v^\mu \,. \tag{1.161}$$

Note that for a filled Landau level, i.e., $\nu = 1$, this exactly reproduces the previously derived expression for the topological Lagrangian (1.27), and just as before the gauge field b parametrizes the current. For the Laughlin states $k = 3, 5, \ldots$ the theory superficially looks very similar, but the higher integer values of k have far reaching consequences: fractional charge and statistics for the quasi-particles, ground-state degeneracy on higher genus surfaces, and chiral bosonic edge modes with k-dependent correlation functions [33].

1.7.3 The Abelian Hierarchy

In experiments on very clean samples, one sees a large number of FQH states [62]. Most of them fit beautifully into a hierarchical scheme [63, 64], where "daughter" states are formed by the condensation of anyonic quasi-particles in a "parent" FQH state, just as the Laughlin states can be thought of as a condensation of the original electrons.

Starting from the just derived topological Lagrangian (1.161) we can deduce what possible topological states can emerge from condensing quasi-holes. To do so, we need a dynamical theory for the holes, that in principle could be obtained by keeping higher order terms in above derivation of the effective theory. This is however difficult so we shall use a heuristic approach.

The basic, and quite reasonable, assumption is that the condensation of the quasiholes can be described by the same procedure as was used above for the electrons. This amounts to taking the following Ginzburg–Landau theory for the quasiholes,

$$\mathcal{L}_{\xi_v} = \phi_v^\star(i\partial_0 - b_0)\phi_v - \frac{1}{2m}|(\mathbf{p}+\mathbf{b})\phi_v|^2 - V(\rho_v)\,, \tag{1.162}$$

where ϕ_v is a bosonic vortex field associated to the quasiholes. We can again introduce a statistical gauge potential a and couple it to ϕ_v, but this time with the coefficient $1/4\pi k_2$ where k_2 is an even integer,

$$\tilde{\mathcal{L}}_{\xi_v} = \phi_v^\star(i\partial_0 + b_0 - a_0)\phi_v - \frac{1}{2m}|(\mathbf{p}+\mathbf{b}-\mathbf{a})\phi_v|^2 - V(\rho_v) + \frac{1}{4\pi k_2}\varepsilon^{\mu\nu\sigma}a_\mu\partial_\nu a_\sigma\,. \tag{1.163}$$

This is a trivial statistical transmutation, which adds the phase $e^{i2\pi k}$ to each particle exchange. In the mean-field approach, it will however have the effect to make the hole condensate thinner, just as in the original electron condensation. We can now, as before, find a mean-field solution where a_μ cancel against b_μ. The equation of motion for the density is obtained by varying b_0 and a_0,

$$\rho_v = -\frac{k}{2\pi}\varepsilon^{ij}\partial_i b_j + \frac{e}{2\pi}B; \qquad \rho_v = \frac{1}{2\pi k_2}\varepsilon^{ij}\partial_i a_j\,. \tag{1.164}$$

Combining this with the requirement $\varepsilon^{ij}\partial_i b_j = \varepsilon^{ij}\partial_i a_j$ imply that the mean-field solution is possible only when $\rho_c = \varepsilon^{ij}\partial_i b_j/2\pi$ equals

$$\rho_c = \frac{1}{2\pi}\frac{eB}{k+\frac{1}{k_2}} = \rho_0\frac{1}{k+\frac{1}{k_2}}\,.$$

To obtain the vortices in the field ϕ_v we can do almost exactly as above, with the result,

$$\begin{aligned}\mathcal{L}_{\text{top}} = &-\frac{k}{4\pi}\varepsilon^{\mu\nu\sigma}b_\mu\partial_\nu b_\sigma - \frac{k_2}{4\pi}\varepsilon^{\mu\nu\sigma}b^2_\mu\partial_\nu b^2_\sigma - \frac{1}{4\pi}\varepsilon^{\mu\nu\sigma}b_\mu\partial_\nu b^2_\sigma \\ &-\frac{1}{4\pi}\varepsilon^{\mu\nu\sigma}b^2_\mu\partial_\nu b_\sigma - \frac{e}{2\pi}\varepsilon^{\mu\nu\sigma}A_\mu\partial_\nu b_\sigma + b_\mu j^\mu_{v1} + b^2_\mu j^\mu_{v2}\,.\end{aligned}$$

To see how this generates a whole hierarchy of states, we consider condensation of the vortices in the previous state and so on. At level n that is with n consecutive condensations we end up with n gauge fields and the Lagrangian

$$\mathcal{L} = -\frac{1}{4\pi}\sum_{\alpha,\beta}K_{\alpha\beta}\varepsilon^{\mu\nu\sigma}b^\alpha_\mu\partial_\nu b^\beta_\sigma - \sum_\alpha\frac{e}{2\pi}t_\alpha A_\mu\varepsilon^{\mu\nu\sigma}\partial_\nu b^\alpha_\sigma + \sum_\alpha l_\alpha b^\alpha_\mu j^\mu\,, \tag{1.165}$$

where the *K-matrix* K and the the *charge vector* t takes the values

$$K_{\alpha\beta} = k_\alpha\delta_{\alpha\beta} - \delta_{\alpha+1,\beta} - \delta_{\alpha,\beta+1}; \qquad t_\alpha = \delta_{\alpha,1}\,, \tag{1.166}$$

and where the vector $\mathbf{l}$ is integer valued and describes all possible the quasi-particles excitations. As we realize from how this was constructed k_α is an even integer except for k_1 which is odd. It is explained in detail in [17] how to extract the topological information from this Lagrangian, and here we just quote the following important results for the filling fraction ν, and the charge (Q_α) and statistics angle (θ_α) for the αth quasi-particle:

$$\nu = \mathbf{t}^T\mathbf{K}^{-1}\mathbf{t}; \qquad Q_\alpha = -e\mathbf{t}^T\mathbf{K}^{-1}\mathbf{l}; \qquad \theta_\alpha = \pi\mathbf{l}^T\mathbf{K}^{-1}\mathbf{l}. \tag{1.167}$$

1.8 Mathematical Background and Proofs

In this section you will find mathematical background that complement the main text of this chapter. You will also find proofs, too technical for the main text, of some of the statements made in the text.

1.8.1 Vector Bundles and Chern Numbers in Quantum Mechanics

Consider a N dimensional subspace $h(\mathbf{b})$ of a Hilbert space $\mathscr{H}$ that varies continuously with the parameters $\mathbf{b}$ in some manifold $\mathbf{b} \in B$. The space

$$E = \bigcup_{\mathbf{b}\in B} \{\mathbf{b}\} \times h(\mathbf{b}) \,,$$

which is a subspace of $B \times \mathscr{H}$, is an example of a fiber bundle. The manifold B is called the base space, and the space $h(\mathbf{b})$ is called the fiber at $\mathbf{b}$. More specifically this is an example of a complex vector bundle, since the fibers are complex vector spaces. In the main text we encounter two distinct cases:

1. A quantum mechanical many-body system on a $2d$ torus with fluxes $\boldsymbol{\phi} = (\phi_x, \phi_y)$ through the two holes. For each value the fluxes there are N degenerate ground-states (recall Fig. 1.2 in Sect. 1.2.1). In this case the base manifold $B \equiv T^2_\phi$ is the flux-torus from Sect. 1.2.1, $\mathscr{H}$ is the many-body Hilbert space and $h(\boldsymbol{\phi})$ is the N-dimensional sub-space consisting of the degenerate ground-states at $\boldsymbol{\phi}$.
2. A systems of non-interacting fermions on a lattice. The base manifold $B \equiv B.Z.$ is the Brillouin-zone torus, $\mathscr{H}$ is the single-particle Hilbert space and $h(\mathbf{k})$ is the space of occupied single-particle states at lattice momentum $\mathbf{k}$.

Complex fiber-bundles are geometrical structures which have certain characteristics which take discrete vales, and which therefore cannot be altered by continuous transformations. Below we will construct one set of such characteristics, the Chern numbers. For the two above examples of fiber bundles, these Chern numbers are related to experimentally measurable transport coefficients.

We will also describe another set of characteristics called the Chern–Simons invariants. Contrary to the Chern numbers they are not topological invariants of the fiber-bundle, but for the two-mentioned examples, they are quantized if certain symmetries are invoked.

Both these quantities are defined in terms of a Berry connection which we will now define.

1.8.1.1 The Berry Connection

The Berry connection $\mathscr{A}_\mu$ is an operator taking $h(\mathbf{b})$ to $h(\mathbf{b}+db^\mu)$; it relates an element $|\mathbf{b}\rangle \in h(\mathbf{b})$ to the element in $h(\mathbf{b}+db^\mu)$ closest to $|\mathbf{b}\rangle$. If $h(\mathbf{b})$ is independent of $\mathbf{b}$ the closest element to $|\mathbf{b}\rangle$ in $h(\mathbf{b}+db^\mu)$ trivially is $|\mathbf{b}\rangle$ itself, while if $h(\mathbf{b})$ varies with $\mathbf{b}$ that is not necessarily the case. The notion of distance in the Hilbert space $\mathscr{H}$ is given by the inner product and the vector closest to $|\mathbf{b}\rangle$ in $h(\mathbf{b}+db^\mu)$ is the orthogonal projection of $|\mathbf{b}\rangle$ into $h(\mathbf{b}+db^\mu)$,

$$P\left(\mathbf{b}+db^\mu\right)|\mathbf{b}\rangle \; . \tag{1.168}$$

The Berry connection is defined by

$$P\left(\mathbf{b}+db^\mu\right)|\mathbf{b}\rangle = \left(\mathbb{1} - i\mathscr{A}_\mu db^\mu\right)|\mathbf{b}\rangle \; , \tag{1.169}$$

where $\mathscr{A}_\mu$ are operators from $h(\mathbf{b})$ to $h(\mathbf{b}+db^\mu)$.

Although this abstract form sometimes is useful we usually have to choose a specific basis to do any calculation. We therefore pick a basis

$$\{|\mathbf{b};\alpha\rangle\}_{\alpha=1,\ldots,N} \; , \tag{1.170}$$

that varies smoothly in some region B and we can write

$$|\mathbf{b}\rangle = \sum_\alpha a_\alpha(\mathbf{b})\,|\mathbf{b};\alpha\rangle \; , \tag{1.171}$$

for some coefficients $a_\alpha(\mathbf{b})$. In this basis the projection operator from $h(\mathbf{b})$ to $h(\mathbf{b}+db^\mu)$ is represented by,

$$\langle\mathbf{b}+db^\mu;\alpha|P\left(\mathbf{b}+db^\mu\right)|\mathbf{b};\beta\rangle = \delta_\alpha^\beta - \left\langle\mathbf{b};\alpha|\frac{\partial}{\partial b^\mu}|\mathbf{b};\beta\right\rangle db^\mu \; , \tag{1.172}$$

(the relation $(\frac{\partial}{\partial b^\mu}\langle\mathbf{b};\alpha|)|\mathbf{b};\beta\rangle) = -\left\langle\mathbf{b};\alpha|\frac{\partial}{\partial b^\mu}|\mathbf{b};\beta\right\rangle$ was used) and we get the expression

$$a_\alpha(\mathbf{b}) - \sum_\beta \left\langle \mathbf{b}; \alpha | \frac{\partial}{\partial b^\mu} | \mathbf{b}; \beta \right\rangle a_\beta(\mathbf{b}) db^\mu \tag{1.173}$$

for the left hand side of the definition (1.169) of the Berry connection. We can then read of

$$\mathscr{A}_{\alpha\,\mu}^{\,\beta} = i \left\langle \mathbf{b}; \alpha | \frac{\partial}{\partial b^\mu} | \mathbf{b}; \beta \right\rangle . \tag{1.174}$$

1.8.1.2 The Berry Field Strength

The Berry connection relates coefficients of vectors in two different Hilbert spaces, one at $h(\mathbf{b})$ and the other at $h(\mathbf{b} + db^\mu)$. Thus, the matrix $\mathscr{A}_{\alpha\,\mu}^{\,\beta}$ can be taken arbitrary since one could change the basis of $h(\mathbf{b})$ independent from the basis of $h(\mathbf{b} + db^\mu)$. The representation, $\mathscr{A}_{\alpha\,\mu}^{\,\beta}$, thus cannot be used to characterise the fiber bundle, we clearly have to look for something else.

The strategy is to consider a closed curve to define a matrix acting in a single Hilbert space $h(\mathbf{b})$. A matrix is a representation of an operator, and the precise form of the matrix is *not* basis independent, the trace however is, and we now show how it can be used to define the Chern numbers.

Consider an infinitesimal loop obtained by first moving the db^μ in the b^μ-direction, then db^ν in the ν-direction, db^μ backward in the μ-direction and finally back to where we started. We start out with the fiber $|\mathbf{b}\rangle$ then take the fiber closest to it in $h(\mathbf{b} + db^\mu)$, i.e.,

$$P\left(\mathbf{b} + db^\mu\right) |\mathbf{b}\rangle \ , \tag{1.175}$$

then take the fiber closest to that in $h(\mathbf{b} + db^\mu + db^\nu)$ etc. until the loop is closed; we end up with

$$P\left(\mathbf{b}\right) P\left(\mathbf{b} + db^\nu\right) P\left(\mathbf{b} + db^\mu + db^\nu\right) P\left(\mathbf{b} + db^\mu\right) |\mathbf{b}\rangle \ . \tag{1.176}$$

The Berry field strength $\mathscr{F}_{\mu\nu}$ is defined from this expression by

$$P\left(\mathbf{b}\right) P\left(\mathbf{b} + db^\nu\right) P\left(\mathbf{b} + db^\mu + db^\nu\right) P\left(\mathbf{b} + db^\mu\right) |\mathbf{b}\rangle = \left(\mathbb{1} - i \mathscr{F}_{\mu\nu} db^\mu db^\nu\right) |\mathbf{b}\rangle \ . \tag{1.177}$$

By Taylor expanding the projectors it is straight forward to write the field strength in terms on the connection,

$$\mathscr{F}_{\mu\nu} = \partial_\mu \mathscr{A}_\nu - \partial_\nu \mathscr{A}_\mu + i \mathscr{A}_\mu \mathscr{A}_\nu - i \mathscr{A}_\nu \mathscr{A}_\mu \ , \tag{1.178}$$

or written as matrices in a specific basis,

$$\mathscr{F}_{\alpha\,\mu\nu}^{\,\beta} = \partial_\mu \mathscr{A}_{\alpha\,\nu}^{\,\beta} - \partial_\mu \mathscr{A}_{\alpha\,\nu}^{\,\beta} + i \mathscr{A}_{\alpha\,\mu}^{\,\gamma} \mathscr{A}_{\gamma\,\nu}^{\,\beta} - i \mathscr{A}_{\alpha\,\nu}^{\,\gamma} \mathscr{A}_{\gamma\,\mu}^{\,\beta} \ , \tag{1.179}$$

where summation over repeated indices is understood. If B is two-dimensional $\mathscr{F}_{\mu\nu}$ has only one non-trivial component and we suppress the lower indices and use the notation

$$\mathscr{F} \equiv \mathscr{F}_{12} = -\mathscr{F}_{21} = \frac{1}{2}\varepsilon^{\mu\nu}\mathscr{F}_{\mu\nu} \,. \tag{1.180}$$

The field strenght $\mathscr{F}_{\mu\nu}$ is an operator that act within a fiber $h(\mathbf{b})$ and the trace of it equals the sum of the diagonal components

$$\mathrm{Tr}[\mathscr{F}_{\mu\nu}] = \mathscr{F}_{\alpha\,\mu\nu}^{\;\alpha} \,, \tag{1.181}$$

in any basis (remember that the repeated index α is summed over). The Berry connection on the other hand is different. $\mathscr{A}_\mu(\mathbf{b})$ is an operator from the fiber $h(\mathbf{b})$ to the fiber $h(\mathbf{b}+b^\mu)$ and an operator between two different spaces have no notion of a trace.[15] We will still write $\mathrm{Tr}[\mathscr{A}_\mu]$ (and similar for products $\mathrm{Tr}[\mathscr{A}_\mu\mathscr{A}_\nu]$ etc.) as a short-hand for $\mathscr{A}_{\alpha\,\mu}^{\;\alpha}$, but you have to remember that this *is a basis dependent expression.* It is instructive to see how this comes about in an explicit calculation: From the definition of the Berry connection (1.169) it follows, that under a coordinate transformation,

$$|\mathbf{b};\alpha\rangle \to \sum_\beta U_{\alpha\beta}(\mathbf{b})\,|\mathbf{b};\beta\rangle \ ,$$

the coefficients of the Berry connection transforms as

$$\mathscr{A}_{\alpha\,\mu}^{\;\beta} \to U_\alpha^{\;\gamma}\mathscr{A}_{\gamma\,\mu}^{\;\delta}U_\delta^{\dagger\,\beta} - iU_\alpha^{\dagger\,\gamma}\partial_\mu U_\gamma^{\;\beta} \,, \tag{1.182}$$

while the field strength just rotates,

$$\mathscr{F}_{\alpha\,\mu\nu}^{\;\beta} \to U_\alpha^{\;\gamma}\mathscr{F}_{\gamma\,\mu\nu}^{\;\delta}U_\delta^{\dagger\,\beta} \,. \tag{1.183}$$

So from the cyclic property one realize that $\mathscr{F}_{\alpha\,\mu\nu}^{\;\alpha}$ is basis independent and equals the trace of the operator $\mathscr{F}_{\mu\nu}$. One the other hand, $\mathrm{Tr}[\mathscr{A}_\mu] \equiv \mathscr{A}_{\alpha\,\mu}^{\;\alpha}$ transforms as

$$\mathrm{Tr}[\mathscr{A}_\mu] \to \mathrm{Tr}[\mathscr{A}_\mu] - iU_\alpha^{\dagger\,\gamma}\partial_\mu U_\gamma^{\;\alpha} \tag{1.184}$$

under a basis transformation.

We will end this section with a useful formula for the trace of $\mathscr{F}_{\mu\nu}$. Note that in the expression for the Berry field-strength (1.179) the two terms without derivatives come with different sign and thus vanishes when traced over, because of the cyclic

[15] You might object since $\mathscr{A}_\mu(\mathbf{b})$ is an operator within the full Hilbert space $\mathscr{H}$, so one should be able to define its trace. That is, in principle, a correct assumption but the trace of $\mathscr{A}_\mu(\mathbf{b})$ does not depend only on the structure of the fibers but of the full Hilbert space $\mathscr{H}$ and it does not necessarily equal $\mathscr{A}_{\alpha\,\mu}^{\;\alpha}(\mathbf{b})$, in a particular basis. The reason is the fact that the basis $\{|\mathbf{b};\alpha\rangle\}$ varies with $\mathbf{b}$; $\mathscr{A}_{\alpha\,\mu}^{\;\beta}(\mathbf{b})$ is the representation of the operator $\mathscr{A}_\mu(\mathbf{b})$ where the bras are written in the basis $\{|\mathbf{b}+b^\mu;\alpha\rangle\}$ and the kets in the basis $\{|\mathbf{b};\alpha\rangle$, see (1.172).

property of the trace,

$$\mathrm{Tr}\left[-i\mathscr{A}_\mu\mathscr{A}_\nu+i\mathscr{A}_\nu\mathscr{A}_\mu\right]\equiv i\mathscr{A}_{\alpha\,\mu}^{\;\gamma}\mathscr{A}_{\gamma\,\nu}^{\;\alpha}-i\mathscr{A}_{\alpha\,\nu}^{\;\gamma}\mathscr{A}_{\gamma\,\mu}^{\;\alpha}=0\,. \tag{1.185}$$

One can therefore write

$$\mathrm{Tr}\left(\mathscr{F}_{\mu\nu}\right)\equiv\mathscr{F}_{\alpha\,\mu_1\nu_1}^{\;\alpha}=-i\sum_\alpha\left(\frac{\partial}{\partial b^\mu}\left\langle\mathbf{b};\alpha\right|\frac{\partial}{\partial b^\nu}\left|\mathbf{b};\alpha\right\rangle-\frac{\partial}{\partial b^\nu}\left\langle\mathbf{b};\alpha\right|\frac{\partial}{\partial b^\mu}\left|\mathbf{b};\alpha\right\rangle\right)\,. \tag{1.186}$$

1.8.1.3 First Chern Number and Chern–Simons Invariant

Let us now assume that B is two-dimensional (or think of a two-dimensional submanifold of a general B), and form,

$$ch_1[\mathscr{F}](S)=\frac{1}{4\pi}\int_B\mathrm{d}^2b\,\varepsilon^{\mu\nu}\mathrm{Tr}[\mathscr{F}_{\mu\nu}]\,. \tag{1.187}$$

This expression defines the first Chern number, ch_1, and the integrand (including the prefactor) is called the first Chern character.

One can in general cannot define a basis for each $h(b)$ that varies continuously for all $b\in B$. For any region topologically equivalent to a subset of $\mathbb{R}^n$ one can define a continuous basis. If B is topologically equivalent to a disc we can define a continuous basis in the whole of B, and we can use the basis dependent expression $\mathscr{A}_{\alpha\,\mu}^{\;\beta}$ throughout B. Then by using $\varepsilon^{\mu\nu}\mathrm{Tr}[\mathscr{A}_\mu\mathscr{A}_\nu]=0$ (remember that $\mathrm{Tr}[\mathscr{A}_\mu\mathscr{A}_\nu]$ is defined by the basis dependent expression $\mathscr{A}_{\alpha\,\mu}^{\;\beta}$) we can write $\varepsilon^{\mu\nu}\mathrm{Tr}[\mathscr{F}_{\mu\nu}]=2\varepsilon^{\mu\nu}\partial_\mu\mathrm{Tr}[\mathscr{A}_\nu]$. This allows us to use Stokes theorem to rewrite the Chern number as

$$\frac{1}{2\pi}\int_{\partial B}dl^\mu\,\mathrm{Tr}[\mathscr{A}_\mu]\,. \tag{1.188}$$

The integrand (again including the prefactor) is called the first Chern–Simons form. Next we assume that B is a sphere[16] and to evaluate the integral in (1.187) we cover the sphere with two caps, the northern hemisphere (α) and the southern hemisphere (β). With this we have

$$\frac{1}{4\pi}\int_B\mathrm{d}^2b\,\varepsilon^{\mu\nu}\mathrm{Tr}[\mathscr{F}_{\mu\nu}]=\frac{1}{4\pi}\int_\alpha\mathrm{d}^2b\,\varepsilon^{\mu\nu}\mathrm{Tr}[\mathscr{F}_{\mu\nu}]+\frac{1}{4\pi}\int_\beta\mathrm{d}^2b\,\varepsilon^{\mu\nu}\mathrm{Tr}[\mathscr{F}_{\mu\nu}]\,. \tag{1.189}$$

Since both the caps are topologically equivalent to the disc, we can in each one of them pick a continuous basis. Using Stokes theorem we get,

[16]The fact that the first Chern number is an integer, proven here, hold for a general closed manifold as e.g., the torus. The proof is more involved since one has to divide B into more regions, but the arguments are analogous to the one for the sphere.

$$\frac{1}{4\pi}\int_B d^2b\,\varepsilon^{\mu\nu}\mathrm{Tr}[\mathscr{F}_{\mu\nu}] = \frac{1}{2\pi}\int_{\partial\alpha} dl^\mu\,\mathrm{Tr}[\mathscr{A}_\mu^{(\alpha)}] - \frac{1}{2\pi}\int_{\partial\alpha} dl^\mu\,\mathrm{Tr}[\mathscr{A}_\mu^{(\beta)}]\,, \quad (1.190)$$

where we used the fact that $\partial\beta$ is the same curve as $\partial\alpha$, but with opposite orientation, and where the superscript α means that $\mathscr{A}_\mu^{(\alpha)}$ is defined with respect to a coordinate system that is continuous in the region α, and similarly for $\mathscr{A}_\mu^{(\beta)}$. On the equator both coordinate systems are continuous and they are thus related by some unitary transformation U,

$$\mathrm{Tr}[\mathscr{A}_\mu^{(\beta)}] = \mathrm{Tr}[\mathscr{A}_\mu^{(\alpha)}] - i\mathrm{Tr}[U^\dagger\partial_\mu U]\,. \quad (1.191)$$

Putting this together we get

$$\frac{1}{4\pi}\int_B d^2b\varepsilon^{\mu\nu}\mathrm{Tr}[\mathscr{F}_{\mu\nu}] = \frac{1}{2\pi i}\int_{\partial\alpha} dl^\mu\mathrm{Tr}[U^\dagger\partial_\mu U]\,. \quad (1.192)$$

The matrix trace is the same in all basis so we can consider it in the basis where U is diagonal

$$U = \mathrm{diag}\left(e^{i\theta_1}, e^{i\theta_2}, \ldots\right)\,, \quad (1.193)$$

to get

$$ch_1(S) = \frac{1}{2\pi i}\int_\gamma dl^\mu\,\mathrm{Tr}[U^\dagger\partial_\mu U] = \frac{1}{2\pi i}\sum_i\int_{\partial\alpha} dl^\mu e^{-i\theta_i}\partial_\mu e^{i\theta_i}\,. \quad (1.194)$$

The ith term in this sum gives the change of the phase angle θ_i accumulated when integrating over $\partial\alpha$, but since this is a closed curve this change has to equal $2\pi n$ for some integer n, which completes the proof that $ch_1(S)$ is an integer.

The formula (1.191) allows for another important conclusion. Since $\mathrm{Tr}[\mathscr{A}_\mu]$ changes by $-i\mathrm{Tr}[U^\dagger\partial_\mu U]$ under a basis transformation and $\int dl^\mu\mathrm{Tr}[U^\dagger\partial_\mu U] = 2\pi ni$ for some integer n, the exponent of 2π times the first Chern–Simons invariant

$$CS_1[\mathscr{A}](\gamma) = \frac{1}{2\pi}\int_\gamma dl^\mu\mathrm{Tr}[\mathscr{A}_\mu]$$

is also a basis independent quantity, meaning that $\exp(iCS_1)$ is well defined. It is however not necessarily quantized as an integer.

1.8.1.4 Relating the Brillouin-Zone and the Flux-Torus Chern Numbers

For a non-interacting fermionic system on lattice with periodic boundary conditions we can define two different fiber bundles. The first is the flux-torus fiber bundle with the flux torus T_ϕ^2 as base space and fibers $h(\boldsymbol{\phi})$ the one-dimensional Hilbert spaces spanned by $|GS; \boldsymbol{\phi}\rangle$ (i.e., the ground-state at flux $\boldsymbol{\phi}$ which are the fluxes encircled by the two independent non-contractible loops on the torus).

The other fiber bundle has the Brillouin-zone as the base space and the fibers at lattice momentum $\mathbf{k}$ are spanned by the single particle wave functions,

$$\psi_{\mathbf{k}\,n}(\mathbf{x}) = e^{i\mathbf{k}\cdot\mathbf{x}} u_{\mathbf{k}\,n}(\mathbf{x}) \ , \tag{1.195}$$

of the N filled bands. The Berry connection over the Brillouin-zone fiber bundle is given by

$$\mathscr{A}_{n\,k_i}^{\ m}(\mathbf{k}) = -i \left\langle \psi_{\mathbf{k}\,n}(\mathbf{x}) | \partial_{k_i} | \psi_{\mathbf{k}\,n}(\mathbf{x}) \right\rangle \equiv \int \mathrm{d}^2x \ \psi^*_{\mathbf{k}\,n}(\mathbf{x}) \partial_{k_i} \psi_{\mathbf{k}\,n}(\mathbf{x}) \ . \tag{1.196}$$

This can also be written as the anti-commutator of the creation and annihilation operators,

$$\mathscr{A}_{n\,k_i}^{\ m}(\mathbf{k}) = -i \left\{ a_{\mathbf{k}\,n}, \partial_{k_i} a^\dagger_{\mathbf{k}\,m} \right\} \tag{1.197}$$

where

$$a^\dagger_{\mathbf{k}\,n} = \int \mathrm{d}^2x \ \psi^\dagger(\mathbf{x}) \psi_{\mathbf{k}\,n}(\mathbf{x}); \qquad a_{\mathbf{k}\,n} = \int \mathrm{d}^2x \ \psi(\mathbf{x}) \psi^*_{\mathbf{k}\,n}(\mathbf{x})) \tag{1.198}$$

and $\psi^\dagger(\mathbf{x})$ is the operator that creates an electron at position $\mathbf{x}$,

$$\left\{ \psi^\dagger(\mathbf{x}), \psi(\mathbf{x}') \right\} = \delta^2(\mathbf{x} - \mathbf{x}') \ . \tag{1.199}$$

The corresponding Berry field-strength is (which we denote by $\mathscr{B}$ to distinguish it from the previously defined flux torus field strength $\mathscr{F}$)

$$\mathscr{B}_{n\,k_i k_j}^{\ m} = \partial_{k_i} \mathscr{A}_{n\,k_j}^{\ m} - \partial_{k_j} \mathscr{A}_{n\,k_i}^{\ m} + i \mathscr{A}_{n\,k_i}^{\ p} \mathscr{A}_{p\,k_j}^{\ m} - i \mathscr{A}_{n\,k_j}^{\ p} \mathscr{A}_{p\,k_i}^{\ m} \ , \tag{1.200}$$

where the repeated index p should be summed over. Since the Brillouin-zone is two-dimensional the field-strength has only one independent component

$$\mathscr{B} \equiv \varepsilon^{ij} \mathscr{B}_{k_i k_j} \ . \tag{1.201}$$

We can then write

$$\mathrm{Tr}[\mathscr{B}] = \varepsilon^{ij} \, \mathrm{Tr}[\partial_{k_i} \mathscr{A}_{k_j}] = -i \sum_n \left\{ a_{\mathbf{k}\,n}, \partial_{k_i} a^\dagger_{\mathbf{k}\,m} \right\} \ , \tag{1.202}$$

for the the defining component of the Brillouin-zone Berry field-strength. Now turn to the other fiber bundle. The fibers are spanned by the ground-states at flux $\boldsymbol{\phi}$,

$$|GS; \boldsymbol{\phi}\rangle = \prod_n \prod_{\mathbf{k} \in B.Z.} a^\dagger_{\mathbf{k}\,n\,\boldsymbol{\phi}} |0\rangle \ , \tag{1.203}$$

where

$$a^\dagger_{\mathbf{k}\,n\,\boldsymbol{\phi}} = \int \mathrm{d}^2x\, \psi^\dagger(\mathbf{x})\psi_{\mathbf{k}\,n\,\boldsymbol{\phi}}(\mathbf{x}) \,, \tag{1.204}$$

and $\psi_{\mathbf{k}\,n\,\boldsymbol{\phi}}(\mathbf{x})$ is the single particle wave function in band n, at lattice momentum $\mathbf{k}$ and at flux $\boldsymbol{\phi}$. The Berry connection is given by

$$\mathscr{A}_{\phi_i} = -i \left\langle GS;\, \boldsymbol{\phi}|\partial_{\phi_i}|GS;\, \boldsymbol{\phi}\right\rangle \tag{1.205}$$

and the defining component of the Berry field strength is

$$\mathscr{F} = -i\varepsilon^{ij}\partial_{\phi_i}\left\langle GS|\partial_{\phi_j}|GS\right\rangle \,. \tag{1.206}$$

Using the commutation realtions

$$\left\{a_{\mathbf{k}\,m\,\boldsymbol{\phi}}, a^\dagger_{\mathbf{k}'\,n\,\boldsymbol{\phi}}\right\} = \delta_{mn}\delta_{\mathbf{k}\,\mathbf{k}'};\quad \left\{a_{\mathbf{k}\,m\,\boldsymbol{\phi}}, a_{\mathbf{k}'\,n\,\boldsymbol{\phi}}\right\} = \left\{a_{\mathbf{k}\,m\,\boldsymbol{\phi}}, a^\dagger_{\mathbf{k}'\,n\,\boldsymbol{\phi}}\right\} \;= 0\,, \tag{1.207}$$

we can write

$$\left\langle GS;\, \boldsymbol{\phi}|\partial_{\phi_i}|GS;\, \boldsymbol{\phi}\right\rangle = \sum_n \int \mathrm{d}^2k \left\{a_{\mathbf{k}\,n\,\boldsymbol{\phi}}, \partial_{\phi_i} a^\dagger_{\mathbf{k}\,n\,\boldsymbol{\phi}}\right\} \,, \tag{1.208}$$

and

$$\mathscr{F}(\phi_i, \phi_j) = -i\varepsilon^{ij}\partial_{\phi_i}\sum_n \int \mathrm{d}^2k \left\{a_{\mathbf{k}\,n\,\boldsymbol{\phi}}, \partial_{\phi_j} a^\dagger_{\mathbf{k}\,n\,\boldsymbol{\phi}}\right\} \,. \tag{1.209}$$

We now pick the gauge potential

$$\mathbf{A} = \tilde{\mathbf{A}} + \frac{\phi_x}{L_x}\hat{x} + \frac{\phi_y}{L_y}\hat{y} \,, \tag{1.210}$$

where the integral of $\tilde{A}_i$ along any of the non-contractible loops on the torus is zero. With this, the Bloch Hamiltonian become

$$H_{Bl} = \frac{\hbar^2}{2m}\left(-i\nabla + e\tilde{\mathbf{A}} + \mathbf{k} + \frac{2\pi}{\phi_0}\left(\phi_x/L_x, \phi_y/L_y\right)\right)^2 + V_{lat} \,. \tag{1.211}$$

Comparing with the Bloch Hamiltonian at zero flux,

$$H_{Bl} = \frac{\hbar^2}{2m}\left(-i\nabla + e\mathbf{A} + \mathbf{k}\right)^2 + V_{lat} \,, \tag{1.212}$$

we realize that we can put $\boldsymbol{\phi} = 0$ and replace

$$\partial_{\phi_i} \to \frac{2\pi}{L_i}\frac{1}{\phi_0}\partial_{k_i} \,, \tag{1.213}$$

which results in

$$\mathscr{F}(\boldsymbol{\phi}) = -i\frac{\varepsilon^{ij}}{\phi_0^2}\int \mathrm{d}^2k\; \partial_{k_i}\left\{a_{\mathbf{k}\,n}, \partial_{k_j} a^{\dagger}_{\mathbf{k}\,n}\right\} = \frac{1}{\phi_0^2}\int \mathrm{d}^2k\; \mathrm{Tr}\,(\mathscr{B}) \,, \tag{1.214}$$

where we used (1.202). Finally integrating over the flux torus concludes the proof that the two Chern numbers are equal.

1.8.2 How to Normalize the Current

Here we shall study the first term $\sim\varepsilon^{\mu\nu\sigma}b_\mu\partial_\nu a_\sigma$ in (1.42) a bit more carefully. Just as the Chern–Simons term this is a topological action, and the two fields a and b have no bulk dynamics. This is however not the full story. If the system is defined on a finite area, $L_x \times L_y$, with periodic boundary conditions, the zero-modes of the fields do acquire dynamics. To understand this, we write the action on the Hamiltonian form,

$$\begin{aligned} S_{BF} &= \frac{k}{2\pi}\int \mathrm{d}^3x\; \varepsilon^{\mu\nu\sigma} b_\mu \partial_\nu a_\sigma \\ &= \frac{k}{2\pi}\int \mathrm{d}^3x\; [\varepsilon^{ij}\dot{a}_i b_j + a_0(\varepsilon^{ij}\partial_i b_j) + b_0(\varepsilon^{ij}\partial_i a_j)] \,. \end{aligned} \tag{1.215}$$

Note that although the Hamiltonian formally vanishes in the $a_0 = b_0 = 0$ gauge, these fields are Lagrange multiplier fields that impose the "Gauss law" constraints $\varepsilon^{ij}\partial_i a_j = \varepsilon^{ij}\partial_i b_j = 0$, which can be solved by,

$$\begin{aligned} a_i &= \partial_i\lambda_a(\mathbf{x}, t) + \frac{2\pi}{L_i}\bar{a}_i(t) \\ b_i &= \partial_i\lambda_b(\mathbf{x}, t) + \frac{2\pi}{L_i}\bar{b}_i(t) \end{aligned} \tag{1.216}$$

where $\bar{a}_i$ and $\bar{b}_i$ are spatially constant, and $\lambda_{a/b}$ are periodic functions on the torus. Inserting this into (1.215) gives the Lagrangian

$$S_{BF} = k\,2\pi\int \mathrm{d}t\; \varepsilon^{ij}\dot{\bar{a}}_i\bar{b}_j \,. \tag{1.217}$$

From this we can read the canonical commutation relations.[17]

[17] In case you do not know how to handle actions that are first order in time derivatives, you can learn in e.g., [65].

$$[\bar{a}_i, \bar{b}_j] = \frac{i}{2\pi k}\varepsilon^{ij}\,. \tag{1.218}$$

The Wilson lines along the cycles of the torus given by,

$$\mathscr{A}_i = e^{i\oint \mathrm{d}x_i\, a_i} = e^{2\pi i \bar{a}} \qquad \mathscr{B}_i = e^{i\oint \mathrm{d}x_i\, b_i} = e^{2\pi i \bar{b}} \tag{1.219}$$

are invariant under the large gauge transformations $\bar{a}_i \to a_i + n_i^a$, and $\bar{b}_i \to b_i + n_i^b$, corresponding to threading unit fluxes though the holes in the torus.

To assume that the charges that couple to the gauge fields a and b are conserved is equivalent to taking the these fields to be compact, which means that field configurations differing by the large gauge transformations are identified. Put differently, the dynamical variables are not the constant $U(1)$ fields $\bar{a}_i$ and $\bar{b}_i$, but the Wilson loop operators $\mathscr{A}_i$ and $\mathscr{B}_i$, which satisfy the commutation relations,

$$\mathscr{A}_l \mathscr{B}_m - e^{\frac{2\pi i}{k}} \mathscr{B}_m \mathscr{A}_l = 0 \quad l \neq m\,. \tag{1.220}$$

For $k = 1$ all operators commute, and there is a unique ground-state, while for $k = 2$ we have the algebra,

$$\mathscr{A}_x \mathscr{B}_y + \mathscr{B}_y \mathscr{A}_x = 0 \quad \text{and} \quad \mathscr{A}_y \mathscr{B}_x + \mathscr{B}_x \mathscr{A}_y = 0\,. \tag{1.221}$$

Each of these algebras have a two dimensional representation (think of the Pauli matrices!) and thus the ground-state is $2 \times 2 = 4$ fold degenerate.

1.8.3 *The Relation Between the Chern Number and the Pontryagin Index*

We will use the notation, $\hat{d}(\mathbf{k}) \cdot \boldsymbol{\sigma}\, \xi_\pm(\mathbf{k}) = \mp\xi_\pm(\mathbf{k})$ and $\xi \equiv \xi_+$ which means that the creation operator for the filled band takes the form

$$a_{\mathbf{k}}^\dagger = \sum_\alpha \xi_-^\alpha(\mathbf{k}) c_{\mathbf{k},\alpha}^\dagger \tag{1.222}$$

and using the definition (1.55) the Berry connection is

$$\mathscr{A}_i = -i\{a_{\mathbf{k}}, \partial_{k_i} a_{\mathbf{k}}^\dagger\} = -i\xi^\dagger(\mathbf{k})\partial_{k_i}\xi(\mathbf{k}) \tag{1.223}$$

Now suppressing the $\mathbf{k}$ dependence and using the short hand notation $\partial_{i/j} \equiv \partial_{k_{i/j}}$ we have,

$$
\begin{aligned}
(\partial_i \xi^\dagger)\partial_j \xi &= \partial_i(\xi^\dagger \hat{d} \cdot \boldsymbol{\sigma})\partial_j \xi = (\partial_i \xi^\dagger)\hat{d} \cdot \boldsymbol{\sigma}\partial_j \xi + \xi^\dagger(\partial_i \hat{d} \cdot \boldsymbol{\sigma})\partial_j \xi \\
&= (\partial_i \xi^\dagger)\hat{d} \cdot \boldsymbol{\sigma}\partial_j \xi + \xi^\dagger(\partial_i \hat{d} \cdot \boldsymbol{\sigma})\partial_j(\hat{d} \cdot \boldsymbol{\sigma}\xi) \\
&= (\partial_i \xi^\dagger)\hat{d} \cdot \boldsymbol{\sigma}\partial_j \xi - \xi^\dagger(\partial_i \hat{d} \cdot \boldsymbol{\sigma})\partial_j \xi + \xi^\dagger(\partial_i \hat{d} \cdot \boldsymbol{\sigma})(\partial_j \hat{d} \cdot \boldsymbol{\sigma})\xi
\end{aligned}
\tag{1.224}
$$

where we in the last equality used $\sigma^a\sigma^b = -\sigma^b\sigma^a + 2\delta^{ab}$ and $\hat{d} \cdot \partial_i \hat{d} = 0$. Making similar manipulations we get,

$$
(\partial_i \xi^\dagger)\partial_j \xi = (\partial_i \xi^\dagger)\hat{d} \cdot \boldsymbol{\sigma}\partial_j \xi + \frac{1}{2}\xi^\dagger(\partial_i \hat{d} \cdot \boldsymbol{\sigma})(\partial_j \hat{d} \cdot \boldsymbol{\sigma})\xi \tag{1.225}
$$

Next winvokee need the spectral decomposition of the Hamiltonian, $\hat{d} \cdot \boldsymbol{\sigma} = \xi\xi^\dagger - \xi_-\xi_-^\dagger$ and the resolution of unity $1 = \xi\xi^\dagger + \xi_-\xi_-^\dagger$ to get the two identities,

$$
\begin{aligned}
(\partial_i \xi^\dagger)\hat{d} \cdot \boldsymbol{\sigma}\partial_j \xi &= (\partial_i \xi^\dagger)(\xi\xi^\dagger - \xi_-\xi_-^\dagger)\partial_j \xi = -\mathscr{A}_i\mathscr{A}_j + (\xi^\dagger\partial_i\xi_-)(\xi_-^\dagger\partial_j\xi) \\
(\partial_i \xi^\dagger)\partial_j \xi &= (\partial_i \xi^\dagger)(\xi\xi^\dagger + \xi_-\xi_-^\dagger)\partial_j \xi = -\mathscr{A}_i\mathscr{A}_j - (\xi^\dagger\partial_i\xi_-)(\xi_-^\dagger\partial_j\xi)
\end{aligned}
\tag{1.226}
$$

where we used $\xi^\dagger\partial_i\xi = -(\partial_i\xi^\dagger)\xi$ etc. Combining (1.225) and (1.226) gives,

$$
(\partial_i \xi^\dagger)\partial_j \xi = -\mathscr{A}_i\mathscr{A}_j + \frac{1}{4}\xi^\dagger(\partial_i \hat{d} \cdot \boldsymbol{\sigma})(\partial_j \hat{d} \cdot \boldsymbol{\sigma})\hat{d} \cdot \boldsymbol{\sigma}\xi \tag{1.227}
$$

Again using the σ matrix algebra, we finally get,

$$
\mathscr{B}(\mathbf{k}) = i\varepsilon_{ij}\partial_i\xi^\dagger\partial_j\xi = -\frac{1}{2}(\partial_i\hat{d} \times \partial_j\hat{d}) \cdot \hat{d} \tag{1.228}
$$

1.8.4 The Parity Anomaly in 2 + 1 Dimensions

The most direct way to extract the effective action (1.68) is to simply calculate the Feynman diagram in Fig. 1.3 with a suitable regularization [36]. An alternative, and quite instructive, way is to pick a specific background field where the Dirac equation can be solved exactly and then invoke gauge and Lorentz invariance to find the general result.

We define the Dirac α-matrices by,

$$
(\beta, \alpha^1, \alpha^2) = (\sigma^3, -\sigma^2, \sigma^1) \tag{1.229}
$$

and will use complex coordinates

$$
z = \sqrt{\frac{eB}{2}}(x + iy) \tag{1.230}
$$

and the notation $\partial = \partial_z$ and $\bar{\partial} = \partial_{\bar{z}}$. In the symmetric gauge, $\mathbf{A} = \frac{B}{2}(-y, x)$, where B is a constant magnetic field. The Hamiltonian for the relativistic massless Landau problem becomes,

$$\begin{aligned} H &= \boldsymbol{\alpha} \cdot (\mathbf{p} - e\mathbf{A}) \\ &= \sqrt{eB} \begin{pmatrix} 0 & -\frac{1}{\sqrt{2}}(\partial - \bar{z}) \\ \frac{1}{\sqrt{2}}(\bar{\partial} + z) & 0 \end{pmatrix} = \sqrt{eB} \begin{pmatrix} 0 & a^\dagger \\ a & 0 \end{pmatrix} \end{aligned} \tag{1.231}$$

where $\mathbf{p} = -i\nabla$, and $[a, a^\dagger] = 1$. Introducing the corresponding number operator state $a^\dagger a \,|n\rangle = n\,|n\rangle$, we easily find the following solutions to the Schrodinger equation,

$$|\Psi_0\rangle = \begin{pmatrix} |0\rangle \\ 0 \end{pmatrix}; \qquad E_0 = 0 \tag{1.232}$$

$$|\Psi_{n\pm}\rangle = \frac{1}{\sqrt{2}} \begin{pmatrix} |n\rangle \\ \pm\,|n-1\rangle \end{pmatrix}; \qquad E_\pm = \pm\sqrt{neB}\,. \tag{1.233}$$

Note that after adding a mass term, $H_m = \beta m$, $|\psi_0\rangle$ is still a solution but with the eigenvalue $E_0 = m$.

The energy levels are however massively degenerate, and in the radial gauge the relevant extra quantum number is the angular momentum. Defining,

$$b^\dagger = a + \sqrt{2}z \tag{1.234}$$

we have $[b, b^\dagger] = [a, a^\dagger] = 1$ with all other commutators vanishing, so the full Schrodinger spectrum is obtained from (1.232), by the replacement

$$|n\rangle \to \frac{(b^\dagger)^k}{\sqrt{k!}} \frac{(a^\dagger)^n}{\sqrt{n!}} |0, 0\rangle\,. \tag{1.235}$$

Next we expand the Dirac field operator in the eigenfunctions $\psi_{n,k\pm} = \langle x\rangle\, \Psi_{k,n\pm}$

$$\hat{\psi}(x) = \sum_{n,k} \left[\psi_{n,k+} e^{-iE_n t} c_{n,k} + \psi_{n,k-} e^{iE_n t} d^\dagger_{n,k} \right] + \sum_k \psi_{0,k} e^{-imt} e \tag{1.236}$$

where we regularized the zero-mode by adding a small mass $m^2 \ll eB$, and introduced the fermionic operators $c_{n,k}$ and $d_{n,k}$ satisfying $\{c^\dagger_{\tilde{n},\tilde{k}}\} = \delta_{n,\tilde{n}}\delta_{k,\tilde{k}}$ etc., and the Majorana operator e which satisfies $e^2 = 0$ and anti-commutes with all other Fermi operators.

For simplicity we considered the $m = 0$ case, but it is not too hard to find the full solution even for $m \neq 0$. The zero-modes we have already found, and the only thing we shall need for the following is that the rest of the spectrum is gapped and symmetric around $E = 0$ which follows from charge conjugation symmetry; fact the energies are $E_n = \sqrt{neB + m^2}$.

The next step is to calculate the vacuum expectation value of the current operator

$$j^\mu = \frac{e}{2}[\bar{\psi}, \gamma^\mu \psi] \tag{1.237}$$

and for simplicity we shall just consider the time component, i.e., the charge. Realling that $\gamma^0 = \beta$, we get

$$j^0 = \rho = \frac{e}{2}(\psi^\dagger \psi - \psi \psi^\dagger) \tag{1.238}$$

and because of the charge conjugation symmetry, only the zero-modes contribute to the expectation value

$$\langle \rho \rangle = \pm \frac{e}{2} \sum_k |\psi_{0,k}|^2 . \tag{1.239}$$

where the positive sign corresponds to a negative m meaning that the (almost) zero-modes are all filled so the contribution comes from the first term in (1.238); the negative sign amounts to these modes all being empty. A direct calculation of the zero-mode wave functions gives,

$$\sum_k |\psi_{0k}|^2 = \sum_k |\sqrt{\frac{eB}{2\pi}}\sqrt{\frac{2^k}{k!}} z^k e^{-|z|^2}|^2 = \sum_k \frac{eB}{2\pi}(2|z|^2)^n e^{-|z|^2} = \frac{eB}{2\pi} \tag{1.240}$$

which just shows that the density of state in the lowest Landau level is the same as in the non-relativistic case. Inserting this in (1.238) gives

$$\langle \rho \rangle = \frac{m}{|m|}\frac{e^2 B}{4\pi} \tag{1.241}$$

and by Lorentz and gauge invariance we get

$$\langle j^\mu \rangle = \frac{m}{|m|}\frac{e^2}{4\pi}\varepsilon^{\mu\nu\sigma}\partial_\nu A_\sigma = -\frac{\delta \Gamma_D[A]}{\delta A_\mu}, \tag{1.242}$$

for any constant electric, or magnetic, field. Integrating this expression to get the effective response action, we get the result (1.68) quoted in the main text. Note that although the Chern–Simons term is not gauge invariant, its variation, given by (1.242), is.

Acknowledgements We thank Maria Hermanns for making useful comments on a preliminary version of these notes, and we also thank Christian Spånslätt for reading and commenting the manuscript. We are very thankful to Marcus Berg and Sergej Moroz, who helped us find many typos and errors.

The nice hand-drawn pictures are provided by Sören Holst and the derivation of the parity anomaly given in Sect. 1.8.4 was shown to THH by the late professor Ken Johnson of MIT.

TKK acknowledge *the Wenner-Gren foundations* and *stiftelsen Olle Engkvist Byggmästare* for financial support during 2019 and 2018 respectively.

References

1. L.D. Landau, Phys. Z. Sowjet. **11**, 26 (1937), http://www.ujp.bitp.kiev.ua/files/journals/53/si/53SI08p.pdf
2. F. Haldane, Phys. Lett. A **93**(9), 464 (1983). https://doi.org/10.1016/0375-9601(83)90631-X
3. M.Z. Hasan, C.L. Kane, Rev. Mod. Phys. **82**, 3045 (2010). https://doi.org/10.1103/RevModPhys.82.3045
4. A. Kitaev, AIP Conf. Proc. **1134**(1), 22 (2009). https://doi.org/10.1063/1.3149495
5. A.P. Schnyder, S. Ryu, A. Furusaki, A.W.W. Ludwig, Phys. Rev. B **78**, 195125 (2008). https://doi.org/10.1103/PhysRevB.78.195125
6. S. Ryu, J.E. Moore, A.W.W. Ludwig, Phys. Rev. B **85**, 045104 (2012). https://doi.org/10.1103/PhysRevB.85.045104
7. M. Levin, X.G. Wen, Phys. Rev. Lett. **96**, 110405 (2006). https://doi.org/10.1103/PhysRevLett.96.110405
8. A. Kitaev, J. Preskill, Phys. Rev. Lett. **96**, 110404 (2006). https://doi.org/10.1103/PhysRevLett.96.110404
9. A.Y. Kitaev, Physics-Uspekhi **44**(10S), 131 (2001), http://stacks.iop.org/1063-7869/44/i=10S/a=S29
10. J. Klassen, X.G. Wen, J. Phys.: Condens. Matter **27**(40), 405601 (2015), http://stacks.iop.org/0953-8984/27/i=40/a=405601
11. S.C. Zhang, Int. J. Mod. Phys. B **06**(01), 25 (1992). https://doi.org/10.1142/S0217979292000037
12. A. Altland, B.D. Simons, *Condensed Matter Field Theory*, 2nd edn. (Cambridge University Press, Cambridge, 2010). https://doi.org/10.1017/CBO9780511789984, https://www.cambridge.org/core/books/condensed-matter-field-theory/0A8DE6503ED868D96985D9E7847C63FF
13. S. Weinberg, *The Quantum Theory of Fields*, vol. 2 (Cambridge University Press, Cambridge, 1996). https://doi.org/10.1017/CBO9781139644174, https://www.cambridge.org/core/books/the-quantum-theory-of-fields/0E0C89894938BE38EE0BCCDB1BC857E5
14. E. Fradkin, *Field Theories of Condensed Matter Physics*, 2nd edn. (Cambridge University Press, Cambridge, 2012). https://doi.org/10.1017/CBO9781139015509, https://www.cambridge.org/core/books/field-theories-of-condensed-matter-physics/EABECB65A0F4F9289B2737A6DD3E6C0D
15. B.A. Bernevig, T.L. Hughes, *Topological Insulators and Topological Superconductors*, 1st edn. (Princeton University Press, Princeton, 2013), http://press.princeton.edu/titles/10039.html
16. X.G. Wen, A. Zee, Phys. Rev. B **46**, 2290 (1992). https://doi.org/10.1103/PhysRevB.46.2290
17. X.G. Wen, Adv. Phys. **44**(5), 405 (1995). https://doi.org/10.1080/00018739500101566
18. T. Hansson, V. Oganesyan, S. Sondhi, Ann. Phys. **313**(2), 497 (2004). https://doi.org/10.1016/j.aop.2004.05.006, http://www.sciencedirect.com/science/article/pii/S0003491604001046
19. G.Y. Cho, J.E. Moore, Ann. Phys. **326**(6), 1515 (2011). https://doi.org/10.1016/j.aop.2010.12.011, http://www.sciencedirect.com/science/article/pii/S0003491610002253
20. K.v. Klitzing, G. Dorda, M. Pepper, Phys. Rev. Lett. **45**, 494 (1980). https://doi.org/10.1103/PhysRevLett.45.494

21. K. Novoselov, Z. Jiang, Y. Zhang, S. Morozov, H. Stormer, U. Zeitler, J. Maan, G. Boebinger, P. Kim, A. Geim, Science **315**(5817), 1379 (2007)
22. Q. Niu, D.J. Thouless, Y.S. Wu, Phys. Rev. B **31**, 3372 (1985). https://doi.org/10.1103/PhysRevB.31.3372
23. R.B. Laughlin, Phys. Rev. B **23**, 5632 (1981). https://doi.org/10.1103/PhysRevB.23.5632
24. D.J. Thouless, M. Kohmoto, M.P. Nightingale, M. den Nijs, Phys. Rev. Lett. **49**, 405 (1982). https://doi.org/10.1103/PhysRevLett.49.405
25. J.E. Avron, R. Seiler, B. Simon, Phys. Rev. Lett. **51**, 51 (1983). https://doi.org/10.1103/PhysRevLett.51.51
26. M. Kohmoto, Ann. Phys. **160**(2), 343 (1985). https://doi.org/10.1016/0003-4916(85)90148-4
27. M. Banerjee, M. Heiblum, V. Umansky, D.E. Feldman, Y. Oreg, A. Stern, Nature **559**(7713), 205 (2018). https://doi.org/10.1038/s41586-018-0184-1
28. A. Chan, T.L. Hughes, S. Ryu, E. Fradkin, Phys. Rev. B **87**, 085132 (2013). https://doi.org/10.1103/PhysRevB.87.085132
29. T. Giamarchi, *Quantum Physics in One Dimension*. International Series of Monographs on Physics (Clarendon Press, Oxford, 2003). https://global.oup.com/academic/product/quantum-physics-in-one-dimension-9780198525004?cc=se&lang=en&
30. M. Blau, G. Thompson, Ann. Phys. **205**(1), 130 (1991). http://dx.doi.org/10.1016/0003-4916(91)90240-9
31. M. Bergeron, G.W. Semenoff, R.J. Szabo, Nuclear Phys. B **437**(3), 695 (1995). https://doi.org/10.1016/0550-3213(94)00503-7
32. M. Stone, Ann. Phys. **207**(1), 38 (1991). https://doi.org/10.1016/0003-4916(91)90177-A
33. X.G. Wen, Int. J. Mod. Phys. B **06**(10), 1711 (1992). https://doi.org/10.1142/S0217979292000840
34. F.D.M. Haldane, Phys. Rev. Lett. **61**, 2015 (1988). https://doi.org/10.1103/PhysRevLett.61.2015
35. X.L. Qi, T.L. Hughes, S.C. Zhang, Phys. Rev. B **78**, 195424 (2008). https://doi.org/10.1103/PhysRevB.78.195424
36. A.N. Redlich, Phys. Rev. D **29**, 2366 (1984). https://doi.org/10.1103/PhysRevD.29.2366
37. H. Nielsen, M. Ninomiya, Phys. Lett. B **130**(6), 389 (1983). https://doi.org/10.1016/0370-2693(83)91529-0
38. C.Z. Chang, J. Zhang, X. Feng, J. Shen, Z. Zhang, M. Guo, K. Li, Y. Ou, P. Wei, L.L. Wang, Z.Q. Ji, Y. Feng, S. Ji, X. Chen, J. Jia, X. Dai, Z. Fang, S.C. Zhang, K. He, Y. Wang, L. Lu, X.C. Ma, Q.K. Xue, Science **340**(6129), 167 (2013). https://doi.org/10.1126/science.1234414, http://science.sciencemag.org/content/340/6129/167
39. K. Fujikawa, Phys. Rev. D **21**, 2848 (1980). https://doi.org/10.1103/PhysRevD.21.2848
40. S. Coh, D. Vanderbilt, A. Malashevich, I. Souza, Phys. Rev. B **83**, 085108 (2011). https://doi.org/10.1103/PhysRevB.83.085108
41. E. Witten, Phys. Lett. B **86**(34), 283 (1979). https://doi.org/10.1016/0370-2693(79)90838-4
42. X.L. Qi, R. Li, J. Zang, S.C. Zhang, Science **323**(5918), 1184 (2009). https://doi.org/10.1126/science.1167747, http://science.sciencemag.org/content/323/5918/1184
43. C.L. Kane, E.J. Mele, Phys. Rev. Lett. **95**, 226801 (2005). https://doi.org/10.1103/PhysRevLett.95.226801
44. C.L. Kane, E.J. Mele, Phys. Rev. Lett. **95**, 146802 (2005). https://doi.org/10.1103/PhysRevLett.95.146802
45. V. Mourik, K. Zuo, S.M. Frolov, S.R. Plissard, E.P.A.M. Bakkers, L.P. Kouwenhoven, Science **336**(6084), 1003 (2012). https://doi.org/10.1126/science.1222360, http://science.sciencemag.org/content/336/6084/1003
46. S. Nadj-Perge, I.K. Drozdov, J. Li, H. Chen, S. Jeon, J. Seo, A.H. MacDonald, B.A. Bernevig, A. Yazdani, Science **346**(6209), 602 (2014). https://doi.org/10.1126/science.1259327, http://science.sciencemag.org/content/346/6209/602
47. A. Kitaev, Ann. Phys. **303**(1), 2 (2003). https://doi.org/10.1016/S0003-4916(02)00018-0
48. L. Fidkowski, A. Kitaev, Phys. Rev. B **81**, 134509 (2010). https://doi.org/10.1103/PhysRevB.81.134509

49. C. Spånslätt, E. Ardonne, J.C. Budich, T.H. Hansson, J. Phys.: Condens. Matter **27**(40), 405701 (2015), http://stacks.iop.org/0953-8984/27/i=40/a=405701
50. S.A. Kivelson, D.S. Rokhsar, Phys. Rev. B **41**, 11693 (1990). https://doi.org/10.1103/PhysRevB.41.11693
51. X.G. Wen, Int. J. Mod. Phys. B **05**(10), 1641 (1991). https://doi.org/10.1142/S0217979291001541
52. A. Balachandran, P. Teotontio-Sobrinho, Int. J. Mod. Phys. A **08**(04), 723 (1993). https://doi.org/10.1142/S0217751X9300028X
53. S. Moroz, A. Prem, V. Gurarie, L. Radzihovsky, Phys. Rev. B **95**, 014508 (2017). https://doi.org/10.1103/PhysRevB.95.014508
54. M. Hermanns, ArXiv e-prints (2008), arxiv.org/abs/0804.1332
55. G.Y. Cho, J.E. Moore, Ann. Phys. **326**(6), 1515 (2011). https://doi.org/10.1016/j.aop.2010.12.011, www.sciencedirect.com/science/article/pii/S0003491610002253
56. D.A. Ivanov, Phys. Rev. Lett. **86**, 268 (2001). https://doi.org/10.1103/PhysRevLett.86.268
57. N. Read, D. Green, Phys. Rev. B **61**, 10267 (2000). https://doi.org/10.1103/PhysRevB.61.10267
58. D.C. Tsui, H.L. Stormer, A.C. Gossard, Phys. Rev. Lett. **48**, 1559 (1982). https://doi.org/10.1103/PhysRevLett.48.1559
59. R.B. Laughlin, Phys. Rev. Lett. **50**, 1395 (1983). https://doi.org/10.1103/PhysRevLett.50.1395
60. A.S. Goldhaber, R. Mackenzie, Phys. Lett. B **214**(3), 471 (1988). http://dx.doi.org/10.1016/0370-2693(88)91397-4
61. S.C. Zhang, T.H. Hansson, S. Kivelson, Phys. Rev. Lett. **62**, 82 (1989). https://doi.org/10.1103/PhysRevLett.62.82
62. W. Pan, H.L. Stormer, D.C. Tsui, L.N. Pfeiffer, K.W. Baldwin, K.W. West, Phys. Rev. Lett. **90**, 016801 (2003). https://doi.org/10.1103/PhysRevLett.90.016801
63. F.D.M. Haldane, Phys. Rev. Lett. **51**, 605 (1983). https://doi.org/10.1103/PhysRevLett.51.605
64. B.I. Halperin, Phys. Rev. Lett. **52**, 1583 (1984). https://doi.org/10.1103/PhysRevLett.52.1583
65. L. Faddeev, R. Jackiw, Phys. Rev. Lett. **60**, 1692 (1988). https://doi.org/10.1103/PhysRevLett.60.1692

Chapter 2
An Introduction to Entanglement Measures in Conformal Field Theories and AdS/CFT

Erik Tonni

2.1 Introduction

Entanglement in quantum systems and many quantities introduced to study is has attracted a lot of research during the last decade, in particular in quantum information theory, condensed matter theory and quantum gravity.

Many excellent reviews have collected all the relevant results about the aspects of entanglement that are relevant for the various communities. In particular, let us mention [1, 2] for quantum field theories (QFT), with particular focus on conformal field theories (CFT), [3] for condensed matter theory and [4, 5] for quantum gravity within the approach based on the gauge/gravity correspondence (AdS/CFT correspondence).

The entanglement entropy is the measure of entanglement that has been mostly studied in the literature because it fully quantifies the bipartite entanglement of pure states. For mixed states a quantity characterising completely the bipartite entanglement is not known and various proposals have been suggested (e.g. the logarithmic negativity).

These introductory lectures focus on the following topics: the entanglement entropies and logarithmic negativity in quantum field theories, with particular emphasis on the analytic results obtained in two dimensional conformal field theories, and the holographic entanglement entropy within the holographic (AdS/CFT) correspondence.

E. Tonni (✉)
SISSA and INFN, via Bonomea 265, 34136 Trieste, Italy
e-mail: etonni@sissa.it

A. Ferraz et al. (eds.), *Strongly Coupled Field Theories for Condensed Matter and Quantum Information Theory*, Springer Proceedings in Physics 239,
https://doi.org/10.1007/978-3-030-35473-2_2

2.2 Bipartite Entanglement

Let us consider a quantum system whose Hilbert space $\mathcal{H}$ can be bipartite, namely it can be written as $\mathcal{H} = \mathcal{H}_1 \otimes \mathcal{H}_2$. The quantum state is characterised by a density matrix ρ and it can be either pure or mixed. In the former case the density matrix is a projector $\rho = |\Psi\rangle\langle\Psi|$ on a vector $|\Psi\rangle$ of the Hilbert space.

Let us first consider pure states. By introducing an orthonormal basis $\{|\Psi_m^{(i)}\rangle, 1 \leqslant m \leqslant d_i\}$ for $\mathcal{H}_i$, a pure state $|\Psi\rangle \in \mathcal{H}$ can be written as

$$|\Psi\rangle = \sum_{a_1=1}^{d_1}\sum_{a_2=1}^{d_2} A_{a_1 a_2} |\Psi_{a_1}^{(1)}\rangle \otimes |\Psi_{a_2}^{(2)}\rangle \tag{2.2.1}$$

where the coefficients $A_{a_1 a_2}$ can be collected into a rectangular matrix $\boldsymbol{A}$. By employing the *singular value decomposition* of a generic complex matrix, $\boldsymbol{A}$ can be written as $\boldsymbol{A} = \boldsymbol{U}\,\boldsymbol{D}\,\boldsymbol{V}^\dagger$, where $\boldsymbol{U}$ and $\boldsymbol{V}$ are unitary matrices of order d_1 and d_2 respectively, while $\boldsymbol{D}$ is a $d_1 \times d_2$ matrix whose elements λ_a can be non vanishing only along a diagonal, hence $1 \leqslant k \leqslant d_0$, where $d_0 = \min(d_1, d_2)$. The unitary matrices $\boldsymbol{U}$ and $\boldsymbol{V}$ can be used to change the local basis of $\mathcal{H}_1$ and $\mathcal{H}_2$ respectively. Writing the state (2.2.1) in terms of the new orthonormal basis $\{|\Phi_m^{(i)}\rangle, 1 \leqslant k \leqslant d_i\}$ for $\mathcal{H}_i$, we get the *Schmidt decomposition*

$$|\Psi\rangle = \sum_{k=1}^{d_0} \lambda_k |\Phi_k^{(1)}\rangle \otimes |\Phi_k^{(2)}\rangle \tag{2.2.2}$$

where λ_k are the Schmidt coefficients. The Schmidt rank $r_0 \leqslant d_0$ is defined as the number of non vanishing Schmidt coefficients. The normalization $\langle\Psi|\Psi\rangle = 1$ is equivalent to the constraint $\sum_{k=1}^{r_0} |\lambda_k|^2 = 1$ on the Schmidt coefficients.

The properties of the bipartite entanglement associated to the pure state $|\Psi\rangle$ and the bipartition $\mathcal{H}_1 \otimes \mathcal{H}_2$ are encoded into the Schmidt coefficients. For instance, when $\lambda_k = 0$ for all $k \neq k_0$, then $|\Psi\rangle = |\Phi^{(1)}\rangle \otimes |\Phi^{(2)}\rangle$ is a product state, hence it is not entangled. In any other case $|\Psi\rangle$ is entangled. When all the Schmidt coefficients have equal size, namely $|\lambda_k| = 1/\sqrt{r_0}$ for all values of k, then $|\Psi\rangle$ is maximally entangled.

The bipartite entanglement of mixed states is more difficult to evaluate. A quantity measuring it is discussed in Sect. 2.4.

2.2.1 *Entanglement Entropies*

Given a state characterised by the density matrix ρ and the bipartition $\mathcal{H}_1 \otimes \mathcal{H}_2$, the *reduced density matrix* ρ_i associated to $\mathcal{H}_i$ is the hermitean operator obtained by tracing over the complementary part of the Hilbert space, i.e.

$$\rho_i \equiv \mathrm{Tr}_{\mathcal{H}_j}\rho \tag{2.2.3}$$

Thus, two reduced density matrices ρ_1 and ρ_2 can be introduced.

The *entanglement entropy* is the Von Neumann entropy associated to the reduced density matrix ρ_i, namely

$$S_i \equiv -\,\mathrm{Tr}(\rho_i \log \rho_i) \tag{2.2.4}$$

An important family of entanglement quantifiers parameterised by an integer $n \geqslant 2$ is given by the *Rényi entropies*

$$S_i^{(n)} \equiv \frac{1}{1-n}\,\log\!\left(\mathrm{Tr}\rho_i^n\right) \tag{2.2.5}$$

A crucial feature of these scalar quantities is that they provide the entanglement entropy (2.2.6) once a proper analytic continuation in n is performed. In particular, the entanglement entropy can be computed as

$$S_i = \lim_{n\to 1} S_i^{(n)} = -\,\partial_n \mathrm{Tr}(\rho_i^n)\big|_{n=1} \tag{2.2.6}$$

The entanglement entropy satisfies important inequalities for any state ρ of the system.

Given the bipartition $\mathcal{H}_1 \otimes \mathcal{H}_2$ of the Hilbert space and adopting the notation $S_i = S(\rho_i)$ the *Araki-Lieb inequality* [6] is given by

$$\left|S(\rho_1) - S(\rho_2)\right| \leqslant S(\rho) \leqslant S(\rho_1) + S(\rho_2) \tag{2.2.7}$$

The second inequality is also called *subadditivity* condition.

Another characteristic feature of the entanglement entropy is given by *strong subadditivity* inequalities [7]

$$\begin{cases} S(\rho_{12}) + S(\rho_{23}) \geqslant S(\rho_2) + S(\rho_{123}) \\ S(\rho_{12}) + S(\rho_{23}) \geqslant S(\rho_1) + S(\rho_3) \end{cases} \tag{2.2.8}$$

where the notation ρ_{12} the reduced density matrix obtained by tracing over $\mathcal{H}_3$. When $\mathcal{H}_2 = \emptyset$, the first inequality of (2.2.8) reduces to the subadditivity condition.

The entanglement entropy fully characterises the bipartite entanglement of pure states.

The density matrix associated to the pure state $|\Psi\rangle$ is the projector $\rho = |\Psi\rangle\langle\Psi|$. By employing the Schmidt decomposition (2.2.2), the reduced density matrix ρ_i can be written as

$$\rho_i = \sum_{a_j=1}^{d_j} \langle\Phi_{a_j}^{(j)}|\,\rho\,|\Phi_{a_j}^{(j)}\rangle = \sum_{k=1}^{r_0} |\lambda_k|^2\, |\Phi_k^{(i)}\rangle\langle\Phi_k^{(i)}| \qquad j \neq i \tag{2.2.9}$$

This expression tells that: (a) the reduced density matrices ρ_i are linear combinations of projectors, hence they describe mixed states; (b) the two reduced density matrices ρ_1 and ρ_2 have the same non vanishing eigenvalues $w_k \equiv |\lambda_k|^2$.

When the entire system is in a pure state $|\Psi\rangle$, the expression (2.2.9) holds for ρ_i, hence the corresponding entanglement entropy can be written in terms of the Schmidt coefficients as follows

$$S_i = -\sum_{k=1}^{r_0} w_k \log w_k \tag{2.2.10}$$

This tells us that $S_1 = S_2$ for pure states, hence (2.2.10) provides the entanglement entropy of the bipartition when the system is in a pure state.

When the pure state $|\Psi\rangle$ is a product state, then $S_1 = S_2 = 0$, while when it is maximally entangled we have that $S_1 = S_2 = \log r_0$.

For pure states we can employ (2.2.9), where $\{|\Phi_k^{(i)}\rangle\}$ an orthonormal basis for $\mathcal{H}_i$, finding

$$\mathrm{Tr}\rho_i^n = \sum_{k=1}^{r_0} w_k^n \tag{2.2.11}$$

Thus, also the Rényi entropies can be determined from the Schmidt coefficients, and this implies that $S_1^{(n)} = S_2^{(n)}$ also for the Rényi entropies.

2.3 Geometric Entanglement Entropies in QFT

When the bipartition of the Hibert space corresponds to a spatial bipartition $A \cup B$ of a slice of the spacetime at constant time, the bipartite entanglement depends also on the geometry of the bipartition itself. In these cases (2.2.5) and (2.2.6) are sometimes called geometric entanglement entropies [8–13] and they are denoted as $S_A^{(n)}$ and $S_B^{(n)}$.

In a Quantum Field Theory, a state can be represented by employing the path integral. Denoting by t_{E} the Euclidean time coordinate and by $\boldsymbol{x} \in \mathbb{R}^{d-1}$ the remaining spatial coordinates, at fixed $t_{\mathrm{E}} = 0$ we have that

$$|\Psi\big(\phi_0(\boldsymbol{x})\big)\rangle = \frac{1}{\mathcal{Z}^{1/2}} \int_{t_{\mathrm{E}}<0} \mathcal{D}\phi \; \delta\big(\phi(t_{\mathrm{E}} = 0, \boldsymbol{x}) - \phi_0(\boldsymbol{x})\big) \, e^{-S[\phi]} \tag{2.3.1}$$

where we have denoted by $\phi_0(\boldsymbol{x})$ the field configuration at $t_{\mathrm{E}} = 0$. Similarly, the path integral representation of $\langle\Psi\big(\phi_0(\boldsymbol{x})\big)|$ involves the integration over $t_{\mathrm{E}} > 0$. Given these expressions, the density matrix $\rho = |\Psi\rangle\langle\Psi|$ can be written as follows

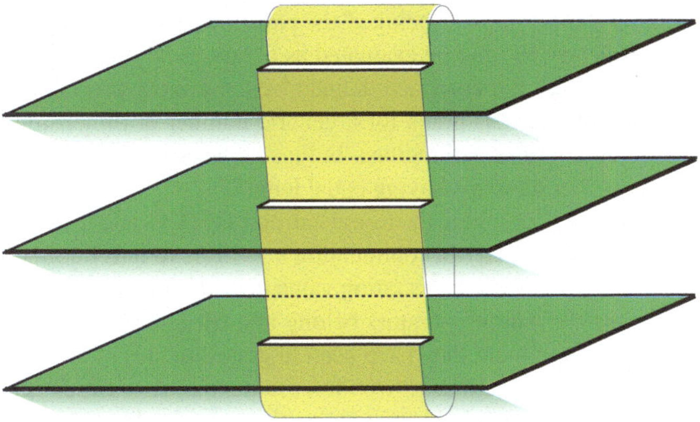

Fig. 2.1 Riemann surface $\mathscr{R}_{N,n}$ corresponding to an interval ($N = 1$) and $n = 3$

$$\rho\big(\phi_+(\boldsymbol{x}), \phi_-(\boldsymbol{x})\big) = \frac{1}{\mathcal{Z}} \int_{t_{\mathrm{E}}<0} \mathcal{D}\phi \int_{t_{\mathrm{E}}>0} \mathcal{D}\phi \; \delta\big(\phi(0^+, \boldsymbol{x}) - \phi_+(\boldsymbol{x})\big) \; \delta\big(\phi(0^-, \boldsymbol{x}) - \phi_-(\boldsymbol{x})\big) \; e^{-S[\phi]} \tag{2.3.2}$$

which can be roughly understood as the continuum limit of a matrix where $\phi_+(\boldsymbol{x})$ and $\phi_-(\boldsymbol{x})$ are the continuum limit of the rows and the columns respectively.

The reduced density matrix is obtained by identifying the fields $\phi_+(\boldsymbol{x})$ and $\phi_-(\boldsymbol{x})$ only for $\boldsymbol{x} \in B$ and performing the path integration corresponding to this region. This gives

$$\rho_A\big(\phi_+(\boldsymbol{x}_A), \phi_-(\boldsymbol{x}_A)\big) = \int_{t_{\mathrm{E}}=0} \mathcal{D}\phi \prod_{\boldsymbol{x}_B} \delta\big(\phi_+(\boldsymbol{x}_B) - \phi_-(\boldsymbol{x}_B)\big) \; \rho\big(\phi_+(\boldsymbol{x}), \phi_-(\boldsymbol{x})\big) \tag{2.3.3}$$

where we have adopted the notation $\boldsymbol{x} = \boldsymbol{x}_B \in B$ and $\boldsymbol{x} = \boldsymbol{x}_A \in A$. Notice that (2.3.3) depends on the field configurations $\phi_+(\boldsymbol{x}_A)$ and $\phi_-(\boldsymbol{x}_A)$, which are defined respectively slightly above and below A along the Euclidean time direction.

Given the path integral representation (2.3.3) for the reduced density matrix, the Rényi entropies can be constructed in a similar way by taking n copies of (2.3.3) and tracing them along A in the proper way. In particular, the field configuration $\phi_+(\boldsymbol{x}_A)$ of the jth copy must be identified with the field configuration $\phi_-(\boldsymbol{x}_A)$ of the $(j+1)$th copy in a cyclic way. This procedure, which holds for any number of spatial dimensions, is described in Fig. 2.1 for the simplest case case where the spacetime has two dimensions and A is an interval.

The net result is that the Rényi entropies can be written in terms of the partition function $\mathcal{Z}_n$ of n copies of the model coupled through the boundary conditions imposed by joining the copies as above. Thus

$$\mathrm{Tr}\, \rho_A^n = \frac{\mathcal{Z}_n}{\mathcal{Z}^n} \tag{2.3.4}$$

which is normalised to $\mathrm{Tr}\,\rho_A = 1$, being $\mathcal{Z}$ the partition function of a single copy of the model. Equivalently, $\mathcal{Z}_n$ can be evaluated as the partition function of the model defined on the n-sheeted hypersurface defined by the joining procedure described above. We remark that (2.3.4) holds for a generic number of dimensions and also when A is made by disconnected regions. In Fig. 2.2 we show an example where A is made by two disjoint intervals having equal length.

Another useful way to compute the Rényi entropies (2.3.4), and then the entanglement entropy through the analytic continuation $n \to 1$, is to introduce some particular operators along ∂A whose expectation value provides the moments $\mathrm{Tr}\rho_A^n$ of the reduced density matrix. These operators belong to a particular class of *twist fields*. In the context of entanglement they have been first used in [13].

When the Euclidean spacetime is two dimensional and $A = (u, v)$ is an interval on the infinite line, ∂A is made by two isolated points. In this case the moments (2.3.4) can be computed also as the following two point function

$$\mathrm{Tr}\rho_A^n = \langle \mathcal{T}_n(u)\, \bar{\mathcal{T}}_n(v) \rangle \tag{2.3.5}$$

where the twist fields $\mathcal{T}_n$ and $\bar{\mathcal{T}}_n$, that are one the inverse of the other one, are located respectively at the left and at the right endpoint of the interval. This procedure allows to treat the case where A is made by N disjoint intervals in a straightforward way. The Riemann surface underlying the partition function $\mathcal{Z}_n$ occurring in (2.3.4) will be denoted by $\mathscr{R}_{N,n}$. For instance, for two disjoint intervals $A = (u_1, v_1) \cup (u_2, v_2)$ we have

$$\mathrm{Tr}\rho_A^n = \langle \mathcal{T}_n(u_1)\, \bar{\mathcal{T}}_n(v_1)\, \mathcal{T}_n(u_2)\, \bar{\mathcal{T}}_n(v_2) \rangle \tag{2.3.6}$$

which can be easily extended to the case where $A = \cup_{i=1}^N (u_i, v_i)$ is made by N disjoint intervals, finding

$$\mathrm{Tr}\rho_A^n = \langle \mathcal{T}_n(u_1)\, \bar{\mathcal{T}}_n(v_1)\, \mathcal{T}_n(u_2)\, \bar{\mathcal{T}}_n(v_2) \rangle \tag{2.3.7}$$

For higher dimensional spacetimes, the twist fields are non local operators whose support is given by a connected component of ∂A, which has codimension two. Thus, for instance, in a four dimensional Euclidean spacetime, when $A = A_1 \cup A_2$ is made by the union of a disk A and an ellipse A_2, the corresponding Rényi entropies can be computed as a two point function of two twist fields whose supports are ∂A_1 and ∂A_2. The twist fields can be understood also as special class of defects [14, 15].

Conformal field theories (CFT) provide an important class of quantum field theories for our purposes because this symmetry allows to get analytic results for the entanglement entropies of some configurations. We mainly focus on two spacetime dimensions because in this case the conformal symmetry is infinite dimensional, while in higher dimensions it has a finite number of generators. The generators of the conformal symmetry in two spacetime dimensions satisfy the Virasoro algebra, which is determined by an important model dependent constant c known as central charge.

2.3.1 Intervals in 2D CFT

In quantum systems defined in one spatial dimension, the simplest configuration to consider is a single interval A when the entire system is in its ground state. In this case the moments of the reduced density matrix ρ_A can be written as (2.3.5). Focussing on Conformal Field Theories, the conformal symmetry can be employed to evaluate (2.3.5). In particular, in [13] it has been shown that the twist fields $\mathcal{T}_n$ and $\bar{\mathcal{T}}_n$ are primary operators. For these operators, conformal symmetry constraints the two point function to depend on the distance $\ell \equiv |u - v|$ between the positions of the operators and on their common conformal dimension. For the twist fields, we have

$$\mathrm{Tr}\rho_A^n = \frac{c_n}{(\ell/\epsilon)^{2\Delta_n}} \tag{2.3.8}$$

where ϵ is the UV cutoff and

$$\Delta_n = \frac{c}{12}\left(n - \frac{1}{n}\right) \tag{2.3.9}$$

is the scaling dimension of the twist fields $\mathcal{T}_n$ and $\bar{\mathcal{T}}_n$, being c is the central charge of the model and c_n is the normalisation constant. The normalisation condition $\mathrm{Tr}\rho_A = 1$ is guarantee by the requiring $c_1 = 1$. Taking the replica limit (2.2.6) for the moments (2.3.8), one obtains [10–12]

$$S_A = \frac{c}{3}\log(\ell/\epsilon) + \text{const} \tag{2.3.10}$$

We remark that the Riemann surface underlying the computation of $\mathcal{Z}_n$ is the two dimensional sphere.

The conformal symmetry allows also to explore the cases where A is an interval in the infinite line while the entire system is at finite temperature $T \equiv 1/\beta$. In this case, one obtains

$$\mathrm{Tr}\rho_A^n = \frac{c_n}{\left[\frac{\beta}{\pi\epsilon}\sinh(\pi\ell/\beta)\right]^{2\Delta_n}} \qquad S_A = \frac{c}{3}\log\left[\frac{\beta}{\pi\epsilon}\sinh(\pi\ell/\beta)\right] + \text{const} \tag{2.3.11}$$

Notice that for $\ell \ll \beta$ the zero temperature results (2.3.8) and (2.3.10) are recovered, as expected. In the large temperature regime $\ell \gg \beta$ and (2.3.11) gives $S_A = \pi c\,\ell/(3\beta) + \dots$, which is the extensive Gibbs entropy for a system of length ℓ.

Another configuration that can be studied analytically by exploiting the conformal symmetry is the one given by an interval of length ℓ in a circumference of length $L > \ell$, when the entire system is at zero temperature. For this configuration the moments of the reduced density matrix read

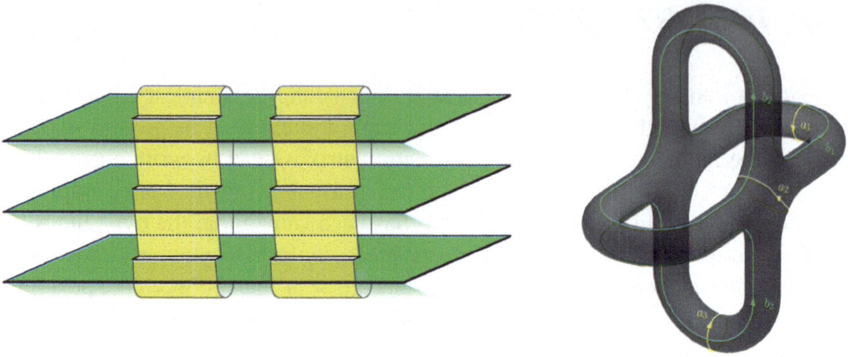

Fig. 2.2 Riemann surface $\mathscr{R}_{N,n}$ corresponding to two disjoint intervals ($N = 2$) and $n = 3$ (left) or $n = 4$ (right) (Figure on the right reprinted with permission from [33])

$$\mathrm{Tr}\rho_A^n = \frac{c_n}{\left[\frac{L}{\pi\epsilon}\sin(\pi\ell/L)\right]^{2\Delta_n}} \qquad S_A = \frac{c}{3}\log\left[\frac{L}{\pi\epsilon}\sin(\pi\ell/L)\right] + \mathrm{const} \tag{2.3.12}$$

In the limit $\ell \gg L$ the results (2.3.8) and (2.3.10) for the interval in the infinite line are recovered, as expected.

In the framework of the two dimensional CFT, also models with physical boundaries, namely boundary CFT (BCFT) can be considered. In these models the boundary conditions imposed on the fields enter in a crucial way to determine the spectrum of the allowed operators. In the context of entanglement, an analysis similar to the one performed to get (2.3.10) allows to study the configuration where A is an interval of length ℓ in a semi-infinite line adjacent to the boundary, i.e. whose endpoint coincides with the initial point of the semi-infinite line. In this case one finds

$$S_A = \frac{c}{6}\log(\ell/\epsilon) + \mathrm{const} \tag{2.3.13}$$

where the subleading term depends on the boundary conditions characterising the BCFT.

When A is made by two or more disjoint intervals in the infinite line and the entire system is in its ground state, the Riemann surface $\mathscr{R}_{N,n}$ obtained through the replica method that occurs in (2.3.4) has higher genus given by $g = (n-1)(N-1)$, where $N \geqslant 1$ is the number of disjoint intervals and $n \geqslant 2$ the number of copies. In the simplest case of $N = 2$ intervals, $\mathscr{R}_{2,n}$ has genus $g = n - 1$ but it is not the most generic Riemann surface having this genus. Indeed, $\mathscr{R}_{N,n}$ obtained through the replica method provide a small subclass of Riemann surfaces (see Fig. 2.2, on the right, where $\mathscr{R}_{2,4}$ is shown). Furthermore, focussing on two dimensional CFTs, the conformal symmetry does not allow to fix entirely the dependence on the positions of the endpoints of the intervals. Thus, these configurations are more difficult to treat analytically.

The simplest case within this class of configurations is given by $A = (u_1, v_1) \cup (u_2, v_2)$, namely two disjoint intervals in the infinite line. In two dimensional CFT, the conformal symmetry allows to write the four point function (2.3.7) in the following form

$$\mathrm{Tr}\rho_A^n = c_n^2 \left(\frac{|u_1 - u_2|\,|v_1 - v_2|}{|u_1 - v_1|\,|u_2 - v_2|\,|u_1 - v_2|\,|u_2 - v_1|} \right)^{2\Delta_n} \mathcal{F}_n(x) \tag{2.3.14}$$

where Δ_n is the scaling dimension (2.3.9) and x is the harmonic ratio of the four endpoints, i.e.

$$x = \frac{(u_1 - v_1)(u_2 - v_2)}{(u_1 - u_2)(v_1 - v_2)} \tag{2.3.15}$$

The function $\mathcal{F}_n(x)$ is model dependent and it is known analytically only in few cases. Furthermore, its dependence on the replica index n can be highly non trivial, making the replica limit very difficult to perform.

For instance, for the two dimensional Ising CFT, which has $c = 1/2$, it has been found that [16, 17]

$$\mathcal{F}_n^{\text{Ising}}(x) = \frac{\sum_e \left|\Theta[e](\mathbf{0}|\tau(x))\right|}{2^{n-1}\left|\Theta(\mathbf{0}|\tau(x))\right|} \tag{2.3.16}$$

being Θ the Riemann theta function with characteristic $\boldsymbol{e}^{\mathrm{t}} = (\varepsilon^{\mathrm{t}}, \delta^{\mathrm{t}})$, that is defined as

$$\Theta[e](z|\Omega) = \sum_{m \in \mathbb{Z}^p} \exp\left[\mathrm{i}\pi(\boldsymbol{m} + \varepsilon)^{\mathrm{t}} \cdot \Omega \cdot (\boldsymbol{m} + \varepsilon) + 2\pi\mathrm{i}(\boldsymbol{m} + \varepsilon)^{\mathrm{t}} \cdot (z + \delta) \right] \tag{2.3.17}$$

where $z \in \mathbb{C}^p/(\mathbb{Z}^p + \Omega\,\mathbb{Z}^p)$ is the independent variable, while ε and $\boldsymbol{\delta}$ are vectors whose entries are either 0 or $1/2$, and ω is a $p \times p$ complex, symmetric matrix with positive imaginary part. In (2.3.16), the sum in the numerator is performed over all the possible characteristics and the elements of the $(n-1) \times (n-1)$ complex and symmetric matrix $\tau(x)$ are given by

$$(\tau_2)_{ij} = \mathrm{i}\frac{2}{n}\sum_{k=1}^{n-1} \sin(\pi k/n)\, \frac{{}_2F_1(k/n, 1 - k/n; 1; 1 - x)}{{}_2F_1(k/n, 1 - k/n; 1; x)} \cos[2\pi k/n(i - j)] \tag{2.3.18}$$

in terms of the hypergeometric function ${}_2F_1$.

The function $\mathcal{F}_n(x)$ can be written analytically also for the free massless boson compactified on a circle and for the free massless Dirac fermion, which have central charge $c = 1$ [16, 17].

For these three simple CFTs, analytic results can be obtained also for the most general case where A is made by N disjoint intervals on the infinite line [18].

The analytic continuation of (2.3.14) with $\mathcal{F}_n$ given by (2.3.16) is difficult to perform. Thus, a numerical method based on the rational interpolation has been employed [19], finding agreement with the corresponding results obtained through spin chains calculations [20] (see also [21]).

2.3.2 *Higher Dimensional CFT*

Evaluating the entanglement entropy S_A of a spatial domain A in a CFT_{d+1} with $d > 1$ is challenging also because the shape of A occurs in a highly non trivial way.

When the CFT is in its ground state and for smooth entangling hypersurfaces ∂A, the following expansion of the entanglement entropy as $\epsilon \to 0$ is obtained

$$S_A = \alpha_1 \frac{\text{Area}(\partial A)}{\epsilon^{d-1}} + \frac{C_3}{\epsilon^{d-3}} + \frac{C_5}{\epsilon^{d-3}} + \cdots - \begin{cases} F_d + o(1) & \text{even } d \\ C_{\text{log}} \log(R/\epsilon) + O(1) & \text{odd } d \end{cases} \tag{2.3.19}$$

where R is a typical length of A (e.g. the radius for a spherical region) the coefficients C_k of the divergent terms depends on the geometry of ∂A. The coefficient of the leading divergences is proportional to the area of the entangling hypersurface and this property is known as *area law* of the entanglement entropy. For odd values of d, notice that a logarithmic divergence occurs. This is very important because its coefficient C_{log} contains the central charges of the underlying CFT. The simplest example is given by $d = 1$, where $C_{\text{log}} = c/3$, being c the central charge of the CFT_2.

Various techniques have been employed to study the coefficients occurring in the expansion (2.3.19). For spherical regions, conformal mappings have be very successful because they lead to important results also for the corresponding reduced density matrix [22]. For generic shapes, a promising approach is based on the computation of the moments of the reduced density matrix as correlation function of proper *defect operators* [14, 15], which are the higher dimensional generalisation of the twist fields.

Configurations where A is made by disjoint regions have been also explored [23].

2.4 Entanglement Negativity

The bipartite entanglement of mixed states, i.e. the entanglement associated of a bipartition of the Hilbert space $\mathcal{H} = \mathcal{H}_1 \otimes \mathcal{H}_2$ when the entire system is in a mixed state, is difficult to quantify and various quantities have been introduced. Unfortunately, many of them require to perform extremization procedures that make very difficult to get quantitative results. Here we focus on the logarithmic negativity because these extremisation procedures to not occur in its evaluation.

In order to define the logarithmic negativity, we must introduce the partial transpose. Considering the density matrix ρ characterising the mixed state in the bipartite Hilbert space $\mathcal{H} = \mathcal{H}_1 \otimes \mathcal{H}_2$, its partial transpose with respect e.g. to $\mathcal{H}_2$, that will be denoted by ρ^{T_2}, is defined by introducing its elements as follows

$$\langle e_i^{(1)} e_j^{(2)} \,|\, \rho^{T_2} \,|\, e_r^{(1)} e_s^{(2)} \rangle = \langle e_i^{(1)} e_s^{(2)} \,|\, \rho \,|\, e_r^{(1)} e_j^{(2)} \rangle \tag{2.4.1}$$

The partial transpose ρ^{T_1} w.r.t. $\mathcal{H}_2$ can be defined in a similar way [24, 25].

It is important to remark that, while ρ is semipositive definite, ρ^{T_2} can have also negative eigenvalues. By considering the trace norm of ρ^{T_2}, we obtain

$$\|\rho^{T_2}\| \equiv \mathrm{Tr}|\rho^{T_2}| = \sum_i |\lambda_i| = \sum_{\lambda_i>0} \lambda_i - \sum_{\lambda_i<0} \lambda_i = 1 - 2\sum_{\lambda_i<0} \lambda_i \equiv 1 + 2\mathcal{N} \tag{2.4.2}$$

where the normalisation condition $\sum_i \lambda_i = 1$ has been employed. In (2.4.2) we have introduced the *negativity* $\mathcal{N}$ as minus the sum of the negative eigenvalues. The *logarithmic negativity* is defined as follows

$$\mathcal{E} = \log \|\rho^{T_2}\| \tag{2.4.3}$$

The logarithmic negativity $\mathcal{E}$ satisfies the conditions required by a measure of the bipartite entanglement for mixed states. For instance, for any bipartition of the Hilbert space we have

$$\mathcal{E}_1 = \mathcal{E}_2 \tag{2.4.4}$$

A useful approach to study the logarithmic negativity is based on a replica limit similar to the one introduced for the entanglement entropy [26, 27]. In particular, let us consider the moments of the partial transpose

$$\mathrm{Tr}(\rho^{T_2})^{n_e} = \sum_i \lambda_i^{n_e} = \sum_{\lambda_i>0} |\lambda_i|^{n_e} + \sum_{\lambda_i<0} |\lambda_i|^{n_e} \tag{2.4.5}$$

$$\mathrm{Tr}(\rho^{T_2})^{n_o} = \sum_i \lambda_i^{n_o} = \sum_{\lambda_i>0} |\lambda_i|^{n_o} - \sum_{\lambda_i<0} |\lambda_i|^{n_o} \tag{2.4.6}$$

where n_e is an even positive integer and n_o is an odd positive integer. Thus, the parity of n plays a crucial role due to the occurrence of negative eigenvalues.

From the last expression in (2.4.5) it is straightforward to realise that the trace norm (2.4.2) can be also obtained through the following analytic continuation

$$\|\rho^{T_2}\| = \lim_{n_e\to 1} \mathrm{Tr}(\rho^{T_2})^{n_e} \tag{2.4.7}$$

implying that the logarithmic negativity (2.4.3) can be computed through the following replica limit

$$\mathcal{E} = \lim_{n_e\to 1} \log\Big(\mathrm{Tr}\,(\rho^{T_2})^{n_e}\Big) \tag{2.4.8}$$

where only the sequence of the moments corresponding to even n's must be considered. The same limit performed on the moments $\mathrm{Tr}(\rho^{T_2})^{n_o}$ gives zero identically, because of the normalisation condition $\mathrm{Tr}\,\rho^{T_2} = 1$. The result is based only on the definition (2.4.3), hence it holds for any quantum system and for any state ρ. The usual technical difficulties related to the analytic continuation occur also in this analysis.

When ρ is pure state, the Schmidt decomposition can be employed to show that

$$\mathrm{Tr}(\rho^{T_2})^n = \begin{cases} \mathrm{Tr}_2\big(\rho_2^{n_e/2}\big)^2 & \text{even } n = n_e \\ \mathrm{Tr}_2\,\rho_2^{n_o} & \text{odd } \; n = n_o \end{cases} \tag{2.4.9}$$

where $\mathrm{Tr}_2 \equiv \mathrm{Tr}_{\mathcal{H}_2}$ and is the reduced density matrix $\rho_2 \equiv \mathrm{Tr}_{\mathcal{H}_1}\,\rho$. Combining (2.4.9) and (2.4.8), it is straightforward to find that for pure states the logarithmic negativity is equal to the Rényi entropy of order $1/2$, i.e.

$$\mathcal{E} = S_2^{(1/2)} \tag{2.4.10}$$

In the remaining part of this section, we will consider partitions of the Hilbert space corresponding to a geometric partitions of a spatial slice of the spacetime. In order to study the bipartite entanglement of a mixed state, one can study a spatial bipartition $A \cup B$ of a system in a pure state (e.g. its ground state) and consider the mixed state defined by the reduced density matrix ρ_A. Then, by introducing a spatial bipartition $A = A_1 \cup A_2$, the partial transpose w.r.t. either A_1 or A_2 can be explored.

For quantum field theories, the twist field approach has been employed to study also the moments of the partial transpose [26, 27]. Then, the logarithmic negativity is obtained through the replica limit (2.4.8).

In two dimensional spacetimes (one spatial dimension), the most intuitive configuration to consider is the one where $A = A_1 \cup A_2$ is made by two disjoint intervals $A_1 = (u_1, v_1)$ and $A_2 = (u_2, v_2)$. The moments of the partial transpose w.r.t. A_2 can be computed through the following four point function of twist fields

$$\mathrm{Tr}\big(\rho_A^{T_2}\big)^n = \langle \mathcal{T}_n(u_1)\,\bar{\mathcal{T}}_n(v_1)\,\bar{\mathcal{T}}_n(u_2)\,\mathcal{T}_n(v_2)\rangle \tag{2.4.11}$$

where the order of the fields along the line is the crucial difference with respect to the moments of the reduced density matrix in (2.3.7). The four point function in (2.4.11) can be also evaluated through the partition function of the model on the Riemann surface shown in Fig. 2.3 in the case of two disjoint intervals having the same length.

From the previous result we can find how to employ the twist fields to evaluate the moments of the partial transpose for the configuration where A_1 and A_2 are adjacent intervals. Indeed, taking the limit $u_2 \to v_1$, for two adjacent intervals $A_1 = (u, w)$ and $A_2 = (w, v)$ we find that

$$\mathrm{Tr}\big(\rho_A^{T_2}\big)^n = \langle \mathcal{T}_n(u)\,\bar{\mathcal{T}}_n^2(w)\,\mathcal{T}_n(v)\rangle \tag{2.4.12}$$

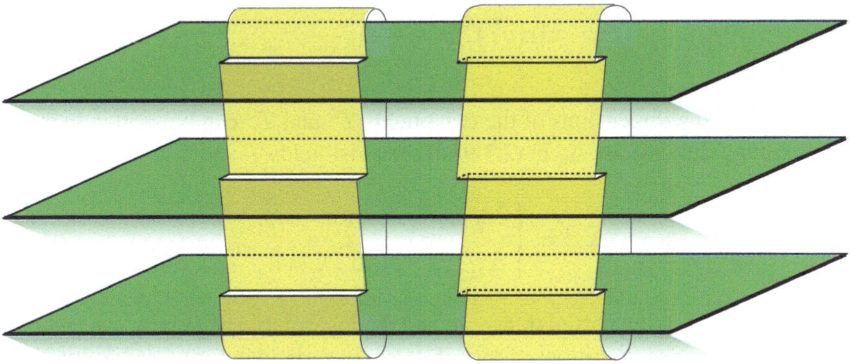

Fig. 2.3 Riemann surface underlying the evaluation of $\mathrm{Tr}(\rho_A^{T_2})^n$ when $A = A_1 \cup A_2$ is made by two disjoint intervals of equal length

where in this correlator the new field $\bar{\mathcal{T}}^2$ occurs.

Finally, we can also address the configuration where the system is in a pure state $\rho = |\psi\rangle\langle\psi|$ and the bipartition of the spatial dimension is given by $A_1 \cup A_2$. This configuration can be studied by taking the limit $B \to \emptyset$ of the previous one made by two adjacent interval. The result reads

$$\mathrm{Tr}(\rho_A^{T_2})^n = \langle \mathcal{T}_n^2(u)\, \bar{\mathcal{T}}_n^2(v)\rangle \tag{2.4.13}$$

We remark that, in quantum field theories, finding the properties of composite fields (like $\mathcal{T}^2$ and $\bar{\mathcal{T}}^2$) from the elementary ones is a highly non trivial task.

In our case the geometrical interpretation of the twist fields allows to conclude that $\mathcal{T}^2$ and $\bar{\mathcal{T}}^2$ implement a jump from the ith copy to the $(i+2)$th copy once one encircles the branch point where either $\mathcal{T}^2$ or $\bar{\mathcal{T}}^2$ is located.

In two or higher spatial dimensions the analysis of the moments of the partial transpose becomes more complicated because the underlying twist fields are non local. Numerical analysis for some free lattice models have been performed in [28, 29].

2.4.1 Intervals in 2D CFT

In two dimensional conformal field theories, the infinite dimensional conformal symmetry allows to get analytic results for some of the correlation function mentioned above. In the following we report some results obtained in [26, 27].

For the spatial bipartition of a pure state, the two point function (2.4.13) occurs and, by employing (2.3.8) and (2.4.9), one finds that

$$\mathrm{Tr}\big(\rho_A^{T_2}\big)^n = \frac{c_n^{(2)}}{|u-v|^{2\Delta_n^{(2)}}} \tag{2.4.14}$$

where the scaling dimensions of the twist fields $\mathcal{T}^2$ and $\bar{\mathcal{T}}^2$ can be written in terms of the scaling dimensions Δ_n in (2.3.9) of the twist fields $\mathcal{T}$ and $\bar{\mathcal{T}}$ as follows

$$\Delta_n^{(2)} = \begin{cases} 2\Delta_{n_e/2} & \text{even}\, n = n_e \\ \Delta_{n_o} & \text{odd} \quad n = n_o \end{cases} \tag{2.4.15}$$

The moments of the partial transpose and the logarithmic negativity can be computed also for some bipartitions of mixed states. In particular, for two adjacent intervals, by employing the general structure of the three point functions in CFT with the proper scaling dimensions given in (2.3.9) and (2.4.15), the replica limit (2.4.8) provides the following universal expression for the logarithmic negativity

$$\mathcal{E} = \frac{c}{4}\log\bigg(\frac{\ell_1\ell_2}{\ell_1+\ell_2}\bigg) + \text{const} \tag{2.4.16}$$

where c is the central charge and the constant term is non universal.

The moments of the partial transpose for the configuration where $A = A_1 \cup A_2$ is made by two disjoint intervals is more complicated and non universale because the four point function (2.4.11) occurs. In particular, since all the fields in the correlator have the same scaling dimension, we have that

$$\mathrm{Tr}\big(\rho_A^{T_2}\big)^n = c_n^2 \bigg(\frac{|u_1-u_2|\,|v_1-v_2|}{|u_1-v_1|\,|u_2-v_2|\,|u_1-v_2|\,|u_2-v_1|}\bigg)^{2\Delta_n} \mathcal{G}_n(x) \tag{2.4.17}$$

where the function $\mathcal{G}_n(x)$ is a model dependent function of the ratio (2.3.15). It can be shown that this function is related to the one introduced in (2.3.14) for the Rényi entropies of two disjoint intervals as follows

$$\mathcal{G}_n(x) = (1-x)^{c(n-1/n)/3}\, \mathcal{F}_n\bigg(\frac{x}{x-1}\bigg) \tag{2.4.18}$$

where we stress that the function $\mathcal{F}_n$ depends on the underlying CFT.

Taking the replica limit (2.4.8) by employing the expressions (2.4.17) and (2.4.18), one finds that the logarithmic negativity of two disjoint intervals in CFT is given by

$$\mathcal{E} = \lim_{n_e\to 1} \mathcal{F}_{n_e}\bigg(\frac{x}{x-1}\bigg) \tag{2.4.19}$$

This analytic continuation is very difficult to find analytically. Furthermore, the function $\mathcal{F}_n$ is known in very few cases. For instance, for the Ising model on the infinite line the function $\mathcal{F}_n$ is given by (2.3.16) and it has been shown that it can be extended also to complex values of x.

In the literature, the moments of the partial transpose have been checked against lattice calculations both in fermionic chains [30–33] and harmonic chains [26, 27]. Then, in some cases numerical extrapolations have been performed [19].

2.5 Holographic Entanglement Entropy

In the context of quantum gravity, the gauge/gravity correspondence (also called AdS/CFT correspondence) provide a crucial arena to explore quantitatively quantum gravity models (coming from string theories as microscopic model) through quantum field theory methods. According to the gauge/gravity duality, the strong coupling regime of a $(d+1)$ dimensional CFT can be described through the semiclassical limit of gravity model where the background is given by an asymptotically $(d+2)$ dimensional Anti de Sitter (AdS) spacetime. For instance, for a CFT_2 with central charge c, the regime described by a semiclassical gravitational model in AdS_3 is given by $c \to \infty$.

The AdS_{d+2} spacetime is a maximally symmetric spacetime with constant negative curvature. In Poincaré coordinates, it is defined by the following metric

$$ds^2 = \frac{L_{AdS}^2}{z^2}\left(-\,dt^2 + dz^2 + dx_1^2 + \cdots + dx_d^2\right) \tag{2.5.1}$$

where L_{AdS} is the radius of AdS_{d+2}. The dual CFT_{d+1} is defined on the boundary of AdS_{d+2}, at $z = 0$.

Considering a constant time slice of a CFT_{d+1} and introducing a spatial bipartition $A \cup B$, the entanglement entropy S_A in the regime of strong coupling can be computed through a gravitational computation in the corresponding asymptotically AdS_{d+2} spacetime. This formula has been found by Ryu and Takayanagi in [34, 35] and it requires to compute the area $\mathcal{A}_A$ of the minimal area surface $\hat{\gamma}_A$ anchored to ∂A (sometimes called entangling hypersurface) and extending into the gravitational asymptotically AdS_{d+2} background. In particular, the holographic entanglement entropy reads

$$S_A = \frac{\mathcal{A}_A}{4G_N} \tag{2.5.2}$$

where G_N is the gravitational Newton constant in $(d+2)$ spacetime dimensions. The covariant prescription for the holographic entanglement entropy, that is crucial to study e.g. the time dependent gravitational backgrounds like the one describing the formation of a black hole, has been proposed in [36].

In AdS_4, that is described by (2.5.1) with $d = 2$, a constant time slice is the Euclidean hyperbolic space $\mathbb{H}_3$. In this case A is a two dimensional spatial region in the boundary at $z = 0$ and its boundary ∂A is a closed curve whenever A is

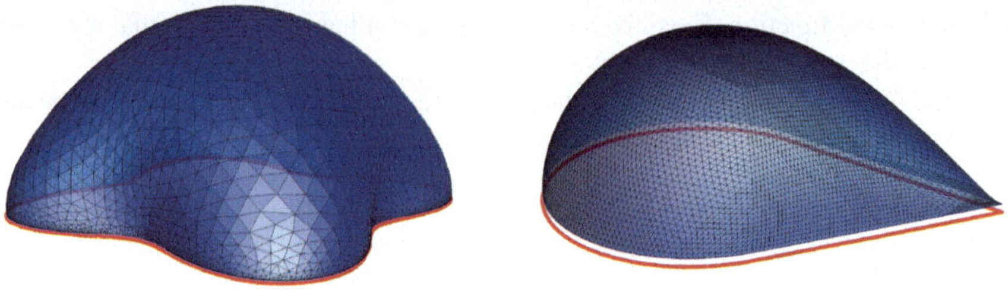

Fig. 2.4 Minimal area surfaces $\hat{\gamma}_A$ in the Euclidean hyperbolic space $\mathbb{H}_3$ (a constant time slice of AdS_4) anchored the entangling curve ∂A (red curves) defined in the boundary. In the left panel the entangling curve is smooth, while in the right panel it contains a corner (Figures reprinted with permission from [41, 52] respectively)

finite. In Fig. 2.4 we show two examples of minimal area surfaces $\hat{\gamma}_A$ anchored to the entangling curve ∂A (red curves in the figure): in the left panel ∂A is smooth, while in the right panel it contains a singular point given by the vertex of a corner.

The area of $\mathcal{A}_A$ is divergent because $\hat{\gamma}_A$ reaches the boundary $z = 0$ of the AdS space, that is a non compact spacetime. Thus, in order to evaluate (2.5.2), first one introduces a small cutoff $\varepsilon > 0$, evaluating the area of the part of $\hat{\gamma}_A$ having $z \geqslant \varepsilon$. Then, the holographic entanglement entropy (2.5.2) can be written as an expansion as $\varepsilon \to 0$ whose coefficients depends on various features of the minimal surface $\hat{\gamma}_A$. For smooth regions A, an expansion like (2.3.19) is obtained, where the coefficients of the divergent terms depend on the geometry of ∂A, i.e. only on the part of $\hat{\gamma}_A$ close to the boundary.

It is important to remark that the definition of the holographic entanglement entropy (2.5.2) includes the condition that $\hat{\gamma}_A$ must be homologous to A, i.e. it must be possible to obtain A by deforming $\hat{\gamma}_A$ in a smooth way inside the spacetime [37]. This condition (called *homology constraint*) is relevant when the asymptotically AdS gravitational background contains a black hole.

The formula (2.5.2) for the holographic entanglement entropy has been tested in various ways. It has been shown in [37] that it satisfies the highly non trivial constraints given by the strong subadditivity condition (2.2.8). Furthermore, also the Araki-Lieb inequality (2.2.7) has been discussed through the holographic formula (2.5.2) [37, 38]. Besides the characteristic features of the entanglement entanglement entropy, the holographic entanglement entropy (2.5.2) satisfies additional properties that could be employed to identify conformal field theories that can have a gravitational dual description. The main example is given by the monogamy of the mutual information. In order to define this property, let us consider three regions A_1, A_2 and A_3 and introduce the tripartite mutual information as

$$I_3(A_1, A_2, A_3) \equiv S_{A_1} + S_{A_2} + S_{A_3} - S_{A_1 \cup A_2} - S_{A_1 \cup A_3} - S_{A_2 \cup A_3} + S_{A_1 \cup A_2 \cup A_3} \tag{2.5.3}$$

The sign of $I_3(A_1, A_2, A_3)$ is not determined by fundamental properties of the entanglement entropy. For the holographic entanglement entropy given by (2.5.2), it has been shown that $I_3(A_1, A_2, A_3) \leqslant 0$ for any choice of the regions [39].

Finding minimal area surfaces analytically and then computing their area can be a very hard task. The difficulty originates both from the shape of the region A and on the type of gravitational background that must be considered. In the simplest case of AdS_{d+2} and for very simple regions like infinite strips and spherical regions, where some symmetry can be employed, $\hat{\gamma}_A$ and their regularised area have been found already in the original references [34, 35]. Other smooth domains and more complicated backgrounds have been explored e.g. in [40–42].

There are other important features of the holographic entanglement entropy (2.5.2) that can be used to characterise the field theory with a holographic dual description. Here we mention the occurrence of transitions in the holographic entanglement entropy of a domain made two or more disjoint components as the distances between the different components change [43–45]. The simplest case is when $A = A_1 \cup A_2$ is made by two disjoint intervals on the line. It is not difficult to realise that the geodesic in AdS_3 connecting the endpoints of an interval on the boundary is given by half circumference. Thus, given $A_1 = (u_1, v_1)$ and $A_2 = (u_2, v_2)$, when the two intervals are very far apart $\hat{\gamma}_A$ is made by two half circumferences anchored to A_1 and A_2; while when they are very close, $\hat{\gamma}_A$ is made by two half circumferences anchored to (u_1, v_2) and (v_1, u_2). These two configurations are geometrically very different and a transition between them occurs at some critical distance.

A more complicated example in AdS_4 is given in Fig. 2.5, where we show the the minimal area surface $\hat{\gamma}_A$ when A is made by three disjoint disks with the same radius whose distances between the centres compared with the radius guarantee the existence of a connected $\hat{\gamma}_A$. Obviously, when the disks are very far apart from each

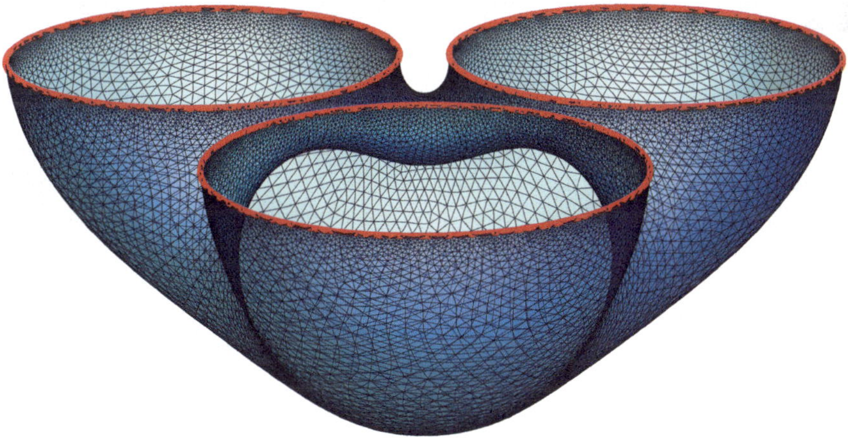

Fig. 2.5 Minimal area surfaces $\hat{\gamma}_A$ in $\mathbb{H}_3$ anchored to ∂A, with A made by the union of three disjoint disks having the same radius (Figure reprinted with permission from [40])

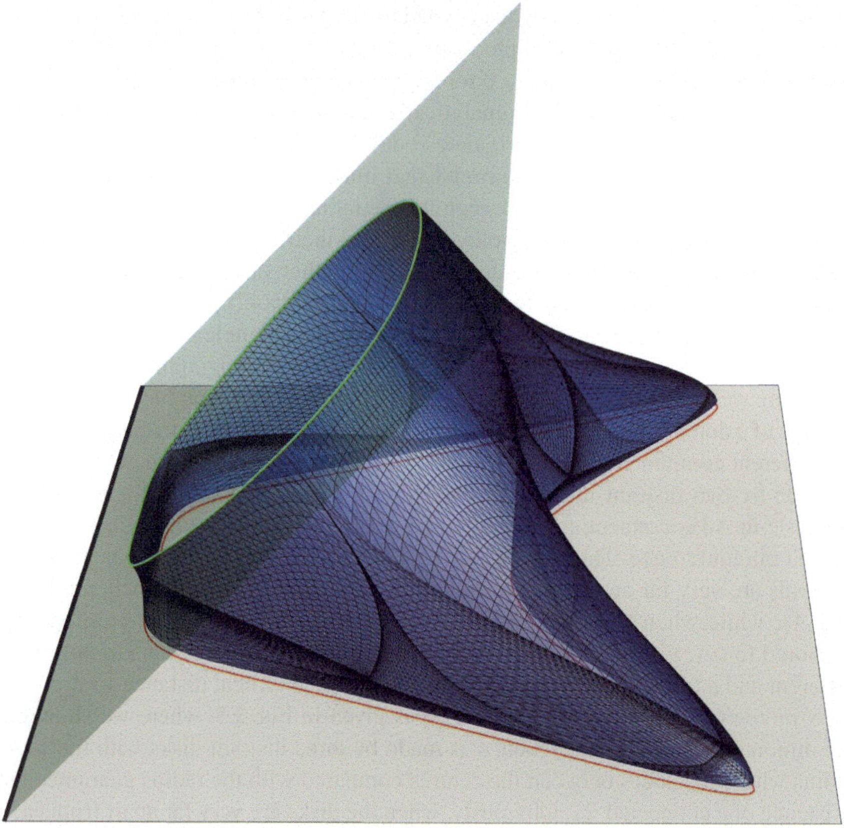

Fig. 2.6 Minimal area surfaces $\hat{\gamma}_A$ in AdS$_4$/BCFT$_3$ anchored to ∂A given by the red curve. The surface intersects orthogonally the green brane delimiting the bulk gravitational dual, that encodes the boundary conditions of the dual BCFT$_3$ whose spatial slice is the grey horizontal half plane (Figure reprinted with permission from [53])

other $\hat{\gamma}_A$ is made be three disconnected surfaces anchored to the boundaries of the three disks. Thus, as the distances change, transitions occur in the tripartite mutual information (2.5.3). The occurrence of these transitions is a characteristic feature of the holographic entanglement entropy (2.5.2).

In the above discussion we have considered only smooth regions. Focussing on the case of AdS$_4$/CFT$_3$ for simplicity, when the entangling curve ∂A is smooth, we have that $\mathcal{A}_A = P_A/\varepsilon - F_A + o(1)$ as ε, being P_A the perimeter of A. When A contains corners, i.e. the entangling curve is singular, also a logarithmic divergence occurs, namely

$$\mathcal{A}_A = \frac{P_A}{\varepsilon} - b \log(P_A/\varepsilon) + O(1) \tag{2.5.4}$$

where the coefficient $b = \sum_i f(\theta_i)$ is a sum over the corners of a function $f(\theta)$ of the opening angle of the corner. The function f is often called corner function. This logarithmic divergence for singular entangling curves occurs also for a generic CFT_3 and the corner function is model dependent. For AdS_4/CFT_3, the corner function has been computed analytically in [46], while for some free theories it has been found in [47, 48].

The holographic entanglement entropy formula (2.5.2) has been employed also in the context of the so called AdS/BCFT correspondence [49–53], where the conformal field theory has a physical boundary whose holographic gravitational dual corresponds to an hypersurface in the bulk that enclose the gravitational bulk. In Fig. 2.6 we show a minimal surface.

Finally, we find worth remarking that the covariant prescription for the holographic entanglement entropy given in [36] has been largely employed to study the holographic evolution of the entanglement entropy in gravitational backgrounds describing the formation of a black hole [54–61].

References

1. P. Calabrese, J. Cardy, Entanglement entropy and conformal field theory. J. Phys. A **42**, 504005 (2009), arXiv:0905.4013 [cond-mat.stat-mech]
2. H. Casini, M. Huerta, Entanglement entropy in free quantum field theory. J. Phys. A **42**, 504007 (2009), arXiv:0905.2562 [hep-th]
3. V. Eisler, I. Peschel, Reduced density matrices and entanglement entropy in free lattice models. J. Phys. A: Math. Theor. **42** 504003 (2009), arXiv:0906.1663 [cond-mat.stat-mech]
4. T. Nishioka, S. Ryu, T. Takayanagi, Holographic entanglement entropy: an overview. J. Phys. A **42**, 504008 (2009), arXiv:0905.0932 [hep-th]
5. M. Rangamani, T. Takayanagi, Holographic entanglement entropy. Lect. Notes Phys. **931**, 1 (2017), arXiv:1609.01287 [hep-th]
6. H. Araki, E. Lieb, Entropy inequalities. Commun. Math. Phys. **18**, 160 (1970)
7. H. Araki, E. Lieb, M.B. Ruskai, Proof of the strong subadditivity of quantum-mechanical entropy. J. Math. Phys. AIP Publishing. **14**(12) (1938)
8. L. Bombelli, R.K. Koul, J. Lee, R.D. Sorkin, A quantum source of entropy for black holes. Phys. Rev. D **34**, 373 (1986)
9. M. Srednicki, Entropy and area. Phys. Rev. Lett. **71**, 666 (1993), arXiv:hep-th/9303048
10. C.G. Callan Jr., F. Wilczek, On geometric entropy. Phys. Lett. B **333**, 55 (1994), arXiv:hep-th/9401072
11. C. Holzhey, F. Larsen, F. Wilczek, Geometric and renormalized entropy in conformal field theory. Nucl. Phys. B **424**, 443 (1994), arXiv:hep-th/9403108
12. G. Vidal, J.I. Latorre, E. Rico, A. Kitaev, Entanglement in quantum critical phenomena. Phys. Rev. Lett. **90**, 227902 (2003), arXiv:quant-ph/0211074
13. P. Calabrese, J.L. Cardy, Entanglement entropy and quantum field theory. J. Stat. Mech. **0406**, P06002 (2004), arXiv:hep-th/0405152
14. L.Y. Hung, R.C. Myers, M. Smolkin, Twist operators in higher dimensions. JHEP **1410**, 178 (2014), arXiv:1407.6429 [hep-th]
15. L. Bianchi, M. Meineri, R.C. Myers, M. Smolkin, Rényi entropy and conformal defects. JHEP **1607**, 076 (2016), arXiv:1511.06713 [hep-th]
16. P. Calabrese, J. Cardy, E. Tonni, Entanglement entropy of two disjoint intervals in conformal field theory. J. Stat. Mech. **0911**, P11001 (2009), arXiv:0905.2069 [hep-th]

17. P. Calabrese, J. Cardy, E. Tonni, Entanglement entropy of two disjoint intervals in conformal field theory II. J. Stat. Mech. **1101**, P01021 (2011), arXiv:1011.5482 [hep-th]
18. A. Coser, L. Tagliacozzo, E. Tonni, On Rényi entropies of disjoint intervals in conformal field theory. J. Stat. Mech. **1401**, P01008 (2014), arXiv:1309.2189 [hep-th]
19. C. De Nobili, A. Coser, E. Tonni, Entanglement entropy and negativity of disjoint intervals in CFT: some numerical extrapolations. J. Stat. Mech. **1506**(6), P06021 (2015), arXiv:1501.04311 [cond-mat.stat-mech]
20. V. Alba, L. Tagliacozzo, P. Calabrese, Entanglement entropy of two disjoint blocks in critical Ising models. Phys. Rev. B **81**, 060411 (2010), arXiv:0910.0706 [cond-mat.stat-mech]
21. M. Fagotti, P. Calabrese, Entanglement entropy of two disjoint blocks in XY chains. J. Stat. Mech. **1004**, P04016 (2010), arXiv:1003.1110 [cond-mat.stat-mech]
22. H. Casini, M. Huerta, R.C. Myers, Towards a derivation of holographic entanglement entropy. JHEP **1105**, 036 (2011), arXiv:1102.0440 [hep-th]
23. J. Cardy, Some results on the mutual information of disjoint regions in higher dimensions. J. Phys. A **46**, 285402 (2013), arXiv:1304.7985 [hep-th]
24. A. Peres, Separability criterion for density matrices. Phys. Rev. Lett. **77**, 1413 (1996)
25. G. Vidal, R.F. Werner, A computable measure of entanglement. Phys. Rev. A **65**, 032314 (2002)
26. P. Calabrese, J. Cardy, E. Tonni, Entanglement negativity in quantum field theory. Phys. Rev. Lett. **109**, 130502 (2012), arXiv:1206.3092 [cond-mat.stat-mech]
27. P. Calabrese, J. Cardy, E. Tonni, Entanglement negativity in extended systems: a field theoretical approach. J. Stat. Mech. **1302**, P02008 (2013), arXiv:1210.5359 [cond-mat.stat-mech]
28. V. Eisler, Z. Zimboras, Entanglement negativity in two-dimensional free lattice models. Phys. Rev. B **93**, 115148 (2016), arXiv:1511.08819 [cond-mat.stat-mech]
29. C. De Nobili, A. Coser, E. Tonni, Entanglement negativity in a two dimensional harmonic lattice: area law and corner contributions. J. Stat. Mech. **1608**(8), 083102 (2016), arXiv:1604.02609 [cond-mat.stat-mech]
30. V. Eisler, Z. Zimboras, On the partial transpose of fermionic Gaussian states. New J. Phys. **17**, 053048 (2015), arXiv:1502.01369 [cond-mat.stat-mech]
31. A. Coser, E. Tonni, P. Calabrese, Partial transpose of two disjoint blocks in XY spin chains. J. Stat. Mech. **1508**(8), P08005 (2015), arXiv:1503.09114 [cond-mat.stat-mech]
32. A. Coser, E. Tonni, P. Calabrese, Towards the entanglement negativity of two disjoint intervals for a one dimensional free fermion. J. Stat. Mech. **1603**(3), 033116 (2016), arXiv:1508.00811 [cond-mat.stat-mech]
33. A. Coser, E. Tonni, P. Calabrese, Spin structures and entanglement of two disjoint intervals in conformal field theories. J. Stat. Mech. **1605**(5), 053109 (2016), arXiv:1511.08328 [cond-mat.stat-mech]
34. S. Ryu, T. Takayanagi, Holographic derivation of entanglement entropy from AdS/CFT. Phys. Rev. Lett. **96**, 181602 (2006), arXiv:hep-th/0603001
35. S. Ryu, T. Takayanagi, Aspects of holographic entanglement entropy. JHEP **0608**, 045 (2006). arXiv:hep-th/0605073
36. V.E. Hubeny, M. Rangamani, T. Takayanagi, A Covariant holographic entanglement entropy proposal. JHEP **0707**, 062 (2007), arXiv:0705.0016 [hep-th]
37. M. Headrick, T. Takayanagi, A Holographic proof of the strong subadditivity of entanglement entropy. Phys. Rev. D **76**, 106013 (2007), arXiv:0704.3719 [hep-th]
38. V.E. Hubeny, H. Maxfield, M. Rangamani, E. Tonni, Holographic entanglement plateaux. JHEP **1308**, 092 (2013), arXiv:1306.4004 [hep-th]
39. P. Hayden, M. Headrick, A. Maloney, Holographic mutual information is monogamous. Phys. Rev. D **87**(4), 046003 (2013), arXiv:1107.2940 [hep-th]
40. P. Fonda, L. Giomi, A. Salvio, E. Tonni, On shape dependence of holographic mutual information in AdS_4. JHEP **1502**, 005 (2015), arXiv:1411.3608 [hep-th]
41. P. Fonda, D. Seminara, E. Tonni, On shape dependence of holographic entanglement entropy in AdS_4/CFT_3. JHEP **1512**, 037 (2015), arXiv:1510.03664 [hep-th]
42. A. Allais, M. Mezei, Some results on the shape dependence of entanglement and Rényi entropies. Phys. Rev. D **914**, 046002 (2015), arXiv:1407.7249 [hep-th]

43. V.E. Hubeny, M. Rangamani, Holographic entanglement entropy for disconnected regions. JHEP **0803**, 006 (2008), arXiv:0711.4118 [hep-th]
44. M. Headrick, Entanglement Renyi entropies in holographic theories. Phys. Rev. D **82**, 126010 (2010), arXiv:1006.0047 [hep-th]
45. E. Tonni, Holographic entanglement entropy: near horizon geometry and disconnected regions. JHEP **1105**, 004 (2011), arXiv:1011.0166 [hep-th]
46. N. Drukker, D.J. Gross, H. Ooguri, Wilson loops and minimal surfaces. Phys. Rev. D **60**, 125006 (1999), arXiv:hep-th/9904191
47. H. Casini, M. Huerta, Universal terms for the entanglement entropy in 2+1 dimensions. Nucl. Phys. B **764**, 183 (2007), arXiv:hep-th/0606256
48. H. Casini, M. Huerta, L. Leitao, Entanglement entropy for a dirac fermion in three dimensions: vertex contribution. Nucl. Phys. B **814**, 594 (2009), arXiv:0811.1968 [hep-th]
49. T. Takayanagi, Holographic dual of BCFT. Phys. Rev. Lett. **107**, 101602 (2011), arXiv:1105.5165 [hep-th]
50. M. Fujita, T. Takayanagi, E. Tonni, Aspects of AdS/BCFT. JHEP **1111**, 043 (2011), arXiv:1108.5152 [hep-th]
51. A. Faraji Astaneh, C. Berthiere, D. Fursaev, S.N. Solodukhin, Holographic calculation of entanglement entropy in the presence of boundaries. Phys. Rev. D **95**(10), 106013 (2017), arXiv:1703.04186 [hep-th]
52. D. Seminara, J. Sisti, E. Tonni, Corner contributions to holographic entanglement entropy in $AdS_4/BCFT_3$. JHEP **1711**, 076 (2017), arXiv:1708.05080 [hep-th]
53. D. Seminara, J. Sisti, E. Tonni, Holographic entanglement entropy in $AdS_4/BCFT_3$ and the Willmore functional. JHEP **1808**, 164 (2018), arXiv:1805.11551 [hep-th]
54. J. Abajo-Arrastia, J. Aparicio, E. Lopez, Holographic evolution of entanglement entropy. JHEP **1011**, 149 (2010), arXiv:1006.4090 [hep-th]
55. V. Balasubramanian et al., Holographic thermalization. Phys. Rev. D **84**, 026010 (2011), arXiv:1103.2683 [hep-th]
56. A. Allais, E. Tonni, Holographic evolution of the mutual information. JHEP **1201**, 102 (2012), arXiv:1110.1607 [hep-th]
57. R. Callan, J.Y. He, M. Headrick, Strong subadditivity and the covariant holographic entanglement entropy formula. JHEP **1206**, 081 (2012), arXiv:1204.2309 [hep-th]
58. V.E. Hubeny, M. Rangamani, E. Tonni, Thermalization of causal holographic information. JHEP **1305**, 136 (2013), arXiv:1302.0853 [hep-th]
59. H. Liu, S.J. Suh, Entanglement growth during thermalization in holographic systems. Phys. Rev. D **89**(6), 066012 (2014), arXiv:1311.1200 [hep-th]
60. V. Keranen, E. Keski-Vakkuri, L. Thorlacius, Thermalization and entanglement following a non-relativistic holographic quench. Phys. Rev. D **85**, 026005 (2012), arXiv:1110.5035 [hep-th]
61. P. Fonda, L. Franti, V. Keränen, E. Keski-Vakkuri, L. Thorlacius, E. Tonni, Holographic thermalization with Lifshitz scaling and hyperscaling violation. JHEP **1408**, 051 (2014), arXiv:1401.6088 [hep-th]

Chapter 3
Entanglement Content of Many-Body States via Concurrence, Negativity and Schmidt Gap

Sougato Bose, Abolfazl Bayat, Henrik Johannesson and Pasquale Sodano

Abstract Quantum entanglement is nearly ubiquitous in equilibrium and non-equilibrium many-body states. Although it has been largely studied through the von Neumann entropy of a subsystem, which quantifies the entanglement between two complementary parts of a many-body system, this is not necessarily the only way. Here we review how some other measures can be fruitful in characterizing the entanglement content of many-body states. For example, we can look at the entangement between two individual spins through the concurrence or between two non-complementary, but in principle large, parts of a many-body system through the negativity. Alternatively, a quantity inspired through entanglement studies, but not itself a measure of entanglement, namely the Schmidt gap, can be effective as an order parameter for phase transitions in which only the entanglement structure of a many-body system changes. We exemplify using equilibrium states of short-range and impurity models and their quantum phase transitions.

S. Bose (✉) · A. Bayat
Department of Physics and Astronomy, University College London,
Gower Street, London WC1E 6BT, UK
e-mail: s.bose@ucl.ac.uk

A. Bayat
e-mail: abolfazl.bayat@ucl.ac.uk

A. Bayat
Institute of Fundamental and Frontier Sciences, University of Electronic Science and Technology of China, Chengdu 610051, China

H. Johannesson
Department of Physics, University of Gothenburg, 41296 Gothenburg, Sweden
e-mail: henrik.johannesson@physics.gu.se

P. Sodano
INFN Sezione di Perugia, Via A. Pascoli, 06123 Perugia, Italy
e-mail: pasquale.sodano01@gmail.com

International Institute of Physics, UFRN, Natal-RN 59078-400, Brazil

A. Ferraz et al. (eds.), *Strongly Coupled Field Theories for Condensed Matter and Quantum Information Theory*, Springer Proceedings in Physics 239,
https://doi.org/10.1007/978-3-030-35473-2_3

3.1 Quantum Entanglement and Its Quantification

An example of an entangled state of two spin-1/2 systems (two qubits) with states $|0\rangle$ and $|1\rangle$ is

$$|\psi^-\rangle = \frac{1}{\sqrt{2}}(|0\rangle|1\rangle - |1\rangle|0\rangle) \tag{3.1}$$

which can never be written down as a pure product of states of the individual systems such as in the form $|\chi\rangle|\phi\rangle$. A general separable (*not* entangled) state of two systems A and B is one given by the equation

$$\rho_{AB} = \sum_i P_i|\psi\rangle\langle\psi|_A \otimes |\phi\rangle\langle\phi|_B. \tag{3.2}$$

Any state which is *not* of the above form is called an entangled state.

Entanglement is a huge area of quantum information science, with several reviews such as [1]. It is a quantifiable, as well as a measurable entity. Simply stating, the "harder" it is to approximate a state as a probability distribution over products of pure states, the "higher" is its entanglement. For example, for the case of two qubits, the state $|\psi^-\rangle$ is the most entangled, whose amount is generally set to unity, while a product state of the form $|\chi\rangle|\phi\rangle$ or mixed states of the form $p|\chi_1\rangle\langle\chi_1|_A \otimes |\phi_1\rangle\langle\phi_1|_B + (1-p)|\chi_2\rangle\langle\chi_2|_A \otimes |\phi_2\rangle\langle\phi_2|_B$ have zero entanglement.

Quantification of entanglement has been accomplished in several ways. The crucial property for an entanglement measure to satisfy is that it cannot be increased between two systems held by distant parties Alice and Bob if they are solely using local operations and classical communications (LOCC). The first measure of entanglement to be introduced was the von Neumann entropy of entanglement [2]. It applies to the entanglement of two systems of arbitrary dimensions (such as when each system is a multi-qubit system), but only for pure states. This is computed by first computing the reduced density matrix ρ_A of system A from the state ρ_{AB} of the total system using the procedure $\rho_A = Tr_B(\rho_{AB})$, where $Tr_B(\,)$ denotes partial tracing over system B. The entanglement of the two systems A and B in the pure state ρ_{AB} is then given by the von Neumann entropy [2]

$$\mathscr{S} = -Tr\{\rho_A \log_2 \rho_A\}, \tag{3.3}$$

which in terms of the eigenvalues η_i of ρ_A, is $\mathscr{S} = -\sum_i \eta_i \log_2 \eta_i$. This measure has been studied and computed for a vast majority of (pure) ground states quantum many-body systems when they are divided into two complementary parts [3]. We will thus not concentrate on the von Neumann entropy in this article.

Here, we would rather concentrate on applying those measures of entanglement to condensed matter systems which are less common. They give entanglement between two non-complementary parts of a many body systems—such as well separated blocks of spins or pairs of individual spins. For the case of two qubits (e.g. spin-1/2 systems) in a mixed state, given by a density matrix ρ, its entanglement can be computed [4]. The procedure is to first compute the matrix

$$\tilde{\rho}_{AB} = \sigma_y \otimes \sigma_y \rho^*_{AB} \sigma_y \otimes \sigma_y \tag{3.4}$$

where the complex conjugate ρ^* of ρ is taken in the basis $|00\rangle$, $|01\rangle$, $|10\rangle$, $|11\rangle$. Then the entanglement can be quantified by a number called concurrence $\mathscr{E}$ given by [4]

$$\mathscr{E} = max\{0, \lambda_1 - \lambda_2 - \lambda_3 - \lambda_4\}, \tag{3.5}$$

where λ_is are the square roots of the eigenvalues of $\rho\tilde{\rho}$ in decreasing order.

However, at times you may have two larger dimensional systems in a mixed state and want to compute their entanglement—for example, for two non-complementary blocks of spins. Then it is best to use another measure of entanglement termed the negativity. The reduced density operators ρ_{AB} carry the information on the entanglement between the blocks A and B. As ρ_{AB} is a mixed state, the block entropy is inappropriate as a measure of the entanglement. We have to use instead the negativity [5, 6] $\mathscr{N} \equiv (\sum_i |a_i| - 1)$ where $|a_i|$ denote the modulus of the eigenvalues of the partial transpose $(\rho_{AB})^{T_A}$ of ρ_{AB} with respect to the subsystem A, i.e., $\langle w_i^A w_j^B | \rho_{AB}^{T_S} | w_k^A w_l^B \rangle = \langle w_k^A w_j^B | \rho_{AB} | w_i^A w_l^B \rangle$ [1]. $\{|w^S\rangle\}$ and $\{|w^E\rangle\}$ are the orthogonal basis states of A and B respectively, chosen by the DMRG procedure. This is a widely used genuine measure of quantum correlations (entanglement monotone [7]) and provides a bound to the fidelity of teleportation with a single copy of the state [8].

Moreover, we will be using another measure, which, although, is not a measure of entanglement, it does indicate, in the broadest coarse grained way, the entanglement present between two complementary halves of a pure state system. This is called the Schmidt gap. We cut the system in two parts, A and B, and write the Schmidt decomposition of a pure state, say the ground state $|GS\rangle$, as

$$|GS\rangle = \sum_k \sqrt{\lambda_k} |A_k\rangle \otimes |B_k\rangle, \quad \lambda_k \geq 0, \tag{3.6}$$

with mutually orthogonal Schmidt basis states $|A_k\rangle$ and $|B_k\rangle$. The density matrix of each part is diagonal in the Schmidt basis,

$$\rho_\alpha = \sum_k \lambda_k |\alpha_k\rangle\langle\alpha_k|, \quad \alpha = A, B. \tag{3.7}$$

with the eigenvalues $\lambda_1 \geq \lambda_2 \geq ...$ in descending order forming the entanglement spectrum (frequently defined as $\{-\ln \lambda_i\}_{i=1,2,...}$ in the literature). Then the Schmidt gap is defined as $\Delta_S = \lambda_1 - \lambda_2$ – where λ_1 and λ_2 are the two largest eigenvalues of the reduced ground state density matrix as defined above.

3.2 Concurrence Between Two Spins of a Many-Body System

As an illustration, we will first study a model of an open ended chain of four spin-1/2 particles coupled through a nearest neighbor isotropic Heisenberg interaction, so that the Hamiltonian is

$$\mathbf{H} = \sum_{i=1}^{3} \sigma^i . \sigma^{i+1} \tag{3.8}$$

The ground state of this chain

$$|GS\rangle = \left(\sqrt{\frac{2}{3}} |\psi^-\rangle|\psi^-\rangle - 0.1494(|00\rangle|11\rangle - |01\rangle|01\rangle - |10\rangle|10\rangle + |11\rangle|00\rangle \right) \tag{3.9}$$

is manifestly an entangled state (just looking at the form of the state is enough to spot that) and for a long time condensed matter physicists have known that the ground states of such systems are indeed entangled. What is more important, though, is the "amount" of entanglement between two spins of the system. This had not been computed till the advent of quantum information. In order to do this, we first obtain the reduced density matrix ρ_{ij} of the spins i and j. From the expression of $|GS\rangle$, it is clear that ρ_{12} for example, is a mixed state with a significant proportion (2/3) of the maximally entangled state $|\psi^-\rangle$. The entanglement between spins i and j is computed as the concurrence $\mathscr{E}$ from the formula given in Sect. 3.1. However, it is worth mentioning here that because of certain symmetries of the Heisenberg model, the concurrence $\mathscr{E}$ reduces to a very simple formula and one need not involve all elements of the reduced density matrix ρ_{ij} for the calculation of concurrence. Note that all the states involved in the expression for $|GS\rangle$ have the same number of zeros. This is a consequence of the commutation of $\mathbf{H}$ with $\sum_i \sigma_z^i$ and holds for all eigenstates of $\mathbf{H}$, and consequently also for their mixtures such as thermal states. It is then easy for the reader to verify (I leave it as an exercise here) that the density matrix ρ_{ij} cannot have any off diagonal terms (or coherence) between spaces with different values of $\sigma_z^i + \sigma_z^j$. In the standard basis $|00\rangle, |01\rangle, |10\rangle, |11\rangle$, ρ_{ij} will thus be of the form

$$\begin{pmatrix} x & 0 & 0 & 0 \\ 0 & y_1 & z & 0 \\ 0 & z^* & y_2 & 0 \\ 0 & 0 & 0 & w \end{pmatrix}$$

For such a simple form of ρ_{ij}, the concurrence is given by the simple formula $\mathscr{E} = 2\max\{0, |z| - \sqrt{xw}\}$.

The concurrence between spins 1 and 2 is found to be 0.866, which is quite high (the highest possible value, for a maximally entangled state, being 1). These are nearest neighbors. On the other hand, no entanglement exists between 2 and 3,

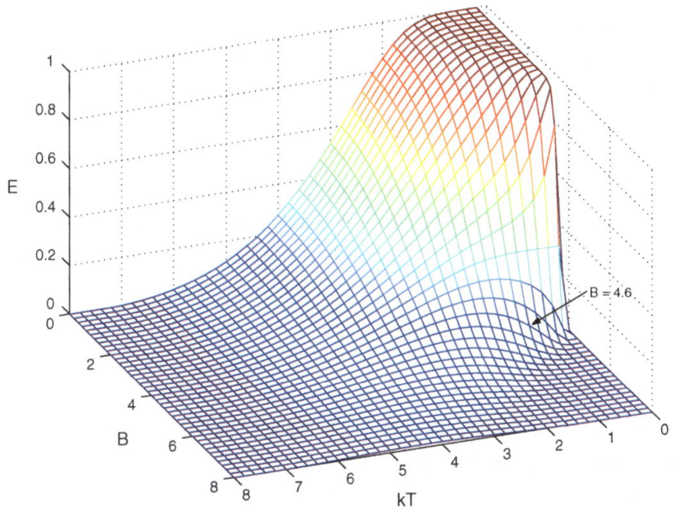

Fig. 3.1 A plot of the entanglement between two spins interacting via exchange in a magnetic field as quantified by their concurrence. It shows that entanglement ca persist at finite temperatures, and can even increase due to heating when the system is under an applied magnetic field. This latter case, while counter-intuitive, can be qualitatively understood as the ordering tendency of an external magnetic field competing with the fluctuations of the temperature to give rise to a finite entanglement

though they are nearest neighbors. The entanglement pattern of the chain is dimerized because of its open ends and there is no entanglement between any of the non-nearest neighbor spins (such as 1 and 3 or 1 and 4). Interestingly, the entanglement between two spins as quantified by concurrence can persist even in systems in a thermal state $\rho(T) = e^{-\mathbf{H}/kT}/Z$ for a system described by a Hamiltonian H at non-zero temperatures (*thermal entanglement*) and moreover, given a can also be tuned by a magnetic field. With respect to the simplest rudimentary case, namely, two spins coupled by an exchange interaction and placed in an external magnetic field,

$$\mathbf{H} = \sum_{i=1}^{N} (B\sigma_z^i + J\boldsymbol{\sigma}^i.\boldsymbol{\sigma}^{i+1}) \tag{3.10}$$

where $\boldsymbol{\sigma}^i = (\sigma_x^i, \sigma_y^i, \sigma_z^i)$ in which $\sigma_{x/y/z}^i$ are the Pauli matrices for the ith spin (we assume cyclic boundary conditions $1 + N = 1$), with $J > 0$ (antiferromagnetic interactions), the entanglement is shown as a function of the magnetic field and temperature in Fig. 3.1 which is taken from [9]. Reference [9], which was the first study of entanglement between individual spins of a spin chain in its natural (thermal/ground) state, also described the entanglement between spins of longer spin chains. Subsequently, the variation of such entanglement was also studied across *quantum phase transitions* [10]—when the ground state of certain quantum many-body systems undergo a sudden "qualitative" change due to the variation of a parameter of the Hamiltonian. Speaking very roughly and qualitatively, the ground state near a quantum phase transition is a highly entangled state because of a competition between different ordering tendencies of different terms of a Hamiltonian. On either side of the transition different tendencies win and impose their order, while at the transition neither can win and only an entangled state can be the lowest energy state. This competition between different ordering tendencies was shown to be captured in terms of the entanglement between two spins i and j, usually nearest and next nearest neighbors, by a peak of the entanglement near the point of a quantum phase transition [11, 12].

3.3 Entanglement Negativity in a Many-Body System: Case Study with the Kondo Model

In the context of spin chains, entanglement negativity was first studied in simultaneous papers [13, 14]. For critical ground states which have no built-in scale, it was found to be *scale invariant* in the sense that it only depended on the *ratio* $\mu = x/L$ of the separation x of two noncomplementary blocks of spins in a spin chain, and the length of the blocks L (length of both the blocks are taken to be equal here). This is shown in Fig. 3.2. A combination of an exponential and a power-law dependence of the entanglement in the ratio μ was found numerically in these papers [13], while subsequently the exponential dependence has also been analytically proved [15]. Entanglement negativity can also be used for quantification of multipartite entanglement in many-body systems which also shows scaling behavior near quantum phase transition [16]. Here we will next concentrate on how the entanglement negativity can be used to extract the Kondo cloud in a spin chain emulation of the Kondo model as discussed in [17].

The simplest Kondo model [18] describes a single impurity spin interacting with the conduction electrons in a metal; the ground state is a highly nontrivial many body state in which the impurity spin is screened by conduction electrons in a large orbital of size ξ, termed as the Kondo cloud. Many physical observables vary on the characteristic length scale ξ, which is a well defined function of the Kondo coupling [18]. Determining the spatial extent of the Kondo cloud has been so far a challenging problem repeatedly addressed by various means [19, 21, 22]. This includes an investigation which introduces a quantity called "impurity entanglement entropy" which, however, is not a bonafide measure of entanglement [19, 20]. Kondo systems are expected to have a more exotic *form* of entanglement than the widely studied spin-spin and complementary block entanglements. Indeed, in Kondo systems, the impurity spin is expected to be mostly entangled with only a specific block of the whole system. This is, of course, merely an intuition which needs to be quantitatively verified with a genuine measure of entanglement: this is indeed a task that negativity can accomplish as we will elucidate in this section following [17].

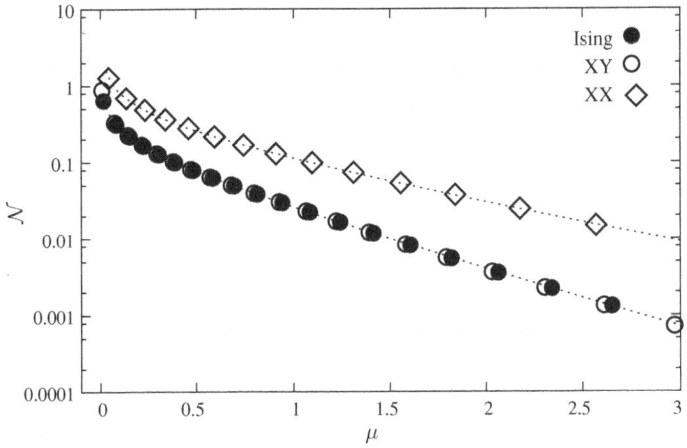

Fig. 3.2 The figure shows the dependence of the entanglement between two blocks of spins in a spin chain as quantified by the entanglement negativity. The entanglement is depicted as a function of the ratio $\mu = L/x$ of each block L and their separation x

It is known [23] that universal low energy long distance physics of this Kondo model arises also in a spin chain when a magnetic impurity is coupled to the end of a gapless Heisenberg anti-ferromagnetic $J_1 - J_2$ spin 1/2 chain, where J_1 (J_2) is the (next) nearest neighbor coupling. This is described by the Hamiltonian

$$H = J'(\sigma_1.\sigma_2 + J_2\sigma_1.\sigma_3) + \sum_{i=2}^{N-1} \sigma_i.\sigma_{i+1} + J_2 \sum_{i=2}^{N-2} \sigma_i.\sigma_{i+2}, \tag{3.11}$$

where the nearest neighbor coupling J_1 has been normalized to 1, $\sigma_i = (\sigma_i^x, \sigma_i^y, \sigma_i^z)$ is a vector of Pauli operators at site i, N is the total length of the chain. The impurity spin, located at one end of the chain, is accounted for by weaker couplings J' to the rest of the system. When J_2 exceeds a critical value, the spin chain enters a gapped dimerized regime and its relation to the Kondo model breaks down. Namely, for $0 \leq J_2 \leq J_2^c = 0.2412$, the spin system is gapless and it supports a Kondo regime [19, 20]. For $J_2 > J_2^c$, the system enters the gapped *dimer regime*, where the ground state takes a dimerised form (Fig. 3.3).

To study the entanglement of the ground state can be accomplished using DMRG as described in [17]. We determine the size of the block A when the entanglement between the impurity and block B is almost zero; by this procedure we measure an Entanglement Healing Length (EHL) L^*, i.e. the length of the block A which is maximally entangled with the impurity. We show that, in the gapless Kondo regime, EHL scales with the strength of the impurity coupling just as the Kondo screening length, ξ, does. Thus, in the gapless regime of the Kondo spin chain, our approach yields a method to detect the Kondo screening length [19, 21, 22] based on a true measure of entanglement. In addition, we *find* that entanglement in the Kondo regime is essentially unchanged if one rescales all the length scales with the EHL L^*.

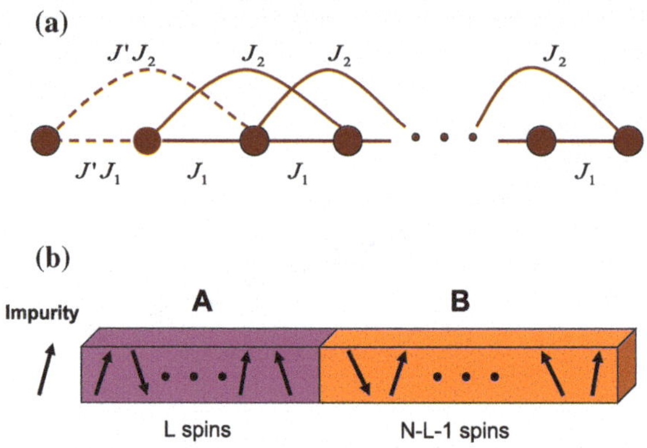

Fig. 3.3 (Color online) **a** Kondo Spin chain with next nearest neighbor Heisenberg interaction with one impurity at one end. **b** The chain is divided into three parts, an impurity, a block A and a block B. Entanglement is computed between the impurity and block B

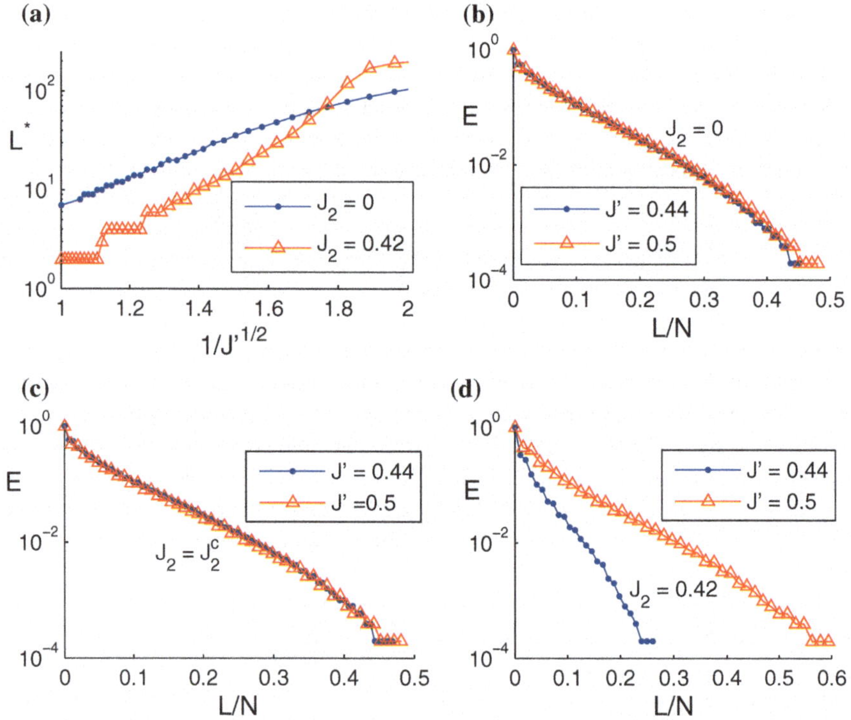

Fig. 3.4 (Color online) **a** L^* versus $1/\sqrt{J'}$ for both Kondo ($J_2 = 0$) and dimer regime ($J_2 = 0.42$). **b** Entanglement versus L/N for fixed $N/L^* = 4$ when $J_2 = 0$. **c** Entanglement versus L/N for fixed $N/L^* = 4$ at the critical point $J_2 = J_2^c$. **d** Entanglement versus L/N for fixed $N/L^* = 4$ in the dimer regime ($J_2 = 0.42$)

We find that there is an EHL L^* so that, for $L > L^*$, the entanglement between the impurity and block B is almost zero: L^* provides us with an estimate of the distance for which the impurity is mostly entangled with the spins contained in block A. For large chains ($N > 200$) in the Kondo regime, one finds that L^* is almost independent of N and depends only on J'. In the Kondo regime, i.e. for $J_2 < J_2^c$, L^* depends on J' just as the Kondo screening length ξ does [19, 20]; for small J', $L^* \propto e^{\alpha/\sqrt{J'}}$, where α is a constant. We plot L^* as a function of $1/\sqrt{J'}$ in Fig. 3.4a. In a semilogarithmic scale, the straight line plot exhibited in the Kondo regime ($J_2 = 0$) shows that L^* may be regarded as the Kondo screening length. Moreover, the nonlinearity of the same plot in the dimer regime ($J_2 = 0.42$), especially for small J', shows that, here, no exponential dependence on $1/\sqrt{J'}$ holds.

We observe also a remarkable scaling of negativity in the Kondo regime. This scaling may be regarded as yet another independent evidence of the fact that L^* is indeed the Kondo length ξ. In general, the entanglement E between the impurity and block B is a function of the three independent variables, J', L and N which, due to the one to one correspondence between J' and L^*, can be written as $E(L^*, L, N)$. We find that, in the Kondo regime, $E = E(N/L^*, L/N)$. To illustrate this, we fix the ratio N/L^* and plot the entanglement in terms of L/N for different values of J' (or equivalently L^*) for $J_2 = 0$ (Fig. 3.4b) and for $J_2 = J_2^c$ (Fig. 3.4c). The complete coincidence of the two plots in Figs. 3.4b and c shows that, in the Kondo regime, the spin chain can be scaled in size without essentially affecting the entanglement as long as L^* is also scaled. In the dimer regime the entanglement stays a function of three independent variables, i.e. $E = E(L^*, L, N)$, and, as shown in Fig. 3.4d, the entanglement does not scale with L^*. In our approach, the Entanglement Healing Length L^* may be evaluated in both the Kondo and the dimer regime: the scaling behavior, as well as the dependence of L^* on J', discriminates then between the very different entanglement properties exhibited by the spin chain Kondo model as J_2 crosses J_2^c.

We defined L^* such that there is no entanglement between the impurity and block B when block A is made of L^* spins. Conventional wisdom based on previous renormalization group analysis suggests that, in both regimes, the impurity and the block A of length L^* form a pure entangled state, while block B is also in a pure state. This is indeed approximately true in the dimer regime (exactly true for $J_2 = 0.5$) but it turns out to be dramatically different in the Kondo regime. To check this, [17] also computed the von Neumann entropy of the block B when block A has L^* spins and found it to be non zero. Thus, the blocks A and B are necessarily entangled in the Kondo regime as there is no entanglement between the impurity and B. In fact, after a distance L^*, the impurity is "screened" i.e, the block B feels as if it is part of a conventional gapless chain and has a diverging von Neumann entropy. The Kondo cloud is maximally entangled with the impurity as well as being significantly entangled with block B. Based on the above, a simple ansatz for the ground state $|GS\rangle$ in the Kondo regime is provided by

$$|GS\rangle = \sum_i \alpha_i \frac{|\uparrow\rangle |L_i^{\uparrow}(J')\rangle - |\downarrow\rangle |L_i^{\downarrow}(J')\rangle}{\sqrt{2}} \otimes |R_i(J')\rangle, \tag{3.12}$$

where α_i are constants, $\{|L_i^{\uparrow}(J')\rangle, |L_i^{\downarrow}(J')\rangle\}$ and $\{|R_i(J')\rangle\}$ are sets of orthogonal states on the cloud and the remaining system, respectively. At the fixed point $J' \to 0$ all spins except the impurity are included in $|L_i^{\uparrow}(J')\rangle$ and $|L_i^{\downarrow}(J')\rangle$. At $J' \to 1$, very few spins are contained in $|L_i^{\uparrow}(J')\rangle$ and $|L_i^{\downarrow}(J')\rangle$ while $\{|R_i(J')\rangle\}$ represents most of the chain.

The above exercise illustrates the efficacy of Negativity as a mixed state entanglement measure to capture emergent structures in many-body states such as the Kondo cloud, and use that to infer variational representations of the states.

3.4 Schmidt Gap for Signalling Quantum Criticality: Case Study with an Impurity Phase Transition

We now move to studying another quantity, the Schmidt gap, as defined in Sect. 3.1, which, although not a bonafide measure of entanglement, is an indicator of the same. It is extremely useful as an indicator of quantum phase transitions which do not have a local order parameter, such as generic impurity quantum phase transitions in which it is only the entanglement structure of the ground state which is re-arranged as one crosses the critical point.

As case study we again turn to a Kondo model, but now with *two* localized spin-1/2 impurities, coupled to the spins of the conduction electrons by an antiferromagnetic Kondo interaction and to each other via a Ruderman–Kittel–Kasuya–Yosida (RKKY) interaction. This is the *two-impurity Kondo model* [25], a theoretical workhorse in the study of impurity quantum phase transitions. When the Kondo interaction dominates, the electron spins screen the impurity spins (similar to the screening of a single spin-1/2 impurity in the ordinary Kondo model), while in the opposite limit the two impurity spins form a local singlet. The crossover between the two regimes sharpens into a quantum phase transition when each impurity is connected to its own distinct reservoir of conduction electrons [26]. The resulting "non-Fermi liquid" response of transport and thermodynamic observables has attracted much attention, with the first experiment reported in 2015 [27]. The lack of an easily identifiable local order parameter which exhibits scaling at the phase transition has triggered a search for alternative markers of the transition, the Schmidt gap being one viable candidate [28].

For a numerical computation of the Schmidt gap, using a Density Matrix Renormalization Group (DMRG) approach, it is convenient to first map the spin sector of the two-impurity Kondo model onto a spin chain, similar to what was done in the previous section for the ordinary Kondo model. One thus obtains a Hamiltonian $H = \sum_{m=L,R} H_m + H_I$, where

$$H_m = J'_m \left(J_1 \sigma_1^m \cdot \sigma_2^m + J_2 \sigma_1^m \cdot \sigma_3^m\right) + J_1 \sum_{i=2}^{N_m - 1} \sigma_i^m \cdot \sigma_{i+1}^m + J_2 \sum_{i=2}^{N_m - 2} \sigma_i^m \cdot \sigma_{i+2}^m, \quad (3.13)$$

$$H_I = J_1 K \sigma_1^L \cdot \sigma_1^R.$$

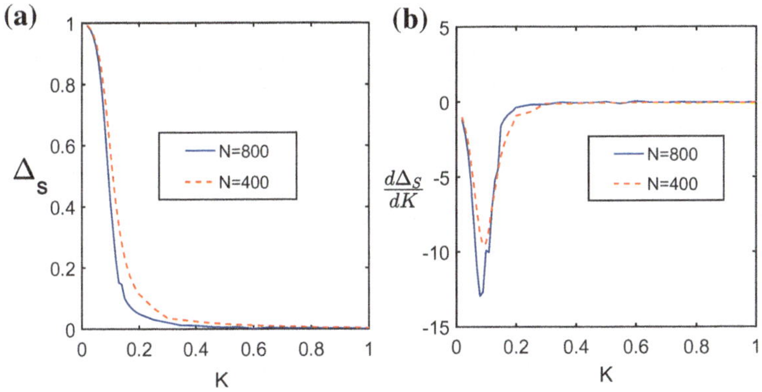

Fig. 3.5 **a** Schmidt gap as a function of the RKKY coupling K for different system sizes. **b** The derivative of the Schmidt gap with respect to the RKKY coupling K which shows non-analytic behavior as the system size increases. The figures are taken from [28]

Here $m = L, R$ labels two chains, denoted "left" and "right" respectively, with σ_i^m the vector of Pauli matrices at site i in chain m, and with J_1 (J_2) nearest- (next-nearest-) neighbor spin-exchange amplitudes. The parameters $J_L' > 0$ and $J_R' > 0$ play the role of antiferromagnetic Kondo couplings, with K the RKKY coupling between the impurity spins σ_1^L and σ_1^R. The total number of sites is thus $N = N_L + N_R$. Similar to the spin chain representation of the ordinary Kondo model, the ratio J_2/J_1 is fine tuned to the dimerization point $(J_2/J_1)_c = 0.2412$ of the spin chain [29], in this way killing off logarithmic corrections to the numerical finite-size data. In Fig. 3.5a we plot the Schmidt gap Δ_S (cf. Sect. 1) as a function of the RKKY coupling K when the system is cut across the bond between the two impurities. In other words, we regard the left (or right) chain as a subsystem of the full composite chain. As seen in the figure, the order-parameter-like profile of Δ_S becomes sharper as the size N of the chain increases, suggesting that Δ_S drops to zero at some critical value, $K = K_c$ in the thermodynamic limit. The interpretation of K_c as a critical point is evidenced by the sharp cusp of $\partial\Delta_S/\partial K$ at K_c as seen in Fig. 3.5b, with the cusp serving as a finite-size precursor of a critical non-analyticity in Δ_S' in the thermodynamic limit. Moreover, as shown in [28], by monitoring how the cusps move as functions of the Kondo coupling J', one finds, for sufficiently large systems, an almost perfect fit to the known exponential scaling of the quantum critical point $K_c \sim \exp(-\alpha/J')$ of the two-impurity Kondo model (with α a positive constant) [30]. It may be worth pointing out that cuts taken far away from the impurities yield Schmidt gaps which depend only weakly on the RKKY coupling, as expected from the local character of the transition [28]. Such cuts are thus less useful for spotting an impurity quantum phase transition.

A finite-size scaling analysis, with the Ansatz

$$\Delta_S = N^{-\beta/\nu} f_{\Delta_S}(|K - K_c|N^{1/\nu}), \tag{3.14}$$

gives further support for using the Schmidt gap Δ_S as a stand-in for a conventional order parameter. Here f_{Δ_S} is a scaling function, with β and δ the critical exponents governing the scaling of Δ_S and the crossover scale ξ [31] at the critical point: $\Delta_S \sim |K - K_c|^{\beta}$ and $\xi \sim |K - K_c|^{-\nu}$ respectively. The choice $\beta = 0.2 \pm 0.05$ and $\nu = 2 \pm 0.1$ yields a near-perfect data collapse when plotting $N^{\beta/\nu}\Delta_S$ as a function of $|K - K_c|N^{1/\nu}$ for different N and different impurity couplings J' [28]. The value $\nu = 2$ agrees with results from conformal field theory [32].

To summarize, DMRG data on the two-impurity Kondo model strongly suggests that the Schmidt gap faithfully captures its critical behavior. It is important to note that while the Schmidt gap is a nonlocal quantity, it can be represented as a superposition of n-point spin correlation functions, and therefore, in principle, be measured. To carry this out in an experiment clearly remains a challenge for the future. In the meantime, it would be interesting to attempt a characterization of impurity quantum phase transitions for other models using the Schmidt gap.

3.5 Entanglement Negativity as a Measurable Entity

In fact, the measure of negativity that we have used in large part of this article to quantify the entanglement, is also a measurable entity and one can accomplish it as long as one has the experimental access to several replicas of a system. In this section we show how to estimate logarithmic negativity using experimentally measurable quantities.

Logarithmic Negativity: For a generic mixed state, logarithmic negativity [5–8] is an entanglement measure. As logarithmic negativity does not rely on usual optimization over Hilbert space, needed in other entanglement measures, is computable efficiently. For a generic mixed state ρ_{AB} which explain the quantum state of two subsystems A and B the logarithmic negativity is defined as It is defined as:

$$\mathscr{E} = \log_2 \left|\rho_{AB}^{T_A}\right| = \log_2 \left|\rho_{AB}^{T_B}\right| = \log_2 \sum_k |\lambda_k| \tag{3.15}$$

with $|\cdot|$ the trace norm, $\rho_{AB}^{T_X}$ the partial transpose with respect to subsystem X, and $\{\lambda_k\}$ the eigenvalues of $\rho_{AB}^{T_X}$. Because of the non-trivial dependence of $\mathscr{E}$ on ρ_{AB}, there is no state-independent observable that can measure it—generally demanding full state tomography. The $\{\lambda_k\}$ are the roots of the characteristic polynomial, $P(\lambda) = \det(\lambda - \rho_{AB}^{T_B}) = \sum_n c_n \lambda^n$, where each c_n is a polynomial function of the partially transposed moments:

$$\mu_m = Tr[(\rho_{AB}^{T_B})^m] = \sum_k \lambda_k^m. \tag{3.16}$$

This means that, $\{\mu_m\}$ contains full information about the spectrum $\{\lambda_k\}$. Interestingly, even though partially transposed density matrices are generically unphysical,

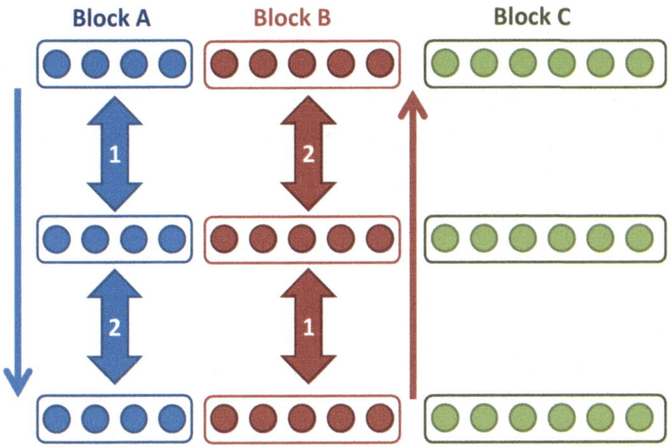

Fig. 3.6 Schematics: Example measurement set-up for the moments, $\mu_m = Tr[(\rho_{AB}^{T_B})^m]$, here for $m = 3$, from which one can extract the logarithmic negativity $\mathcal{E}$ between A and B. The generic mixedness of ρ_{AB} could arise from entanglement with environment C. Here the subsystems contain N_A, N_B and N_C particles respectively. The scheme involves three copies of the original system, and two counter propagating sets of measurements on A and B, ordered by the shown numbers, with direction depicted by the filled arrows

measurement of their moments is possible. In fact, there area few proposals in the literature for computing the the moments μ_m using m copies of the system. This includes, replica techniques [15] proposed for conformal field theory models, exploiting controlled swap operators between the copies [33, 34] and counter propagating swap measurements [35]. In the following, we provide a brief review of how to measure the moments based on the scheme of [35].

Measuring the Moments of $\rho_{AB}^{T_B}$: Here, we show that any moment can be measured using only SWAP-operators between the *individual* constituents of the m copies of the state ρ_{AB}, namely $\rho_{AB}^{\otimes m} = \bigotimes_{c=1}^{m} \rho_{A_c B_c}$. This general set-up is shown in Fig. 3.6, where the mixedness of ρ_{AB} arises from possible entanglement with a third system C, such that $\rho_{AB} = Tr_C|\Psi_{ABC}\rangle\langle\Psi_{ABC}|$ with $|\Psi_{ABC}\rangle$ being a pure tripartite state. The first step is to write the matrix power as an expectation of a permutation operator on the partially transposed copies:

$$\mu_m = Tr[(\bigotimes_{c=1}^{m} \rho_{A_c B_c}^{T_{B_c}}) P^m] \\ = Tr[(\bigotimes_{c=1}^{m} \rho_{A_c B_c})(P^m)^{T_B}] \,, \quad (3.17)$$

where P^m is any linear combination of cyclic permutation operators of order m and the second line makes use of the identity $Tr(\rho_{AB}^{T_B} O) = Tr(\rho_{AB} O^{T_B})$, valid for any operator O. A schematic of the equality in (3.17) for $m = 3$ is shown in Fig. 3.6. For spin lattices, our choice of P^m to measure the moments μ_m results to the following steps in practice:

1. prepare m copies of the state ρ_{AB};
2. sequentially measure a 'forward' sequence of adjacent swaps, $S_A^{c,c+1}$ between neighbouring copies of system A from $c = 1$ to $m - 1$;
3. sequentially measure a 'backward' sequence of adjacent swaps, $S_B^{c,c-1}$ between neighbouring copies of system B from $c = m$ to 2;
4. repeat these steps in order to yield an expectation value.

This procedure is also depicted for $m = 3$ in Fig. 3.6. For each $m \leq 2$ we require $\mathcal{O}(N_A + N_B)$ measurements. This is in stark contrast to tomography, which generically for qubit systems requires $2^{2(N_A+N_B)}$ measurement settings. It is also worth emphasizing the difference between this procedure and other operational methods for measuring Renyi entropies [36–38]. First of all, Renyi entropies only quantify entanglement for pure states, and cannot be used in the more general mixed state scenario. Secondly, while for entropies the operations are only performed on a single subsystem, here, one performs both 'forward' and 'backward' operations on two subsystems at once, as explained above.

Estimating the logarithmic negativity from the moments of μ_m: To estimate the logarithmic negativity, a precise knowledge of all λ_k is not required. Since $-\frac{1}{2} \leq \lambda_k \leq 1$ for all k [39] and $\sum_k \lambda_k = 1$, generically, the magnitude of the moments quickly decreases with m, with the first few carrying the most information. This is crucial to help computing the logarithmic negativity with measuring only very few moments. One approach using only the even moments has been proposed in the quantum field theory literature [15] by exploiting numerical extrapolation. However, this method neglects the odd moments and generally requires a large number of moments and thus copies. In [35], it has been shown that the moments required, $\{\mu_m : m \leq M\}$, to accurately estimate the entanglement can number as few as $M = 3$. This is achieved by avoiding reconstruction of the spectrum or state and instead employing machine learning to directly map moments to logarithmic negativity. Note that μ_0 is simply the dimension of the systems Hilbert space, while $\mu_1 = 1$ in all cases. Additionally, it can be easily shown that μ_2 is equal to the purity of the state $= Tr[\rho_{AB}^2]$, and as such, $M \geq 3$ is needed to extract any information about $\mathcal{E}$. Therefore, using $M = 3$ copies is optimal in terms of resources.

Machine Learning Entanglement: Machine learning is as a key tool for modeling an unknown non-linear map between sets of data. In the supervised machine learning setup, one trains a model with a set of known inputs and their corresponding outputs. Once trained, the model can then be used to predict the unknown output of new input data. We follow the machine learning algorithm of the [35] for estimating the logarithmic negativity from the information contained in the moments, μ_m. The moments μ_m are taken as the input and the logarithmic negativity $\mathcal{E}$ as the output for training a deep neural networks [40, 41]. Training is performed by taking a large set of states for which μ_m and $\mathcal{E}$ can be computed on a classical computer. This model can then be used to predict $\mathcal{E}$ from a set of experimentally measured moments.

Training the neural network with random states: From an entanglement perspective, relevant states in condensed matter physics can be classified as either area-law, or volume-law. In the first case, the entanglement of a subsystem A with the rest is proportional to the number of qubits along their boundary. In the second, this entanglement is instead proportional to N_A, the number of qubits in A. Area-law states arise as low energy eigenstates of local gapped Hamiltonians, with logarithmic corrections in critical systems. Volume-law states however, are associated with the eigenstates found in the mid-spectrum, and as such arise in non-equilibrium dynamics. In order to train a neural network, a set of suitable training states, including both volume and area law quantum states, are required for which both the moments and logarithmic negativities are known. To encompass both area- and volume-law states, we consider two classes of states $|\Psi_{ABS}\rangle$: (i) random generic pure states (R-GPS), e.g. sampled from the Haar measure, which typically have volume-law entanglement [42, 43]; (ii) random matrix product states (R-MPS) with fixed bond dimension, which satisfy an area-law by construction [44]. In order to generate a training set with a wide range of entanglement features, subsystem sizes, and mixedness, we perform the following procedure:

1. For a fixed number of qubits N, take either a R-GPS, or R-MPS with bond dimension D.
2. Take different tri-partitions such that $N = N_A + N_B + N_C$, and for each calculate μ_m and $\mathscr{E}$ for ρ_{AB}.
3. Repeat for different random instances, while separately varying N and D.

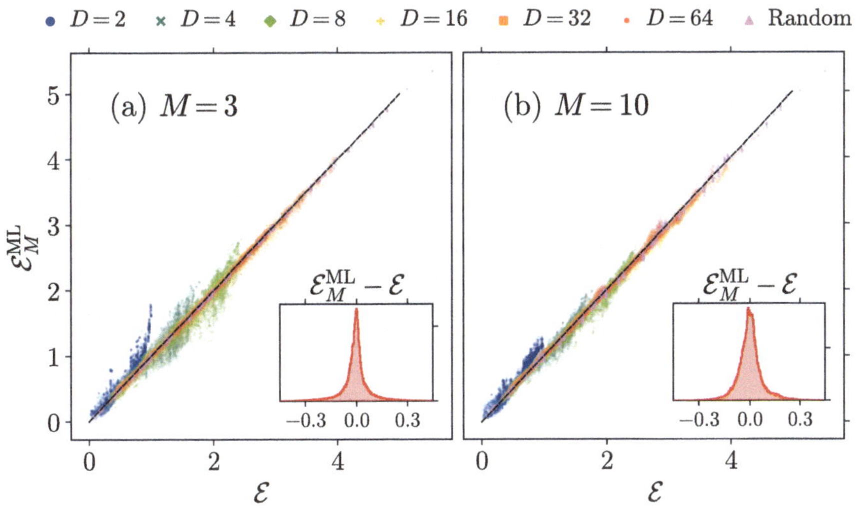

Fig. 3.7 **Machine learning entanglement.** Estimated logarithmic negativity $\mathscr{E}_M^{\mathrm{ML}}$, using a machine learning versus actual logarithmic negativity $\mathscr{E}$, for the same set of random states described in the main text. Training and prediction is performed using the moments μ_m generated from: **a** $M = 3$ copies; **b** $M = 10$ copies. The respective insets show the distribution of error, $\mathscr{E}_M^{\mathrm{ML}} - \mathscr{E}$. The figure is taken from the [35]

To check the performance of the neural network, we take the set of random states and split it in two, one half for training the neural network model, and the other as 'unseen' test data. In Fig. 3.7a we plot the machine learning model's predictions, $\mathscr{E}_M^{\mathrm{ML}}$, for the test data, using only $M = 3$ copies, in which a high degree of accuracy is achieved. In the inset of Fig. 3.7a, we plot a histogram of the errors $\mathscr{E}_M^{\mathrm{ML}} - \mathscr{E}$, which displays a very sharp peak at zero error with standard deviation ~ 0.09 A further improvement, particularly in outliers, is achieved by increasing the number of copies M to 10, see Fig. 3.7b, where the error standard deviation decreases to ~ 0.07. Regardless, the machine learning method is already very accurate for extracting entanglement using only three copies.

3.6 Conclusions

We have reviewed here the study of entanglement in many-body systems from an angle which is slightly different from the bulk of the literature. Most of the literature has concentrated on the entanglement between two complementary blocks in a many-body system or a quantum field theory. On the other hand, if we were to examine entanglement between two non-complementary parts of a system, this strategy will not work—in those cases concurrence and, for arbitrary dimensional systems, negativity, has to be employed. We have exemplified that study with examples: thermal entanglement and evidencing the Kondo cloud as an "entanglement cloud" around the impurity spin. We have also presented a strategy of how negativity can actually be measured in a many-body system using replicas in the laboratory, and how machine-learning can be used to make this more efficient. On a different vein, we have also exemplified how the Schmidt gap, which is not an entanglement measure, but a somewhat broad-brush indicator of entanglement, can be used to identify impurity quantum phase transitions.

Acknowledgements We acknowledge discussions carried out at the Workshop on Strongly Coupled Field Theories for Condensed Matter and Quantum Information Theory, IIP, Natal, Brazil. Much of the original work presented was funded by: EPSRC grant EP/K004077/1; Swedish Research Council grant no. 621-2011-3942 and STINT grant IG 2011-2028; the ERC grant PACOMANEDIA; Ministry of Science, Technology and Innovation of Brazil; CNPq Bolsa de Produtividade em Pesquisa.

References

1. R. Horodecki, P. Horodecki, M. Horodecki, K. Horodecki, Rev. Mod. Phys. **81**, 865 (2009)
2. S. Popescu, D. Rohrlich, Phys. Rev. A **56**, R3319(R) (1997)
3. P. Calabrese, J. Cardy, B. Doyon, J. Phys. A **42**, 500301 (2009)
4. W.K. Wootters, Phys. Rev. Lett. **80**, 2245 (1998)
5. K. Zyczkowski, P. Horodecki, A. Sanpera, M. Lewenstein, Phys. Rev. A **58**, 883 (1998)
6. J. Lee, M. Kim, Y. Park, S. Lee, J. Mod. Opt. **47**, 2151 (2000)

7. M.B. Plenio, Phys. Rev. Lett. **95**, 090503 (2005)
8. G. Vidal, R.F. Werner, Phys. Rev. A **65**, 032314 (2002)
9. M.C. Arnesen, S. Bose, V. Vedral, Phys. Rev. Lett. **87**, 017901 (2001)
10. S. Sachdev (Cambridge University Press, Cambridge, 2001)
11. A. Osterloh, L. Amico, G. Falci, R. Fazio, Nature **416**, 608 (2002)
12. T.J. Osborne, M.A. Nielsen, Phys. Rev. A **66**, 032110 (2002)
13. H. Wichterich, J. Molina-Vilaplana, S. Bose, Phys. Rev. A **80**, 010304(R) (2009)
14. S. Marcovitch, A. Retzker, M.B. Plenio, B. Reznik, Phys. Rev. A **80**, 012325 (2009)
15. P. Calabrese, J. Cardy, E. Tonni, Phys. Rev. Lett. **109**, 130502 (2012)
16. A. Bayat, Phys. Rev. Lett. **118**, 036102 (2017)
17. A. Bayat, P. Sodano, S. Bose, Phys. Rev. B **81**, 064429 (2010)
18. I. Affleck, Lecture Notes. Les Houches (2008), arXiv:0809.3474
19. E.S. Sorensen, M.S. Chang, N. Laflorencie, I. Affleck, J. Stat. Mech., P08003 (2007); E. S. Sorensen, M. S. Chang, N. Laflorencie, I. Affleck, J. Stat. Mech. L01001 (2007)
20. N. Laflorencie, E.S. Srensen, M.S. Chang, I. Affleck, Phys. Rev. Lett. **96**, 100603 (2006)
21. H. Frahm, A.A. Zvyagin, J. Cond. Matt. **9**, 9939 (1997)
22. R.G. Pereira, N. Laflorencie, I. Affleck, B.I. Halperin, Phys. Rev. B **77**, 125327 (2008)
23. N. Laflorencie, E.S. Sorensen, I. Affleck, J. Stat. Mech. **P0**, 2008 (2007)
24. M. Vojta, Phil. Mag. **86**, 1807 (2006)
25. C. Jayaprakash, H.-R. Krishnamurthy, J. Wilkins, Two-impurity Kondo problem. Phys. Rev. Lett. **47**, 737–740 (1981)
26. G. Zaránd, C.-H. Chung, P. Simon, M. Vojta, Phys. Rev. Lett. **97**, 166802 (2006)
27. A. Spinelli, M. Gerrits, R. Toskovic, B. Bryant, M. Ternes, A.F. Otte, Nat. Commun. **6**, 10046 (2015)
28. A. Bayat, H. Johannesson, S. Bose, P. Sodano, Nat. Commun. **5**, 3784 (2014)
29. K. Okamoto, K. Nomura, Phys. Lett. A **169**, 433 (1992)
30. B.A. Jones, C.M. Varma, Phys. Rev. B **40**, 324 (1989)
31. E. Sela, A.K. Mitchell, L. Fritz, Phys. Rev. Lett. **106**, 147202 (2011)
32. I. Affleck, A.W.W. Ludwig, B.A. Jones, Phys. Rev. B **52**, 9528 (1995)
33. H.A. Carteret, Phys. Rev. Lett. **94**, 040502 (2005)
34. K. Bartkiewicz, P. Horodecki, K. Lemr, A. Miranowicz, K. Zyczkowski, Phys. Rev. A **91**, 032315 (2015)
35. J. Gray, L. Banchi, A. Bayat, S. Bose, Phys. Rev. Lett. **121**, 150503 (2018)
36. A. Daley, H. Pichler, J. Schachenmayer, P. Zoller, Phys. Rev. Lett. **109**, 020505 (2012)
37. L. Banchi, A. Bayat, S. Bose, Phys. Rev. B **94**, 241117 (2016)
38. D.A. Abanin, E. Demler, Phys. Rev. Lett. **109**, 020504 (2012)
39. S. Rana, Phys. Rev. A **87**, 054301 (2013)
40. R. Rojas, *Neural Networks: A Systematic Introduction* (Springer Science and Business Media, Berlin, 2013)
41. J. Schmidhuber, Neural Netw. **61**, 85 (2015)
42. S. Popescu, A.J. Short, A. Winter, Nat. Phys. **2**, 754 (2006)
43. A. Hamma, S. Santra, P. Zanardi, Phys. Rev. Lett. **109**, 040502 (2012)
44. U. Schollwöck, Ann. Phys. **326**, 96 (2011)

Chapter 4
Generalized Entanglement Entropy in New Spin Chains

Fumihiko Sugino and Vladimir Korepin

Abstract Motzkin and Fredkin spin chains are new spin chain models exhibiting extraordinary amount of entanglement entropy scaling as a square root of the volume in spite of local interactions. This is a distinguished feature of the models from other spin chains with entanglement of at most logarithmic scaling. We first compute generalized entanglement entropy, called as the Rényi entropy, of these models. The Rényi entropy has further importance, since it can provide the whole spectrum of an entangled subsystem. We find non-analytic behavior of the Rényi entropy that can be regarded as a phase transition never seen in any other spin chain investigated so far. This new phase transition indicates unique and rich structures of the entanglement spectrum.

4.1 Introduction

Entanglement is one of the most characteristic features of quantum mechanics, which provides correlations to objects unable to be explained in classical mechanics. In case that a given full system S is divided into two subsystems A and B, the reduced density matrix of A is defined by tracing out the degrees of freedom of B in the density matrix of the full system ρ:

$$\rho_A = \mathrm{Tr}_B\, \rho, \tag{4.1}$$

where Tr_B means the trace over the Hilbert space belonging to B. Even if ρ is a pure state, i.e., can be expressed as the form $\rho = |\psi\rangle\langle\psi|$ for some state $|\psi\rangle$, ρ_A is

F. Sugino (✉)
Center for Theoretical Physics of the Universe, Institute for Basic Science (IBS), Expo-ro 55, Yuseong-gu, Daejeon 34126, Republic of Korea
e-mail: fusugino@gmail.com

V. Korepin
C.N.Yang Institute for Theoretical Physics, Stony Brook University, Stony Brook, NY 11794, USA
e-mail: korepin@gmail.com

A. Ferraz et al. (eds.), *Strongly Coupled Field Theories for Condensed Matter and Quantum Information Theory*, Springer Proceedings in Physics 239,
https://doi.org/10.1007/978-3-030-35473-2_4

no longer so in general and takes a form like $\rho_A = c_1 |\psi_1\rangle\langle\psi_1| + c_2 |\psi_2\rangle\langle\psi_2| + \cdots$ (c_i's are positive numbers summed to 1) that is called a mixed state. Entanglement is normally measured by the von Neumann entanglement entropy (EE):

$$S_A = -\mathrm{Tr}\,(\rho_A \ln \rho_A), \tag{4.2}$$

which vanishes for the pure states but not for the mixed states. ρ_A carries information of interactions between A and B, some of which can be read off through (4.2). We can say that difference of the behavior of (4.2) reflects difference of dynamical property of the system.

Rényi entropy [16] is a generalization of the EE and defined by

$$S_{A,\alpha} = \frac{1}{1-\alpha} \ln \mathrm{Tr}\,\rho_A^\alpha, \tag{4.3}$$

where α is a positive number not equal to 1. It is easy to see that (4.3) reduces to (4.2) in the limit $\alpha \to 1$. The Rényi entropy has further importance than the EE, because the whole spectrum (entanglement spectrum) of ρ_A or equivalently of the entanglement Hamiltonian

$$H_{\mathrm{ent},A} = -\ln \rho_A \tag{4.4}$$

can be obtained once the Rényi entropy is known as a function of α. In terms of (4.4), the Rényi entropy takes a form analogous to the "thermal free energy" with the "temperature" $1/\alpha$:

$$S_{A,\alpha} = \frac{1}{1-\alpha} \ln \mathrm{Tr}\, e^{-\alpha H_{\mathrm{ent},A}}. \tag{4.5}$$

In this article, we compute the EE and Rényi entropy for highly entangled spin chains to be introduced in the next section, and find a new phase transition with respect to the parameter α. Figure 4.1 shows a phase diagram for the EE and Rényi entropy of the models with respect to α.

4.2 Motzkin and Fredkin Spin Chains

Let us consider ground states of quantum many-body systems with local interactions, Normally, their EEs are proportional to the area of the boundaries of A and B, which is called as an "area law" [9]. In gapped systems, it can be naturally understood from that the correlation length is finite and relevant interactions to the EE are localized around the boundaries. There is a mathematical proof only for the case of gapped one-dimensional systems [10]. In gapless systems described by the $(1+1)$-dimensional conformal field theories (CFTs), we can see violation of the area law [6, 11, 12]:

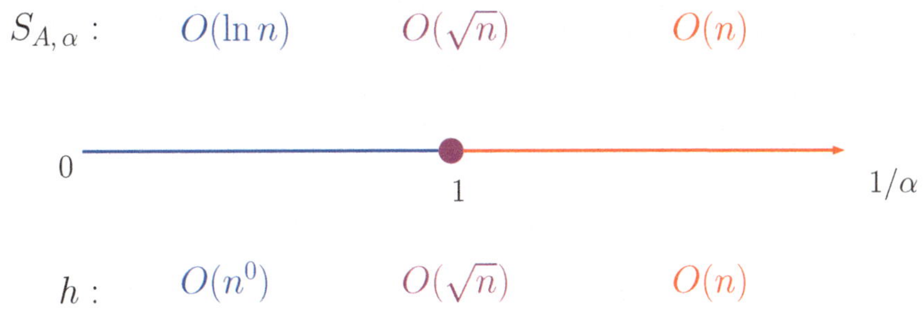

Fig. 4.1 Phase diagram for the EE and Rényi entropy of the models with respect to α. $S_{A,\alpha} = O(n)$ for $0 < \alpha < 1$ ("high temperature"), whereas $S_{A,\alpha} = O(\ln n)$ for $\alpha > 1$ ("low temperature"). At the transition point $\alpha = 1$, the EE behaves as $O(\sqrt{n})$. At the bottom, height of paths dominantly contributing to the EE and Rényi entropy is shown

$$S_A \sim \frac{c + \bar{c}}{6} \ln L, \tag{4.6}$$

where c and $\bar{c}$ are central charges of the holomorphic and anti-holomorphic sectors of the CFT, and L denotes the length scale of the subsystem A. Equation (4.6) is the asymptotic form for L large. In one spatial dimension the boundary is a (zero-dimensional) point, which implies that the area law means S_A asymptotically approaching to a finite constant as L grows. In addition, it was pointed out in spacetime physics that black holes have entropy proportional to the area of their horizons from an analogy to thermodynamics [1, 2]. A possibility has been discussed that the entropy of black holes can be interpreted as an EE satisfying the area law [3, 7, 19].

In gapless systems in D spatial dimensions, the EE had been believed to violate the area law as $S_A = O(L^{D-1} \ln L)$ over decades. Namely, the violation would be at most logarithmic. Recent discovery of a highly entangled spin chain by Movassagh and Shor [14], called as Motzkin spin chain, is striking in this sense. In spite that the Hamiltonian of the model consists of local (nearest neighbor) interactions, the EE exhibits the square-root violation of the area law: $S_A = O(\sqrt{L})$, which is much larger than logarithmic. A different spin chain whose EE behaves similarly, called as the Fredkin spin chain, has been constructed by Salberger and one of the authors (V.K.) [17] (see also [8]). In the remaining of this section, we introduce these two spin chains.

4.2.1 Fredkin Spin Chain

The Fredkin spin chain [17] of length $2n$ has up and down quantum spin degrees of freedom with multiplicity (called as color) s at each of the lattice sites $\{1, 2, \ldots, 2n\}$. We express the up- (down-)spin state with color k at the site i as $|u_i^k\rangle$ ($|d_i^k\rangle$) with

$k = 1, \ldots, s$. The Hamiltonian is defined as the sum of projection operators and has local interactions ranging up to next-to-nearest neighbors:

$$H_{\mathrm{F},s} = \sum_{j=1}^{2n-2} \sum_{k_1,k_2,k_3=1}^{s} \left\{ \left| U^{k_1,k_2,k_3}_{j,j+1,j+2} \right\rangle \left\langle U^{k_1,k_2,k_3}_{j,j+1,j+2} \right| + \left| D^{k_1,k_2,k_3}_{j,j+1,j+2} \right\rangle \left\langle D^{k_1,k_2,k_3}_{j,j+1,j+2} \right| \right\}$$
$$+ \sum_{j=1}^{2n-1} \sum_{k \neq \ell} \left\{ \left| u^k_j, d^\ell_{j+1} \right\rangle \left\langle u^k_j, d^\ell_{j+1} \right| + \frac{1}{2} \left(\left| u^k_j, d^k_{j+1} \right\rangle - \left| u^\ell_j, d^\ell_{j+1} \right\rangle \right) \left(\left\langle u^k_j, d^k_{j+1} \right| - \left\langle u^\ell_j, d^\ell_{j+1} \right| \right) \right\}$$
$$+ \sum_{k=1}^{s} \left\{ \left| d^k_1 \right\rangle \left\langle d^k_1 \right| + \left| u^k_{2n} \right\rangle \left\langle u^k_{2n} \right| \right\}, \tag{4.7}$$

where

$$\left| U^{k_1,k_2,k_3}_{j,j+1,j+2} \right\rangle = \frac{1}{\sqrt{2}} \left(\left| u^{k_1}_j, u^{k_2}_{j+1}, d^{k_3}_{j+2} \right\rangle - \left| u^{k_1}_j, d^{k_2}_{j+1}, u^{k_3}_{j+2} \right\rangle \right), \tag{4.8}$$

$$\left| D^{k_1,k_2,k_3}_{j,j+1,j+2} \right\rangle = \frac{1}{\sqrt{2}} \left(\left| u^{k_1}_j, d^{k_2}_{j+1}, d^{k_3}_{j+2} \right\rangle - \left| d^{k_1}_j, u^{k_2}_{j+1}, d^{k_3}_{j+2} \right\rangle \right). \tag{4.9}$$

4.2.1.1 Ground State and Dyck Walks

For colorless case ($s = 1$), the up- and down-spin states can be represented as arrows in the (x, y)-plane pointing to $(1, 1)$ (up-step) and $(1, -1)$ (down-step), respectively. Each spin configuration of the chain corresponds to a length-$2n$ walk consisting of the up- and down-steps. The Hamiltonian has a unique ground state of zero energy, which is superposition of states with equal weight. Each of the states is identified with each path of length-$2n$ Dyck walks that is random walks starting at the origin, ending at $(2n, 0)$, and not allowing paths to enter $y < 0$ region.

For s-color ($s > 1$) case, the above identification goes on with additional color degrees of freedom. Namely, each spin configuration of the chain corresponds to a length-$2n$ walk consisting of the up- and down-steps with color. The ground state is unique, and corresponds to length-$2n$ colored Dyck walks in which the color of each up-step should be matched with that of the subsequent down-step at the same height. The other is the same as the colorless case.

The ground state is given by

$$\left| P_{\mathrm{F},\,2n,\,s} \right\rangle = \frac{1}{\sqrt{N_{\mathrm{F},\,2n,\,s}}} \sum_{w \in P_{\mathrm{F},\,2n,\,s}} |w\rangle\,, \tag{4.10}$$

where $P_{\mathrm{F},\,2n,\,s}$ denotes the formal sum of length-$2n$ colored Dyck walks, w runs over monomials appearing in $P_{\mathrm{F},\,2n,\,s}$, and $N_{\mathrm{F},\,2n,\,s}$ stands for the number of the length-$2n$ colored Dyck walks:

$$N_{\mathrm{F},\,2n,\,s} = s^n\, N_{\mathrm{F},\,2n} = \frac{s^n}{n+1} \binom{2n}{n}. \tag{4.11}$$

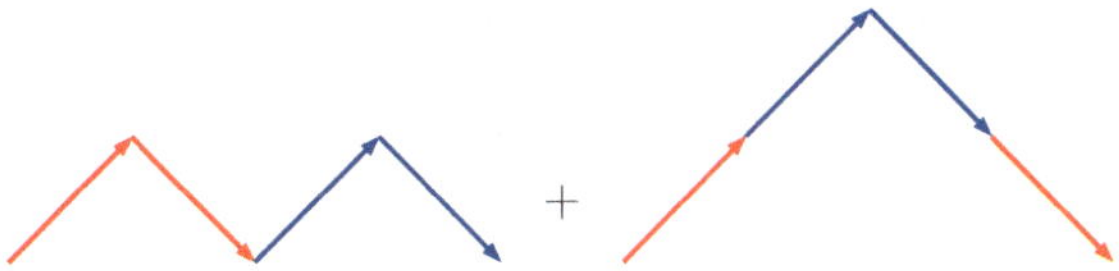

Fig. 4.2 Colored Dyck walks corresponding to the two states of the summand in (4.12). Colors k and ℓ are represented by red and blue

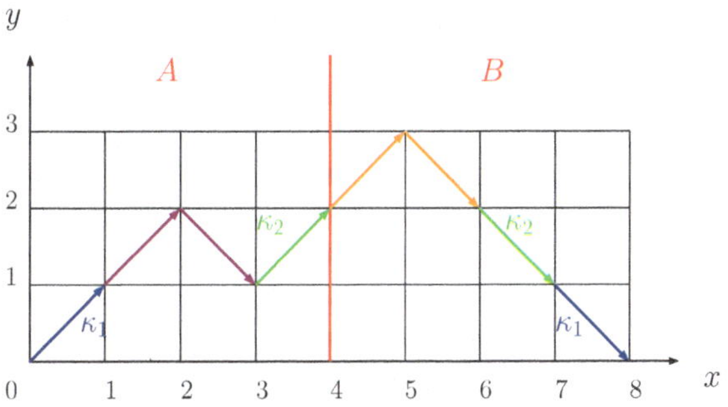

Fig. 4.3 A path in case of $2n = 8$, $r = 0$ and $h = 2$. Colors of violet and orange are matched in A itself and B itself, respectively. On the other hand, colors of blue (κ_1) and green (κ_2) are unmatched in A or B alone and matched across the boundary of A and B

$N_{\mathrm{F},\,2n}$ denotes the number of colorless Dyck walks of length $2n$, which is equal to the n-th Catalan number C_n. Note that $N_{\mathrm{F},\,2n,\,s}$ can be obtained by setting all the u^k and d^k to 1 in $P_{\mathrm{F},\,2n,\,s}$. For example, $2n = 4$ case reads

$$\left|P_{\mathrm{F},\,4,\,s}\right\rangle = \frac{1}{\sqrt{2s^2}} \sum_{k,\ell=1}^{s} \left\{ \left|u_1^k, d_2^k, u_3^\ell, d_4^\ell\right\rangle + \left|u_1^k, u_2^\ell, d_3^\ell, d_4^k\right\rangle \right\}. \tag{4.12}$$

The two states of the summand are drawn as colored Dyck walks in Fig. 4.2.

We take the first $(n + r)$ sites as the subsystem A (and the remaining $(n - r)$ sites as B), and compute the EE and Rényi entropy in case of $n \pm r = O(n)$ (r takes (positive or negative) integers). This gives a generalization of our previous paper of the half-chain ($r = 0$) case [20]. Let $P^{(0\to h)}_{\mathrm{F},\,n+r,\,s}$ be a part of colored Dyck paths belonging to A that is from the origin to $(n + r,\ h)$ in the $(x,\,y)$-plane. The height h takes non-negative integers. Similarly, the remaining part belonging to B is denoted by $P^{(h\to 0)}_{\mathrm{F},\,n-r,\,s}$. Each path in $P^{(0\to h)}_{\mathrm{F},\,n+r,\,s}$ has h unmatched up-steps that are supposed to be matched across the boundary with h unmatched down-steps in $P^{(h\to 0)}_{\mathrm{F},\,n-r,\,s}$. We write as $\tilde{P}^{(0\to h)}_{\mathrm{F},\,n+r,\,s}(\{\kappa_m\})$ ($\tilde{P}^{(h\to 0)}_{\mathrm{F},\,n-r,\,s}(\{\kappa_m\})$) the paths with the color degrees of freedom of the unmatched up- (down-) steps fixed to $\kappa_1, \ldots, \kappa_h$, where κ_m denotes the color of unmatched up- or down step connecting the heights $m - 1$ and m. An example of a path in case of $2n = 8$, $r = 0$ and $h = 2$ is depicted in Fig. 4.3.

The numbers of the paths are given by

$$N^{(0\to h)}_{\mathrm{F},\,n+r,\,s} = P^{(0\to h)}_{\mathrm{F},\,n+r,\,s}\Big|_{u^k=d^k=1} = s^{\frac{n+r+h}{2}} N^{(h)}_{\mathrm{F},\,n}, \qquad \tilde{N}^{(0\to h)}_{\mathrm{F},\,n+r,\,s} = s^{-h} N^{(0\to h)}_{\mathrm{F},\,n+r,\,s} = s^{\frac{n+r-h}{2}} N^{(h)}_{\mathrm{F},\,n+r} \tag{4.13}$$

with

$$N^{(h)}_{\mathrm{F},\,n+r} = \frac{1+(-1)^{n+r+h}}{2}\,\frac{h+1}{\frac{n+r+h}{2}+1}\binom{n+r}{\frac{n+r+h}{2}}. \tag{4.14}$$

By considering the reverse of paths, it is easy to see $N^{(h\to 0)}_{\mathrm{F},\,n-r,\,s} = N^{(0\to h)}_{\mathrm{F},\,n-r,\,s}$ and $\tilde{N}^{(h\to 0)}_{\mathrm{F},\,n-r,\,s} = \tilde{N}^{(0\to h)}_{\mathrm{F},\,n-r,\,s}$. The ground state is decomposed as a linear combination of tensor products of two states belonging to A and B (Schmidt decomposition):

$$\left|P_{\mathrm{F},\,2n,\,s}\right\rangle = \sum_{h=0}^{n-|r|}\sum_{\kappa_1=1}^{s}\cdots\sum_{\kappa_h=1}^{s}\sqrt{p^{(h)}_{\mathrm{F},\,n+r,n-r,\,s}}\,\left|\tilde{P}^{(0\to h)}_{\mathrm{F},\,n+r,\,s}(\{\kappa_m\})\right\rangle\otimes\left|\tilde{P}^{(h\to 0)}_{\mathrm{F},\,n-r,\,s}(\{\kappa_m\})\right\rangle. \tag{4.15}$$

Here,

$$\left|\tilde{P}^{(0\to h)}_{\mathrm{F},\,n+r,\,s}(\{\kappa_m\})\right\rangle = \frac{1}{\sqrt{\tilde{N}^{(0\to h)}_{\mathrm{F},\,n+r,\,s}}}\sum_{w\in\tilde{P}^{(0\to h)}_{\mathrm{F},\,n+r,\,s}(\{\kappa_m\})}|w\rangle\,, \tag{4.16}$$

$$\left|\tilde{P}^{(h\to 0)}_{\mathrm{F},\,n-r,\,s}(\{\kappa_m\})\right\rangle = \frac{1}{\sqrt{\tilde{N}^{(0\to h)}_{\mathrm{F},\,n-r,\,s}}}\sum_{w\in\tilde{P}^{(h\to 0)}_{\mathrm{F},\,n-r,\,s}(\{\kappa_m\})}|w\rangle\,, \tag{4.17}$$

and

$$p^{(h)}_{\mathrm{F},\,n+r,n-r,\,s} = \frac{\tilde{N}^{(0\to h)}_{\mathrm{F},\,n+r,\,s}\tilde{N}^{(0\to h)}_{\mathrm{F},\,n-r,\,s}}{N_{\mathrm{F},\,2n,\,s}} = s^{-h}\,\frac{N^{(h)}_{\mathrm{F},\,n+r}N^{(h)}_{\mathrm{F},\,n-r}}{N_{\mathrm{F},\,2n}}. \tag{4.18}$$

From the density matrix of the ground state $\rho = \left|P_{\mathrm{F},\,2n,\,s}\right\rangle\left\langle P_{\mathrm{F},\,2n,\,s}\right|$ with (4.15), the reduced density matrix is obtained as

$$\rho_A = \mathrm{Tr}_B\,\rho = \sum_{h=0}^{n-|r|}\sum_{\kappa_1=1}^{s}\cdots\sum_{\kappa_h=1}^{s}p^{(h)}_{\mathrm{F},\,n+r,n-r,\,s}\left|\tilde{P}^{(0\to h)}_{\mathrm{F},\,n+r,\,s}(\{\kappa_m\})\right\rangle\left\langle\tilde{P}^{(0\to h)}_{\mathrm{F},\,n+r,\,s}(\{\kappa_m\})\right|, \tag{4.19}$$

where we used the orthonormal property:

$$\left\langle\tilde{P}^{(h\to 0)}_{\mathrm{F},\,n-r,\,s}(\{\kappa_m\})\,\middle|\,\tilde{P}^{(h'\to 0)}_{\mathrm{F},\,n-r,\,s}(\{\kappa'_m\})\right\rangle = \delta_{h,h'}\delta_{\kappa_1,\kappa'_1}\cdots\delta_{\kappa_h,\kappa'_h}. \tag{4.20}$$

Since ρ_A is a diagonal form, the EE (4.2) and the Rényi entropy (4.3) are recast as

$$S_{\mathrm{F},\,A} = -\sum_{h=0}^{n-|r|}s^h\,p^{(h)}_{\mathrm{F},\,n+r,n-r,\,s}\ln p^{(h)}_{\mathrm{F},\,n+r,n-r,\,s}, \tag{4.21}$$

$$S_{\mathrm{F},\,A,\alpha} = \frac{1}{1-\alpha}\ln\sum_{h=0}^{n-|r|}s^h\left(p^{(h)}_{\mathrm{F},\,n+r,n-r,\,s}\right)^{\alpha}. \tag{4.22}$$

Note $p^{(h)}_{\mathrm{F},\,n+r,n-r,\,s}$ does not depend on $\kappa_1,\ldots,\kappa_h$ and the sums $\sum_{\kappa_1=1}^{s}\cdots\sum_{\kappa_h=1}^{s}$ yield the factor s^h.

4.2.1.2 Asymptotic Form of $p^{(h)}_{\mathrm{F},\,n+r,n-r,\,s}$

We plug (4.11) and (4.14)–(4.18) and evaluate its asymptotic behavior. For $n,\ n\pm h\gg 1$, use of Stirling's formula $x!\sim\sqrt{2\pi}\,x^{x+\frac{1}{2}}\,e^{-x}\left[1+O(x^{-1})\right]$ $(x\to\infty)$ leads to

$$\begin{aligned}
p^{(h)}_{\mathrm{F},\,n+r,n-r,\,s} \sim\ & \frac{1+(-1)^{n+r+h}}{2}\,\frac{8}{\sqrt{\pi}}\,s^{-h}\left(\frac{n}{(n+r)(n-r)}\right)^{3/2}(h+1)^2\\
&\times\exp\left[-\frac{n+r+h+3}{2}\ln\frac{n+r+h}{n+r}-\frac{n+r-h+1}{2}\ln\frac{n+r-h}{n+r}\right.\\
&\qquad\left.-\frac{n-r+h+3}{2}\ln\frac{n-r+h}{n-r}-\frac{n-r-h+1}{2}\ln\frac{n-r-h}{n-r}\right]\\
&\times\left[1+O(n^{-1})\right].
\end{aligned}\tag{4.23}$$

In case of $h\le O(n^{1/2})$, we can further expand the logarithms in the exponential in (4.23) to find

$$\begin{aligned}
p^{(h)}_{\mathrm{F},\,n+r,n-r,\,s} \sim\ & s^{-h}\,\frac{1+(-1)^{n+r+h}}{2}\,\frac{8}{\sqrt{\pi}}\left(\frac{n}{(n+r)(n-r)}\right)^{3/2}(h+1)^2\,e^{-\frac{n\,(h+1)^2}{(n+r)(n-r)}}\\
&\times\left[1+O(n^{-1})\right]
\end{aligned}\tag{4.24}$$

4.2.2 *Motzkin Spin Chain*

The Motzkin spin chain [14] has additional spin degrees of freedom (we call zero-spin) at each site compared with the Fredkin spin chain. We express the up- and down-spin states with color $k=1,\ldots,s$ and the zero-spin at the site i as $\left|u_i^k\right\rangle$, $\left|d_i^k\right\rangle$ and $|0_i\rangle$, respectively. The Hamiltonian of the Motzkin spin chain of length $2n$ is also defined by the sum of projection operators:

$$\begin{aligned}
H_{\mathrm{M},s} =\ & \sum_{j=1}^{2n-1}\sum_{k=1}^{s}\left\{\left|U^k_{j,j+1}\right\rangle\left\langle U^k_{j,j+1}\right| + \left|D^k_{j,j+1}\right\rangle\left\langle D^k_{j,j+1}\right| + \left|F^k_{j,j+1}\right\rangle\left\langle F^k_{j,j+1}\right|\right\}\\
&+\sum_{j=1}^{2n-1}\sum_{k\neq\ell}\left|u^k_j,d^\ell_{j+1}\right\rangle\left\langle u^k_j,d^\ell_{j+1}\right| + \sum_{k=1}^{s}\left\{\left|d^k_1\right\rangle\left\langle d^k_1\right| + \left|u^k_{2n}\right\rangle\left\langle u^k_{2n}\right|\right\},
\end{aligned}\tag{4.25}$$

where

$$\left|U^k_{j,j+1}\right\rangle = \frac{1}{\sqrt{2}}\left(\left|0_j, u^k_{j+1}\right\rangle - \left|u^k_j, 0_{j+1}\right\rangle\right), \tag{4.26}$$

$$\left|D^k_{j,j+1}\right\rangle = \frac{1}{\sqrt{2}}\left(\left|0_j, d^k_{j+1}\right\rangle - \left|d^k_j, 0_{j+1}\right\rangle\right), \tag{4.27}$$

$$\left|F^k_{j,j+1}\right\rangle = \frac{1}{\sqrt{2}}\left(\left|0_j, 0_{j+1}\right\rangle - \left|u^k_j, d^k_{j+1}\right\rangle\right), \tag{4.28}$$

and the interactions are among nearest neighbors.

4.2.2.1 Ground State and Motzkin Walks

The Hamiltonian has the unique ground state at zero-energy. By the same identification of the spins and 2D steps as before with the additional zero-spin corresponding to the arrow $(1, 0)$ (flat-step), for colorless case ($s = 1$) the ground state is expressed by the equal-weight superposition of length-$2n$ Motzkin walks, which are random walks consisting of up-, down- and flat-steps, starting at the origin, ending at $(2n, 0)$ and forbidding paths to enter $y < 0$ region. For s-color case ($s > 1$), the color assigned to each up-step should be matched with that of the subsequent down-step at the same height, which is the same as in the Fredkin spin chain.

The ground state is expressed as

$$\left|P_{\mathrm{M},\,2n,\,s}\right\rangle = \frac{1}{\sqrt{N_{\mathrm{M},\,2n,\,s}}} \sum_{w\in P_{\mathrm{M},\,2n,\,s}} |w\rangle\,, \tag{4.29}$$

where $N_{\mathrm{M},\,2n,\,s}$ in the normalization factor is the number of the length-$2n$ colored Motzkin walks given by

$$N_{\mathrm{M},\,2n,\,s} = \sum_{\rho=0}^{n} \binom{2n}{2\rho} s^{n-\rho}\, N_{\mathrm{F},\,2n-2\rho}, \tag{4.30}$$

where 2ρ stands for the number of the flat-steps. For example,

$$\begin{aligned}\left|P_{\mathrm{M},\,4,\,s}\right\rangle = \frac{1}{\sqrt{1+6s+2s^2}}\Bigg[&\left|0_1, 0_2, 0_3, 0_4\right\rangle \\ &+\sum_{k=1}^{s}\Big\{\left|u^k_1, d^k_2, 0_3, 0_4\right\rangle + \left|0_1, u^k_2, d^k_3, 0_4\right\rangle + \left|0_1, 0_2, u^k_3, d^k_4\right\rangle \\ &\quad + \left|u^k_1, 0_2, d^k_3, 0_4\right\rangle + \left|0_1, u^k_2, 0_3, d^k_4\right\rangle + \left|u^k_1, 0_2, 0_3, d^k_4\right\rangle\Big\} \\ &+\sum_{k,\ell=1}^{s}\Big\{\left|u^k_1, d^k_2, u^\ell_3, d^\ell_4\right\rangle + \left|u^k_1, u^\ell_2, d^\ell_3, d^k_4\right\rangle\Big\}\Bigg].\end{aligned} \tag{4.31}$$

We repeat similar computational steps to what was done in the Fredkin case, and obtain

$$S_{\mathrm{M},\,A} = -\sum_{h=0}^{n-|r|} s^h\, p^{(h)}_{\mathrm{M},\,n+r,n-r,\,s} \ln p^{(h)}_{\mathrm{M},\,n+r,n-r,\,s}, \tag{4.32}$$

$$S_{\mathrm{M},\,A,\alpha} = \frac{1}{1-\alpha} \ln \sum_{h=0}^{n-|r|} s^h \left(p^{(h)}_{\mathrm{M},\,n,n,\,s}\right)^{\alpha} \tag{4.33}$$

with

$$p^{(h)}_{\mathrm{M},\,n+r,n-r,\,s} = s^{-h} \frac{N^{(0\to h)}_{\mathrm{M},\,n+r,\,s} N^{(0\to h)}_{\mathrm{M},\,n-r,\,s}}{N_{\mathrm{M},\,2n,\,s}}, \tag{4.34}$$

$$N^{(0\to h)}_{\mathrm{M},\,n\pm r,\,s} = \sum_{\rho=0}^{n\pm r-h} \binom{n\pm r}{\rho} N^{(h)}_{\mathrm{F},\,n\pm r-\rho}\ s^{\frac{n\pm r-\rho+h}{2}}. \tag{4.35}$$

4.2.2.2 Asymptotic Form of $p^{(h)}_{\mathrm{M},\,n+r,n-r,\,s}$

We rewrite (4.35) as

$$N^{(0\to h)}_{\mathrm{M},\,m,\,s} = s^h (h+1) \sum_{\rho=0}^{m-h} \frac{1+(-1)^{m-\rho+h}}{2} C_{m,h,\rho}, \tag{4.36}$$

where

$$C_{m,h,\rho} \equiv \frac{m!\, s^{\frac{m-\rho-h}{2}}}{\rho! \left(\frac{m-\rho-h}{2}\right)! \left(\frac{m-\rho+h}{2}+1\right)!}, \tag{4.37}$$

and m takes $n+r$ or $n-r$. Use of Stirling's formula for $m,\ m-\rho\pm h \gg 1$ yields

$$\begin{aligned} C_{m,h,\rho} \sim{}& \frac{2^{m-\rho+1}}{\pi} \frac{m^{m+\frac{1}{2}} s^{\frac{m-\rho-h}{2}}}{\rho^{\rho+\frac{1}{2}} (m-\rho)^{m-\rho+2}} \\ &\times \exp\left[-\frac{m-\rho-h+1}{2} \ln \frac{m-\rho-h}{m-\rho} - \frac{m-\rho+h+3}{2} \ln \frac{m-\rho+h}{m-\rho}\right] \\ &\times \left[1+O(m^{-1})\right]. \end{aligned} \tag{4.38}$$

We evaluate the sum of ρ in (4.36) by the saddle point method. The saddle point equation

$$C_{m,h,\rho} = C_{m,h,\rho+2}, \tag{4.39}$$

reads

$$(4s-1)\xi^2+2\xi+\eta^2-1=0+O(m^{-1}) \tag{4.40}$$

with $\xi=\rho/m$ and $\eta=h/m$, which is solved by

$$\rho=\rho_0+O(m^0), \qquad \rho_0\equiv\frac{m}{4s-1}\left[-1+\sqrt{4s-(4s-1)\frac{h^2}{m^2}}\right]. \tag{4.41}$$

After expanding (4.38) around the saddle point as $\rho=\rho_0+x$ $(x\ll m)$, we have

$$\begin{aligned}C_{m,h,\rho}\sim\ &\frac{s^{-h/2}}{2\pi s}\frac{m^{m+\frac{1}{2}}}{\rho_0^{m+\frac{5}{2}}}\left(\frac{m-\rho_0-h}{m-\rho_0+h}\right)^{\frac{h+1}{2}}\\ &\times\exp\left[-\left\{\frac{1}{8s\rho_0^2}\sqrt{4sm^2-(4s-1)h^2}+O(m^{-2})\right\}x^2\right]\\ &\times\exp\left[\frac{1}{4s\rho_0^2}\left(2m-2(m+1)\rho_0-h\right)x+O\left(x^3/m^2,x^4/m^3,\ldots\right)\right].\end{aligned} \tag{4.42}$$

The sum in (4.36) can be carried out by converting it to an integral as

$$\sum_{\rho=0}^{m-h}\frac{1+(-1)^{m-\rho+h}}{2}(\cdots)\rightarrow\frac{1}{2}\int_{-\infty}^{\infty}dx(\cdots), \tag{4.43}$$

where we perform Gaussian integrations after expanding the exponential in the last line of (4.42). As a result, contribution from the last line amounts to a factor $\left[1+O(m^{-1})\right]$. We finally find

$$\begin{aligned}N^{(0\to h)}_{\mathrm{M},\,m,\,s}\sim\ &\frac{s^{h/2}}{\sqrt{2\pi s}}\frac{m^{m+\frac{1}{2}}}{\rho_0^{m+\frac{3}{2}}}\frac{h+1}{\left[4sm^2-(4s-1)h^2\right]^{1/4}}\left(\frac{m-\rho_0-h}{m-\rho_0+h}\right)^{\frac{h+1}{2}}\\ &\times\left[1+O(m^{-1})\right].\end{aligned} \tag{4.44}$$

$N_{\mathrm{M},\,2n,\,s}$ is obtained by setting $m=2n$ and $h=0$ in (4.44), and then $p^{(h)}_{\mathrm{M},\,n+r,n-r,\,s}$ through the relation (4.34). When h is at most $O(n^{1/2})$, the expression simplifies to

$$\begin{aligned}p^{(h)}_{\mathrm{M},\,n+r,n-r,\,s}\sim\ &s^{-h}\sqrt{\frac{2}{\pi\sigma^3}}\left(\frac{n}{(n+r)(n-r)}\right)^{3/2}(h+1)^2\,e^{-\frac{1}{2\sigma}\frac{n\,(h+1)^2}{(n+r)(n-r)}}\\ &\times\left[1+O(n^{-1})\right]\end{aligned} \tag{4.45}$$

with $\sigma\equiv\frac{\sqrt{s}}{2\sqrt{s}+1}$.

4.3 Rényi Entropy of Colorless Fredkin Spin Chain

First, let us compute asymptotic behavior of the Rényi entropy (4.22) as $n \to \infty$ for colorless case ($s = 1$). In the sum $\sum_{h=0}^{n} \left(p^{(h)}_{\mathrm{F},\, n+r, n-r,\, s=1}\right)^{\alpha}$ with (4.23), there is a saddle point $h_* = O(\sqrt{n})$. which justifies use of (4.24).

After converting the sum to an integral:

$$\sum_{h=0}^{n-|r|} \frac{1+(-1)^{n+r+h}}{2} (\cdots) \to \frac{1}{2}\sqrt{n} \int_0^{\infty} \mathrm{d}x \, (\cdots) \tag{4.46}$$

with $x = \frac{h}{\sqrt{n}}$, we obtain

$$\begin{aligned} \sum_{h=0}^{n} \left(p^{(h)}_{\mathrm{F},\, n+r, n-r,\, s=1}\right)^{\alpha} &\sim \left(\frac{8}{\sqrt{\pi}(1-r^2)^{3/2}}\right)^{\alpha} \frac{1}{2} n^{\frac{1-\alpha}{2}} \\ &\times \left(\int_0^{\infty} \mathrm{d}x \, x^{2\alpha} e^{-\frac{\alpha}{1-u^2}x^2} - \int_0^{n^{-1/2}} \mathrm{d}x \, x^{2\alpha} e^{-\frac{\alpha}{1-u^2}x^2}\right), \end{aligned} \tag{4.47}$$

where we set $u \equiv r/n$. The first integral in the parenthesis is elementary and the second one is evaluated as $O(n^{-\alpha-\frac{1}{2}})$. The result becomes

$$\begin{aligned} S_{\mathrm{F},\, A,\, \alpha} = &\frac{1}{2} \ln \frac{(n+r)(n-r)}{n} + \frac{1}{1-\alpha} \ln \Gamma\left(\alpha + \frac{1}{2}\right) \\ &- \frac{1}{2(1-\alpha)} \left\{(1+2\alpha) \ln \alpha + \alpha \ln \frac{\pi}{64} + \ln 16\right\} \\ &+ (\text{terms vanishing as } n \to \infty). \end{aligned} \tag{4.48}$$

This grows logarithmically as $n \to \infty$, and the $\alpha \to 1$ limit gives the EE:

$$S_{\mathrm{F},\, A} = \frac{1}{2} \ln \frac{(n+r)(n-r)}{n} + \frac{1}{2} \ln \frac{\pi}{4} + \gamma - \frac{1}{2} + (\text{terms vanishing as } n \to \infty) \tag{4.49}$$

with γ being the Euler constant. This is consistent with the result obtained in [17].

4.4 Rényi Entropy of Colorless Motzkin Spin Chain

After similar calculation to the previous case, the result reads

$$S_{\mathrm{M},\,A,\alpha} = \frac{1}{2}\ln\frac{(n+r)(n-r)}{n} + \frac{1}{1-\alpha}\ln\Gamma\left(\alpha+\frac{1}{2}\right)$$
$$-\frac{1}{2(1-\alpha)}\left\{(1+2\alpha)\ln\alpha + \alpha\ln\frac{\pi}{24} + \ln 6\right\}$$
$$+(\text{terms vanishing as } n\to\infty). \tag{4.50}$$

Again, this grows logarithmically as $n \to \infty$, and reproduces the EE obtained in [4, 13] in the $\alpha \to 1$ limit:

$$S_{\mathrm{M},\,A} = \frac{1}{2}\ln\frac{(n+r)(n-r)}{n} + \frac{1}{2}\ln\frac{2\pi}{3} + \gamma - \frac{1}{2} + (\text{terms vanishing as } n\to\infty). \tag{4.51}$$

4.5 Rényi Entropy of s-Color Fredkin Spin Chain

We compute asymptotic behavior of the Rényi entropy (4.22) for colored case ($s > 1$). The expression (4.23) leads to

$$\sum_{h\geq 0} s^h \left(p^{(h)}_{\mathrm{F},\,n+r,n-r,\,s}\right)^{\alpha}$$
$$\simeq \left(\frac{8}{\sqrt{\pi}}\right)^{\alpha}\left(\frac{n}{(n+r)(n-r)}\right)^{\frac{3}{2}\alpha} e^{\alpha(n+r+2)\ln(n+r)+\alpha(n-r+2)\ln(n-r)}$$
$$\times\sum_{h\geq 0}\frac{1+(-1)^{n+r+h}}{2}\,e^{f_{\mathrm{F}}(h)} \times \left[1+O(n^{-1})\right] \tag{4.52}$$

with

$$f_{\mathrm{F}}(h) \equiv (1-\alpha)(\ln s)\,h$$
$$-\frac{\alpha}{2}(n+r+h+3)\ln(n+r+h) - \frac{\alpha}{2}(n+r-h+1)\ln(n+r-h)$$
$$-\frac{\alpha}{2}(n-r+h+3)\ln(n-r+h) - \frac{\alpha}{2}(n-r-h+1)\ln(n-r-h)$$
$$+2\alpha\ln(h+1). \tag{4.53}$$

We discuss two cases $0 < \alpha < 1$ and $\alpha > 1$ separately.

4.5.1 $0 < \alpha < 1$ Case

We evaluate the sum in (4.52) in the saddle point method for large n. The saddle point equation $f_F(h) = f_F(h+2)$ reads

$$2\frac{1-\alpha}{\alpha}\ln s = \ln\frac{(n+h)^2-r^2}{(n-h)^2-r^2} + O(n^{-1}), \tag{4.54}$$

and its solution is

$$h_* = n\xi_* + O(n^0), \qquad \xi_* \equiv \coth\theta - \sqrt{\frac{1}{\sinh^2\theta}+u^2} \tag{4.55}$$

with

$$\theta = \frac{1-\alpha}{\alpha}\ln s, \qquad u \equiv \frac{r}{n}. \tag{4.56}$$

In (4.55) we chose the "$-$" branch, because the "+" branch leads to h_* larger than n for $u = 0$ that is unphysical. Notice that the value of h_* is $O(n)$, and use of (4.23) instead of (4.24) is consistent. We obtain

$$f_F''(h_*) = -\frac{\alpha}{n}\frac{\sinh^2\theta}{\xi_*}\sqrt{\frac{1}{\sinh^2\theta}+u^2} \times \left[1+O(n^{-1})\right], \tag{4.57}$$

$$\begin{aligned} f_F(h_*) &= -(1-\alpha)\ln s - \alpha(n+2)\ln\left(n^2\frac{2\xi_*}{\sinh\theta}\right) \\ &\quad -\frac{\alpha}{2}nu\ln\frac{\xi_*\sqrt{\frac{1}{\sinh^2\theta}+u^2}+u(1+u)}{\xi_*\sqrt{\frac{1}{\sinh^2\theta}+u^2}-u(1-u)} + O(n^{-1}), \end{aligned} \tag{4.58}$$

and evaluate the sum as an integral around the saddle point:

$$\begin{aligned} &\sum_{h=0}^{n-|r|}\frac{1+(-1)^{n+r+h}}{2}(h+1)^{2\alpha}\,e^{f_F(h)} \\ &= (h_*+1)^{2\alpha}e^{f_F(h_*)}\,\frac{1}{2}\int_{-\infty}^{\infty}dx\,e^{\frac{1}{2}f_F''(h_*)x^2}\times\left[1+O(n^{-1})\right] \\ &= h_*^{2\alpha}\,e^{f_F(h_*)}\,\frac{1}{2}\sqrt{\frac{2\pi}{-f_F''(h_*)}}\times\left[1+O(n^{-1})\right]. \end{aligned} \tag{4.59}$$

Combining (4.52) and (4.59), we end up with

$$
\begin{aligned}
S_{A,\alpha} &= \frac{1}{1-\alpha}\ln\sum_{h\geq 0} s^h\left(p^{(h)}_{\mathrm{F},n+r,n-r,s}\right)^\alpha \\
&= n\frac{\alpha}{1-\alpha}\left\{\ln\left[(1+u)^{1+u}(1-u)^{1-u}\frac{\sinh\theta}{2\xi_*}\right]\right. \\
&\qquad\left. -\frac{u}{2}\ln\frac{\xi_*\sqrt{\frac{1}{\sinh^2\theta}+u^2}+u(1+u)}{\xi_*\sqrt{\frac{1}{\sinh^2\theta}+u^2}-u(1-u)}\right\} \\
&\quad+\frac{1+\alpha}{2(1-\alpha)}\ln\left(n(1-u^2)\right)-\ln s+\frac{1}{2}\ln\pi+\frac{\alpha}{1-\alpha}\ln\left(2\sinh^2\theta\right) \\
&\quad+\frac{1}{2(1-\alpha)}\ln\frac{\xi_*}{2\alpha(1-u^2)}-\frac{1}{1-\alpha}\ln\left[\sinh\theta\left(\frac{1}{\sinh^2\theta}+u^2\right)^{1/4}\right] \\
&\quad+O(n^{-1}).
\end{aligned}
\tag{4.60}
$$

The Rényi entropy grows linearly as the volume in large n. Note that we cannot take $\alpha \to 1$ or $s \to 1$ limit in (4.60). In the limit, θ becomes vanish and ξ_* also goes to zero, which makes invalid the computation so far for $h_* = O(n)$. Namely, the $n \to \infty$ limit does not commute with $\alpha \to 1$ or $s \to 1$ limit.

For $r = 0$ ($u = 0$) case in which the subsystems A and B are the same length n, (4.60) reduces to

$$
\begin{aligned}
S_{A,\alpha} &= n\frac{2\alpha}{1-\alpha}\ln\cosh\frac{\theta}{2}+\frac{1+\alpha}{2(1-\alpha)}\ln n \\
&\quad-\ln s+\frac{1}{2}\ln\frac{\pi}{4}-\frac{1}{2(1-\alpha)}\ln\alpha-\frac{1}{1-\alpha}\ln\cosh\frac{\theta}{2} \\
&\quad+\frac{2\alpha}{1-\alpha}\ln\sinh\theta+O(n^{-1}),
\end{aligned}
\tag{4.61}
$$

which coincides with the result in [20].

4.5.2 $\alpha > 1$ Case

In this case, the summand of $\sum_{h\geq 0} s^h\left(p^{(h)}_{\mathrm{F},n+r,n-r,s}\right)^\alpha$ contains a factor $s^{-(\alpha-1)h}$ which damps exponentially as h grows. Since there is no other factor compensating this damping, we can say that $h \lesssim \frac{1}{(\alpha-1)\ln s} = O(n^0)$ is relevant to the sum. Hence, use of (4.24) leads to

$$
\begin{aligned}
\sum_{h=0}^{n-|r|} s^h\left(p^{(h)}_{\mathrm{F},n+r,n-r,s}\right)^\alpha &\sim \left(\frac{8}{\sqrt{\pi}}\right)^\alpha\left(\frac{n}{\beta}\right)^{-\frac{3}{2}\alpha} s^{\alpha-1}\sum_{h\geq 1}\frac{1-(-1)^{n+r+h}}{2}s^{-(\alpha-1)h}h^{2\alpha} \\
&\quad\times\left[1+O(n^{-1})\right],
\end{aligned}
\tag{4.62}
$$

where

$$\beta \equiv \frac{1}{1-u^2}, \qquad u \equiv \frac{r}{n}, \tag{4.63}$$

and the factor $e^{-\frac{\alpha\beta}{n}(h+1)^2}$ was further simplified as $1 + O(n^{-1})$. The sum cannot be converted to an integral, but should be treated as it is. By introducing the Lerch transcendent

$$\Phi(z, s, a) = \sum_{k=0}^{\infty} \frac{z^k}{(k+a)^s}, \tag{4.64}$$

the final result is expressed as follows:

$$\begin{aligned} S_{\mathrm{F},\,A,\,\alpha} = & \frac{3\alpha}{2(\alpha-1)} \ln \frac{(n+r)(n-r)}{n} + \frac{\alpha}{2(\alpha-1)} \ln \frac{\pi}{32^2} \\ & - \frac{1}{\alpha-1} \ln \Phi\left(s^{-2(\alpha-1)}, -2\alpha, \frac{1}{2}\right) + O(n^{-1}) \end{aligned} \tag{4.65}$$

for $n+r$ even, and

$$\begin{aligned} S_{\mathrm{F},\,A,\,\alpha} = & \frac{3\alpha}{2(\alpha-1)} \ln \frac{(n+r)(n-r)}{n} - \ln s + \frac{\alpha}{2(\alpha-1)} \ln \frac{\pi}{32^2} \\ & - \frac{1}{\alpha-1} \ln \Phi\left(s^{-2(\alpha-1)}, -2\alpha, 0\right) + O(n^{-1}) \end{aligned} \tag{4.66}$$

for $n+r$ odd. Both of the expressions have the common logarithmic term of the volume, whereas the difference appears in $O(n^0)$ terms. At $r=0$, these results reduce to what is obtained in [20]. Again we cannot take $\alpha \to 1$ or $s \to 1$ limit in (4.65) or (4.66), because the damping by $s^{-(\alpha-1)h}$ ceases in the limit and the computation will become invalid.

Qualitatively, the Lerch transdents in (4.65) and (4.66) behave as

$$\Phi\left(s^{-2(\alpha-1)}, -2\alpha, \frac{1}{2}\right) \sim \frac{s^{\alpha-1}\Gamma(2\alpha+1, (\alpha-1)\ln s)}{(2(\alpha-1)\ln s)^{2\alpha+1}}, \tag{4.67}$$

$$\Phi\left(s^{-2(\alpha-1)}, -2\alpha, 0\right) \sim \frac{\Gamma(2\alpha+1)}{(2(\alpha-1)\ln s)^{2\alpha+1}}, \tag{4.68}$$

where $\Gamma(z, x)$ is the incomplete Gamma function:

$$\Gamma(z, x) \equiv \int_x^{\infty} \mathrm{d}t\, t^{z-1} e^{-t}. \tag{4.69}$$

4.5.3 EE of s-Color Fredkin Spin Chain

In the expression of the EE (4.21) with (4.23) or (4.24), the first factor s^h in the summand cancels with s^{-h} in $p^{(h)}_{\mathrm{F},\,n+r,n-r,\,s}$, and the summand does not contain exponentially growing factor with h. Then, the saddle point of h in (4.21) is $O(\sqrt{n})$, which justifies the use of (4.24). By converting the sum to an integral with $h = \sqrt{n}\,x$, we have

$$S_{\mathrm{F},\,A} = -\frac{4}{\sqrt{\pi}\,(1-u^2)^{3/2}} \int_{n^{-1/2}}^{\infty} \mathrm{d}x\, x^2 e^{-\beta x^2} \ln\left[s^{1-\sqrt{n}\,x} \frac{8n^{-1/2}}{\sqrt{\pi}\,(1-u^2)^{3/2}}\, x^2 e^{-\beta x^2}\right] \times \left[1 + O(n^{-1})\right] \tag{4.70}$$

with (4.63). We divide the integral $\int_{n^{-1/2}}^{\infty}$ to $\int_0^{\infty} - \int_0^{n^{-1/2}}$, where the second integral is evaluated as $O\left(\frac{\ln n}{n^{3/2}}\right)$ and negligible. After computing the first integral, we obtain

$$S_{\mathrm{F},\,A} = (2\ln s)\sqrt{\frac{(n+r)(n-r)}{\pi n}} + \frac{1}{2}\ln\frac{(n+r)(n-r)}{n} + \frac{1}{2}\ln\frac{\pi}{4} + \gamma - \frac{1}{2} - \ln s + (\text{terms vanishing as } n \to \infty). \tag{4.71}$$

This is consistent with the result in [17], and coincides with the computation in [20] at $r = 0$.

4.5.4 Phase Transition

The Rényi entropy $S_{\mathrm{F},\,A,\,\alpha}$ shows different asymptotic behavior for $0 < \alpha < 1$ and $\alpha > 1$. It grows linearly with the volume for the former case, whereas logarithmically for the latter. From the point of view of (4.5), the result can be interpreted as a phase transition occurring at $\alpha = 1$.

As we saw in the computation, Dyck walks with large height $h = O(n)$ dominantly contribute to the $S_{\mathrm{F},\,A,\,\alpha}$ in "high temperature" region $0 < \alpha < 1$, which leads to the volume law behavior. On the other hand, Dyck walks with low height $h = O(n^0)$ dominate in "low temperature" $\alpha > 1$, which does not give qualitative change to the behavior of the colorless case. The transition point itself forms a phase, where the EE behaves as a square root of the volume. Main contribution to (4.71) comes from $h = O(\sqrt{n})$. These are summarized in Fig. 4.1.

4.6 Rényi Entropy of s-Color Motzkin Spin Chain

In this section, we compute large-n behavior of the Rényi entropy of s-color Motzkin spin chain ($s > 1$). Plugging (4.34) and (4.44) to (4.33), the sum we should evaluate becomes

$$\sum_{h=0}^{n-|r|} s^h \left(p^{(h)}_{\mathrm{M},\,n+r,n-r,\,s}\right)^\alpha \sim \left(\frac{1}{\sqrt{\pi}\, s^{1/4}}\right)^\alpha \frac{(2n)^{\frac{3}{2}\alpha}\,(n+r)^{(n+r+\frac{1}{2})\alpha}\,(n-r)^{(n-r+\frac{1}{2})\alpha}}{\left(2\sqrt{s}+1\right)^{(2n+\frac{3}{2})\alpha}} \times \sum_{h=0}^{n-|r|} e^{f_{\mathrm{M}}(h)} \times \left[1+O(n^{-1})\right], \tag{4.72}$$

where

$$\begin{aligned} f_{\mathrm{M}}(h) \equiv{}& (1-\alpha)(\ln s)h - \alpha\left(n+r+\frac{3}{2}\right)\ln\rho_{0+} - \alpha\left(n-r+\frac{3}{2}\right)\ln\rho_{0-} \\ &+\alpha\frac{h+1}{2}\left\{\ln(n+r-\rho_{0+}-h)+\ln(n-r-\rho_{0-}-h)\right. \\ &\left.\qquad -\ln(n+r-\rho_{0+}+h)-\ln(n-r-\rho_{0-}+h)\right\} \\ &-\frac{\alpha}{4}\ln\left[4s(n+r)^2-(4s-1)h^2\right]-\frac{\alpha}{4}\ln\left[4s(n-r)^2-(4s-1)h^2\right] \\ &+2\alpha\ln(h+1). \end{aligned} \tag{4.73}$$

$\rho_{0\pm}$ is given by ρ_0 in (4.41) with m replaced by $n \pm r$. We discuss the two cases of $0 < \alpha < 1$ and $\alpha > 1$ separately.

4.6.1 $0 < \alpha < 1$ *Case*

We try to evaluate (4.72) by the saddle point method. Noting the relation

$$h^2 = (n \pm r - \rho_{0\pm})^2 - 4s\rho_{0\pm}^2, \tag{4.74}$$

the saddle point equation reads

$$\frac{1+\sqrt{1-\frac{4s\rho_{0+}^2}{(n+r-\rho_{0+})^2}}}{1-\sqrt{1-\frac{4s\rho_{0+}^2}{(n+r-\rho_{0+})^2}}}\;\frac{1+\sqrt{1-\frac{4s\rho_{0-}^2}{(n-r-\rho_{0-})^2}}}{1-\sqrt{1-\frac{4s\rho_{0-}^2}{(n-r-\rho_{0-})^2}}} = s^{2\frac{1-\alpha}{\alpha}} + O(n^{-1}). \tag{4.75}$$

It seems however technically difficult to solve (4.75) analytically. Let us consider the case $r = 0$, in which the subsystems A and B have the same length n. Then,[1] $\rho_{0\pm}$ reduces to ρ_0, and (4.75) becomes

$$\frac{1+\sqrt{1-\frac{4s\rho_0^2}{(n-\rho_0)^2}}}{1-\sqrt{1-\frac{4s\rho_0^2}{(n-\rho_0)^2}}} = s^{\frac{1-\alpha}{\alpha}} + O(n^{-1}). \tag{4.76}$$

This is solved by

$$h_* = n\frac{s^{\frac{1}{2\alpha}} - s^{1-\frac{1}{2\alpha}}}{s^{\frac{1}{2\alpha}} + s^{1-\frac{1}{2\alpha}} + 1} + O(n^0), \tag{4.77}$$

$$\rho_0|_{h=h_*} = \frac{n}{s^{\frac{1}{2\alpha}} + s^{1-\frac{1}{2\alpha}} + 1} + O(n^0), \tag{4.78}$$

which give

$$f''_{\rm M}(h_*) = -\frac{2\alpha}{n}\frac{\left(s^{\frac{1}{2\alpha}} + s^{1-\frac{1}{2\alpha}} + 1\right)^2}{s^{\frac{1}{2\alpha}} + s^{1-\frac{1}{2\alpha}} + 4s} \times \left[1 + O(n^{-1})\right], \tag{4.79}$$

$$f_{\rm M}(h_*) = -(2n+2)\alpha \ln n + (2n+3)\alpha \ln\left(s^{\frac{1}{2\alpha}} + s^{1-\frac{1}{2\alpha}} + 1\right) - (1-\alpha)\ln s$$
$$-\frac{\alpha}{2}\ln\left[1 + 4\frac{\left(2s^{\frac{1}{2\alpha}}+1\right)\left(2s^{1-\frac{1}{2\alpha}}+1\right)}{\left(s^{\frac{1-\alpha}{2\alpha}} - s^{-\frac{1-\alpha}{2\alpha}}\right)^2}\right] + O(n^{-1}). \tag{4.80}$$

Evaluating the sum for $r = 0$, we end up with

$$\begin{aligned}
S_{{\rm M},\,A,\,\alpha} &= \frac{1}{1-\alpha}\ln\sum_{h=0}^{n} s^h\left(p^{(h)}_{{\rm M},n,n,s}\right)^\alpha \\
&= n\frac{2\alpha}{1-\alpha}\ln\left[\sigma\left(s^{\frac{1-\alpha}{2\alpha}} + s^{-\frac{1-\alpha}{2\alpha}} + s^{-1/2}\right)\right] + \frac{1+\alpha}{2(1-\alpha)}\ln n \\
&+\frac{1}{2}\ln\pi - \frac{1}{1-\alpha}\ln\left(s\sqrt{\alpha}\right) + \frac{1}{2(1-\alpha)}\ln\left(s^{\frac{1}{2\alpha}} + s^{1-\frac{1}{2\alpha}} + 4s\right) + \frac{3\alpha}{2(1-\alpha)}\ln(2\sigma) \\
&+\frac{3\alpha-1}{1-\alpha}\ln\left(s^{\frac{1}{2\alpha}} + s^{1-\frac{1}{2\alpha}} + 1\right) - \frac{\alpha}{2(1-\alpha)}\ln\left[1 + 4\frac{\left(2s^{\frac{1}{2\alpha}}+1\right)\left(2s^{1-\frac{1}{2\alpha}}+1\right)}{\left(s^{\frac{1-\alpha}{2\alpha}} - s^{-\frac{1-\alpha}{2\alpha}}\right)^2}\right] \\
&+(\text{terms vanishing as } n\to\infty)
\end{aligned} \tag{4.81}$$

[1]Although the analysis for the case $r = 0$ is already given in [20], we briefly present the derivation to make this article self-contained as much as possible.

with $\sigma = \frac{\sqrt{s}}{2\sqrt{s}+1}$. The Rényi entropy has asymptotic behavior proportional to the volume as we saw in the s-color Fredkin case (4.60). The coefficient of $\ln n$ term coincides with that in (4.61), which implies to show some universal property.

4.6.2 $\alpha > 1$ Case

As in the Fredkin case, due to the exponential damping factor $s^{-(\alpha-1)h}$ in the sum $\sum_{h=0}^{n} s^h \left(p^{(h)}_{\mathrm{M},\, n+r, n-r, s}\right)^{\alpha}$, we can regard h as a quantity at most $O(n^0)$, which justifies the use of (4.45).

With the notations (4.63) and (4.64), the sum is recast as

$$\sum_{h=0}^{n-|r|} s^h \left(p^{(h)}_{\mathrm{M},n+r,n-r,s}\right)^{\alpha} \simeq \left(\frac{4}{\sqrt{\pi}}\right)^{\alpha} \left(\frac{n}{\beta}\right)^{-\frac{3}{2}\alpha} s^{-1+\alpha}\, \Phi\left(s^{-(\alpha-1)}, -2\alpha, 0\right) \times \left[1 + O(n^{-1})\right]. \tag{4.82}$$

We thus find

$$S_{\mathrm{M},\,A,\,\alpha} = \frac{3\alpha}{2(\alpha-1)} \ln\left(2\sigma \frac{(n+r)(n-r)}{n}\right) - \ln s + \frac{\alpha}{2(\alpha-1)} \ln\frac{\pi}{16} - \frac{1}{\alpha-1} \ln \Phi\left(s^{-(\alpha-1)}, -2\alpha, 0\right) + O(n^{-1}), \tag{4.83}$$

which reduces to the result in [20] at $r = 0$. Qualitative behavior of $\Phi\left(s^{-(\alpha-1)}, -2\alpha, 0\right)$ is evaluated as

$$\Phi\left(s^{-(\alpha-1)}, -2\alpha, 0\right) \sim \frac{\Gamma(2\alpha+1)}{((\alpha-1)\ln s)^{2\alpha+1}}. \tag{4.84}$$

4.6.3 EE of s-Color Motzkin Spin Chain

After similar computation to the s-color Fredkin case, we have

$$S_{\mathrm{M},\,A} = (2\ln s)\sqrt{\frac{2\sigma}{\pi}\frac{(n+r)(n-r)}{n}} + \frac{1}{2}\ln\frac{(n+r)(n-r)}{n} + \frac{1}{2}\ln(2\pi\sigma) + \gamma - \frac{1}{2} - \ln s + (\text{terms vanishing as } n \to \infty). \tag{4.85}$$

Again this reduces to the result of the half-chain case [20] at $r = 0$. As mentioned there, in the computation we should not approximate $h + 1$ to h in (4.45) to obtain the $O(n^0)$ term correctly.

4.6.4 Phase Transition

Similar to the s-color Fredkin case, the Rényi entropy $S_{\mathrm{M},\,A,\,\alpha}$ has different asymptotic behavior for $0 < \alpha < 1$ and $\alpha > 1$ – linear of the volume and logarithm of the volume. Motzkin walks with large height $h = O(n)$ dominantly contribute to the $S_{\mathrm{M},\,A,\,\alpha}$ in "high temperature" region $0 < \alpha < 1$, leading to the volume law behavior, whereas Motzkin walks with low height $h = O(n^0)$ dominate in "low temperature" $\alpha > 1$, qualitatively same as the colorless case. The transition point itself consists of a phase, where the EE behaves as a square root of the volume. Main contribution to (4.85) comes from height $h = O(\sqrt{n})$. Figure 4.1 give a summary of the result.

4.7 Discussion

We have analytically computed the EE and the Rényi entropy of two blocks of unequal length ($n + r$ and $n - r$) in highly entangled Motzkin and Fredkin spin chains. In case of $0 < \alpha < 1$ in the colored Motzkin spin chain, we presented the result of half chains ($r = 0$) due to technical difficulty.

For colored cases, we found non-analyticity in the expression of the Rényi entropy at $\alpha = 1$, which is a totally new phase transition never seen before in any spin chain. We investigated for two blocks of unequal length but both of $n + r$ and $n - r$ being $O(n)$. It is interesting to extend our analysis to the case of two blocks of different orders of size and to see to what extent the phase transition persists. Finally, it will be intriguing to perform similar computation to cousins of the models [5, 15, 18, 21–23].

References

1. J.M. Bardeen, B. Carter, S.W. Hawking, Commun. Math. Phys. **31**, 161 (1973)
2. J.D. Bekenstein, Phys. Rev. D **7**, 2333 (1973)
3. L. Bombelli, R.K. Koul, J. Lee, R.D. Sorkin, Phys. Rev. D **34**, 373 (1986)
4. S. Bravyi, L. Caha, R. Movassagh, D. Nagaj, P.W. Shor, Phys. Rev. Lett. **109**, 207202 (2012)
5. L. Caha, D. Nagaj, arXiv:1805.07168 [quant-ph]
6. P. Calabrese, J. Cardy, J. Phys. A **42**, 504005 (2009)
7. C.G. Callan Jr., F. Wilczek, Phys. Lett. B **333**, 55 (1994)
8. L. Dell'Anna, O. Salberger, L. Barbiero, A. Trombettoni, V.E. Korepin, Phys. Rev. B **94**, 155140 (2016)
9. J. Eisert, M. Cramer, M.B. Plenio, Rev. Mod. Phys. **82**, 277 (2010) and arXiv:0808.3773 [quant-ph]
10. M.B. Hastings, J. Stat. Mech. Theory Exp. **P08024** (2007)
11. C. Holzhey, F. Larsen, F. Wilczek, Nucl. Phys. **B [FS] 424**, 443–467 (1994)
12. V.E. Korepin, Phys. Rev. Lett. **92**, 096402 (2004)
13. R. Movassagh, J. Math. Phys. **58**, 031901 (2017)
14. R. Movassagh, P.W. Shor, Proc. Natl. Acad. Sci. **113**, 13278–13282 (2016)

15. P. Padmanabhan, F. Sugino, V. Korepin, arXiv:1804.00978 [quant-ph]
16. A. Rényi, *Probability Theory* (Elsevier, Amsterdam, 1970)
17. O. Salberger, V. Korepin, Rev. Math. Phys. **29**, 1750031 (2017)
18. O. Salberger, T. Udagawa, Z. Zhang, H. Katsura, I. Klich, V. Korepin, J. Stat. Mech. Theory Exp. **1706**, 063103 (2017)
19. M. Srednicki, Phys. Rev. Lett. **71**, 666 (1993)
20. F. Sugino, V. Korepin, Int. J. Mod. Phys. B **32**(28), 1850306 (2018)
21. F. Sugino, P. Padmanabhan, J. Stat. Mech. Theory Exp. **1801**, 013101 (2018)
22. Z. Zhang, A. Ahmadain, I. Klich, Proc. Natl. Acad. Sci. **114**, 5142–5146 (2017)
23. Z. Zhang, I. Klich, J. Phys. A **50**, 425201 (2017)

Chapter 5
Topological Kondo Effect

Francesco Buccheri and Reinhold Egger

Abstract We review recent theoretical progress in the understanding of the topological Kondo effect in Coulomb-blockaded Majorana devices and generalizations thereof. The central ingredient in Majorana devices is the so-called Majorana box which encodes a spin-$1/2$ degree of freedom in the Majorana subspace that can be addressed by electron cotunneling processes. In particular, after explaining the basic physics of the topological Kondo effect in a Majorana box connected to a set of normal-conducting leads, we discuss the Josephson current-phase relation in a superconducting multi-terminal setup where the central junction is again formed by a Majorana box but the leads are phase-biased superconductors. For large Kondo temperature, one finds that the competition between two-channel Kondo physics and the gap opening in the leads results in a 6π periodicity of the current-phase relation. This periodicity is due to the fractionalized charge excitations characterizing the non-Fermi liquid two-channel Kondo fixed point. We also explore generalizations of this Majorana-based topological Kondo setup to platforms hosting parafermionic excitations.

5.1 Introduction

An important aim of contemporary condensed matter physics is to understand the physics of devices containing Majorana bound states (MBSs) [1–4], or even more exotic setups hosting parafermionic zero modes (i.e., fractionalized Majorana states) [5, 6]. Besides the inherent fundamental interest in such systems, a set of localized zero modes can encode quantum states featuring non-Abelian braiding statistics. Such states may be useful for topologically protected quantum information

F. Buccheri (✉) · R. Egger
Institute for Theoretical Physics, Heinrich-Heine-University Düsseldorf, 40225 Düsseldorf, Germany
e-mail: buccheri@hhu.de

R. Egger
e-mail: egger@hhu.de

A. Ferraz et al. (eds.), *Strongly Coupled Field Theories for Condensed Matter and Quantum Information Theory*, Springer Proceedings in Physics 239,
https://doi.org/10.1007/978-3-030-35473-2_5

processing applications [5, 7]. While experimental signatures of MBSs have already been reported in a variety of experimental platforms [8–24], condensed-matter realizations of parafermions (PFs) are still in their infancy; for recent progress towards possible PF implementations, see [25, 26]. The physics expected to be seen in PF devices has been addressed by a large number of theoretical works [6, 27–43]. Theoretical PF constructions often rely on competing gap-inducing mechanisms acting on the edge states of topologically ordered 2D phases, e.g., bilayer fractional quantum Hall (FQH) systems [36], proximitized fractional topological insulators [27], and proximitized FQH systems with filling factor $\nu = 2/3$ [33] or $\nu = 1/(2k+1)$ (with integer k) [27, 29]. Eventually this may allow one to realize Fibonacci anyons which are capable of topologically protected universal quantum computation [33, 40].

For a Coulomb-blockaded superconducting island containing more than two MBSs (the so-called Majorana box [44]), an effective impurity spin operator is encoded by pairs of spatially separated MBSs. When normal leads are coupled to the MBSs, this spin is screened through cotunneling processes, culminating in the topological Kondo effect (TKE). The theory of the TKE has been laid out in a series of recent articles [45–67]. It exhibits local quantum criticality associated with a non-Fermi liquid fixed point. This physics should be experimentally observable at temperatures T below the Kondo temperature T_K, see below for a definition. Unlike conventional overscreened multi-channel Kondo systems [68–71], the TKE is intrinsically stable against anisotropies. Majorana devices could thus realize multi-channel Kondo effects without delicate fine tuning of parameters.

Theoretical [72–74] and experimental [15] work has pointed out the important role of Coulomb charging effects in a floating mesoscopic topological superconductor (TS), i.e., in a Majorana box. Since charge degrees of freedom are gapped out and, in addition, several quasiparticle poisoning channels are blocked [75, 76] by a large box charging energy E_C, the Majorana subsector of the Hilbert space tends to be stabilized by charging effects. Moreover, charging effects suppress sequential tunneling in favor of long-ranged cotunneling events. The latter effectively connect different leads—coupled to the Majorana box through tunnel contacts—in a phase-coherent fashion. For that reason, Majorana boxes are also discussed as promising ingredients for topologically protected quantum information processing applications [44, 76–79].

Consider now a Majorana box operated under Coulomb valley conditions, where we assume that $M \geq 3$ normal-conducting (and due to the MBS spin polarization also effectively spinless [1]) leads are individually tunnel-coupled to different MBSs on the box. In that case, the Majorana subsector of the Hilbert space will be equivalent to an effective quantum impurity spin with orthogonal SO(M) symmetry [45]. (The orthogonal symmetry, as opposed to the conventional unitary one, can be traced back to the reality condition $\gamma = \gamma^\dagger$ imposed on Majorana fermion operators [50].) For the minimal case $M = 3$, this impurity spin corresponds to a standard spin-1/2 operator, where different spin components are nonlocally represented by distinct Majorana bilinears on the box, see below. The lead space also has an SO(M) symmetry, and therefore cotunneling processes effectively implement an exchange coupling of the lead spin density at the central junction (the box). Through this exchange coupling, the effective impurity spin undergoes screening which in the end drives the system to a *robust* non-Fermi liquid fixed point [45]. This fixed point is similar to the overscreened multi-channel Kondo fixed point [68].

One can even include Coulomb interactions in the leads within this scenario. Taking $M \geq 3$ spinless Luttinger liquid leads with interaction parameter g, where $g = 1$ for the noninteracting case and $g < 1$ for repulsive interactions [68, 80], the linear TKE conductance between leads j and k has the form [47, 50]

$$G_{jk}^{\text{TKE}} = \frac{2ge^2}{h}\left[1 - (T/T_K)^{2\Delta_M - 2} + \cdots\right]\left(\frac{1}{M} - \delta_{jk}\right), \tag{5.1}$$

where the scaling dimension $\Delta_M = 2g(M-1)/M$ refers to the leading irrelevant operator at the strong-coupling fixed point. Equation (5.1) holds for $\Delta_M > 1$ and in the asymptotic low-temperature regime $T \ll T_K$. The non-Fermi-liquid nature of the TKE fixed point is visible in the power-law T dependence with rational (non-integer) exponent. Moreover, it is worth stressing that the conductance tensor becomes completely isotropic, regardless of possible anisotropies in the bare tunnel couplings. By observing the conductances in (5.1), experiments could detect strong evidence for nonlocality due to MBSs. Let us give an example: For $g = 1$ and $M = 3$, the $T = 0$ conductance between leads 1 and 2 has the (large) value $G_{12}^{\text{TKE}} = 2e^2/3h$. After decoupling lead 3 by a suitable gate voltage change, the TKE is destroyed and only an exponentially small conductance G_{12} due to residual cotunneling is expected [74]. This offers a very distinct and hard-to-fake Majorana signature since the MBS coupled to lead 3 is located far away from the MBSs coupled to leads 1 and 2. Recent works have studied various other aspects and variations of the above setup, including systems of coupled Majorana boxes [67].

In this article, after reviewing the basic TKE physics, we discuss two recent generalizations of the above setup. The first generalization concerns the Josephson effect for a Majorana box connected to *superconducting* instead of normal leads, as we schematically illustrate in Fig. 5.1. Most theoretical works for Majorana systems contacted by superconducting electrodes have addressed cases without TKE [82–86]. Below we summarize the results of [62] for the superconducting version of the TKE, where a nontrivial competition between superconductivity and TKE arises because no lead states below the superconducting gap Δ are available for screening the box spin. The corresponding single-channel spin-1/2 Kondo variant is of Fermi-liquid type and can be realized for two superconducting leads connected to an Anderson impurity (spinful quantum dot) [87]. The latter case has been studied in great detail over the past decades, both theoretically [88–96] and experimentally [97–102]. It is known that a local quantum phase transition for $\Delta/T_K \simeq 1$ separates the so-called 0-phase (realized at small Δ/T_K) from the so-called π-phase (at large Δ/T_K). Remarkably, almost the entire crossover in the free energy of the Josephson junction was found to be described by universal scaling functions of Δ/T_K. In the deep 0-phase, the Kondo resonance persists and yields the current-phase relation of a fully transparent superconducting junction [96], while in the π-phase the Kondo effect is almost completely quenched and one finds a negative supercurrent.

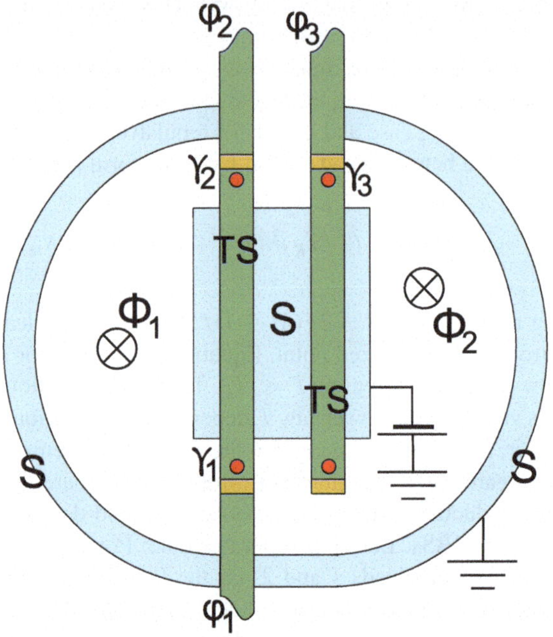

Fig. 5.1 TKE setup for the case of $M = 3$ superconducting leads. As concrete platform, one may imagine two long parallel InAs or InSb nanowires with superconducting shells. These proximitized nanowires define 1D TS segments (green) hosting MBSs at their ends (filled red circles, shown only on the Majorana box). The Majorana box has Majorana operators γ_j, where the two central TS segments are connected through a floating conventional s-wave superconductor (blue island). The box has a large charging energy E_C. Short non-proximitized wire sections (yellow) serve as gate-tunable tunnel contacts. Superconducting leads are realized by contacting the outer nanowire sections with large conventional superconducting electrodes (blue). Using the shown loop geometry, one may study the supercurrents I_j as function of the superconducting phases $(\varphi_1, \varphi_2, \varphi_3)$ by means of tuning the magnetic fluxes $\Phi_{1,2}$, cf. [81]

For the Majorana device in Fig. 5.1, [62] has shown that an even richer interplay between superconductivity and (now multi-channel) Kondo screening physics takes place. The TKE symmetry group is however changed by even a tiny gap Δ due to the proliferation of crossed Andreev reflection processes. For $\Delta \ll T_K$ and $M = 3$ attached leads, an exact strong-coupling calculation along the lines of [103–105] predicts that two-channel Kondo physics is responsible for a 6π periodicity of the Josephson effect, with a critical current of order $I_c \approx e\Delta^2/\hbar T_K$. This periodicity directly implies charge fractionalization in units of $e^* = 2e/3$ for the elementary transfer processes across the junction. One thus has to wind the phase by 6π in order to transfer a Cooper pair. On the other hand, for $\Delta \gg T_K$, one obtains the 4π-periodic current-phase relation of a parity-conserving topological Josephson junction [1–3]. In view of the rapid experimental progress on MBSs in semiconductor-superconductor devices, these predictions could be put to experimental tests, e.g., using techniques used for observing the 4π Josephson effect [106].

As second generalization of the Majorana-based TKE scenario, we here describe what happens for fractional MBSs, i.e., parafermions (PF). References [42, 43] have introduced a PF box device generalizing the Majorana box to a setup with *parafermionic zero modes*. This PF box can in principle be realized by using opposite-spin FQH edge states proximitized by alternating SC and FM domains, closely following the proposals in [27, 29] but taking into account the charging energy E_C of the box; recent experiments reported progress along these lines [25, 26]. The PF model then replaces the normal-conducting (Luttinger liquid) leads behind (5.1) by chiral FQH edge states, which are described by chiral Luttinger liquid theory and host fractional quasiparticles [68, 80, 107, 108]. Importantly, such edge states are experimentally under control for more than two decades [109–114]. In order to see whether PFs will also establish a–possibly modified–Kondo effect, we study whether the TKE conductance tensor in (5.1) will be modified in the PF scenario. In fact, as discussed below, [43] argues that such a PF generalization does not yield a Kondo Hamiltonian which is invariant under the action of a continuous group. For $M = 3$ edges, the quantum impurity spin defined on the PF box transforms in a representation of the SU(n) group, where $n = 2/\nu$ for filling factor $\nu = 1/(2k+1)$. Remarkably, the effective low-energy Hamiltonian does, however, not have this symmetry. Technically, the reason behind this fact is that one cannot perform rotations in lead space because each FQH lead necessitates different Klein factors. Nevertheless, for $M \geq 3$ chiral edges attached to the PF box, one finds a non-trivial strong-coupling regime where the conductance tensor exhibits an almost identical behavior as in (5.1). The PF generalization therefore constitutes a new multi-terminal quantum junction setups for Luttinger liquids that is distinct from all previously studied cases [115–127], even though transport exhibits qualitative similarities to the TKE [46, 47, 50] and to the junctions considered in [125, 127]. Although both the physical realizations and the transport characteristics differ, all three problems share key features. In particular, the system is driven to a strong coupling regime, where at the fixed point incoming currents are universally distributed between all outgoing channels, without producing any shot noise.

The remainder of this article is structured as follows. In Sect. 5.2, we provide a basic introduction to the standard TKE with normal-conducting leads. In Sect. 5.3, we then discuss the case of superconducting leads where the TKE competes with the gap opening in the leads [62]. In Sect. 5.4, a PF generalization of the TKE setup is discussed [43]. Finally, in Sect. 5.5, we offer some conclusions. We often put $\hbar = e = k_B = 1$, and sometimes also set the Fermi velocity $v_F = 1$.

5.2 TKE Basics

We now turn to a theoretical description of the setup in Fig. 5.1. For convenience, we directly allow for a superconducting gap in the leads, even though the results considered in this section are for $\Delta = 0$. The M leads attached to the Majorana box are described as semi-infinite TS wires. The effectively spinless low-energy Hamiltonian for the leads is given by [1]

$$H_{\text{leads}} = \sum_{j=1}^{M} \int_0^\infty \mathrm{d}x\, \Psi_j^\dagger(x) \left[-i\partial_x \sigma_z + \Delta_j e^{-i\varphi_j \sigma_z} \sigma_y \right] \Psi_j(x), \tag{5.2}$$

where Δ_j is the absolute value and φ_j the phase of the respective proximity gap, Pauli matrices $\sigma_{x,y,z}$ and unity σ_0 act in Nambu space, and the spinors $\Psi_j = (\psi_{j,R}, \psi_{j,L}^\dagger)^T$ are written in terms of right/left-moving fermion operators with boundary condition $\psi_{j,R}(0) = \psi_{j,L}(0)$. We shall only discuss results for identical gaps, $\Delta_j = \Delta$, but see [62]. Notably, we do not assume conventional s-wave superconductors as leads since different pairing symmetries for the box and the leads would imply a supercurrent blockade [82], where only above-gap quasiparticle Josephson transport is possible [83–86]. Fortunately, leads with effective p-wave pairing symmetry may be implemented in a natural way, see Fig. 5.1. For $\Delta = 0$, we will then arrive at the canonical TKE model first studied by Béri and Cooper in 2012 [45].

At the position of the contact to the box, which is taken at $x = 0$ for each lead, the respective boundary fermion ψ_j is then coupled by a tunnel amplitude t_j to the Majorana operator $\gamma_j = \gamma_j^\dagger$ on the box, with anticommutator $\{\gamma_j, \gamma_k\} = \delta_{jk}$. We assume that only energy scales well below the proximity gap Δ_{box} on the box are probed, where Δ_{box} and Δ (for simplicity, we take $\Delta < \Delta_{\text{box}}$) are formally independent parameters. Above-gap quasiparticles on the box are neglected throughout. For large charging energy E_C, charge quantization implies a parity constraint for the box MBSs and tends to suppress certain quasiparticle poisoning events. At the same time, the ground state of the isolated box is degenerate for all $M > 2$, where the Majorana bilinears $i\gamma_j\gamma_k$ represent the box quantum impurity spin [45, 51]. The projection to the Hilbert space sector with quantized box charge yields [45]

$$H_{\text{EC}} = \sum_{j \neq k}^{M} \lambda_{jk} \psi_j^\dagger(0) \psi_k(0) \gamma_k \gamma_j, \tag{5.3}$$

where the dimensionless exchange couplings $\lambda_{jk} = 2t_j t_k^* / E_C$ describe elastic cotunneling between leads $j \leftrightarrow k$.

For the case of interest in this section, $\Delta = 0$, [45–47] have demonstrated that $H_{\text{leads}} + H_{\text{EC}}$ gives a TKE of $\text{SO}_2(M)$ symmetry. The Kondo temperature is given by

$$T_K = E_C e^{-\pi/[2(M-2)\bar{\lambda}]}, \quad \bar{\lambda} = \frac{1}{M(M-1)} \sum_{j \neq k} \lambda_{jk}, \tag{5.4}$$

where the group relation $\text{SO}_2(3) \sim \text{SU}_4(2)$ implies a four-channel Kondo effect for $M = 3$ [103]. The low-energy regime of this problem can be studied in an essentially exact manner by bosonization, exploiting the Klein-Majorana fusion trick [46, 47]. One then arrives, e.g., at the TKE conductance tensor in (5.1).

Before turning to generalizations of this basic TKE scenario, we briefly discuss how one may realize this model for the simplest nontrivial case $M = 3$, cf. Fig. 5.1. The floating box is defined by connecting two parallel TS wires by an s-wave superconductor. Nanowires can be fabricated with an epitaxial superconducting shell, where a magnetic field simultaneously drives both wires into the TS phase [1]. We assume that the TS sections on the box are so long that overlap between different Majorana states is negligible. Non-proximitized wire parts yield gate-tunable tunnel barriers, and leads are defined by the outer TS wires in Fig. 5.1. Using available Majorana wires [15], it appears possible to realize the Kondo regime [45, 50, 51, 64].

5.3 Superconducting Leads: The Josephson Current-Phase Relation

We now consider the case $\Delta \neq 0$ in (5.2), i.e., superconducting electrodes. Here the competition between Kondo physics and superconductivity is controlled by the ratio Δ/T_K. We note in passing that for $\varphi_j = 0$, the above model describes junctions of off-critical anisotropic spin chains [48, 49, 60, 61]. In a loop geometry with magnetic fluxes [81], one can change the phase differences between TS leads and measure the current-phase relation.

In our effective field-theoretical description, the free energy F of the system is written in terms of functional integral over the Grassmann-Nambu spinor fields (see (5.2)) and the localized Majorana fields described above. Specifically, the partition function reads

$$Z = e^{-\beta F} = \int \mathscr{D}(\Psi_j, \gamma_j) e^{-S} \,. \tag{5.5}$$

The action $S = S_{\text{leads}} + S_{\text{box}} + S_{\text{EC}}$ consists of three terms, describing the dynamics of the leads, of the Majorana box and the cotunneling of fermions through the box. The box action is

$$S_{\text{box}} = \frac{1}{2} \int \mathrm{d}\tau \sum_j \gamma_j \partial_\tau \gamma_j \tag{5.6}$$

and the interaction of the leads in the box is written from (5.3) as $S_{\text{EC}} = \int \mathrm{d}\tau\, H_{\text{EC}}(\tau)$.

It is convenient to integrate out all fermion modes in the leads away from the boundary at $x = 0$, in order to write S_{leads} in terms of a single boundary fermionic degrees of freedom $\Psi_j = \Psi_j(\tau)$ for each chain. It reads

$$S_{\text{leads}} = -\frac{1}{2} \int_0^\beta \mathrm{d}\tau \mathrm{d}\tau' \sum_j \bar{\Psi}_j(\tau) G_j^{-1}(\tau - \tau') \Psi_j(\tau'). \tag{5.7}$$

at inverse temperature β. The Fourier transform of the boundary Green's function $G_j(\tau)$ is then [85]

$$G_j(\omega) = -i\ \mathrm{sgn}(\omega)\sqrt{1+\frac{\Delta_j^2}{\omega^2}}\ \sigma_0 + \frac{\Delta_j e^{-i\varphi_j\sigma_z}}{i\omega}\sigma_x . \tag{5.8}$$

The supercurrent I_j exiting lead j and entering the box is computed from the phase derivative of the free energy in (5.5), $I_j = (2e/\hbar)\partial_{\varphi_j} F$. The current conservation at the junction is encoded in the sum rule

$$\sum_j I_j = 0 . \tag{5.9}$$

A simple and instructive limit is the so-called atomic limit, defined by $\Delta \gg T_K$. In other words, it is the limit in which pairing correlations are dominant with respect to the Kondo physics. The gap is so wide, that there are no low-energy quasiparticles in the low-temperature dynamics. Also in the leads, therefore, the only degrees of freedom are effectively the Majorana modes localized on the boundary of the TS. In this limit, (5.8) takes the simple form

$$G_j(\omega) \simeq \frac{\Delta_j}{i\omega}(\sigma_0 + \sigma_x) \tag{5.10}$$

in which we have chosen, without loss of generality, a gauge which removes the phases φ_j from the bulk order parameter. The price to pay, of course, is to have complex tunneling amplitudes. From (5.10), one can see that the effective action can be written in terms of the single boundary Majorana combination

$$\eta_j = \frac{\psi_j + \psi_j^\dagger}{\sqrt{2\Delta_j}} \tag{5.11}$$

as

$$S_{\text{leads}} \simeq \frac{1}{2}\sum_j \int d\tau\ \eta_j \partial_\tau \eta_j . \tag{5.12}$$

The effective low-energy Hamiltonian can be written from (5.3) in the form

$$H_{\text{eff}} = \frac{1}{2}\sum_{j\neq k} \sqrt{\Delta_j \Delta_k}\lambda_{jk} e^{i(\varphi_j-\varphi_k)/2}\eta_j\eta_k\gamma_k\gamma_j . \tag{5.13}$$

All mutually commuting products $2i\eta_j\gamma_j = \sigma_j = \pm 1$ are conserved. In fact, the set $\{\sigma_j\}$ determines a fermion parity sector for all tunnel contacts. Setting all $\Delta_j = \Delta$, we arrive at the 4π-periodic supercurrents

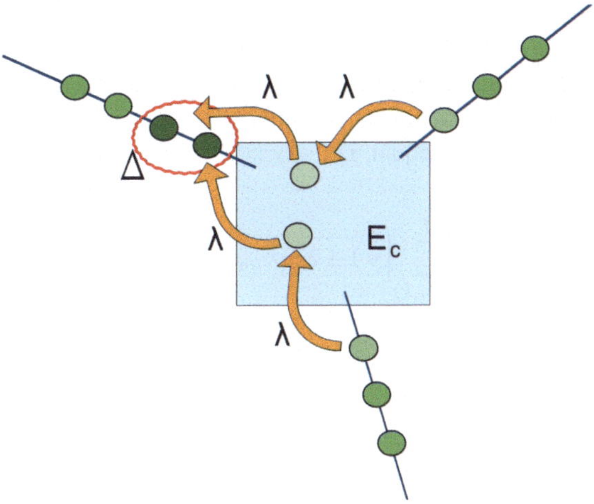

Fig. 5.2 TKE with superconducting leads: multiple tunneling through the island generates crossed Andreev reflection processes, see main text. Only the two participating MBSs are shown

$$I_j = \frac{e\Delta}{\hbar}\sum_{k\neq j}\lambda_{jk}\sigma_j\sigma_k \sin\left(\frac{\varphi_j-\varphi_k}{2}\right), \tag{5.14}$$

We note that (5.14) also describes topological Josephson junctions with featureless tunnel contacts [1], which can be traced back to the fact that the Kondo effect is suppressed in the atomic limit because low-energy quasiparticles in the leads, which are necessary to screen the box spin, are not available below the gap.

5.3.1 Renormalization Group (RG) Analysis

We now study the running of the coupling constants λ_{jk}, using perturbation theory up to second order. First of all, we notice that, even when absent at zeroth order, crossed Andreev reflection will be generated at this order. This correspond to the simultaneous cotunnelling of two fermions from two different leads, creating an additional Cooper pair on the third lead as visualized in Fig. 5.2. The conjugated process, i.e., the splitting of a Cooper pair, with the subsequent tunneling of two electrons to different leads, is also generated. Together, these are modeled by the additional term

$$H_{\rm CAR} = \sum_{j<k}\kappa_{jk}\psi_j^\dagger(0)\psi_k^\dagger(0)\gamma_k\gamma_j + \text{h.c.} \tag{5.15}$$

This implies that we have to take into account also the running of the crossed Andreev reflection amplitudes $\kappa_{jk} = \kappa_{kj}$. These are generally complex-valued, since the order parameters of the three leads have different phases. We remark that Δ is also a running parameter in our RG scheme and that it is possible to define the dimensionless parameter $\delta(\ell) = e^{\ell}\Delta/D$, with the bandwidth $D \simeq \min(E_C, \Delta_{\rm box})$ acting as a natural UV cutoff. The flow has a natural endpoint as soon as $\delta(\ell) \simeq 1$.

It is always possible to choose a gauge in which the phase dependence appears in $H_{\rm EC} + H_{\rm CAR}$ only. Considering the case in which all the superconducting gap parameters are equal, i.e., $\Delta_j = \Delta$ for all j, we obtain the RG equations

$$\frac{d\lambda_{jk}}{d\ell} = \frac{2}{\pi} \sum_{m\neq(j,k)}^{M} \left[\sqrt{1+\delta^2}\left(\lambda_{jm}\lambda_{mk} + \kappa_{jm}\kappa^*_{mk}\right) + \delta\left(\lambda_{jm}\kappa^*_{mk} + \kappa_{jm}\lambda_{mk}\right)\right], \quad (5.16)$$

$$\frac{d\kappa_{jk}}{d\ell} = \frac{2}{\pi} \sum_{m\neq(j,k)}^{M} \left[\delta\left(\lambda_{jm}\lambda^*_{mk} + \kappa_{jm}\kappa_{mk}\right) + \sqrt{1+\delta^2}\left(\lambda_{jm}\kappa_{mk} + \kappa_{jm}\lambda^*_{mk}\right)\right],$$

with $j \neq k$.

The phases of the coupling constants will also flow under renormalization. First, let us consider the phase-unbiased situation, in which all $\varphi_j = 0$. The RG equations (5.16) can be solved analytically by defining the real symmetric matrices $\Lambda^{(\pm)}_{jk} = \lambda_{jk} \pm \kappa_{jk}$. The flow Eq. (5.16) are now decoupled as

$$\frac{d\Lambda^{(\pm)}_{jk}}{d\ell} = \frac{2}{\pi}(\sqrt{1+\delta^2} \pm \delta) \sum_{m\neq(j,k)}^{M} \Lambda^{(\pm)}_{jm}\Lambda^{(\pm)}_{mk}, \quad (5.17)$$

which are equivalent to those for the TKE, apart from the presence of the gap parameter δ. Analogously to [45–47, 50], even if the initial conditions for $\Lambda^{(\pm)}_{jk}$ are chosen to be anisotropic with respect to the pair of indices j, k, the matrices scale towards a unique value $\Lambda^{(\pm)}_{jk}(\ell) \to \Lambda_{\pm}(\ell)[1 - \delta_{jk}]$. Rephrasing, anisotropies are irrelevant perturbations.

The flow can be solved for any initial value, but the solution has a simple form for isotropic initial conditions, $\Lambda_{\pm}(0) = \bar{\lambda}$. The coupling $\bar{\lambda}$ can then be taken as the average defined in (5.4). From (5.17), one has

$$\Lambda_{\pm}(\ell) = \lambda(\ell) \pm \kappa(\ell) = \frac{\bar{\lambda}}{1 - \frac{2(M-2)\bar{\lambda}}{\pi}\mathscr{F}_{\pm}(\ell)}, \quad (5.18)$$

in which the functions

$$\mathscr{F}_{\pm}(\ell) = \left[\sqrt{1+\delta^2} + \ln\sqrt{\frac{\sqrt{1+\delta^2}-1}{\sqrt{1+\delta^2}+1}} \pm \delta\right]_{\delta(0)}^{\delta(\ell)}. \quad (5.19)$$

appear. We observe that both $\Lambda_{\pm}$ grow for small values of the flow parameter ℓ, but the fate of the couplings will depend on the bare value of Δ as well. For $\Delta \gg T_K$, with T_K in (5.4), the dimensionless parameter δ reaches the value of unity before

$\Lambda_{\pm}$ can catch up. In this regime, the topological Kondo effect is suppressed by the strong electron-electron pairing and one can effectively consider the leads to be in the atomic limit.

In the opposite regime $\Delta \ll T_K$, the situation $\Lambda_+ \gg \Lambda_-$ is quickly reached and the coupling Λ_+ shows a divergence, while Λ_- remains bounded. The energy scale where $\Lambda_+(\ell)$ enters the strong-coupling regime can be estimated from (5.18) as

$$T_+ \approx T_K + \Delta \left(1 - e^{-\frac{\pi}{2(M-2)\lambda}}\right) . \tag{5.20}$$

We see from the above equation that a small pairing interaction increases the Kondo temperature. In this regime, the difference between the λ and the κ couplings is much smaller than their sum. For an arbitrarily small pairing Δ, finite CAR couplings κ are generated and they flow to strong coupling together with the λ. One can then tentatively write a strong coupling Hamiltonian in which the two sets of variables are equal and verify that the localized Majorana modes only couple to half of the channels in the wires, which corresponds to a $SO_1(M)$ Kondo fixed point. Therefore, any finite Δ destabilizes the $SO_2(M)$ Kondo fixed point and induces a flow to a stable fixed point with symmetry group $SO_1(M)$. For $M = 3$, this has been shown in [60], where the relation $SO_1(3) \sim SU_2(2)$ implies a two-channel (instead of the $\Delta = 0$ four-channel [45, 103]) Kondo fixed point.

We now study the effect of having different phases $\varphi_j \neq 0$ of the superconducting order parameter in the different leads. The RG flow thus only depends on the gauge-invariant phase differences $\varphi_j - \varphi_k$. In order to retain simplicity, we discuss the limit $\Delta \ll T_K$ for $M = 3$ leads. The RG equations (5.16) acquire a phase dependence, which can then be studied numerically. The flow of the absolute value is not altered by the phase bias, but, as the strong-coupling regime is approached, $\lambda_{jk}(\ell) \to \lambda(\ell)e^{i(\varphi_j-\varphi_k)/2}$ and $\kappa_{jk}(\ell) \to \kappa(\ell)e^{i\theta_{jk}}$, having isolated the phase dependence. The phases of the CAR parameter are $\theta_{jk} = (\varphi_j + \varphi_k)/2 - \varphi_0$, where the symmetry under exchange of the legs and gauge invariance determine the value of the center-of-mass phase $\varphi_0 = (\varphi_1 + \varphi_2 + \varphi_3)/3$ at the stable strong-coupling fixed point.

5.3.2 *Physics at Low Temperatures: The Topological Kondo Effect*

Having established which couplings are expected to dominate the low-temperature physics, we focus here on $M = 3$ and $\Delta \ll T_K$. In analogy to [105], we will consider perturbations around the two-channel Kondo fixed point [52, 104] and argue that they do not destabilize this fixed point.

The Hamiltonian equation (5.7) can be written in terms of a single chiral fermion via the "unfolding" transformation

$$\Phi_j(x>0) = \psi_{j,R}(x)\ , \qquad \Phi_j(x<0) = \psi_{j,L}(-x). \tag{5.21}$$

The real and imaginary part of the fermion field define two Majorana fields $\Phi_j(x) = [\eta_j(x) + i\xi_j(x)]/\sqrt{2}$ for $j = 1, 2, 3$. Using these definitions, out of the boundary degrees of freedom, we can define the spin-1/2 operators

$$\mathbf{S} = -\frac{i}{2}\boldsymbol{\gamma} \times \boldsymbol{\gamma}\ , \qquad \mathbf{S}_\eta = -\frac{i}{2}\eta(0) \times \eta(0)\ , \qquad \mathbf{S}_\xi = -\frac{i}{2}\xi(0) \times \xi(0) \tag{5.22}$$

where we have used the vector notation $\gamma = (\gamma_1, \gamma_2, \gamma_3)^T$, as well as $\eta = (\eta_1, \eta_2, \eta_3)^T$ and $\xi = (\xi_1, \xi_2, \xi_3)^T$. The operators in (5.22) satisfy the Pauli algebra.

We argued that, once the RG flow reaches the strong-coupling regime, we have $\Lambda_+ \gg \Lambda_-$. Using (5.18) this implies that $\lambda \approx \kappa$ for every pair of legs. Having also determined the phases of the renormalized couplings, we find it convenient to use the gauge transformation $\psi_{j,R/L}(x) \to e^{i(\varphi_j - \varphi_0)/2}\psi_{j,R/L}(x)$, which removes the phase factors from $H_{\mathrm{EC}} + H_{\mathrm{CAR}}$. Substituting into the interaction Hamiltonian and making use of the definitions in (5.22), one has the residual interaction which perturbs the Kondo fixed point in the form

$$H_{\mathrm{EC}} + H_{\mathrm{CAR}} = \Lambda_+ \mathbf{S} \cdot \mathbf{S}_\eta + \Lambda_- \mathbf{S} \cdot \mathbf{S}_\xi\ , \tag{5.23}$$

in which we have considered again the limit $\Delta \ll \Lambda_\pm$. We recognize here a competing Kondo interaction proportional to Λ_-. Whenever $\Delta \sim \Lambda_- \approx 0$, we have a single spin-spin interaction between the island and half of the fermionic degrees of freedom of the wire, which describes precisely the two-channel Kondo problem.

Analogously to the the description of [105], one can study the low-temperature physics. In particular, the strong coupling fixed point is characterized by a change in the boundary condition for the Majorana field η, which obeys twisted boundary conditions, obtained by the substitution

$$\boldsymbol{\eta}(x) \to \mathrm{sgn}(x)\boldsymbol{\eta}(x)\ . \tag{5.24}$$

Conversely, the other component of the complex bulk fermions still obeys $\xi(x) = \xi(-x)$. In terms of the original fermions, these boundary conditions correspond to perfect Andreev reflection at the boundary, $\psi_{j,R}(0) = -\psi^\dagger_{j,L}(0)$.

The Kondo screening of the impurity effective spin can be modeled via the replacement $S \to i\sqrt{T_K}\gamma_0\boldsymbol{\eta}(0)$, as in [105]. Here, the Majorana operator $\gamma_0 = i\xi_1(0)\xi_2(0)\xi_3(0)$ switches between the states of the two-dimensional low-energy manifold in which the imaginary part of boundary fermions on the legs are constrained. Its time-ordered ($\mathcal{T}$) correlation functions are $\langle \mathcal{T}\gamma_0(\tau)\gamma_0(0)\rangle = \frac{1}{2}\mathrm{sgn}(\tau)$. The leading irrelevant operator around the Kondo fixed point, describing the competition of the kinetic term of the lead Hamiltonian against the tunneling, can then be written as

$$H_{\mathrm{LIO}} = 2\pi T_K^{-1/2}\gamma_0\eta_1(0)\eta_2(0)\eta_3(0) \tag{5.25}$$

and has scaling dimension $d = 3/2$. The coupling of the island to the real part of the boundary fermions on the leads, proportional to Λ_- in (5.23), is also an irrelevant operator with dimension $d = 3/2$.

We now discuss the bulk pairing $\propto \Delta$ in the leads and compute the Josephson current. We will find a fractional dependence on the relative superconducting phases, which is detectable by probing the system with a finite frequency electric field, as in [106].

In order to achieve our goal, the boundary Green's functions of the field combinations $[\psi_{j,R}(0) - \psi^\dagger_{j,L}(0)]/\sqrt{2}$ at the strong coupling fixed point are needed. Given that "half" of the boundary spin is producing the screening on the Majorana island, the strong-coupling theory consists of decoupled leads where the η Majorana fermions obey twisted boundary conditions. Following [85], one obtains the Majorana correlation functions at the boundary ($x = 0^+$) in the form

$$\begin{aligned} \langle \mathscr{T} \eta_j(\tau) \xi_k(0) \rangle &= -i\delta_{jk} \Delta \cos(\varphi_j - \varphi_0) f(\tau), \\ \langle \mathscr{T} \eta_j(\tau) \eta_k(0) \rangle &= \langle \mathscr{T} \xi_j(\tau) \xi_k(0) \rangle = -\delta_{jk} \partial_\tau f(\tau), \end{aligned} \tag{5.26}$$

where the function

$$f(\tau) = \int \frac{d\omega}{2\pi} \frac{1 - e^{-\sqrt{\omega^2+\Delta^2}/T_K}}{\sqrt{\omega^2 + \Delta^2}} \cos(\omega\tau) \tag{5.27}$$

has been employed. The Josephson current can now be computed from the relation

$$\frac{2e}{\hbar} \frac{\partial}{\partial \varphi_\alpha} \lim_{T\to 0} F(T) \tag{5.28}$$

in which the free energy $F(T)$ is evaluated to second order in perturbation theory as

$$F^{(2)} = -\frac{1}{2} \int d\tau \langle \mathscr{T} H'(\tau) H'(0) \rangle \, . \tag{5.29}$$

Using (5.26) and Wick's theorem, one arrives at the Josephson current

$$I_j(\varphi_1, \varphi_2, \varphi_3) = \frac{I_0}{3} \sum_{k\neq j}^{3} \left[3 \sin \varphi_{j,k} + \sin \frac{\varphi_{j,p} + \varphi_{k,p}}{3} - \sin \frac{\varphi_{k,j} + \varphi_{p,j}}{3} \right], \tag{5.30}$$

where $p \neq j, k$. This is a function of phase differences $\varphi_{a,b} = \varphi_a - \varphi_b$ only. The critical current I_c, is set by

$$I_0 = \zeta \frac{e\Delta^2}{\hbar T_K}, \qquad \zeta = \frac{\Lambda_-^2 f^3(0)}{3} \, . \tag{5.31}$$

In the non-topological version of the Kondo effect, a perturbative calculation of the critical current in the pairing parameter has been presented in [88]. We remark, in comparison, the additional suppression factor $\Delta/T_K \ll 1$ due to the residual unscreened spin encoded by γ_0. Current conservation (5.9) is satisfied, which can be checked by inspection.

Noticeably, the phase dependence of the current (5.30) has a 6π-periodic component, alongside a 2π-periodic component, in sharp contrast with (5.14), obtained in the atomic limit. In general, the 6π periodicity coexists with 2π and 4π effects. Given that the particle number and the phase of the superconducting order parameter are conjugated variables, one can argue that the elementary charge degrees of freedom, in proximity of the strong coupling fixed point, carry fractional charge $e^* = 2e/3$. Note that, for any finite Δ, the low-temperature fixed point is a two-channel instead of a four-channel Kondo problem, hence this value of e^* differs from the one for normal leads probed by shot noise [50, 64].

5.4 Parafermionic Version

We now turn our attention to a parafermionic generalization of the TKE. The first ingredient is naturally a set of localized parafermionic modes, which constitutes a PF box. Subsequently, we need to engineer external leads which can probe the fractional quasiparticles and couple them to the localized modes.

5.4.1 The Hamiltonian

Two FQH puddles are created with opposite spin orientation. We assume that only one massless mode is running along the edge and, in particular, we consider the series $\nu = 1/(2k+1)$. These effectively one-dimensional modes are described by the Hamiltonian [68, 80, 107, 108]

$$H_{\mathrm{edge}} = \frac{v}{4\pi} \int \mathrm{d}x \, (\partial_x \hat{\phi})^2, \tag{5.32}$$

where v is the edge velocity. The bosonic field ϕ satisfies

$$[\hat{\phi}(x), \hat{\phi}(x')] = i\pi \, \mathrm{sgn}(x - x') \tag{5.33}$$

Counterpropagating edge modes from different FQH puddles can be gapped by proximitizing them either with a superconductor (SC), which introduces a relevant perturbation pinning the ϕ field, or with a ferromagnet (FM), which instead tends to pin the conjugated field θ in the respective segment.

Low-energy excitations can only originate from operators, which create localized domain walls between adjacent domains. The low-energy Hilbert space, in particular, is spanned by the states $|Q_{\text{tot}}, Q_1 \text{mod } 2, \ldots, Q_{N-1} \text{mod } 2\rangle$, where Q_{tot} is the total charge of the proximitizing SCs and the FQH edges within the PF box. Importantly, Q_{tot} can assume fractional values differing by multiples of ν. Moreover, given that the number of Cooper pairs is undetermined in a superconductor, Q_{tot} is only defined module 2. Similar to the Majorana case, the operators creating a domain wall switch between states of a zero-energy manifold, i.e., create stable zero-energy modes. These are the PF operators $\hat{\alpha}_j$, obeying the Z_n PF algebra with index $n = 2/\nu$,

$$\hat{\alpha}_j \hat{\alpha}_k = \omega^{\text{sgn}(k-j)} \hat{\alpha}_k \hat{\alpha}_j, \quad \omega = e^{2\pi i/n} = e^{i\pi\nu}. \tag{5.34}$$

The construction of the PF box is then completed by applying a backgate voltage to the device. The Hamiltonian of the floating device, finally, is complemented by the Coulomb charging energy depending on the total charge Q_{tot},

$$H_{\text{box}} = E_C \left(\hat{Q}_{\text{tot}} - q_0 \right)^2. \tag{5.35}$$

The parameter q_0 depends on the applied voltage and can be tuned in such a way that the box has quantized ground-state charge, given by the value of Q_{tot} closest to q_0. When the charging energy E_C is the largest energy scale, indeed, only virtual transitions to states with a different values of Q_{tot} are allowed. Special values of q_0 correspond to charge degenerate points, but we will not consider such cases here. This recipe has been outlined in [27, 29, 42], where more details on these localized parafermionic modes can be found.

We now construct the external leads, which serve as probing leads in transport studies. As shown in Fig. 5.3, it is possible to insert Ohmic contacts between different segments of the edges of the FQH droplet, so that the boundary degree of freedom in each segment is dynamically independent from each of the others. We denote the bosonic field as $\hat{\phi}_j$, with a subscript $j = 1, \ldots, M$ denoting the edge portion. The commutation relations between these chiral boson fields already incorporate Klein factors [68] since (5.36) follows from a single-edge commutation relation by imagining that all leads actually belong to one long edge.

$$[\hat{\phi}_j(x), \hat{\phi}_k(x')] = i\pi \left[\delta_{jk} \text{sgn}(x - x') + \text{sgn}(k - j) \right], \tag{5.36}$$

The dynamics of the chiral bosons is encoded in the free edge Hamiltonian

$$H_{\text{edge}} = \sum_{j=1}^{M} \frac{\nu}{4\pi} \int_{-L/2}^{+L/2} dx (\partial_x \hat{\phi}_j)^2, \tag{5.37}$$

Anisotropies in these velocities do not cause physical effects since they can be absorbed by a renormalization of cotunneling amplitudes. Since we are not inter-

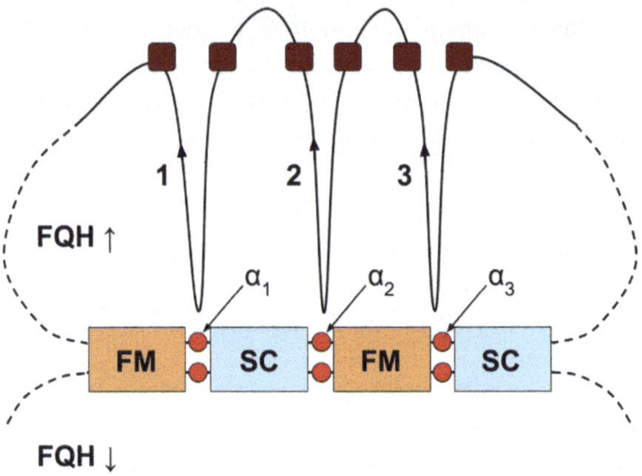

Fig. 5.3 Schematic setup of the parafermion box for the generalization of the topological Kondo effect. Proximitizing FM and SC domains to a pair of FQH edges with opposite spin opens spin gaps or superconducting gaps in the different regions. At each interface, pair of parafermion operators $\hat{\alpha}_j$ are created (with opposite spin). In this realization, the different superconducting domains are connected in order to establish a common coherent phase. A FQH drop can be shaped in such a way that the gapless edges are close to the localized parafermions, such that tunneling becomes possible. Dynamical independence of the edges is ensured by Ohmic contacts, the brown squares in the figure above

ested in finite-size effects in the leads, we will also assume $L \to \infty$. The fractional quasiparticle operator can then be expressed as vertex operator of the respective chiral boson field,

$$\hat{\psi}_j(x) \sim e^{i\sqrt{\nu}\hat{\phi}_j(x)} , \tag{5.38}$$

see [68, 80, 107, 108]. Using (5.36), the fractional quasiparticle operators (5.38) obey the algebra

$$\hat{\psi}_j(x)\hat{\psi}_k(x') = e^{-i\pi\nu \text{sgn}(k-j) - i\pi\nu\delta_{jk}\text{sgn}(x-x')}\hat{\psi}_k(x')\hat{\psi}_j(x). \tag{5.39}$$

Using the construction of the parafermion modes in [27], one can also show the commutation relation

$$\left[\hat{\phi}_j(x), \hat{\alpha}_k\right] = -\pi\sqrt{\nu} , \tag{5.40}$$

from which one also has

$$\hat{\psi}_j(x)\hat{\alpha}_k = e^{-i\pi\nu}\hat{\alpha}_k\hat{\psi}_j(x) . \tag{5.41}$$

We can now write down a tunneling Hamiltonian for the fractional quasiparticles. According to [27, 42], for a point-like tunnel contact at $x = 0$, it is given by

$$H_{\text{tun}} = \sum_{j=1}^{M} \eta_j \hat{\psi}_j(0)\hat{\alpha}_j^\dagger + \text{h.c.} \tag{5.42}$$

where the amplitudes η_j are generally complex-valued and describe tunneling of quasiparticles between the respective edge ($\hat{\psi}_j$) through the FQH bulk to the PF box via $\hat{\alpha}_j$. The relations (5.39) and (5.40) imply that all terms in (5.42) commute with each other. The Klein factors, which are needed to ensure proper statistical phase relations between different edges [68], are fully taken into account by (5.36), (5.34), (5.39) and (5.41).

Due to the large charging energy E_C in (5.35), the ground state of the PF box has a fixed value of Q_{tot} and the dominant low-energy processes are cotunneling of fractional quasiparticles from one edge to the box and from the box to some other edge. By projecting the full Hamiltonian, $H = H_{\text{edge}} + H_{\text{box}} + H_{\text{tun}}$, to the charge ground-state sector of the PF box via a Schrieffer–Wolff transformation [80, 131], one obtains the effective Hamiltonian $H \to H_{\text{eff}} = H_{\text{edge}} + H_{\text{cot}}$, with the the cotunneling term H_{cot} in the form

$$\begin{aligned} H_{\text{cot}} = & - \sum_{j,k=1;\, j\neq k}^{M} \lambda_{jk} \hat{\psi}_j^\dagger(0)\hat{\psi}_k(0)\hat{\alpha}_j\hat{\alpha}_k^\dagger \\ & - \sum_j |\eta_j|^2 \left(U_+^{-1} \hat{\psi}_j^\dagger(0)\hat{\psi}_j(0) + U_-^{-1} \hat{\psi}_j(0)\hat{\psi}_j^\dagger(0) \right), \end{aligned} \tag{5.43}$$

where the notation

$$\lambda_{jk} = \eta_j^* \eta_k \left(U_+^{-1} + U_-^{-1} \right) \tag{5.44}$$

for the cotunneling amplitude from lead k to lead j has been used. Here, U_+ and U_- denote the energy cost for adding one quasiparticle to the box or or for removing one quasiparticle from it, respectively. Besides (5.43), a term describing the simultaneous tunneling of two quasi-particles onto/off the island is obtained [131], but such a term would change the PF box electric charge by $\pm 2e^*$ and thus can be neglected after the projection to the low-energy subspace. The complex phases of η_j can, without loss of generality, be gauged away via the gauge transformation

$$\hat{\phi}_j(x) \to \hat{\phi}_j(x) + \text{const} . \tag{5.45}$$

After such a shift of the bosonic fields, all λ_{jk} in (5.44) are real, non-negative and symmetric in the leg indexes, i.e., $\lambda_{kj} = \lambda_{jk} > 0$. The total electric charge on the PF box is explicitly preserved by (5.43), as well as the total Z_n charge described in [6]. Summarizing, we have obtained the parafermionic generalization of the topological Kondo model as the effective Hamiltonian

$$H_{\text{eff}} = \sum_{j=1}^{M} \frac{\nu}{4\pi} \int_{-\infty}^{\infty} dx\, (\partial_x \hat{\phi}_j)^2 - \sum_{j\neq k}^{M} \lambda_{jk} \hat{\psi}_j^\dagger(0) \hat{\psi}_k(0) \hat{\alpha}_j \hat{\alpha}_k^\dagger, \tag{5.46}$$

together with the commutation relations in (5.36), (5.34), (5.39), and (5.41).

We remark that, from (5.41), the localized parafermions are the Majorana operators of the previous section when the filling $\nu = 1$. Therefore, the quantum impurity spin operators in (5.46) reproduces then the definition of **S** in (5.22) and generate the so(M) algebra [45, 50]. In fact, (5.46) reduces to the TKE Hamiltonian. However, for $\nu < 1$, the PF bilinears $\hat{\alpha}_j \hat{\alpha}_k^\dagger$ appearing in (5.46) do not constitute a closed Lie algebra. On the other hand, it is always possible to close the su$\left(n^{[(M-1)/2]}\right)$ algebra, where $[x]$ is the integer part of x, by adding powers and products of PF bilinears. For comparison with the TKE (i.e., $n = 2$), let us note that the algebra su(d), with $d = 2^{[(M-1)/2]}$, contains the expected subalgebra so(M) for $M \geq 7$. In the TKE, Majorana bilinears generate this subalgebra, and leads are only coupled to this subalgebra [45]. For $M < 7$, on the contrary, the Hilbert space dimension is $d < M$, meaning that the system forms a representation of so(M) which is smaller than the fundamental one. The quoted value for d follows by noting that one needs at least $[(M+1)/2]$ PF pairs on the box to couple to M external edges (leads), where constraining the total Z_n charge of the box is equivalent to removing one PF pair, cf. [42]. In particular, for $M = 3$, the algebra su(n) is generated by the set of operators

$$\left\{ \hat{\alpha}_1^{k_1} \hat{\alpha}_2^{k_2} \hat{\alpha}_3^{-k_1-k_2} \right\}, \tag{5.47}$$

where integers k_j are defined modulo $n = 2/\nu$ as $\hat{\alpha}_j^n = 1$, and we have used the convention $\hat{\alpha}_j^{-k} \equiv (\hat{\alpha}_j^\dagger)^k$ for $k > 0$. The fact that the dimension of the quantum impurity representation space is n follows from the PF representation in [6, 133] together with the Z_n-charge conservation constraint. For $\nu = 1/3$, we obtain the 35 generators of su(6) plus the identity from (5.47). The bilinears $\hat{\alpha}_j \hat{\alpha}_k^\dagger$ appearing in the Hamiltonian (5.46), however, constitute only a subset out of these algebra generators.

Taken all this into account, (5.46) has the form of a quantum impurity problem, in which M independent chiral edges interact with localized fractional quasiparticles in the origin. The cotunneling between the edges causes transitions between the different states of the PF box ground state manifold via the PF bilinears $\sim \hat{\alpha}_j \hat{\alpha}_k^\dagger$.

5.4.2 RG Equations

We now argue that, analogously to the Majorana box, the coupling to the PF box will make the boundary conditions for the chiral edges unstable, driving the system towards a non Fermi-liquid low-temperature fixed point. The operator product expansion of the vertex operators [132] can be used to derive RG equations up to second order in perturbation theory. Importantly, the new operator

$$H' = \left(\hat{\psi}_k^\dagger(0)\right)^2 \hat{\psi}_j(0)\hat{\psi}_m(0)\hat{\alpha}_k^2\hat{\alpha}_j^\dagger\hat{\alpha}_m^\dagger, \tag{5.48}$$

of scaling dimension $\Delta' = 3\nu$, will be generated after contraction of two tunneling terms. This is marginal for $\nu = 1/3$, and irrelevant for larger filling fractions and for $\nu = 2/3$ in particular. Given that the cotunneling terms $\sim \lambda_{jk}$ in (5.46) have smaller scaling dimension $\Delta_{jk} = \nu$, we neglect in the following this family of terms. Contractions of cotunneling operators can only produce PF bilinears of the type $\sim \hat{\alpha}_j\hat{\alpha}_k^\dagger$, which are in turn already present in (5.46).

Defining the variable

$$\tilde{\lambda}_{jk} = \frac{\tau_c^{1-\nu}}{\sqrt{2}v^\nu}\lambda_{jk}, \tag{5.49}$$

where τ_c denotes a ultraviolet cutoff in imaginary time, the RG equations take the same form as those for the Kondo coupling in the Majorana-based TKE with Luttinger liquid parameter $g = 1/\nu$ [46, 47],

$$\frac{d\tilde{\lambda}_{jk}}{d\ell} = (1-\nu)\tilde{\lambda}_{jk} + \sum_{m\neq(j\neq k)}^{M} \tilde{\lambda}_{jm}\tilde{\lambda}_{mk}. \tag{5.50}$$

For simplicity, let us first consider all the cotunneling amplitudes to be equal $\tilde{\lambda}_{jk} = \tilde{\lambda}(1-\delta_{jk})$. Then, (5.50) is simplified as

$$\frac{d\tilde{\lambda}}{d\ell} = (1-\nu)\tilde{\lambda} + (M-2)\tilde{\lambda}^2, \tag{5.51}$$

and its solution can be readily written as

$$\tilde{\lambda}(\ell) = \frac{\tilde{\lambda}(0)e^{(1-\nu)\ell}}{1 + \frac{M-2}{1-\nu}\tilde{\lambda}(0)[1 - e^{(1-\nu)\ell}]}. \tag{5.52}$$

We can see that $\tilde{\lambda}(\ell)$ grows under renormalization. A Kondo temperature can be defined as the energy scale for which (5.52) diverges, the system therefore entering a regime which is nonperturbative in the Kondo coupling. One finds

$$T_K \simeq E_C \left(\frac{(M-2)\tilde{\lambda}(0)}{1-\nu}\right)^{1/(1-\nu)}. \tag{5.53}$$

Comparing with the Kondo temperature characterizing the Majorana box, one sees that the dependence on the bare cotunneling coupling $\tilde{\lambda}(0)$ is now a power law, in contrast with (5.4) [45].

We will not tackle here the analysis of the tentative strong coupling theory and of its stability. A thorough study of the low-temperature fixed point and a derivation of the conductance tensor can be found in [43]. Here, we limit ourselves to remark the similarities, both in the Hamiltonian and in the RG equations, with the Kondo and the Topological Kondo models, which let us expect interesting nonperturbative physics at low temperature.

5.5 Conclusions

In this article, we have provided a summary of the present activities devoted to topological Kondo physics in Majorana devices. In particular, we have stressed two recent noteworthy developments, namely the Josephson effect for a multi-channel topological Kondo impurity and a parafermionic extension of the topological Kondo setup. The former problem could be realized using a Majorana box device with superconducting leads. The different periodicities in the atomic and in the strong-coupling limit (4π vs 6π for three leads) are particularly interesting. Moreover, we believe that there could be a so-far unexplored quantum phase transition at $\Delta \approx T_K$. This point requires a detailed numerical study. As second example, we have examined recent progress on a parafermionic extension of the basic TKE setting. We hope that experiments can soon start to catch up with the rapid theoretical developments.

Acknowledgements We thank A. Altland, B. Béri, Y. Gefen, K. Snizhko, P. Sodano, A. Tsvelik, and A. Zazunov for collaborations on the works reported above.

References

1. J. Alicea, Rep. Prog. Phys. **75**, 076501 (2012)
2. M. Leijnse, K. Flensberg, Semicond. Sci. Techn. **27**, 124003 (2012)
3. C.W.J. Beenakker, Annu. Rev. Condens. Matt. Phys. **4**, 113 (2013)
4. R.M. Lutchyn, E.P.A.M. Bakkers, L.P. Kouwenhoven, P. Krogstrup, C.M. Marcus, Y. Oreg, arXiv:1707.04899
5. C. Nayak, S.H. Simon, A. Stern, M. Freedman, S. Das Sarma, Rev. Mod. Phys. **80**, 1083 (2008)
6. J. Alicea, P. Fendley, Annu. Rev. Condens. Matter Phys. **7**, 119 (2016)
7. S. Das Sarma, M. Freedman, C. Nayak, Npj Quantum Inf. **1**, 15001 (2015)
8. V. Mourik, K. Zuo, S.M. Frolov, S.R. Plissard, E.P.A. Bakkers, L.P. Kouwenhoven, Science **336**, 1003 (2012)
9. M.T. Deng, C.L. Yu, G.Y. Huang, M. Larsson, P. Caroff, H.Q. Xu, Nano Lett. **12**, 6414 (2012)
10. A. Das, Y. Ronen, Y. Most, Y. Oreg, M. Heiblum, H. Shtrikman, Nature Phys. **8**, 887 (2012)
11. L.P. Rokhinson, X. Liu, J.K. Furdyna, Nature Phys. **8**, 795 (2012)
12. S. Nadj-Perge, I.K. Drozdov, J. Li, H. Chen, S. Jeon, J. Seo, A.H. MacDonald, B.A. Bernevig, A. Yazdani, Science **346**, 602 (2014)
13. M. Ruby, F. Pientka, Y. Peng, F. von Oppen, B.W. Heinrich, K.J. Franke, Phys. Rev. Lett. **115**, 197204 (2015)
14. H.H. Sun, K.W. Zhang, L.H. Hu, C. Li, G.Y. Wang, H.Y. Ma, Z.A. Xu, C.L. Gao, D.D. Guan, Y.Y. Li, C. Liu, D. Qian, Y. Zhou, L. Fu, S.C. Li, F.C. Zhang, J.F. Jia, Phys. Rev. Lett. **116**, 257003 (2016)

15. S.M. Albrecht, A.P. Higginbotham, M. Madsen, F. Kuemmeth, T.S. Jespersen, J. Nygård, P. Krogstrup, C.M. Marcus, Nature **531**, 206 (2016)
16. M.T. Deng, S. Vaitiekenas, E.B. Hansen, J. Danon, M. Leijnse, K. Flensberg, J. Nygård, P. Krogstrup, C.M. Marcus, Science **354**, 1557 (2016)
17. Ö. Gül, H. Zhang, F.K. de Vries, J. van Veen, K. Zuo, V. Mourik, S. Conesa-Boj, M.P. Nowak, D.J. van Woerkom, M. Quintero-Pérez, M.C. Cassidy, A. Geresdi, S. Koelling, D. Car, S.R. Plissard, E.P.A.M. Bakkers, L.P. Kouwenhoven, Nano Lett. **17**, 2690 (2017)
18. H. Zhang, Ö. Gül, S. Conesa-Boj, M. Nowak, M. Wimmer, K. Zuo, V. Mourik, F.K. de Vries, J. van Veen, M.W.A. de Moor, J.D.S. Bommer, D.J. van Woerkom, D. Car, S.R. Plissard, E.P.A.M. Bakkers, M. Quintero-Perez, M.C. Cassidy, S. Koelling, S. Goswami, K. Watanabe, T. Taniguchi, L.P. Kouwenhoven, Nat. Commun. **8**, 16025 (2017)
19. S.M. Albrecht, E.B. Hansen, A.P. Higginbotham, F. Kuemmeth, T.S. Jespersen, J. Nygård, P. Krogstrup, J. Danon, K. Flensberg, C.M. Marcus, Phys. Rev. Lett. **118**, 137701 (2017)
20. F. Nichele, A.C.C. Drachmann, A.M. Whiticar, E.C.T. O'Farrell, H.J. Suominen, A. Fornieri, T. Wang, G.C. Gardner, C. Thomas, A.T. Hatke, P. Krogstrup, M.J. Manfra, K. Flensberg, C.M. Marcus, Phys. Rev. Lett. **119**, 136803 (2017)
21. H.J. Suominen, M. Kjaergaard, A.R. Hamilton, J. Shabani, C.J. Palmstrøm, C.M. Marcus, F. Nichele, Phys. Rev. Lett. **119**, 176805 (2017)
22. S. Gazibegovich, D. Car, H. Zhang, S.C. Balk, J.A. Logan, M.W.A. de Moor, M.C. Cassidy, R. Schmits, D. Xu, G. Wang, P. Krogstrup, R.L.M. Op het Veld, J. Shen, D. Bouman, B. Shojaei, D. Pennachio, J.S. Lee, P.J. van Veldhoven, S. Koelling, M.A. Verheijen, L.P. Kouwenhoven, C.J. Palmstrøm, E.P.A.M. Bakkers, Nature **548**, 434 (2017)
23. B.E. Feldman, M.T. Randeria, J. Li, S. Jeon, Y. Xie, Z. Wang, I.K. Drozdov, B. Andrei Bernevig, A. Yazdani, Nat. Phys. **13**, 286 (2017)
24. R.S. Deacon, J. Wiedenmann, E. Bocquillon, F. Domínguez, T.M. Klapwijk, P. Leubner, C. Brüne, E.M. Hankiewicz, S. Tarucha, K. Ishibashi, H. Buhmann, L.W. Molenkamp, Phys. Rev. X **7**, 021011 (2017)
25. Y. Ronen, Y. Cohen, D. Banditt, M. Heiblum, V. Umansky, Nat. Phys. **14**, 411 (2018)
26. T. Wu, Z. Wan, A. Kazakov, Y. Wang, G. Simion, J. Liang, K.W. West, K. Baldwin, L.N. Pfeiffer, Y. Lyanda-Geller, L.P. Rokhinson, Phys. Rev. B **97**, 245304 (2018)
27. N.H. Lindner, E. Berg, G. Refael, A. Stern, Phys. Rev. X **2**, 041002 (2012)
28. M. Cheng, Phys. Rev. B **86**, 195126 (2012)
29. D.J. Clarke, J. Alicea, K. Shtengel, Nat. Commun. **4**, 1348 (2013)
30. M. Burrello, B. van Heck, E. Cobanera, Phys. Rev. B **87**, 195422 (2013)
31. A. Vaezi, Phys. Rev. B **87**, 035132 (2013)
32. F. Zhang, C.L. Kane, Phys. Rev. Lett. **113**, 036401 (2014)
33. R.S.K. Mong, D.J. Clarke, J. Alicea, N.H. Lindner, P. Fendley, C. Nayak, Y. Oreg, A. Stern, E. Berg, K. Shtengel, M.P.A. Fisher, Phys. Rev. X **4**, 011036 (2014)
34. D.J. Clarke, J. Alicea, K. Shtengel, Nat. Phys. **10**, 877 (2014)
35. M. Barkeshli, Y. Oreg, X.L. Qi, arXiv:1401.3750
36. M. Barkeshli, X.L. Qi, Phys. Rev. X **4**, 041035 (2014)
37. J. Klinovaja, D. Loss, Phys. Rev. Lett. **112**, 246403 (2014)
38. J. Klinovaja, A. Yacoby, D. Loss, Phys. Rev. B **90**, 155447 (2014)
39. M. Cheng, R.M. Lutchyn, Phys. Rev. B **92**, 134516 (2015)
40. J. Alicea, A. Stern, Phys. Scr. **T164**, 14006 (2015)
41. Y. Kim, D.J. Clarke, R.M. Lutchyn, Phys. Rev. B **96**, 041123 (2017)
42. K. Snizhko, R. Egger, Y. Gefen, Phys. Rev. B **97**, 081405(R) (2018)
43. K. Snizhko, F. Buccheri, R. Egger, Y. Gefen, Phys. Rev. B **97**, 235139 (2018)
44. S. Plugge, A. Rasmussen, R. Egger, K. Flensberg, New J. Phys. **19**, 012001 (2017)
45. B. Béri, N.R. Cooper, Phys. Rev. Lett. **109**, 156803 (2012)
46. A. Altland, R. Egger, Phys. Rev. Lett. **110**, 196401 (2013)
47. B. Béri, Phys. Rev. Lett. **110**, 216803 (2013)
48. N. Crampé, A. Trombettoni, Nucl. Phys. B **871**, 526 (2013)
49. A.M. Tsvelik, Phys. Rev. Lett. **110**, 147202 (2013)

50. A. Zazunov, A. Altland, R. Egger, New J. Phys. **16**, 015010 (2014)
51. A. Altland, B. Béri, R. Egger, A.M. Tsvelik, Phys. Rev. Lett. **113**, 076401 (2014)
52. E. Eriksson, C. Mora, A. Zazunov, R. Egger, Phys. Rev. Lett. **113**, 076404 (2014)
53. M.R. Galpin, A.K. Mitchell, J. Temaismithi, D.E. Logan, B. Béri, N.R. Cooper, Phys. Rev. B **89**, 045143 (2014)
54. E. Eriksson, A. Nava, C. Mora, R. Egger, Phys. Rev. B **90**, 245417 (2014)
55. O. Kashuba, C. Timm, Phys. Rev. Lett. **114**, 116801 (2015)
56. F. Buccheri, H. Babujian, V.E. Korepin, P. Sodano, A. Trombettoni, Nucl. Phys. B **896**, 52 (2015)
57. S. Plugge, A. Zazunov, E. Eriksson, A.M. Tsvelik, R. Egger, Phys. Rev. B **93**, 104524 (2016)
58. L. Herviou, K. Le Hur, C. Mora, Phys. Rev. B **94**, 235102 (2016)
59. F. Buccheri, G.D. Bruce, A. Trombettoni, D. Cassettari, H. Babujian, V.E. Korepin, P. Sodano, New J. Phys. **18**, 075012 (2016)
60. D. Giuliano, P. Sodano, A. Tagliacozzo, A. Trombettoni, Nucl. Phys. B **909**, 135 (2016)
61. D. Giuliano, G. Campagnano, A. Tagliacozzo, Eur. Phys. J. B **89**, 251 (2016)
62. A. Zazunov, F. Buccheri, P. Sodano, R. Egger, Phys. Rev. Lett. **118**, 057001 (2017)
63. Z.Q. Bao, F. Zhang, Phys. Rev. Lett. **119**, 187701 (2017)
64. B. Béri, Phys. Rev. Lett. **119**, 027701 (2017)
65. L.A. Landau, E. Sela, Phys. Rev. B **95**, 035135 (2017)
66. K. Michaeli, L.A. Landau, E. Sela, L. Fu, Phys. Rev. B **96**, 205403 (2017)
67. M. Gau, S. Plugge, R. Egger, Phys. Rev. B **97**, 184506 (2018)
68. A.O. Gogolin, A.A. Nersesyan, A.M. Tsvelik, *Bosonization and Strongly Correlated Systems* (Cambridge University Press, Cambridge UK, 1998)
69. R.M. Potok, I.G. Rau, H. Shtrikman, Y. Oreg, D. Goldhaber-Gordon, Nature **446**, 167 (2007)
70. Z. Iftikhar, S. Jezouin, A. Anthore, U. Gennser, F.D. Parmentier, A. Cavanna, F. Pierre, Nature **526**, 233 (2015)
71. A.J. Keller, L. Peeters, C.P. Moca, I. Weymann, D. Mahalu, V. Umansky, G. Zarand, D. Goldhaber-Gordon, Nature **526**, 237 (2015)
72. L. Fu, Phys. Rev. Lett. **104**, 056402 (2010)
73. A. Zazunov, A.L. Yeyati, R. Egger, Phys. Rev. B **84**, 165440 (2011)
74. R. Hützen, A. Zazunov, B. Braunecker, A.L. Yeyati, R. Egger, Phys. Rev. Lett. **109**, 166403 (2012)
75. D. Aasen, M. Hell, R.V. Mishmash, A. Higginbotham, J. Danon, M. Leijnse, T.S. Jespersen, J.A. Folk, C.M. Marcus, K. Flensberg, J. Alicea, Phys. Rev. X **6**, 031016 (2016)
76. S. Plugge, L.A. Landau, E. Sela, A. Altland, K. Flensberg, R. Egger, Phys. Rev. B **94**, 174514 (2016)
77. L.A. Landau, S. Plugge, E. Sela, A. Altland, S.M. Albrecht, R. Egger, Phys. Rev. Lett. **116**, 050501 (2016)
78. T. Karzig, C. Knapp, R.M. Lutchyn, P. Bonderson, M.B. Hastings, C. Nayak, J. Alicea, K. Flensberg, S. Plugge, Y. Oreg, C.M. Marcus, M.H. Freedman, Phys. Rev. B **95**, 235305 (2017)
79. D. Litinski, M.S. Kesselring, J. Eisert, F. von Oppen, Phys. Rev. X **7**, 031048 (2017)
80. A. Altland, B. Simons, *Condensed Matter Field Theory*, 2nd edn. (Cambridge University Press, Cambridge UK, 2010)
81. E. Strambini, S. D'Ambrosio, F. Vischi, F.S. Bergeret, Yu.V. Nazarov, F. Giazotto, Nat. Nanotech. **11**, 1055 (2016)
82. A. Zazunov, R. Egger, Phys. Rev. B **85**, 104514 (2012)
83. Y. Peng, F. Pientka, Y. Vinkler-Aviv, L.I. Glazman, F. von Oppen, Phys. Rev. Lett. **115**, 266804 (2015)
84. P.A. Ioselevich, P.M. Ostrovsky, M.V. Feigel'man, Phys. Rev. B **93**, 125435 (2016)
85. A. Zazunov, R. Egger, A.L. Yeyati, Phys. Rev. B **94**, 014502 (2016)
86. F. Setiawan, W.S. Cole, J.D. Sau, S. Das Sarma, Phys. Rev. B **95**, 020501 (2017)
87. A. Martín-Rodero, A.L. Yeyati, Adv. Phys. **60**, 899 (2011)
88. L.I. Glazman and K.A. Matveev, Pis'ma Zh. Eksp. Teor. Fiz. **49**, 570 (1989) [JETP Lett. **49**, 659 (1989)]

89. A. Golub, Phys. Rev. B **54**, 3640 (1996)
90. A.V. Rozhkov, D.P. Arovas, Phys. Rev. Lett. **82**, 2788 (1999)
91. E. Vecino, A. Martín-Rodero, A.L. Yeyati, Phys. Rev. B **68**, 035105 (2003)
92. F. Siano, R. Egger, Phys. Rev. Lett. **93**, 047002 (2004)
93. M.S. Choi, M. Lee, K. Kang, W. Belzig, Phys. Rev. B **70**, 020502 (2004)
94. C. Karrasch, A. Oguri, V. Meden, Phys. Rev. B **77**, 024517 (2008)
95. D.J. Luitz, F.F. Assaad, T. Novotný, C. Karrasch, V. Meden, Phys. Rev. Lett. **108**, 227001 (2012)
96. A. Zazunov, S. Plugge, R. Egger, Phys. Rev. Lett. **121**, 207701 (2018)
97. A.Y. Kasumov, R. Deblock, M. Kociak, B. Reulet, H. Bouchiat, I.I. Khodos, Y.B. Gorbatov, V.T. Volkov, C. Journet, M. Burghard, Science **284**, 1508 (1999)
98. J.A. van Dam, Y.V. Nazarov, E.P.A.M. Bakkers, S. De Franceschi, L.P. Kouwenhoven, Nature **442**, 7103 (2006)
99. J.P. Cleuziou, W. Wernsdorfer, V. Bouchiat, T. Ondarcuhu, M. Monthioux, Nat. Nanotech. **1**, 53 (2006)
100. H. Ingerslev Jørgensen, T. Novotný, K. Grove-Rasmussen, K. Flensberg, P.E. Lindelof, Nano Lett. **7**, 2441 (2007)
101. A. Eichler, R. Deblock, M. Weiss, C. Karrasch, V. Meden, C. Schönenberger, H. Bouchiat, Phys. Rev. B **79**, 161407 (2009)
102. R. Delagrange, D.J. Luitz, R. Weil, A. Kasumov, V. Meden, H. Bouchiat, R. Deblock, Phys. Rev. B **91**, 241401 (2015)
103. M. Fabrizio, A.O. Gogolin, Phys. Rev. B **50**, 17732(R) (1994)
104. I. Affleck, A.W.W. Ludwig, Phys. Rev. B **48**, 7297 (1993)
105. P. Coleman, L.B. Ioffe, A.M. Tsvelik, Phys. Rev. B **52**, 6611 (1995)
106. E. Bocquillon, R.S. Deacon, J. Wiedenmann, P. Leubner, T.M. Klapwijk, C. Brüne, K. Ishibashi, H. Buhmann, L.W. Molenkamp, Nat. Nanotech. **12**, 137 (2017)
107. X.G. Wen, Phys. Rev. B **44**, 5708 (1991)
108. C.L. Kane, M.P.A. Fisher, Phys. Rev. Lett. **68**, 1220 (1992)
109. V.J. Goldman, B. Su, Science **267**, 1010 (1995)
110. V.J. Goldman, Surf. Sci. **361–362**, 1 (1996)
111. R. de Picciotto, M. Reznikov, M. Heiblum, V. Umansky, G. Bunin, D. Mahalu, Nature **389**, 162 (1997)
112. L. Saminadayar, D.C. Glattli, Y. Jin, B. Etienne, Phys. Rev. Lett. **79**, 2526 (1997)
113. V.J. Goldman, Phys. E **1**, 15 (1997)
114. I.J. Maasilta, V.J. Goldman, Phys. Rev. B **55**, 4081 (1997)
115. C. Nayak, M.P.A. Fisher, A.W.W. Ludwig, H.H. Lin, Phys. Rev. B **59**, 15694 (1999)
116. S. Chen, B. Trauzettel, R. Egger, Phys. Rev. Lett. **89**, 226404 (2002)
117. C. Chamon, M. Oshikawa, I. Affleck, Phys. Rev. Lett. **91**, 206403 (2003)
118. X. Barnabé-Thériault, A. Sedeki, V. Meden, K. Schönhammer, Phys. Rev. Lett. **94**, 136405 (2005)
119. M. Oshikawa, C. Chamon, I. Affleck, J. Stat. Mech.: Theor. Exp. **P0**, 2006 (2008)
120. S. Das, S. Rao, D. Sen, Phys. Rev. B **74**, 045322 (2006)
121. C.-Y. Hou, C. Chamon, Phys. Rev. B **77**, 155422 (2008)
122. A. Agarwal, S. Das, S. Rao, D. Sen, Phys. Rev. Lett. **103**, 026401 (2009)
123. D. Giuliano, P. Sodano, Nucl. Phys. B **811**, 395 (2009)
124. B. Bellazzini, P. Calabrese, M. Mintchev, Phys. Rev. B **79**, 085122 (2009)
125. A. Altland, Y. Gefen, B. Rosenow, Phys. Rev. Lett. **108**, 136401 (2012)
126. A. Rahmani, C.-Y. Hou, A. Feiguin, M. Oshikawa, C. Chamon, I. Affleck, Phys. Rev. B **85**, 045120 (2012)
127. A. Altland, Y. Gefen, B. Rosenow, Phys. Rev. B **92**, 085124 (2015)
128. H. Yi, C.L. Kane, Phys. Rev. B **57**, R5579(R) (1998)
129. H. Yi, Phys. Rev. B **65**, 195101 (2002)
130. S. Ganeshan, M. Levin, Phys. Rev. B **93**, 075118 (2016)
131. J.R. Schrieffer, P.A. Wolff, Phys. Rev. **149**, 491 (1966)
132. J. Cardy, *Scaling and Renormalization in Statistical Physics* (Cambridge University Press, UK, 1996)
133. Z.Y. Dong, S.L. Yu, J.X. Li, Phys. Rev. B **96**, 245114 (2017)

Chapter 6
Holographic Kondo Models

Johanna Erdmenger

Abstract These lecture notes are devoted to studying the Kondo problem from the perspective of gauge/gravity duality. This duality is a major recent development within theoretical physics. It maps strongly coupled quantum systems to weakly coupled gravity theories and thus provides a new approach to their description. The Kondo model as originally proposed by J. Kondo in 1961 played a decisive role in the development of major concepts in quantum field theory, such as the renormalization group and the use of conformal symmetry. It describes describes a spin impurity interacting with a free electron gas: At low energies, the impurity is screened and there is a logarithmic rise of the resistivity. In quantum field theory, this amounts to a negative beta function for the impurity coupling and the theory flows to a non-trivial IR fixed point. In these lectures we construct and examine a variant of the Kondo model within gauge/gravity duality. The motivation is twofold: On the one hand, the model may be used for calculating observables for the case of a spin impurity interacting with a strongly correlated electron gas. On the other hand, the models allows for new insights into the working mechanisms of gauge/gravity duality. For constructing the gravity dual, we consider a version of the Kondo model with $SU(N)$ spin at large N, in which the ambient electrons are strongly coupled even before the interaction with the impurity is switched on. We present the brane construction which motivates a gravity dual Kondo model and use this model to calculate the impurity entanglement entropy. The resistivity has a power-law behaviour in this model. We also study quantum quenches, and discuss the relation to the Sachdev-Ye-Kitaev model.

J. Erdmenger (✉)
Institut für Theoretische Physik und Astrophysik, Julius-Maximilians-Universität Würzburg, Würzburg, Germany
e-mail: erdmenger@physik.uni-wuerzburg.de

A. Ferraz et al. (eds.), *Strongly Coupled Field Theories for Condensed Matter and Quantum Information Theory*, Springer Proceedings in Physics 239,
https://doi.org/10.1007/978-3-030-35473-2_6

6.1 Introduction

Dualities are special relations between theories in physics as given by a Hamiltonian or Lagrangian. Two different theories describing the same physical system are *dual* to each other. A familiar example is the duality between the Thirring model and the sine-Gordon model. The Thirring model describes fermions in $1+1$ dimensions with a quartic interaction. As discovered by Coleman [1], it may be mapped to the bosonic sine-Gordon model via bosonization. Both theories thus describe the same physics.

Gauge/gravity duality, as first realized by the AdS/CFT correspondence of Maldacena [2], is a very special duality in the sense that it relates a gravity theory to a gauge theory, i.e. a quantum field theory without gravity. This new relation implies new questions about the nature of gravity itself: How is gravity related to quantum physics? It is equivalent to a non-gravity theory at least in this special context-does this imply that it is non-fundamental? This is an open question which we will not explore in detail here. Nevertheless we note that gauge/gravity duality opens up new issues about the nature of gravity. It is important to emphasize in this context that so far the known examples of gauge/gravity duality involve gravity theories with negative cosmological constant, different from the theory describing our Universe in which the cosmological constant is extremely small but positive.

The best understood example of gauge/gravity duality is the AdS/CFT correspondence. For a quantum field theory in $3+1$ dimensions, it maps $\mathcal{N}=4$ $SU(N)$ supersymmetric Yang–Mills theory to supergravity on the space $\text{AdS}_5 \times S^5$, where AdS stands for Anti-de Sitter space and S^5 for the five-dimensional sphere. Anti-de Sitter space is a hyperbolic space with a negative cosmological constant and a boundary. The $\mathcal{N}=4$ supersymmetric quantum field theory is a conformal field theory, i.e. its coupling is not renormalized and the theory is invariant unter local scale transformations. It may be viewed as being defined on the boundary of the $4+1$-dimensional Anti-de Sitter space.

The equivalence between the two theories in the AdS/CFT correspondence may be made plausible by two arguments: First, the holographic principle [3, 4], and second, the fact that the symmetries of both theories coincide. The holographic principle states that the information contained in a volume is stored on its boundary. More precisely, in the context of semiclassical considerations for quantum gravity, the holographic principle states that the information stored in a spatial volume V_d is encoded in its boundary area A_{d-1}, measured in units of the Planck area l_p^{d-1}. This principle is realized for black holes for instance, for which according to the famous result of Bekenstein [5], their entropy scales with the area of its horizon. Secondly, in a string theory approach it may be seen that the symmetries under which the fields of the quantum field theory involved transform is realized geometrically in the dual gravity solution.

Applications of gauge/gravity duality. The fact that gauge/gravity duality relates strongly coupled quantum field theories to weakly coupled classical gravity theories provides a new approach to calculating observables in these strongly coupled quantum field theories. Generically, such theories are hard to study since there is no universal approach for calculating observables in them. This is crucially different from weakly coupled quantum field theories, for which perturbation theory is the method of choice and provides very accurate results. An example for an approach to strongly coupled theories are advanced numerical techniques such as Monte Carlo methods, in which space-time is discretized. This approach is very successful in calculating observables such as bound state masses and determining the structure of the phase diagram. However, it is afflicted by the sign problem which renders the description of transport properties very complicated, in particular at finite temperature and density. It is thus desirable to have an alternative approach at hand which allows for comparison. Gauge/gravity duality provides such an approach.

Strongly coupled quantum field theories appear in all areas of physics, including particle and condensed matter physics. Weakly coupled theories may successfully be described in a quasiparticle approach. Quasiparticles are quantum excitation in one-to-one correspondence with the states in the corresponding free (non-interacting) theory. In strongly-coupled systems however, this map is no longer present. In general, the excitations in these systems are collective modes of the individual degrees of freedom. Gauge/gravity duality provides an elegant way of describing these modes by mapping them to *quasinormal modes* of the gravity theory. These modes are complex eigenfrequencies of the fluctuations about the gravity background: Their real part is related to the mass of the fluctuations and their complex part to the decay width.

Before we proceed, it is important to stress that to the present day, gauge/gravity duality is a conjecture which has not been proved. The proof is hard in particular since it would require a non-perturbative understanding of string theory in a curved space background, which is not available so far.

Holographic Kondo model. As an example of how to generalize the original example of the AdS/CFT correspondence to more general cases of gauge/gravity duality, we will study in this lecture how to obtain a gravity dual of the well-known Kondo model of condensed matter physics.

The original Kondo model [6] describes the interaction of a free electron gas with a localized magnetic spin impurity. A crucial feature is that at low energies, the impurity is screened by the electrons. The Kondo model is in agreement with experiments involving metals with magnetic impurities, as it correctly predicts a logarithmic rise of the resistivity as the temperature approaches zero.

The significance of the Kondo model goes far beyond its origin as a model for metals with magnetic impurities. In particular, it played a crucial role in the development of the renormalization group (RG). The impurity coupling in the Kondo model has a negative beta function and perturbation theory breaks down at low energies, a property it shares with quantum chromodynamics (QCD). In some respects the Kondo model may thus be viewed as a toy model for QCD. Moreover, the Kondo model corresponds to a boundary RG flow connecting two RG fixed points. These correspond to a UV and a IR CFT, respectively. CFT techniques have proved very useful in studying the Kondo model, as reviewed in [7]. Moreover, the Kondo model has a large N limit in which it may be exactly solved using the Bethe ansatz [8, 9].

The holographic Kondo model we will introduce below differs from the original condensed matter model in that the ambient electrons are strongly coupled among themselves even before the interaction with the magnetic impurity is turned on. Moreover, the impurity is an $SU(N)$ spin with $N \to \infty$. The ambient degrees of freedom are dual to a gravity theory in an AdS_3 geometry at finite temperature. The impurity degrees of freedom are dual to an AdS_2 subspace. As we will see in detail below, the dual gravity model corresponds to a holographic RG flow dual to a UV fixed point perturbed by a marginally relevant operator, which flows to an IR fixed point. In addition, in the IR a condensate forms, such that the model has some similarity to a holographic superconductor [10]. For this model, we may calculate spectral functions and compare their shape to what is expected for the original Kondo model. This may be relevant for the physics of quantum dots. Including the backreaction of the impurity geometry on the ambient geometry allows to calculate the entanglement entropy. Quantum quenches of the Kondo coupling may also be considered.

Related sets of lecture notes including discussions of the holographic Kondo model by the same author may be found in [11, 12]. Detailed information on gauge/gravity duality, the AdS/CFT correspondence and its applications may be found for instance in the books [13–17].

6.2 AdS/CFT Correspondence

6.2.1 Statement of the Correspondence

Let us begin by considering the best understood example of gauge/gravity duality, the AdS/CFT correspondence. Here, 'AdS' stands for 'Anti-de Sitter space' and and 'CFT' for 'conformal field theory. The Dutch physicist Willem de Sitter was a friend of Einstein. The prefix 'Anti' refers to the fact that a crucial sign changes from plus to minus. In fact, Anti-de Sitter space is a hyperbolic space with a negative cosmological constant.

In this example a four-dimensional CFT, $\mathcal{N} = 4$ $SU(N)$ Super Yang–Mills theory, is conjectured to be dual to gravity in the space $AdS_5 \times S^5$. This was proposed along with other examples for AdS/CFT by Maldacena in his seminal paper [2] in 1997. As we will see, the two theories have the same amount of degrees of freedom per unit volume and the same global symmetries. We will first state the duality and then explain it in detail. The AdS/CFT correspondence states that

$\mathcal{N} = 4$ Super Yang–Mills (SYM) theory
with gauge group $SU(N)$ and Yang–Mills coupling constant g_{YM}
is dynamically equivalent to
IIB superstring theory
with string length $l_s = \sqrt{\alpha'}$ and coupling constant g_s
on $AdS_5 \times S^5$ with radius of curvature L, and N units of $F_{(5)}$ flux on S^5.

The two free parameters on the field-theory side, i.e. g_{YM} and N, are related to the free parameters g_s and $L/\sqrt{\alpha'}$ on the string theory side by

$$g_{\mathrm{YM}}^2 = 2\pi g_s \qquad \text{and} \qquad 2g_{\mathrm{YM}}^2 N = L^4/\alpha'^2.$$

For understanding this duality and its motivation in detail, let us first recall some properties of the ingredients involved. We begin with the field theory side and introduce conformal field theories and $\mathcal{N} = 4$ supersymmetry.

6.2.2 Prerequisites for AdS/CFT

6.2.2.1 Conformal Symmetry

An essential aspect for the AdS/CFT correspondence is that the quantum field theory involved is a conformal field theory (CFT). Such a theory consists of fields that transform covariantly under conformal coordinate transformation. These leave angles invariant (locally) and in flat d-dimensional spacetime are defined by the following transformation law of the metric,

$$\mathrm{d}x'_\mu \mathrm{d}x'^\mu = \Omega^{-2}(x)\mathrm{d}x_\mu \mathrm{d}x^\mu . \tag{6.1}$$

Infinitesimally, with $\Omega(x) = 1 - \sigma(x)$ and $x'^\mu = x^\mu + v^\mu(x)$, this gives rise to the conformal Killing equation

$$\partial_\mu v_\nu + \partial_\nu v_\mu = 2\sigma(x)\eta_{\mu\nu} , \qquad \sigma(x) = \frac{1}{d}\partial \cdot v . \tag{6.2}$$

In $d = 2$ dimensions, this reduces to the Cauchy–Riemann equations, which are solved by any holomorphic function. This implies that in $d = 2$, conformal symmetry is infinite dimensional and thus leads to an infinite number of conserved quantities. In more than two dimensions however, conformal symmetry is finite dimensional and the only solutions to the conformal Killing equation (6.2) are

$$v^\mu(x) = a^\mu + \omega^\mu{}_\nu x^\nu + \lambda x^\mu + b^\mu x^2 - 2(b \cdot x)x^\mu ; \quad \omega_{\mu\nu} = -\omega_{\nu\mu} , \sigma(x) = \lambda - 2b \cdot x . \tag{6.3}$$

In $d > 2$, the conformal Killing vector $v_\mu(x)$ is at most quadratic in x. It contains translations (of zeroth order in x), rotations and scale transformations (both linear in x) and special conformal transformations (quadratic in x). The scalar λ, the vectors a_μ and b_μ and the antisymmetric matrix $\omega_{\mu\nu}$ contain a total of

$$1 + 2d + d(d-1)/2 = (d+1)(d+2)/2 \tag{6.4}$$

free parameters. In Euclidean signature, the symmetry group generated by these transformations is $SO(d+1, 1)$, while in Lorentzian signature, it is $SO(d, 2)$. Let us examine the algebra associated to the infinitesimal transformations (6.3) with parameters $(a^\mu, \omega^{\mu\nu}, \lambda, b^\mu)$ for the Lorentzian case. The generator for translations is the momentum operator P_μ. The generator for Lorentz transformations is denoted by $L_{\mu\nu}$. The generator for scale transformations is D and the generator for special conformal transformations is K_μ. The conformal algebra consists of the Poincaré algebra supplemented by the relations

$$[L_{\mu\nu}, K_\rho] = i(\eta_{\mu\rho}K_\nu - \eta_{\nu\rho}K_\mu)\,, \qquad [D, P_\mu] = iP_\mu\,, \tag{6.5}$$

$$\left[D, K_\mu\right] = -iK_\mu\,, \qquad [D, L_{\mu\nu}] = 0\,, \qquad [K_\mu, K_\nu] = 0\,, \tag{6.6}$$

$$\left[K_\mu, P_\nu\right] = -2i(\eta_{\mu\nu}D - L_{\mu\nu})\,. \tag{6.7}$$

For the representations we postulate

$$[D, \phi(0)] = -i\Delta\phi(0) \tag{6.8}$$

for any field $\phi(x)$. This implies

$$\phi(x) \to \phi'(x') = \lambda^{-\Delta}\phi(x) \tag{6.9}$$

for $x \to x' = \lambda x$. Δ is the scaling dimension of the field ϕ. For an infinitesimal transformation this gives

$$\delta_D\phi \equiv [D, \phi(x)] = -i\Delta\phi(x) - ix^\mu\partial_\mu\phi(x)\,, \tag{6.10}$$

with similar relation for the other conformal transformations $\delta_P\phi$, $\delta_L\phi$, $\delta_K\phi$.

For organising the representations, it is useful to define the *quasiprimary* fields which satisfy

$$[K_\mu, \phi(0)] = 0\,. \tag{6.11}$$

This defines the fields of lowest scaling dimension in an irreducible representation of the conformal algebra. All other fields in this multiplet, the conformal *descendents* of ϕ, are obtained by acting with P_μ on the quasiprimary fields.

The infinitesimal transformations $\delta\phi$ give rise to the *conformal Ward identities*

$$\sum_{i=1}^{n} \langle \phi_1(x_1) \ldots \delta\phi_i(x_i) \ldots \phi_n(x_n) \rangle = 0 \,. \tag{6.12}$$

For scalar conformal fields this implies

$$\langle \phi_1(x_1)\phi_2(x_2) \rangle = \begin{cases} \frac{c}{(x_1-x_2)^{2\Delta}} & \text{if } \Delta_1 = \Delta_2 = \Delta \,, \\ 0 & \text{otherwise.} \end{cases} \tag{6.13}$$

For fields with spin, the conformal transformation acts also on the spacetime indices and reads

$$\delta_v \mathcal{O}(x) = -\mathcal{L}_v \mathcal{O}(x) \,, \qquad \mathcal{L}_v \equiv v \cdot \partial_x + \frac{\Delta}{d} \partial \cdot v - \frac{i}{2} \partial^{[\mu} v^{\nu]} L_{\mu\nu} \,, \tag{6.14}$$

for an operator $\mathcal{O}(x)$ of arbitrary spin. The Lorentz generator $L_{\mu\nu}$ acts on the spin indices. For these operators, the conformal correlation functions are more involved. However, conformal symmetry still fixes them up to a small number of independent contributions.

6.2.2.2 $\mathcal{N} = 4$ Supersymmetry

The $\mathcal{N} = 4$ $SU(N)$ Super Yang–Mills theory has some very special properties which are at the origin of it possessing a gravity dual. First of all, it was shown [18, 19] that this theory is conformally invariant even when quantised; its beta function vanishes to all orders in perturbation theory and also non-perturbative contributions are absent. A further important property is that this theory has a global $SU(4)$ symmetry, which is isomorphic to $SO(6)$. We will see that both the $SO(4, 2)$ conformal symmetry as well as $SU(4)$ are also realized as isometries in the dual gravity theory.

For the $\mathcal{N} = 4$ theory, the global $SU(4)$ symmetry is realized as an R symmetry of the supersymmetry algebra. This algebra has four supersymmetry generators which satisfy the anticommutation relations

$$\{Q^a_\alpha, \bar{Q}_{b\dot{\beta}}\} = 2\sigma^\mu{}_{\alpha\dot{\beta}} P_\mu \delta^a{}_b \,, \qquad a = 1, 2, 3, 4 \,, \tag{6.15}$$

with $\sigma^\mu = (\mathbb{1}, \boldsymbol{\sigma})$ and $\boldsymbol{\sigma}$ the three Pauli matrices. Equation (6.15) is invariant under $SU(4)$ rotations. This algebra may be combined with the conformal algebra into a superconformal algebra. This requires the introduction of further fermionic generators, the special superconformal generators S^a_α that satisfies

$$\{S^a_\alpha, \bar{S}_{b\dot{\beta}}\} = 2\sigma^\mu{}_{\alpha\dot{\beta}} K_\mu \delta^a{}_b \,, \qquad a = 1, 2, 3, 4, \tag{6.16}$$

with K_μ the generator of special conformal transformations. We note that the anticommutation relation for the generators S^a_α (6.16) is formally similar to the one for

Table 6.1 Supermultiplet of $\mathcal{N}=4$ supersymmetry

Fields		$SU(4)$ rep.
Gauge field	A_μ	**1**
Complex fermions	λ^a_α	**4**
Real scalars	X^i	**6**

the generators Q^a_α given by (6.15), with the momentum operator P_μ replaced by the special conformal transformations K_μ. The operators P_μ, $L_{\mu\nu}$, D, K_μ together with the Q^a_α, S^a_α form the superconformal algebra associated to the superconformal group $SU(2,2|4)$.

The elementary fields of $\mathcal{N}=4$ Super Yang–Mills theory are organized in a single multiplet of $SU(4)$, as shown in Table 6.1. The $SU(N)$ gauge field is a singlet of $SU(4)$. Moreover, the supermultiplet involves four complex Weyl fermions λ^a_α in the fundamental representation **4** of $SU(4)$ and six real scalars X^i in the representation **6** of $SU(4)$. Note that due to the supersymmetry, both the Weyl fermions and the scalars are in the adjoint representation of the gauge group $SU(N)$ since they are in the same multiplet as the gauge field.

The action of $\mathcal{N}=4$ Super Yang–Mills theory reads

$$S=\mathrm{tr}\int d^4x\left(-\frac{1}{2g_{\mathrm{YM}}{}^2}F_{\mu\nu}F^{\mu\nu}-i\sum_{a=1}^{4}\bar{\lambda}^a\bar{\sigma}^\mu D_\mu\lambda_a-\sum_{i=1}^{6}D_\mu\phi^i D^\mu\phi^i\right.$$
$$\left.+g_{\mathrm{YM}}\sum_{a,b,i}C^{ab}{}_i\lambda_a[\phi^i,\lambda_b]+g_{\mathrm{YM}}\sum_{a,b,i}\bar{C}_{iab}\bar{\lambda}^a[\phi^i,\bar{\lambda}^b]+\frac{g_{\mathrm{YM}}{}^2}{2}\sum_{i,j}[\phi^i,\phi^j]^2\right), \quad (6.17)$$

with gg_{YM} the Yang–Mills coupling. The C^{ab}_i are Clebsch–Gordan coefficients that couple two **4** representations to one **6** representation of the algebra of $SU(4)_R$. We note that in addition to the kinetic terms, this action contains interactions between three and four gauge fields via the non-abelian gauge-field commutators in $F^{\mu\nu}$, as well as Yukawa interaction terms between two fermions and a scalar, and a quartic scalar interaction.

6.2.2.3 Large *N* Limit

The large N limit plays an essential role for the AdS/CFT correspondence. It corresponds to a saddle point approximation. As realized by 't Hooft in 1974 [20], the perturbative expansion of fields in the adjoint representation of the $SU(N)$ gauge group may be reorganized using a double-line notation.

A field ϕ in the adjoint representation may be written as

$$\phi=\phi^A T^A \Leftrightarrow (\phi)^i{}_j=\phi^A(T^A)^i{}_j\,, \quad (6.18)$$

Fig. 6.1 Double-line propagator

where the T^A are the $N^2 - 1$ generators of $SU(N)$. These are matrices with indices i, j. If ϕ is a scalar field in $3 + 1$ dimensions, then its propagator in configuration space is given by

$$\langle \phi^i{}_j(x) \phi^k{}_l(y) \rangle = \delta^i{}_l \delta^k{}_j \frac{g^2}{4\pi^2 (x-y)^2}, \tag{6.19}$$

where g is a typical coupling in the theory. The Kronecker deltas enter from the $SU(N)$ completeness relation

$$\sum_{A=1}^{N^2-1} (T^A)^i{}_j (T^A)^k{}_l = \delta^i{}_l \delta^k{}_j - \frac{1}{N} \delta^i{}_j \delta^k{}_l, \tag{6.20}$$

in which the second term is suppressed for $N \to \infty$. The double-line propagator for (6.19) is shown in Fig. 6.1.

For scalar fields, g in (6.19) may be the coupling of a cubic interaction term; a quartic interaction term may then enter with coefficient g^2. In Yang–Mills theory, g will be the gauge coupling. It will turn out to be extremely useful to define the *'t Hooft coupling*

$$\lambda = g^2 N. \tag{6.21}$$

Let us now count how the contributions corresponding to Feynman diagrams scale with N and with λ. Note that in the normalization for the propagators chosen in (6.19), the vertices scale as $1/g^2$. Also, the sum over traces of indices contributes a factor of N for every closed loop. Assembling all the ingredients, we find that the Feynman diagrams scale as

$$f(\lambda, N) \sim N^{V-E+F} \lambda^{E-V} = N^{\chi} \lambda^{E-V}, \tag{6.22}$$

where V, E and F are the numbers of vertices, edges and faces of the surfaces created by the Feynman diagrams, respectively. χ is the Euler characteristic given by

$$\chi = V - E + F = 2 - 2G, \tag{6.23}$$

with G the genus of the surface. We see that the leading order in N is given by $G = 0$, i.e. by planar diagrams. We note that double-line Feynman diagrams are similar to string-theory diagrams, with strings splitting and joining. This provides a hint that large N quantum field theories are related to string theories. In the simple example with scalar fields considered here, it is not possible to determine exactly which string theory is given by the collection of large N field-theory Feynman diagrams.

The AdS/CFT correspondence however provides a map between well-defined field theories and string theories.

6.2.2.4 AdS Spaces

Anti-de Sitter (AdS) spaces play an important role in the AdS/CFT correspondence. This has several reasons: First of all, the isometries of AdS space in $d+1$ dimensions form the group $SO(d,2)$, which corresponds to the conformal group of a CFT in d dimensions. Moreover, AdS space has a constant negative curvature and a boundary at which we may imagine this CFT to be defined.

The embedding of $(d+1)$-dimensional AdS space into $(d+2)$-dimensional flat Minkowski spacetime is provided by the surface satisfying

$$X_1{}^2 + X_2{}^2 + \cdots + X_d{}^2 - X_0{}^2 - X_{d+1}{}^2 = -L^2 \,, \tag{6.24}$$

where $X_0, X_1, \ldots X_{d+1}$ are the coordinates of $(d+2)$-dimensional Minkowski space. L is referred to as the *AdS radius*. We note that in Lorentzian signature, the symmetry of the isometries of AdS_{d+1} is thus $SO(d,2)$, which coincides with the symmetry of a CFT_d, i.e. a conformal field theory in d dimensions with Lorentzian signature. In Euclidean signature, the sign in front of $X_0{}^2$ becomes a plus and the symmetry is $SO(d+1,1)$.

The boundary of AdS_{d+1} is located at the limit of all coordinates X_M becoming asymptotically large. For large X_M, the hyperboloid given by (6.24) approaches the light-cone in $\mathbb{R}^{d,2}$, given by

$$-X_0{}^2 + \sum_{i=1}^{d} X_i{}^2 - X_{d+1}{}^2 = 0 \,. \tag{6.25}$$

The boundary corresponds to the set of all lines on the light cone given by (6.25) which originate from the origin of $\mathbb{R}^{d,2}$, i.e. $0 \in \mathbb{R}^{d,2}$. This space corresponds to a conformal compactification of Minkowski space.

A set of coordinates that solves (6.24) is

$$\begin{aligned} X^0 &= L \cosh\rho \, \cos\tau \,, \\ X^{d+1} &= L \cosh\rho \, \sin\tau \,, \\ X^i &= L\, \Omega_i \, \sinh\rho \,, \qquad \text{for } i = 1, \ldots, d \,, \end{aligned} \tag{6.26}$$

where Ω_i with $i = 1, \ldots, d$ are angular coordinates satisfying $\sum_i \Omega_i^2 = 1$.The remaining coordinates take the ranges $\rho \in \mathbb{R}_+$ and $\tau \in [0, 2\pi[$. The coordinates (ρ, τ, Ω_i) are referred to as *global coordinates* of AdS_{d+1}. It is convenient to introduce a new coordinate θ by $\tan\theta = \sinh\rho$. Then the metric associated to the parametrization (6.26) becomes that of the Einstein static universe $\mathbb{R} \times S^d$,

$$\mathrm{d}s^2 = \frac{L^2}{\cos^2\theta}\left(-\,\mathrm{d}\tau^2 + \mathrm{d}\theta^2 + \sin^2\theta\,\mathrm{d}\Omega_{d-1}^2\right). \tag{6.27}$$

Since $0 \le \theta < \frac{\pi}{2}$, this metric covers half of $\mathbb{R} \times S^d$.

It is often useful to consider a metric in local coordinates on AdS_{d+1}. This is obtained from the parametrization, with $\mathbf{x} = (x^1, \ldots, x^{d-1})$,

$$\begin{aligned} X_0 &= \frac{L^2}{2r}\left(1 + \frac{r^2}{L^4}(\mathbf{x}^2 - t^2 + L^2)\right), \\ X_i &= \frac{r x_i}{L} \quad \text{for } i \in \{1, \ldots, d-1\}, \\ X_d &= \frac{L^2}{2r}\left(1 + \frac{r^2}{L^4}(\mathbf{x}^2 - t^2 - L^2)\right), \\ X_{d+1} &= \frac{rt}{L}. \end{aligned} \tag{6.28}$$

This covers only one half of the AdS spacetime since $r > 0$. The corresponding metric is referred to as *Poincaré metric* and reads

$$\mathrm{d}s^2 = \frac{L^2}{r^2}\mathrm{d}r^2 + \frac{r^2}{L^2}\eta_{\mu\nu}\mathrm{d}x^\mu \mathrm{d}x^\mu . \tag{6.29}$$

The boundary is located at $r \to \infty$. The embedding of the Poincaré patch into global AdS is shown in Fig. 6.2.

Note that the Ricci scalar and cosmological constant for Anti-de Sitter space are both negative,

$$R = -\frac{d(d+1)}{L^2}, \qquad \Lambda = -\frac{d(d-1)}{2L^2}. \tag{6.30}$$

A further choice of coordinates is obtained by introducing the coordinate $z \equiv L^2/r$, for which the Poincaré metric (6.29) becomes

$$\mathrm{d}s^2 = \frac{L^2}{z^2}\left(\mathrm{d}z^2 + \eta_{\mu\nu}\mathrm{d}x^\mu \mathrm{d}x^\nu\right). \tag{6.31}$$

In this case, the boundary is located at $z \to 0$. Note that in this limit, there is a coordinate singularity but the space remains regular since the curvature remains finite.

6.2.3 String Theory Origin of the AdS/CFT Correspondence

In full generality, the Maldacena conjecture [2] states that $\mathcal{N} = 4$ $SU(N)$ Super Yang–Mills theory is dual to type IIB string theory on $\mathrm{AdS}_5 \times S^5$ for all values of

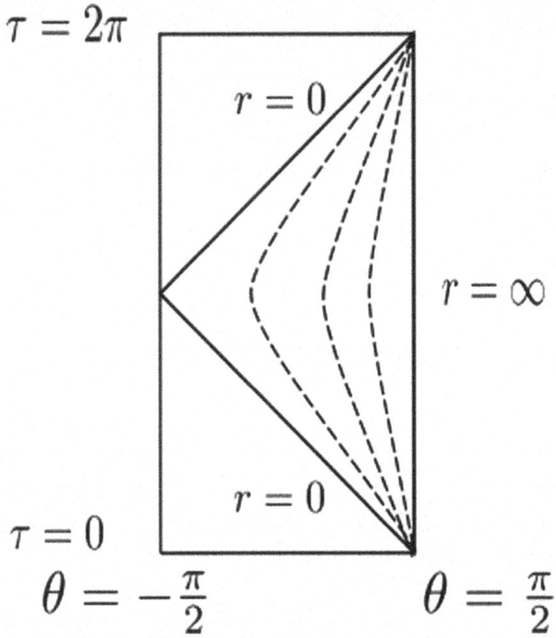

Fig. 6.2 Within AdS_2, the poincaré coordinates cover the triangular region shown. The dashed lines correspond to fixed constant values of r. The boundary is at $r = \infty$. θ and τ are as defined in (6.27)

N and λ. While this is a very beautiful idea, performing actual explicit calculations for testing this proposal requires to consider particular low-energy limits which we will discuss in detail. This is due to the fact that quantum string theory on curved backgrounds has not yet been formulated. This is also a reason why it is hard to provide an actual proof for the AdS/CFT proposal.

6.2.3.1 Motivating AdS/CFT from String Theory

As a particular limit, we consider weakly coupled string theory with string coupling $g_s \ll 1$, keeping $L/\sqrt{\alpha'}$ fixed. The leading order is the classical string theory with $g_s = 0$, which means to only tree-level string diagrams are taken into account. On the CFT side, since $g_{\rm YM}^2 = 2\pi g_s$ this implies $g_{YM}^2 = \lambda/N \to 0$. This in turn implies that $N \to \infty$ since $\lambda = L^4/(2\alpha'^2)$ remains finite. We are thus considering the 't Hooft limit. The duality conjectured in this limit, where λ is fixed but may be small, and the dual field theory contains classical strings, is often referred to as the *strong form* of the AdS/CFT correspondence. There is also the *weak form* of AdS/CFT in which additionally, λ is taken to be very large such that the CFT involved becomes strongly coupled. In this case, the strongly coupled CFT is mapped to a classical gravity theory of pointlike particles, since $\alpha' = \ell_s^2$ (with ℓ the string length) becomes asymptotically small. The gravity theory involved is type IIB supergravity in the

Table 6.2 Different forms of the AdS/CFT correspondence

	$\mathcal{N}=4$ SYM theory	IIB theory on $AdS_5 \times S^5$
Strongest form	any N and λ	Quantum string theory, $g_s \neq 0, \alpha' \neq 0$
Strong form	$N \to \infty$, λ fixed but arbitrary	Classical string theory, $g_s \to 0, \alpha' \neq 0$
Weak form	$N \to \infty$, λ large	Classical supergravity, $g_s \to 0, \alpha' \to 0$

Table 6.3 Embedding of N coincident D3–branes in flat ten-dimensional spacetime

	0	1	2	3	4	5	6	7	8	9
N $D3$	•	•	•	•	–	–	–	–	–	–

example considered. Type IIB supergravity admits D3-brane solutions. The possible limits of the AdS/CFT correspondence are collected together in Table 6.2.

Let us now consider D3-branes to motivate the weak form of the AdS/CFT correspondence. These branes may be viewed from two different perspectives: The open and the closed string perspective. It is crucial for the correspondence that in the low-energy limit where only massless degrees of freedom contribute, open strings give rise to gauge theories while closed strings give rise to gravity theories.

Open string perspective. We begin with the open string perspective on D3-branes. For $g_s N \ll 1$, D-branes may be visualised as higher-dimensional charged objectd on which open strings may end. The 'D' stands for Dirichlet boundary condition. Consider a stack of N D3-branes embedded in $9+1$ flat spacetime dimensions. (Recall that in $9+1$ dimensions, superstring theory is anomaly free and thus consistent.) Neumann and Dirichlet boundary conditions are imposed on the string modes according to Table 6.3.

For N coincident D3-branes, the open strings are described by a *Dirac-Born-Infeld (DBI) action* with gauge group $U(N)$, with integration over the $3+1$-dimensional worldvolume of the branes. In flat ten-dimensional space, the DBI action is given by

$$S_{\mathrm{DBI}} = - T_3 \,\mathrm{tr} \int \mathrm{d}^4 x e^{-\varphi} \sqrt{-\det(P[g] + 2\pi\alpha' F)} + \text{fermionic partners}\,, \tag{6.32}$$

where $T_3 \equiv 2/((2\pi)^3 \alpha'^2 g_s)$ is the brane tension, φ is the dilaton, and $P[g]$ is the pullback of the metric to the worldvolume of the branes. F is the field strength tensor of the gauge field associated to the brane charge. We now consider low-energy excitations with $E \ll \alpha'^{-1/2}$, such that only massless excitations are taken into account. In this limit, the DBI action reduces to

$$S_{\mathrm{DBI}} = -\frac{1}{2\pi g_s} \operatorname{tr} \int d^4x \left(\frac{1}{2} F_{\mu\nu} F^{\mu\nu} + \sum_{i=1}^{6} \partial^\mu \phi^i \partial_\mu \phi^i - \pi g_s \sum_{i,j=1}^{6} [\phi^i, \phi^j]^2 \right) + \text{fermions} + \mathcal{O}(\alpha') , \tag{6.33}$$

where the six scalars $\phi^i = \phi^{iA} T^A$ in the adjoint representation of $U(N)$ arise from the pull-back of the metric to the world-volume of the N D3-branes. They are given by $X^{i+3} = 2\pi\alpha'\phi^i$ with the $X^i + 3$ the coordinates in the directions perpendicular to the brane.

The total action for the D3-branes is

$$S_{D3} = S_{\mathrm{DBI}} + S_{\mathrm{closed}} + S_{\mathrm{int}} , \tag{6.34}$$

where S_{closed} describes the closed string excitations in the ten-dimensional space and S_{int} the interaction between open and closed string modes. In the low-energy limit $\alpha' \to 0$, the open strings decouple from any closed string excitations in the $9+1$-dimensional space: In (6.34), S_{closed} becomes a free theory of massless metric fluctuations, and S_{int} goes to zero. In this limit we are thus left with the low-energy modes in the DBI action as given by (6.33), plus free massless gravity excitations about flat space. The low-energy modes described by the DBI action coincide with the field-theory action of $\mathcal{N} = 4$ Super Yang–Mills theory as given by (6.17),

$$\lim_{\alpha' \to 0} S_{\mathrm{DBI}} = S_{\mathcal{N}=4\,\mathrm{SYM}} , \tag{6.35}$$

subject to identifying $2\pi g_s = g_{YM}^2$. We thus recover the action of $\mathcal{N} = 4$ Super Yang–Mills theory in this limit. By modding out the center of the gauge group, we may reduce the $U(N)$ gauge symmetry to $SU(N)$. Note that the limit taken is $\alpha' \to 0$ while keeping $u = r/\alpha'$ fixed, with r any length scale. This is referred to as the *Maldacena limit.*

Closed string perspective. We now turn to the closed string perspective on D-branes. In the limit $g_s N \gg 1$, the N D3-branes may be viewed as massive extended charged objects sourcing the fields of type IIB supergravity. Closed strings will propagate in this background. The supergravity solution of N D3-branes preserving $SO(3, 1) \times SO(6)$ symmetry in $9+1$ dimensions is given by

$$\begin{aligned} ds^2 &= H(r)^{-1/2} \eta_{\mu\nu} dx^\mu dx^\nu + H(r)^{1/2} \delta_{ij} dy^i dy^j , \\ e^{\varphi(r)} &= g_s , \qquad C_{(4)} = \left(1 - H(r)^{-1}\right) dx^0 \wedge dx^1 \wedge dx^2 \wedge dx^3 + \ldots , \end{aligned} \tag{6.36}$$

with $\mu\nu \in \{0, 1, 2, 3\}$ and $i, j \in \{1, 2, \ldots, 6\}$. Here, $r^2 = y_1^2 + y_2^2 + \cdots + y_6^2$ and the terms denoted by the dots ... in the expression for the four-form $C_{(4)}$ ensure self-duality of $F_{(5)} = dC_{(4)}$, i.e. the five-form given by the exterior derivative of $C_{(4)}$. Inserting the ansatz (6.36) into the Einstein equations of motion in $9+1$ dimensions, we find that $H(r)$ must be harmonic, i.e.

$$\triangle H(r) = 0\,, \quad \text{for } r \neq 0\,, \tag{6.37}$$

with $\triangle$ the Laplace operator in six Euclidean dimensions. The Laplace equation is solved by

$$H(r) = 1 + \left(\frac{L}{r}\right)^4 . \tag{6.38}$$

We will determine L below.

Similarly to the open string case considered before, we now investigate low-energy limits within the closed string perspective. First we note that asymptotically for $r \to \infty$, we have $H(r) \to 1$, i.e. asymptotically for large r we recover flat $9+1$-dimensional space. On the other hand, there is the *near-horizon limit* in which $r \ll L$. Then, $H(r) \sim L^4/r^4$ and the D3-brane metric becomes

$$\begin{aligned} \mathrm{d}s^2 &= \frac{r^2}{L^2}\eta_{\mu\nu}\mathrm{d}x^\mu \mathrm{d}x^\nu + \frac{L^2}{r^2}\delta_{ij}\mathrm{d}y^i\mathrm{d}y^j\,, \\ &= \frac{L^2}{z^2}\left(\eta_{\mu\nu}\mathrm{d}x^\mu \mathrm{d}x^\nu + \mathrm{d}z^2\right) + L^2\mathrm{d}\Omega_5^2\,, \end{aligned} \tag{6.39}$$

where in the second line we define the new radial coordinate $z \equiv L^2/r$ and introduced polar coordinates on the space spanned by the six y^i coordinates, $\mathrm{d}y^i\mathrm{d}y^i = \mathrm{d}r^2 = r^2\mathrm{d}\Omega_5^2$ with $\mathrm{d}\Omega_5^2$ the angular element on S^5. We see that in the near-horizon limit, the D3-brane metric becomes $\mathrm{AdS}_5 \times S^5$!

L, i.e. the radius of both the AdS_5 and the S^5, may be determined from string theory. For this we note that the flux of $F_{(5)}$ through the S^5 has to be quantized. The sphere S^5 surrounds the six Euclidean dimensions perpendicular to the D3-branes at infinity. The charge Q of the D3-branes is determined by

$$Q = \frac{1}{16\pi G_{10}} \int\limits_{S^5} {}^*F_{(5)}\,. \tag{6.40}$$

The charge has to coincide with the number of D-branes, i.e. $Q = N$. This implies the important relation

$$L^4 = 4\pi g_s N \alpha'^2\,, \tag{6.41}$$

since $16\pi G_{10} = 2\kappa_{10}^2 = (2\pi)^7\alpha'^4 g_s^2$.

For stating the correspondence, we note that asymptotically, we observe two kinds of closed strings: Those in flat space at $r \to \infty$, and those in the near-horizon region. Both kinds decouple in the low-energy limit. For an observer at infinity, the energy of fluctuations in the near-horizon region is redshifted,

$$E_\infty \sim \frac{r}{L}E_r \to 0\,. \tag{6.42}$$

Recall that $\sqrt{\alpha'}$ is fixed, but $r \ll L$. This implies that for an observer at infinity, the energy of fluctuations in the near-horizon region is very small. We thus have two types of massless excitations: Massless modes in flat space at $r \to \infty$ and the modes in the near-horizon region, which appear as massless too.

Combining open and closed string perspectives. The AdS/CFT correspondence is now motivated by identifying the massless modes in the open and closed string perspectives. First we note that as discussed above, both in the open and closed string pictures there are massless modes corresponding to free gravity in flat $9+1$-dimensional space. Moreover, in the open string picture further massless modes are given by the Lagrangian of $3+1$-dimensional $\mathcal{N}=4$ $SU(N)$ Super Yang–Mills theory. On the other hand, in the closed string picture we have gravity in the near-horizon region, which is given by IIB supergravity on $\text{AdS}_5 \times S^5$. Identifying these second types of massless modes in the open and closed string pictures gives rise to the AdS/CFT conjecture.

As a final remark in this section, we note that in the near-horizon limit of the closed string picture, it is not possible to locate the D3-branes. In particular, it is not correct to state that they sit at $r=0$. Rather, the D3-brane is a solitonic solution to 10d supergravity which extends over all values of r and which gives rise to $\text{AdS}_5 \times S^5$ in the near-horizon limit.

6.2.3.2 Field-Operator Map

The argument given in Sect. 6.2.3 motivates the conjectured duality between a quantum field theory and a gravity theory. The map between these two theories may be refined to a one-to-one map between individual operators, i.e. between gauge invariant operators in $\mathcal{N}=4$ $SU(N)$ Super Yang–Mills theory and classical gravity fields in $\text{AdS}_5 \times S^5$. Each pair is given by identifying entries transforming in the same representation of the superconformal group $SU(2,2|4)$. The most prominent example are the 1/2 BPS or chiral primary operators in the $[0, \Delta, 0]$ representation of the algebra of $SU(4)$. Here, the three entries are the Dynkin labels, with Δ the conformal dimension of the corresponding operator.[1] The corresponding gauge invariant field theory operators are

$$\mathcal{O}_\Delta(x) = \text{Str}\left(X^{(i_1}(x)X^{i_2}(x)\ldots X^{i_\Delta)}(x)\right) = C^\Delta_{i_1\ldots i_\Delta}\text{tr}\left(X^{(i_1}(x)X^{i_2}(x)\ldots X^{i_\Delta)}(x)\right), \tag{6.43}$$

with the elementary real scalar fields X^i as in (6.17). Str denotes the symmetrized trace over the indices (a,b) of the $SU(N)$ representation matrices $T^{A}{}^b_a$. The symmetrization involves the totally symmetric $SU(4)$ rank Δ tensor representation $C^\Delta_{i_1\ldots i_\Delta}$. An important property of the 1/2 BPS operators is that their two- and three-point functions in $\mathcal{N}=4$ Super Yang–Mills theory are not renormalized and thus independent of the 't Hooft coupling λ. The perturbative small λ results for these

[1] A review of the group theory concepts mentioned here may for instance be found in Appendix B of [13].

two- and three-point functions may then directly be compared to their counterparts calculated from the gravity side, which apply to large λ. However, since these correlation functions are independent of λ, an exact matching of the field theory and gravity results is expected and was indeed obtained in explicit computations [21, 22]. This provides a non-trivial test of the AdS/CFT proposal.

To obtain the corresponding fields on the supergravity side of the correspondence, a Kaluza–Klein reduction is performed on S^5, i.e. the fields in ten dimensions are expanded in spherical harmonics on S^5,

$$\phi(x, z, \Omega_5) = \sum_{l=0}^{\infty} \phi^l(x, z) Y^l(\Omega_5),$$
$$\Box_{S^5} Y^l(\Omega_5) = -\frac{1}{L^2} l(l+4) Y^l(\Omega_5). \tag{6.44}$$

This calculation was already performed in 1985 in [23]. From the Kaluza–Klein modes of the supergravity metric and five-form, we may construct five-dimensional scalars $s^l(x, z)$ that are in the same representation $[0, \Delta, 0]$ as the field-theory operators $\mathcal{O}^\Delta$ if $l = \Delta$. These scalars satisfy

$$\Box_{AdS_5} s^l(z, x) = -\frac{1}{L^2} l(l-4) s^l(x, z). \tag{6.45}$$

Asymptotically, near the AdS boundary at $z \to 0$, the solutions to this equation satisfy

$$s^l(z, x) \sim s^l_{(0)} z^{4-\Delta} + \langle \mathcal{O} \rangle z^\Delta + \text{subleading terms}. \tag{6.46}$$

According to [24], the leading term $s^l_{(0)}$ may be identified with a source for the 1/2 BPS operator $\mathcal{O}^l$, while the subleading term involves the vacuum expectation value of this operator.

For writing the AdS/CFT conjecture in terms of an equation, we add sources for any gauge invariant composite operators to the CFT action,

$$S' = S - \int \mathrm{d}^4 x \phi_{(0)}(x) \mathcal{O}(x). \tag{6.47}$$

Wick rotating to Euclidean time, the generating functional for these operators then reads

$$Z[\phi_{(0)}] = e^{-W[\phi_{(0)}]} = \left\langle \exp\left(\int \mathrm{d}^d x \phi_{(0)}(x) \mathcal{O}(x) \right) \right\rangle_{\text{CFT}}. \tag{6.48}$$

The AdS/CFT conjecture may then be stated as

$$W[\phi_{(0)}] = S_{\text{SUGRA}}[\phi] \Big|_{\lim\limits_{z \to 0} (\phi(x,z) z^{\Delta-4}) = \phi_{(0)}(x)}. \tag{6.49}$$

Fig. 6.3 Witten diagram for a three-point function

The boundary values of the supergravity fields are identified with the sources of the dual field theory. Within AdS/CFT, the operator sources of the CFT become dynamical classical fields propagating into the AdS space in one dimension higher. Note also that AdS/CFT has elements of a saddle point approximation since the CFT functional is given by a classical action on the gravity side. This is expected in the large N limit which also amounts to a saddle point approximation.

From the proposal (6.49) we may calculate connected Green's functions in the CFT by taking functional derivatives with respect to the sources on both sides of this equation. On the field theory side we have

$$\langle \mathcal{O}_1(x_1)\dots\mathcal{O}_n(x_n)\rangle = -\frac{\delta^n W}{\delta\phi^1_{(0)}(x_1)\dots\delta\phi^n_{(0)}(x_n)}\Bigg|_{\phi^i_{(0)}=0}. \tag{6.50}$$

Using (6.49) we may thus calculate CFT correlation functions from the propagation of the source fields through AdS space. Since the gravity action is classical, only tree diagrams contribute. The classical propagators on the gravity side are given by the Green's functions of the operator $\Box_{\mathrm{AdS}_5}$, while the vertices are obtained from higher order terms in the Kaluza–Klein reduction of the ten-dimensional gravity fields on S^5. The corresponding Feynman diagrams are referred to as Witten diagrams [25]. These are usually drawn as a circle depicting the boundary of AdS space, with the interior of the circle corresponding to the AdS bulk space. An example for a Witten diagram leading to a three-point function is shown in Fig. 6.3. Here, each of the three lines in the bulk of AdS corresponds to bulk-to-boundary propagator, i.e. to the appropriate Green's function of $\Box_{\mathrm{AdS}_5}$ with one endpoint at the boundary. For scalar operators, the bulk-to-boundary propagator is given by

$$K_\Delta(z_0,\mathbf{z};\mathbf{x}) = \frac{\Gamma(\Delta)}{\pi^{d/2}\Gamma\left(\Delta-\frac{d}{2}\right)}\left(\frac{z_0}{z_0^2+(\mathbf{z}-\mathbf{x})^2}\right)^{\Delta} \tag{6.51}$$

in Euclidean AdS space with five-dimensional coordinates $z \equiv (z_0,\mathbf{z})$ with z_0 the radial coordinate and $\mathbf{z}$ the four coordinates parallel to the boundary. For the second coordinate, $x_0 = 0$ since x is located at the boundary. The index Δ corresponds to the dimension of the dual scalar operator. Moreover, the vertex in the Witten diagram corresponds to a cubic coupling obtained from the Kaluza–Klein reduction of the type IIB supergravity action on S^5. For four-point functions or even higher correlation functions, there are contributions involving bulk-to-bulk propagators that link two vertices in the bulk of AdS space. The calculation of two- and three point functions of 1/2 BPS operators in $\mathcal{N} = 4$ Super Yang–Mills theory and in IIB supergravity on $AdS_5 \times S^5$ provides an impressive test of the AdS/CFT conjecture: The results for the three-point function in field theory and gravity coincide, subject to an appropriate normalization using the expressions for the two-point function [21, 22].

6.2.4 Finite Temperature

Let us now consider how the AdS/CFT correspondence may be generalized to quantum field theory at finite temperature. In fact, there is a natural way to proceed, which is based on the following. In thermal equilibrium, quantum field theories may be described in the imaginary time formalism. This means that the ensemble average of an operator at temperature T is given by

$$\langle \mathcal{O} \rangle_\beta = \mathrm{tr}\left(\frac{\exp(-\beta H)}{Z} \mathcal{O} \right), \quad Z = \mathrm{tr}\exp(-\beta H), \tag{6.52}$$

where $\beta = 1/(k_B T)$ and we set $k_B = 1$. H is the Hamiltonian of the theory considered. Formally, β corresponds to an imaginary time, $t = i\tau$. An important point is that the analyticity properties of thermal Green's functions require $\tau \in [0, \beta]$. This implies that the imaginary time τ is compactified on a circle.

Let us consider the gravity dual thermodynamics of $\mathcal{N} = 4$ Super Yang–Mills theory on $I\!R^3$. We note that the compactification of the time direction breaks supersymmetry, since antiperiodic boundary conditions have to be imposed on the fermions present in the field theory Lagrangian.

The essential point for constructing the gravity dual is that on the gravity side, the field theory described above is identified with the thermodynamics of black D3-branes in Anti-de Sitter space. The solitonic solution for these branes is given by the metric

$$\mathrm{d}s^2 = H(r)^{-1/2}\left(-f(r)\mathrm{d}t^2 + \mathrm{d}\mathbf{x}^2\right) + H(r)^{1/2}\left(\frac{\mathrm{d}r^2}{f(r)} + r^2 \mathrm{d}\Omega_5{}^2\right), \tag{6.53}$$

$$f(r) = 1 - \left(\frac{r_H}{r}\right)^4, \quad H(r) = 1 + \frac{L^4}{r^4}, \tag{6.54}$$

The blackening factor $f(r)$ vanishes at the Schwarzschild horizon r_h of the black brane. The difference between a black brane and a black hole is that the black brane is infinitely extended in the spatial $\mathbf{x}$ directions, which span $I\!R^3$. Setting $z = L^2/r$, Wick rotating to imaginary time and taking the near-horizon limit as before, this gives

$$ds^2 = \frac{L^2}{z^2}\left(\left(1 - \frac{z^4}{z_H^4}\right)d\tau^2 + d\mathbf{x}^2 + \frac{1}{1 - \frac{z^4}{z_H^4}}dz^2\right) + L^2 d\Omega_5^2\,, \tag{6.55}$$

with z_H the Schwarzschild radius. As for a black hole, we note that $g_{\tau\tau} \to 0, g_{zz} \to \infty$ for $z \to z_H$. Let us now introduce a further variable

$$z = z_H\left(1 - \frac{\rho^2}{L^2}\right). \tag{6.56}$$

Here, ρ is a measure for the distance from the horizon at z_H, outside the black hole. We expand about the horizon. To lowest order in ρ, the (τ, z) contribution to the Euclidean metric becomes

$$ds^2 \simeq \frac{4\rho^2}{z_H^2}d\tau^2 + d\rho^2\,. \tag{6.57}$$

With $\phi \equiv 2\tau/z_H$, this becomes $ds^2 = d\rho^2 + \rho^2 d\phi^2$. For regularity at $\rho = 0$, we have to impose that ϕ is periodic with period 2π, such that we have a plane rather than a conical singularity. This implies that τ becomes periodic with period $\Delta\tau = \pi z_H$. From the field-theory side we know that $\Delta\tau = \beta = 1/T$, which implies

$$z_H = \frac{1}{\pi T}. \tag{6.58}$$

Thus the field-theory temperature is identified with the Hawking temperature of the black brane!

We may now compute the field-theory thermal entropy from the Bekenstein-Hawking entropy of the black brane [26]. In general, the Bekenstein-Hawking entropy is given by the famous result

$$S_{\rm BH} = \frac{A_{d-1}}{4G_{d+1}}, \tag{6.59}$$

where A_{d-1} is the area of the black brane horizon and G_{d+1} is the Newton constant. For a black D3-brane, the horizon area is given by

$$\begin{aligned} A_3 &= \int d^3x \sqrt{g_{3d}\Big|_{z=z_H}} \cdot \mathrm{Vol}(S^5)\,, \qquad g_{3d} = g_{11}g_{22}g_{33} = \frac{L^6}{z^6} \\ &= \pi^6 L^8 T^3 \mathrm{Vol}(\mathbb{R}^3)\,, \end{aligned} \tag{6.60}$$

where we used the useful formulae $\mathrm{Vol}(S^5) = \pi^3 L^5$,

$$G_5 = \frac{G_{10}}{\mathrm{Vol}(S^5)} = \frac{\pi L^3}{2N^2}, \tag{6.61}$$

$2\kappa_{10} = 16\pi G_{10} = (2\pi)^7 \alpha'^4 g_s^2$ and $L^4 = 4\pi g_s N \alpha'^2$. Combining all results, we find

$$S_{\mathrm{BH}} = \frac{\pi^2}{2} N^2 T^3 \mathrm{Vol}(\mathbb{R}^3). \tag{6.62}$$

This result, valid at strong coupling, differs just by its prefactor from the free field theory result

$$S_{\mathrm{free}} = \frac{2\pi^2}{3} N^2 T^3 \mathrm{Vol}(\mathbb{R}^3). \tag{6.63}$$

We note that the result at strong coupling is small by a factor of $3/4$.

6.3 Kondo Model Within Field Theory and Condensed Matter Physics

We now turn to the discussion of models for magnetic impurities. We begin by considering the original model of Kondo [6], which describes the interaction of a free electron gas with a $SU(2)$ spin impurity. The electrons are also in the spin 1/2 representation of a second $SU(2)$. Using field-theory language, the corresponding Hamiltonian may be written as

$$H = \frac{v_F}{2\pi} i\psi^\dagger \partial_x \psi + \frac{v_F}{2} \lambda_K \delta(x) \mathbf{J} \cdot \mathbf{S}. \tag{6.64}$$

Here, v_F is the Fermi velocity, and $\mathbf{S}$ is the magnetic impurity satisfying

$$\left[S^a, S^b\right] = i\varepsilon^{abc} S^c, \tag{6.65}$$

which takes values in the internal $SU(2)$ spin space. The spin impurity interacts with the electron current

$$J^a = \psi^\dagger \sigma^a \psi, \tag{6.66}$$

with σ^a the Pauli matrices. The Hamiltonian consists of a kinetic term for the electrons and an interaction localized at the site of the impurity. Hence the interaction term involves a delta distribution.

The Kondo model is simplified in the s-wave approximation, where the problem becomes spherically symmetric. We thus introduce polar coordinates (r, θ, ϕ). The dependence on the two angles becomes trivial and we are left with a $1+1$-dimensional theory in the space spanned by (r, t). The radial coordinate r runs from zero to infinity. The impurity sits at the origin and provides a boundary condition.

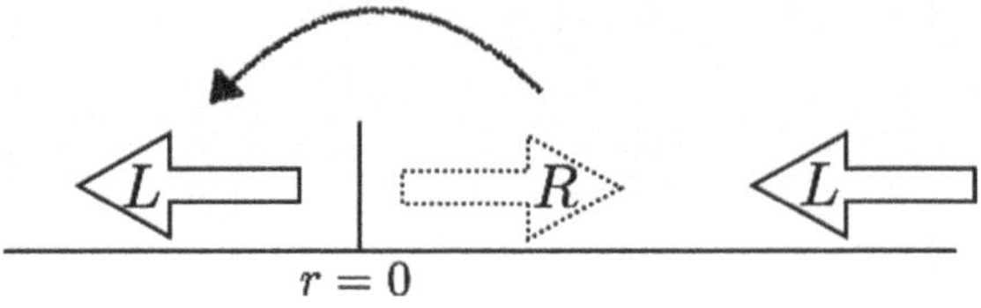

Fig. 6.4 Analytic continuation to negative values of r. The right-movers become left-movers travelling at negative values of r

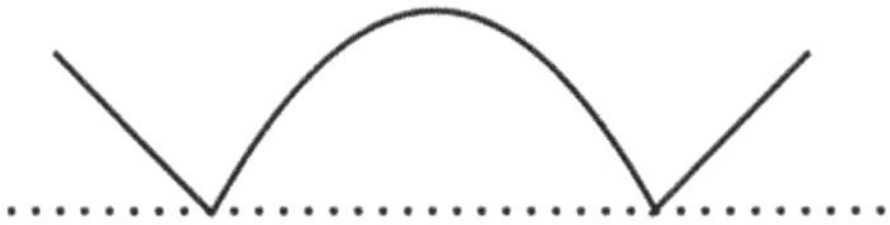

Fig. 6.5 One-loop Feynman graph contributing to the renormalization of the Kondo coupling, with an electron (solid line) scattering off the impurity (dashed line)

The electrons separate into left- and right movers. It is now convenient to analytically continue r to negative values. Then, the previous right-movers become left-movers travelling at negative values of r, i.e. $\psi_R(r) \to \psi_L(-r)$, as shown in Fig. 6.4.

The Hamiltonian (6.64) was proposed and solved perturbatively by Jun Kondo [6]. To first order in perturbation theory, the quantum correction to the resistivity is

$$\rho(T) = \rho_0 \left[\lambda_K + \nu \lambda_K^2 \ln \frac{D}{T} + \cdots \right]^2 , \tag{6.67}$$

where ν is the density of states and D a UV cut-off, for instance the bandwidth. The corresponding Feynman graph is shown in Fig. 6.5. This correction explains the experimental result for a logarithmic rise at low temperatures. From a theoretical perspective, we note that perturbation theory breaks down at a temperature scale

$$T_K = D \exp\left(-\frac{1}{\nu \lambda_K} \right) , \tag{6.68}$$

which defines the *Kondo temperature* T_K. At this scale, the first order perturbative correction is of the same order as the zeroth order term, which implies that perturbation theory breaks down.

For the coupling itself, the first order perturbative correction gives the beta function

$$\beta(\lambda_K)_{\text{one-loop}} = T \frac{\mathrm{d}\lambda_K}{\mathrm{d}T} = -\nu \lambda_K^2 . \tag{6.69}$$

So the beta function is negative. This is analogous to the gauge beta function in QCD, which is also negative—a property associated with asymptotic freedom in the UV.

By analogy, we see that the Kondo temperature T_K plays a similar role as the scale Λ_{QCD} in QCD, at which perturbation theory breaks down.

A resummation of (6.69) leads to the effective coupling

$$\lambda_{\text{eff}}(T) = \frac{\lambda_K}{1 - \nu\lambda_K \ln(D/T)} . \tag{6.70}$$

$\lambda_{\text{eff}}(T)$ diverges at $T \sim T_K = D\exp(-1/(\nu\lambda_K))$. In the IR for $T \to 0$, the theory has a strongly coupled fixed point where the effective coupling vanishes. In fact, the impurity is *screened*: The impurity spin forms a singlet with the electron spin,

$$|\psi\rangle = \frac{1}{\sqrt{2}} (| \Uparrow\downarrow\rangle - | \Downarrow\uparrow\rangle) . \tag{6.71}$$

This is reminiscent of the formation of meson bound states in QCD.

The theories at the UV and IR fixed points of the flow are described by boundary conformal field theories (bCFT). Using the analytic continuation described above, In the UV, the theory is free, and we may impose the boundary condition $\psi_L(0) = \psi_R(0)$ for the left- and right moving electrons introduced above. In the IR however, due to the screening it costs energy to add a further electron to the singlet at $r = 0$. The probability for an electron to be at $r = 0$ in the ground state is zero. This observation is encoded in the antisymmetric boundary condition $\psi_R(0) = -\psi_L(0)$. Within bCFT, the Kondo model was analyzed extensively by Affleck and Ludwig [27], making non-trivial use of the appropriate representations of the conformal and the spin Kac–Moody algebra.

Both the UV and the non-trivial IR fixed point of the Kondo RG flow may be described using CFT techniques. Essentially, the interaction may be translated into a boundary condition at $r = 0$. Let us sketch this approach, considering a general $SU(N)$ spin group instead of the $SU(2)$ considered above, as well as k species (also called channels or flavours) of electrons. In the UV, the boundary condition relating the left- and right movers is just $\psi_L(0) = \psi_R(0)$. In the IR, a bound state involving the impurity spin forms, which is a singlet when $N = k = 2$. This implies that it costs energy to add another electron at $r = 0$, and the probability of finding another electron there is zero. This is described by an antisymmetric wave function as provided by the boundary condition $\psi_L(0) = -\psi_R(0)$.

It may be shown [7] that by introducing the currents

$$J_{\text{charge}} =: \psi^{\dagger\alpha i}\psi_{\alpha i} : , \qquad J^a_{\text{spin}} =: \psi^{\dagger\alpha i} T^{a\beta}_{\alpha} \psi_{\beta i} : , \qquad J^A_{\text{channel}} =: \psi^{\dagger\alpha i} \tau^{Aj}_i \psi_{\alpha j} : , \tag{6.72}$$

where the colon denotes normal ordering, $T^{a\beta}_{\alpha}$ are $SU(N)$ generators and τ^{Aj}_i are $SU(k)$ generators, the Kondo Hamiltonian may be written as

$$H = \frac{1}{2\pi(N+k)} J^a_{\text{spin}} J^a_{\text{spin}} + \frac{1}{2\pi(k+N)} J^A_{\text{channel}} J^A_{\text{channel}} + \frac{1}{4\pi Nk} (J_{\text{charge}})^2 + \lambda_K \delta(r) S^a J^a_{\text{spin}} \,. \tag{6.73}$$

In the IR, by writing

$$\mathcal{J}^a_{\text{spin}} = J^a_{\text{spin}} + \lambda_K \delta(r) S^a \,, \tag{6.74}$$

the interaction term may be absorbed into a new current $\mathcal{J}^a_{\text{spin}}$. Written in terms of this new current, the Hamiltonian again reduces to the Hamiltonian of the free theory without interaction. The interaction is thus absorbed and replaced by the non-trivial boundary condition discussed above.

At the conformal fixed points, the spin, channel and charge currents may be expanded in a Laurent series,

$$J^a(z) = \sum_{n \in \mathbb{Z}} z^{-n-1} J^a_n \,. \tag{6.75}$$

The mode expansions then satisfy Kac–Moody algebras,

$$[J^a_n, J^b_m] = i f^{abc} J^c_{n+m} + \frac{n}{2} k \delta^{ab} \delta_{m+n,0} \,, \tag{6.76}$$

as shown here for the spin current with $SU(N)_k$ symmetry, where k denotes the level of the Kac–Moody algebra. Similarly, for the channels we have a $SU(k)_N$ symmetry. The total symmetry of the model is $SU(N)_k \times SU(k)_N \times U(1)$. The representations of the two Kac–Moody algebras are fused in a tensor product. The two different boundary conditions in the UV and in the IR lead to different representations and thus operator spectra for the total theory.

In the simplest example when the spin is $s = 1/2$ and there is only one species of electrons, $k = 1$, then in the IR a singlet forms. More generally, a singlet is present when $2s = k$, which is referred to as *critical* screening. When $k < 2s$, however, the impurity has insufficient channels to screen the impurity completely, and there is a residual spin of size $|s - k/2|$. This is referred to as *underscreening*. On the other hand, when $k > 2s$ there are too many electron species for a critical screening of the spin, which leads to non-Fermi liquid behaviour, a situation called *overscreening*.

6.4 Large N Kondo Model

As was found by condensed matter physicists in the eighties [28, 29], the Kondo model simplifies considerably when the rank N of the spin group is taken to infinity. In this limit, the interaction term $\mathbf{J} \cdot \mathbf{S}$ reduces to a product $\mathcal{O}\mathcal{O}^\dagger$ involving as scalar operator $\mathcal{O}$, and the screening corresponds to the condensation of $\mathcal{O}$. For comparison to gauge/gravity duality, it will be useful to consider this large N solution in which the Kondo screening appears as a condensation process in $0+1$ dimensions. In the large N limit, a phase transition is possible in such low dimensions since long-range fluctuations are suppressed. Moreover, there is an alternative large N solution of the Kondo model using the Bethe ansatz [8, 9].

The large N limit of the Kondo model involves $N \to \infty$, $\lambda \to 0$ with λN fixed. The vector large N limit of the Kondo model provides information about the spectrum, thermodynamics and transport properties everywhere along the RG flow, even away from the fixed points. $1/N$ corrections may be calculated.

We consider totally antisymmetric representations of $SU(N)$ given by a Young tableau consisting of one column with q boxes, $q < N$. We write the spin in terms of Abrikosov pseudo-fermions χ, which means that we consider

$$S^a = \chi^{\dagger i} T^a{}_i{}^j \chi_j \,, \qquad a = 1, 2, \ldots, N^2 - 1 \,, \tag{6.77}$$

with χ in the fundamental representation of $SU(N)$. A state in the impurity Hilbert space is obtained by acting on the vacuum state with q of the $\chi^\dagger$. This gives rise to a totally antisymmetric tensor product with rank q. Since (6.77) is invariant under phase rotations of the χ's, there is an additional new $U(1)$ symmetry. This implies that we need to impose a constraint since considering the χ's instead of S^a should not introduce any new degrees of freedom. We impose

$$\chi^\dagger \chi = q \,, \tag{6.78}$$

i.e. the charge density of the Abrikosov fermions is given by the size of the totally antisymmetric representation. Together with the fermions ψ of the Kondo model, we have a $SU(N)$ singlet operator

$$\mathcal{O}(t) \equiv \psi^\dagger \chi \,, \qquad \Delta_{\mathcal{O}} = \frac{1}{2} \,. \tag{6.79}$$

Now in the large N limit, the Kondo interaction $\mathbf{J} \cdot \mathbf{S}$ simplifies considerably as follows. We make use of the Fierz identity (6.20). For the Kondo interaction this implies

$$\lambda \delta(x) J^a S^a = \lambda \delta(x) (\psi^\dagger T^a \psi)(\chi^\dagger T^a \chi) = \frac{1}{2} \lambda \delta(x) \left(\mathcal{O} \mathcal{O}^\dagger - \frac{q}{N} (\psi^\dagger \psi) \right) , \tag{6.80}$$

where for sufficiently small q we may neglect the last term in the limit $N \to \infty$.

In the large N limit, the Kondo coupling is thus the coupling of a 'double-trace' deformation $\mathcal{O}\mathcal{O}^\dagger$, with two separately gauge invariant operators $\mathcal{O}$ and $\mathcal{O}^\dagger$. This is similar to double-trace operators where two separately gauge-invariant operators are multiplied to each other. For operators involving fields in the adjoint representation, traces have to be taken to generate gauge-invariant operators. Here however, $\mathcal{O}$ is gauge invariant without trace, since both ψ and χ are in the fundamental of $SU(N)$. The operator $\mathcal{O}\mathcal{O}^\dagger$ is of engineering dimension one. As defect operator, it is marginally relevant, i.e. it is marginal at the classical level, but quantum corrections make it relevant.

In the large N limit, the solution of the field-theory saddle point equations reveals a second order mean-field phase transition in which $\mathcal{O}$ condenses: There is a critical temperature T_c above which $\langle \mathcal{O} \rangle = 0$ and below which $\langle \mathcal{O} \rangle \neq 0$. The critical temperature T_c is slightly smaller than the Kondo temperature T_K and may be calculated analytically. The condensate spontaneously breaks the $U(1)$ symmetry of the χ fermions. $1/N$ corrections smoothen this transition to a cross-over.

At large N, the Kondo model thus has similarity with superconductivity that is triggered by a marginally relevant operator. This observation provides a guiding principle for constructing a gauge/gravity dual of the large N Kondo model.

6.5 Gravity Dual of the Kondo Model

The motivation of establishing a gravity dual of the Kondo model is twofold: On the one hand, this provides a new application of gauge/gravity duality of relevance to condensed matter physics. On the other hand, this provides a gravity dual of a well-understood field theory model with an RG flow, which may provide new insights into the working mechanisms of the duality. It is important to note that our holographic Kondo model will have some features that are distinctly different from the well-known field theory Kondo model described above. Most importantly, the $1+1$-dimensional electron gas will be strongly coupled even before considering interactions with the impurity. This has some resemblance with a Luttinger liquid coupled to a spin impurity. Moreover, the $SU(N)$ spin symmetry will be gauged. The holographic Kondo model has provided insight into the entanglement entropy of this system. Moreover, quenches of the Kondo coupling in the holographic model provide a new geometric realization of the formation of the Kondo screening cloud. It is conceivable that further work will also lead to new insight into the Kondo lattice that involves a lattice of magnetic impurities. The Kondo lattice is a major unsolved problem within condensed matter physics. Preliminary results in this direction that were obtained using holography may be found in [30]. Further holographic studies of holographic Kondo models include [31].

6.5.1 Brane Construction for a Holographic Kondo Model

Here we aim at constructing a holographic Kondo model realizing similar features to the ones of the large N field theory Kondo model described in the previous section, including a RG flow triggered by a double-trace operator [32]. For this purpose, consider an appropriate configuration of D-branes which allows us to realize the field theory operators needed.

Table 6.4 Brane configuration for a holographic Kondo model

	0	1	2	3	4	5	6	7	8	9
N D3	X	X	X	X						
1 D7	X	X			X	X	X	X	X	X
1 D5	X				X	X	X	X	X	

The field theory involves fermionic fields ψ in $1+1$ dimensions in the fundamental representation of $SU(N)$, as well as Abrikosov fermion fields χ localized at the $0+1$-dimensional defect. These transform in the fundamental representation of $SU(N)$ as well. From these we will construct the required operators. For the brane configuration we will use probe branes, which means that a small number of coincident branes are embedded into a D3-brane background, neglecting the backreaction on the geometry. For a holographic Kondo model, a suitable choice of probe branes consists of D7- and D5-branes embedded as shown in Table 6.4. Fields in the fundamental representation are obtained from strings stretching between the D3- , D5- and D7-branes. The D7-brane probe extends in $1+1$ dimensions of the worldvolume of the D3-branes. As we discuss below, strings stretching between the D3- and D7-branes give rise to chiral fermions, which we identify with the electrons of the Kondo model. On the other hand, since the D5-brane only shares the time direction with the D3-branes, the D3–D5 strings give rise to the $0+1$ dimensional Abrikosov fermions.

We note that in a in absence of the D5-branes, the D3/D7-brane system has eight ND directions, such that half of the original supersymmetry is preserved. However, the D5/D7-system has only two ND directions, such that supersymmetry is broken. This leads to the presence of a tachyon potential and a condensation as required for the large N Kondo model. The tachyon, a complex scalar field Φ, is identified as the gravity dual of the operator $\mathcal{O} = \psi^\dagger \chi$.

As discussed in [33, 34], the D7-brane gives rise to an action

$$S_7 = \frac{1}{\pi} \int \mathrm{d}^2 x \, \psi_L^\dagger (i\partial_- - A_-)\psi_L \qquad (6.81)$$

of chiral fermions which are coupled to the $\mathcal{N}=4$ supersymmetric gauge theory in $3+1$ dimensions. A_- is a restriction of a component of the $\mathcal{N}=4$ Super Yang–Mills gauge field to the subspace of the fermions. These fermions are in the fundamental representation of the gauge group $SU(N)$. For simplicity, from now on we drop the label L for left-handed. The gauge field A_- is a component of the $\mathcal{N}=4$ theory gauge field on the $1+1$-dimensional subspace spanned by the D7-brane. We identify the ψ_L with the electrons of the Kondo model.

Similarly, for the Abrikosov fermions χ we obtain from the D3/D5-brane system the action

$$S_5 = \int \mathrm{d}t \, \chi^\dagger (i\partial_t - A_t - \Phi_9)\chi \, . \qquad (6.82)$$

Table 6.5 Field-operator map for the holographic Kondo model

Operator		Gravity field
Electron current $J^{\mu} = \bar{\psi}\gamma^{\mu}\psi$	$\Leftrightarrow$	Chern-Simons gauge field A in AdS_3
Charge density $q = \chi^{\dagger}\chi$	$\Leftrightarrow$	2d gauge field a in AdS_2
Operator $\mathcal{O} = \psi^{\dagger}\chi$	$\Leftrightarrow$	2d complex scalar Φ in AdS_2

Here, Φ_9 is the adjoint scalar of $\mathcal{N} = 4$ Super Yang–Mills theory whose eigenvalues represent the positions of the D3-branes in the x^9 direction. In (6.82), both A_t and Φ_9 are restricted to the subspace of the χ fields. Note that unlike the original Kondo model, the $SU(N)$ spin symmetry is gauged in this approach. Also, the background $\mathcal{N} = 4$ theory is strongly coupled in the gravity dual approach and provides strong interactions between the electrons.

Let us now turn to the gravity dual of this configuration. The N D3-branes provide an $AdS_5 \times S^5$ supergravity background as before. The probe D7-brane wraps an $AdS_3 \times S^5$ subspace of this geometry, while the probe D5-branes wraps $AdS_2 \times S^4$. The Dirac-Born-Infeld action for the D5-brane contains a gauge field a_{μ} on the AdS_2 subspace spanned by (t, r), with t the time coordinate and r the radial coordinate in the AdS geometry. The a_t component of this gauge field is dual to the charge density of the Abrikosov fermions, $q = \chi^{\dagger}\chi$. The D7-brane action contains a Chern–Simons term for a gauge field A_{μ} on AdS_3. As noted before, the D5–D7 strings lead to a complex scalar tachyon field.

We may thus establish the holographic dictionary for the operators of the field-theory large N Kondo model. This is listed in Table 6.5. The electron current in $1 + 1$ dimensions is dual to the Chern–Simons field in $2 + 1$ dimensions. The Abrikosov fermion charge density q in $0 + 1$ dimensions is dual to the gauge field component a_t in $1 + 1$ dimensions. Finally, the operator $\mathcal{O} = \psi^{\dagger}\chi$ in $0 + 1$ dimensions is dual to the complex scalar field Φ in $1 + 1$ dimensions.

The brane picture has allowed us to neatly establish the required holographic dictionary. Unfortunately, it is extremely challenging to derive the full action describing the brane construction given. In particular, the exact form of the tachyon potential is not known.

For making progress towards describing a variant of the Kondo model holographically, we thus turn to a simplified model consisting of a Chern–Simons field in AdS_3 coupled to a Yang–Mills gauge field and a complex scalar in AdS_2. This simplification still allows us to use the holographic dictionary established above. The information we lose though is about the full field content of the strongly coupled field theory. On the other hand, this simplifield model allows for explicit calculations of observables such as two-point functions and the impurity entropy, as we discuss below. It is instructive to compare the results of these calculations with features of the field-theory large N Kondo model, as we shall see.

The simplified model we consider is

$$S = \frac{1}{8\pi G_N} \int \mathrm{d}z\mathrm{d}x\mathrm{d}t\sqrt{-g}\,(R - 2\Lambda) - \frac{N}{4\pi} \int_{\mathrm{AdS}_3} A \wedge dA$$
$$- N \int \mathrm{d}x\mathrm{d}t\sqrt{-g}\left(\frac{1}{4}\mathrm{tr} f^{mn} f_{mn} + (D^m \Phi)^\dagger (D_m \Phi) - V(\Phi)\right) . \tag{6.83}$$

Here, z is the radial AdS coordinate, x is the spatial coordinate along the boundary and t is time. The defect sits at $x = 0$. The first term is the standard Einstein–Hilbert action with negative cosmological constant Λ. The second term is a Chern–Simons term involving the gauge field A_μ dual to the electron current J^μ. We take A_μ to be an Abelian gauge field, which implies that we consider only one flavour of electrons, or—in condensed matter terms—only one channel. f_{mn} is the field strength tensor of the gauge field a_m with $m \in \{t, z\}$, which we take to be Abelian too. Its time component a_t is dual to the charge density $\chi^\star\chi$, which at the boundary takes the value $Q = q/N$ with q the dimension of the antisymmetric prepresentation of the spin impurity. D_m is a covariant derivative given by $D_m = \partial_m + iA_m\Phi - ia_m\Phi$. For the complex scalar, we assume its potential to take the simple form

$$V(\Phi^\dagger\Phi) = M^2\Phi^\dagger\Phi . \tag{6.84}$$

We write the complex field as $\Phi = \phi \exp i\delta$ with $\phi = |\Phi|$. We choose M^2 in such a way that $\Phi^\dagger\Phi$ is a relevant operator in the UV limit. It becomes marginally relevant when perturbing about the fixed point. Moreover, for the time being we consider the matter fields as probes, such that they do not influence the background geometry. For this background geometry we take the solution to the gravity equations of motion which corresponds to the AdS BTZ black hole, i.e.

$$\mathrm{d}s^2_{\mathrm{BTZ}} = \frac{1}{z}\left(\frac{1}{h(z)}\mathrm{d}z^2 - h(z)\mathrm{d}t^2\right) ,$$
$$h(z) = 1 - \frac{z^2}{z_h^2} , \tag{6.85}$$

where we set the AdS radius to one, $L = 1$, and z_h is related to the temperature by

$$T = \frac{1}{2\pi z_h} . \tag{6.86}$$

The non-trivial equations of motion for the matter fields are given by

$$\partial_z A_x = 4\pi\delta(x)\sqrt{g}g^{tt}a_t\phi^2 ,$$
$$\partial_z(\sqrt{-g}g^{zz}g^{tt}\partial_z a_t) = 2\sqrt{-g}g^{tt}a_t\phi^2 ,$$
$$\partial_z(\sqrt{-g}g^{zz}\partial_z\phi) = \sqrt{-g}g^{tt}a_t^2\phi + \sqrt{-g}M^2\phi . \tag{6.87}$$

The three-dimensional gauge field A_μ is non-dynamical, but will be responsible for a phase shift similar to the one observed in the field-theory Kondo model.

Above the critical temperature T_c where $\mathcal{O}$ dual to the scalar field condenses, we have $\phi = 0$. Then, asymptotically near the boundary, we have $a_t(z) \sim \frac{Q}{z} + \mu$, where μ is a chemical potential for the spurious $U(1)$ symmetry rotating the χ's. The charge density is given by $\chi^\dagger \chi = NQ$, with $Q = q/N$.

For generating the Kondo RG flow, we need to turn on the marginally relevant 'double-trace' operator $\mathcal{O}\mathcal{O}^\dagger$. We choose the mass M in the potential such that the field $\phi(z)$ is at the Breitenlohner–Freedman stability bound [35]. The asymptotic behaviour of $\phi(z)$ near the boundary is then

$$\phi(z) = \alpha z^{1/2} \ln(\Lambda z) - \beta z^{1/2} + \mathcal{O}(z^{3/2} \ln(\Lambda z)) . \tag{6.88}$$

Following [36, 37], the gravity dual of a double-trace perturbation is obtained by imposing a linear relation between α and β,

$$\alpha = \kappa \beta . \tag{6.89}$$

We choose α to correspond to a source for the operator $\mathcal{O}$, while β is related to is vacuum expectation value. The physical coupling $\phi(z)$ should be a RG invariant, i.e. invariant under changes of the cut-off Λ. This implies

$$\kappa = \frac{\kappa_0}{1 + \kappa_0 \ln(\Lambda_0/\Lambda)} . \tag{6.90}$$

At finite temperature, we obtain the analogous result

$$\kappa_T = \frac{\kappa_0}{1 + \kappa_0 \ln(\Lambda z_h)} \tag{6.91}$$

This expression for the coupling κ_T diverges at the temperature

$$T_K = \frac{1}{2\pi} \Lambda e^{1/\kappa_0} , \tag{6.92}$$

where T_K is the *Kondo temperature*. A similar behaviour is observed in the condensed matter Kondo models. Moreover, this behaviour bears some similarity to QCD, where the coupling becomes strong at a scale Λ_{QCD}, below which bound states provide the natural description of the degrees of freedom. Of course, in the holographic Kondo model there are *two* couplings, one between the electrons themselves and secondly the Kondo coupling κ_T. While the first is strong along the entire flow, κ_T diverges at the Kondo temperature and then becomes small again at lower temperatures, where the condensate forms.

For determining the physical properties of the model considered, we have to resort to numerics to solve the equations of motion (6.87). We find a mean-field phase transition as expected for a large-N theory, as shown in Fig. 6.6. In the screened

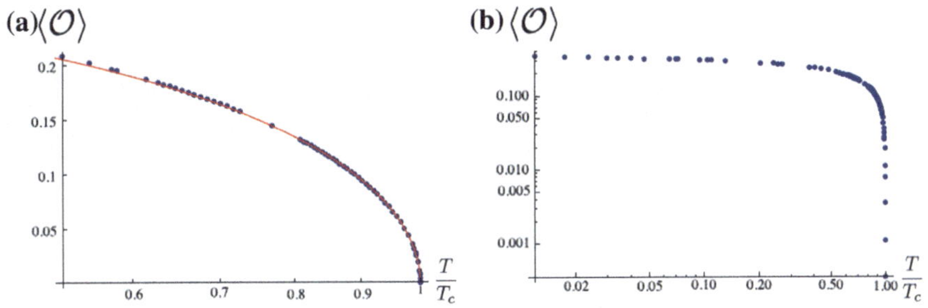

Fig. 6.6 Expectation value of the operator $\mathcal{O} = \psi^\dagger \chi$ as function of the temperature. Below T_c, a condensate forms. **a** Close to the transition temperature, displaying that the phase transition is mean-field; **b** Log-log plot showing a larger temperature range. The VEV appears to approach a constant at low temperatures, however further stabilisation by a quartic potential contribution is expected to be required in the limit $T \to 0$. Figures from [32]

phase, a condensate of the operator $\mathcal{O} = \psi^\dagger \chi$ forms. We note that for very small temperatures, the numerical solution of the equations of motion becomes extremely time-consuming and thus our results are less accurate in this regime. We expect that in the limit $T \to 0$, to obtain a stable constant solution for $\langle \mathcal{O} \rangle$ requires to add a quartic term to the potential (6.84).

Our holographic model allows for a geometrical description of the screening mechanism in the dual strongly-coupled field theory. For this we consider the electric flux $\mathscr{F}$ of the AdS_2 gauge field $a_t(z)$. At the boundary of the holographic space, this flux encodes information about the impurity spin representation,

$$\lim_{z \to 0} \mathscr{F} = \lim_{z \to 0} \sqrt{-g} f^{zt} = a_t'(z)|_{z \to 0} = Q \,, \tag{6.93}$$

with $Q = q/N$ and q as in (6.78). When $\phi = 0$, this flux is a constant and takes the same value at the black hole horizon. However for $T < T_c$, the non-trivial profile $\phi(z)$ draws electric charge away from $a_t(z)$, reducing the electric flux at the horizon. This implies that the effective number of impurity degrees of freedom is reduced, which corresponds to screening. This is shown in Fig. 6.7 which shows the flux $\mathscr{F}_{z \to z_h}$ at the horizon as a function of temperature. The numerical solution of the equations of motion yields a decreasing flux when the temperature is decreased.

The temperature dependence of the resistivity may be obtained by an analysis of the leading irrelevant operator at the IR fixed point, i.e. by perturbing about the IR fixed point by this operator. This gives $\rho(T) \propto T^\gamma$ with $\gamma \in \mathbb{R}$ a real number. A similar behaviour occurs also in Luttinger liquids [38]. The model thus does not reproduce the logarithmic rise of the resistivity with decreasing temperature observed in the original Kondo model. This behaviour is expected since the model is at large N and the ambient electrons are strongly coupled.

Let us emphasize again the differences between the holographic Kondo model considered here and the large N Kondo model of condensed matter physics: Here, the electrons are strongly coupled among themselves even before coupling them to

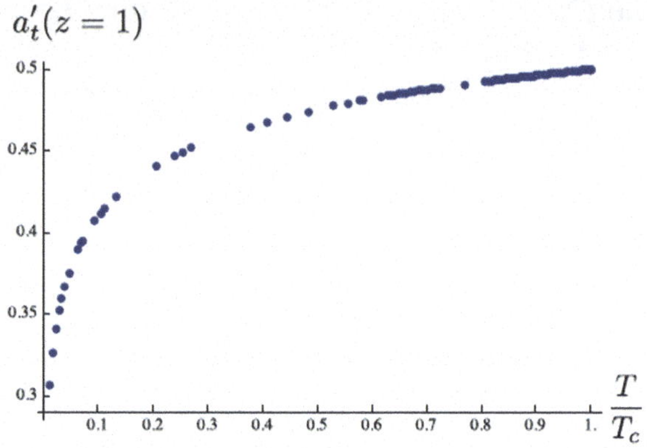

Fig. 6.7 Electric flux through the boundary of AdS_2 at the black hole horizon. This is a measure for the number of degrees of freedom. Its decrease at low temperatures indicates that the impurity is screened. For $T/T_c \lesssim 0.2$, the decrease is only logarithmic. The radial variable is normalized such that $z = 1$ at the horizon. Figure from [32]

the spin defect. The system thus has two couplings: the electron-electron coupling which is always large, and the Kondo coupling to the defect that triggers the RG flow. Moreover, we point out that in our model, the $SU(N)$ symmetry is gauged, while it is a global symmetry in the condensed matter models.

To conclude, let us consider different applications of the holographic Kondo model we introduced. These involve three aspects: the impurity entropy, quantum quenches and correlation functions.

6.6 Applications of the Holographic Kondo Model

6.6.1 Entanglement Entropy

The concept of holographic entanglement entropy introduced by Ryu and Takayanagi in 2006 has proved to be an important ingredient to the holographic dictionary [39], opening up new relations between gauge/gravity duality and quantum information. In general, the entanglement entropy is defined for two Hilbert spaces $\mathscr{H}_A$ and $\mathscr{H}_B$. In the AdS/CFT correspondence, it is useful to consider A and B to be two disjunct space regions in the CFT. Defining the reduced density matrix to be

$$\rho_A = \text{tr}_B \rho \,, \tag{6.94}$$

where ρ is the density matrix of the entire space, the entanglement entropy is given by its von Neumann entropy

$$S = -\text{tr}_A \rho_A \ln \rho_A \,. \tag{6.95}$$

The entanglement entropy bears resemblance with the black hole entropy since it quantifies the lost information hidden in B. Ryu and Takayanagi proposed the

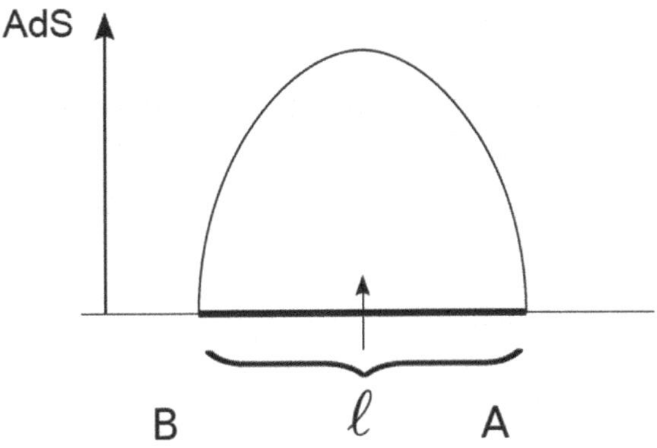

Fig. 6.8 The impurity entropy in the holographic Kondo model is obtained from the entanglement entropy. The entanglement area is a line of length ℓ in the dual field theory. The holographic minimal surface is a geodesic. For the impurity entropy, the entanglement entropy in absence of the defect is subtracted from the one in presence of the defect

holographic dual of the entanglement entropy to be

$$S = \frac{\text{Area}\gamma_A}{4G_{d+1}}, \tag{6.96}$$

where G_{d+1} is the Newton constant of the dual gravity space and γ_A is the area of the minimal bulk surface whose boundary coincides with the boundary of region A. For a field theory in $1+1$ dimensions, the region A may be taken to be a line of length ℓ, and the bulk minimal surface γ_A becomes a bulk geodesic joining the two endpoints of this line, as shown for the holographic Kondo model in Fig. 6.8. We note that for a $1+1$-dimensional CFT at finite temperature, with the BTZ black hole as gravity dual, it is found both in the CFT [40] and on the gravity side [39] that the entanglement entropy for a line of length ℓ is given by

$$S_{\text{BH}}(\ell) = \frac{c}{3}\ln\left(\frac{1}{\pi\varepsilon T}\sinh(2\pi\ell T)\right), \tag{6.97}$$

with ε a cut-off parameter.

For the Kondo model, a useful quantity to consider is the *impurity entropy* which is given by the difference of the entanglement entropies in presence and in absence of the magnetic impurity,

$$S_{\text{imp}} = S_{\text{impurity present}} - S_{\text{impurity absent}}. \tag{6.98}$$

In the previous sections, we considered the probe limit of the holographic Kondo model, in which the fields on the AdS_2 defect do not backreact on the AdS_3 geometry. However, including the backreaction is necessary in order to calculate the effect of the defect on the Ryu–Takayanagi surface. A simple model that achieves this [41, 42] consists of cutting the $2+1$-dimensional geometry in two halves at the defect at $x = 0$ and joining these back together subject to the *Israel junction condition* [43]

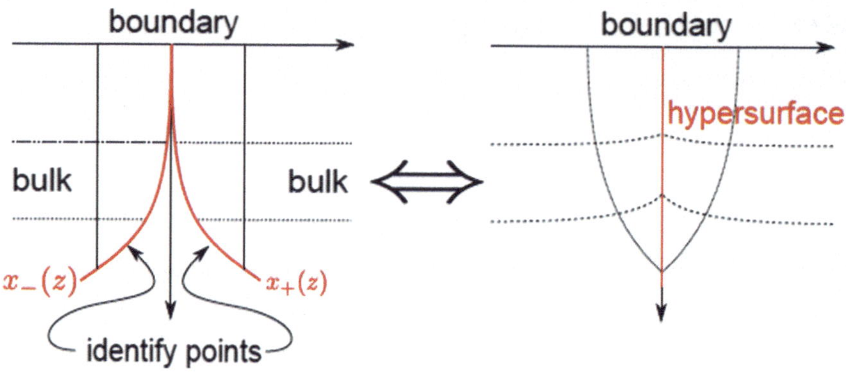

Fig. 6.9 Cutting and joining of two halves of the AdS BTZ geometry subject to the Israel junction at the defect. Figure by Mario Flory

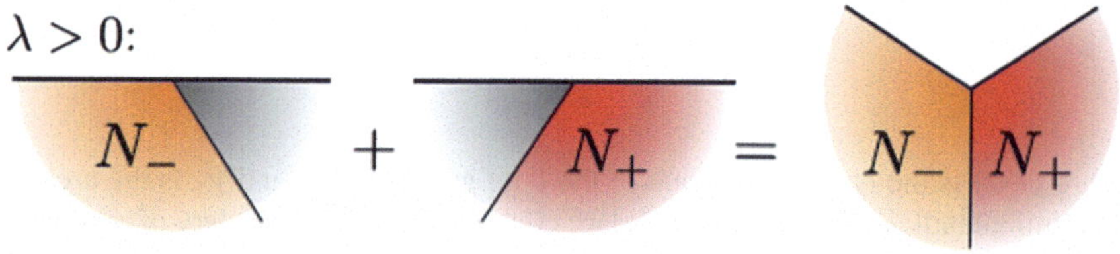

Fig. 6.10 Geometry in a vicinity of the backreacting defect brane at positive brane tension. The horizontal black line corresponds to the boundary of the deformed AdS space, as in Fig. 6.9. The volume is increased in a given region around the defect as compared to the case when the brane tension vanishes. This will lead to a longer geodesic for a given entanglement interval and thus to a non-zero positive impurity entropy. Figure by Mario Flory

$$K_{\mu\nu} - \gamma_{\mu\nu} K = -\frac{\kappa_G}{2} T_{\mu\nu} , \tag{6.99}$$

This procedure is shown in Fig. 6.9. We refer to the joining hypersurface as 'brane'. In (6.99), γ and K are the induced metric and extrinsic curvature at the joining hypersurface extending in (t, z) directions. $T_{\mu\nu}$ is the energy-momentum tensor for the matter fields a and Φ at the defect, and κ_G is the gravitational constant with $\kappa_G^2 = 8\pi G_N$.

The matter fields Φ and a lead to a non-zero tension on the brane, which varies with the radial coordinate. The higher the tension on this brane, the longer the geodesic joining the two endpoints of the entangling interval will be, as shown in Fig. 6.10. A numerical solution of the Israel junction condition reveals that the brane tension decreases with decreasing temperature, which leads to a shorter geodesic. This in turn leads to a decrease of the impurity entropy (6.98). This decreases is expected and in agreement with the screening of the impurity degrees of freedom.

In the holographic Kondo model, the brane is actually curved since the brane tension depends on the radial coordinate. For large entangling regions ℓ, we may

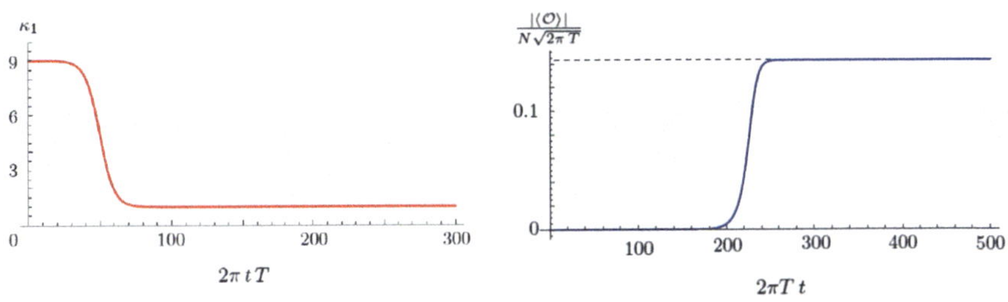

Fig. 6.11 Left: Quench of the 'double-trace' Kondo coupling from the unscreened to the screened phase. Right: Reaction of the system to this quench: A condensate forms. There are no oscillations about the new equilibrium configuration. Figure from [46]

approximate the impurity entropy to linear order by noting that the length decrease of the Ryu–Takayanagi geodesic γ_A translates into a decrease of the entangling region ℓ itself. To linear order, this implies that the entangling region is given by $\ell + D$ in the UV and by ℓ in the IR, for $D \ll \ell$. Using (6.97) we may thus write for the difference of the impurity between its UV and IR values

$$\begin{aligned} \Delta S_{\text{imp}} &= S_{\text{BH}}(\ell + D) - S_{\text{BH}}(\ell) \\ &\simeq D \cdot \partial_\ell S_{\text{BH}}(\ell) = \frac{2\pi D T}{3} \coth(2\pi \ell T)\,. \end{aligned} \tag{6.100}$$

It is a non-trivial result that subject to identifying the scale D with the *Kondo correlation length* of condensed matter physics, $D \propto \xi_K$, then the result agrees with previous field-theory results for the Kondo impurity entropy [44, 45].

6.6.2 *Quantum Quenches*

A quantum quench corresponds to introducing a time dependence of the Kondo coupling. On the gravity side, this implies that the equations of motion become partial differential equations (PDEs), since both the dependence on the AdS radial coordinate and on time are relevant. Quenches of the holographic 'double trace' Kondo coupling κ_T were considered in [46]. Figure 6.11 shows a quench from the unscreened to the screened phase. The system reacts to this quench of the coupling by forming a condensate. There is a certain time lapse before this happens. It is also noteworthy that the reaction is overdamped, i.e. there are no oscillations around the new equilibrium value. This behaviour follows from the structure of the *quasinormal modes*, i.e. the eigenmodes of the gravity system. The leading eigenmode is purely imaginary in this system. This is in agreement with the behaviour of the correlation functions discussed in the next section.

6.6.3 Correlation Functions

AdS/CFT allows to calculate retarded Green's functions by adapting the methods presented in Sect. 6.2.3.2 to Lorentzian signature [47]. The required causal structure is obtained by imposing infalling boundary conditions on the gravity field fluctuations at the black hole horizon. Moreover, a careful regularization using the methods of holographic regularization [48] is essential. This approach was used in [49, 50] to calculate spectral functions for the Kondo operator $\mathscr{O} = \psi^\dagger \chi$ of (6.79). Spectral functions are generally obtained from the retarded Green's function by virtue of

$$\rho(\omega) = -2\,\mathrm{Im}\,G_R(\omega)\,. \tag{6.101}$$

The spectral function measures the number of degrees of freedom present at a given energy. The results for the holographic Kondo model obtained in [49, 50] are shown in Fig. 6.12.

Above the critical temperature, these spectral functions show a *spectral asymmetry* related to a *Fano resonance* [51]. In the holographic case, this asymmetry is characteristic of the interaction between the ambient strongly coupled CFT and the localized impurity degrees of freedom. A similar spectral asymmetry also appears in the condensed-matter large N Kondo model (which involves free electrons) at vanishing temperature [52]. In the screened phase, the holographic spectral function displayed in Fig. 6.12 is antisymmetric, consistent with the relation

$$\omega_P \propto -i|\langle\mathscr{O}\rangle|^2 \tag{6.102}$$

between the condensate and the leading pole ω_P in the retarded Green's function. This relation is also satisfied by the condensed matter large-N Kondo model involving free electrons [53].

A similar spectral asymmetry also arises in the context of the Sachdev-Ye-Kitaev (SYK) model that received a lot of attention recently [54, 55]. In fact, the original variant of this model due to Sachdev and Ye [54] involves Weyl fermions, as opposed to the Majorana fermions of the SYK model. This Sachdev-Ye may be obtained from the Ising model by the same mechanism as discussed in (6.77) above, i.e. by writing the Ising spin in terms of a bilinear of auxiliary fermions. In this case, the Ising model is given by

$$H_S = -\frac{1}{\sqrt{N}}\sum_{A<B} J_{A,B} S^{aA} S^{aB}\,, \qquad S^a = \chi^\dagger T^a \chi\,, \tag{6.103}$$

where the A, B label the different sites of the Ising lattice, and the index a refers to spin space as in (6.77). We see that inserting the fermion bilinear expression for S^a into the Ising model will give rise to a four-fermion model. Indeed, as explained in [54, 56], reducing (6.103) to a single-site model by averaging over disorder, and taking the large N limit, gives rise to the Sachdev-Ye model

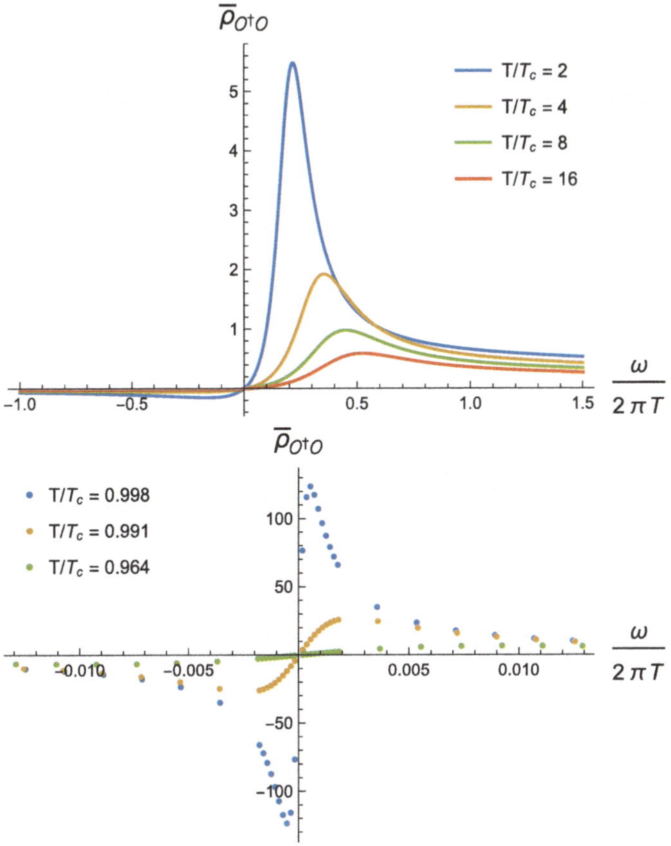

Fig. 6.12 Spectral function $\rho(\omega)$ for the Kondo operator $\mathcal{O}$ at the defect, as function of the frequency ω. **a** Left: In the unscreened phase above T_c. The spectral function corresponds to a Fano resonance with a spectral asymmetry. **b** Right: In the screened phase below T_c. The spectral function is antisymmetric. The Green's functions' poles leading to the extrema in $\rho(\omega)$ are determined by the size of the condensate for $\mathcal{O}$. Figures from [49]

$$H_{SY} = \frac{1}{(2N)^{3/2}} \sum_{i,j,k,l=1}^{N} \overline{J_{ij,kl}}\, \chi^{\dagger i} \chi^{j} \chi^{\dagger k} \chi^{l} - \mu \sum_{i} \chi^{\dagger i} \chi^{i} , \tag{6.104}$$

where the second term involving the chemical potential μ is added to fix the representation q of the spin impurity. As discussed in [57], the Sachdev-Ye model also displays a spectral asymmetry. This asymmetry is of an analogous form to the one found above for the holographic Kondo model. In [57], it is shown that the spectral asymmetry in the Sachdev-Ye model may be mapped to the entropy of a black hole in AdS_2 space. A similar mechanism is expected to be at work in the holographic Kondo model introduced above.

6.7 Conclusion and Outlook

The holographic Kondo model demonstrates nicely how the original concept of the AdS/CFT conjecture may be applied to more involved configurations, in this case involving a marginally relevant perturbation by a 'double-trace' operator and a condensation process. It also demonstrates that holographic models may be linked to previous results, in this case the large N Kondo model of condensed matter physics. On the other hand, they also add new features, in this case the coupling of the magnetic impurity to a strongly coupled electron system, leading in particular to new features in quantum quenches and in the spectral function.

The AdS/CFT correspondence and gauge/gravity duality are undoubtedly one of the most exciting developments in physics within the last twenty years. As discussed, new avenues are opening up and are expected to lead to further important discoveries in the future.

Acknowledgements I am very grateful to my collaborators Mario Flory, Carlos Hoyos, Max-Niklas Newrzella, Andy O'Bannon, Ioannis Papadimitriou, Jonas Probst and Jackson Wu.

References

1. S. Coleman, Quantum sine-Gordon equation as the massive Thirring model. Phys. Rev. **D 11**, 2088–2097 (1975). https://doi.org/10.1103/PhysRevD.11.2088
2. J.M. Maldacena, The large N limit of superconformal field theories and supergravity. Int. J. Theor. Phys. **38**, 1113–1133 (1999). arXiv:hep-th/9711200 [hep-th]. [Adv. Theor. Math. Phys. **2**, 231 (1998)]
3. G.'t. Hooft, Dimensional reduction in quantum gravity. Conf. Proc. **C930308**, 284–296 (1993). arXiv:gr-qc/9310026 [gr-qc]
4. L. Susskind, The world as a hologram. J. Math. Phys. **36**, 6377–6396 (1995). arxiv: hep--th/9409089
5. J.D. Bekenstein, Black holes and entropy. Phys. Rev. **D7**, 2333–2346 (1973). https://doi.org/10.1103/PhysRevD.7.2333
6. J. Kondo, Resistance minimum in dilute magnetic alloys. Prog. Theor. Phys. **32**(1), 37–49 (1964)
7. I. Affleck, Conformal field theory approach to the Kondo effect. Acta Phys. Polon. **B26**, 1869–1932 (1995). arXiv:cond-mat/9512099 [cond-mat]
8. N. Andrei, Diagonalization of the Kondo Hamiltonian. Phys. Rev. Lett. **45**, 379–382 (1980)
9. P. Wiegmann, Towards an exact solution of the Anderson model. Phys. Lett. A **80**, 163–167 (1980)
10. S.A. Hartnoll, C.P. Herzog, G.T. Horowitz, Building a holographic superconductor. Phys. Rev. Lett. **101**, 031601 (2008). arXiv:0803.3295 [hep-th]
11. J. Erdmenger, Introduction to gauge/gravity duality. PoS **TASI2017**, 001 (2018). https://doi.org/10.22323/1.305.0001. arXiv:1807.09872 [hep-th]
12. N. Andrei et al., Boundary and defect CFT: open problems and applications. arXiv:1810.05697 [hep-th]
13. M. Ammon, J. Erdmenger, *Gauge/Gravity Duality: Foundations and Applications* (Cambridge University Press, Cambridge, 2015)
14. H. Nastase, *Introduction to the ADS/CFT Correspondence* (Cambridge University Press, 2015)

15. M. Natsuume, AdS/CFT duality user guide. textitLect. Notes Phys. **903**, 1–294 (2015).https://doi.org/10.1007/978-4-431-55441-7, arXiv:1409.3575 [hep-th]
16. J. Casalderrey-Solana, H. Liu, K. Mateos, D. Rajagopal, U. Wiedemann, *Gauge/String Duality, Hot QCD and Heavy Ion Collisions* (Cambridge University Press, Cambridge, 2014). https://doi.org/10.1017/CBO9781139136747
17. J. Zaanen, Y.-W. Sun, Y. Liu, K. Schalm, *Holographic Duality in Condensed Matter Physics* (Cambridge University Press, Cambridge, 2015)
18. P.S. Howe, K.S. Stelle, P.C. West, A class of finite four-dimensional supersymmetric field theories. Phys. Lett. **124B**, 55–58 (1983). https://doi.org/10.1016/0370-2693(83)91402-8
19. P.S. Howe, K.S. Stelle, P.K. Townsend, Miraculous ultraviolet cancellations in supersymmetry made manifest. Nucl. Phys. **B236**, 125–166 (1984). https://doi.org/10.1016/0550-3213(84)90528-5
20. G. 't Hooft, A planar diagram theory for strong interactions. Nucl. Phys. **B72**, 461 (1974); 337(1973)]. https://doi.org/10.1016/0550-3213(74)90154-0
21. D.Z. Freedman, S.D. Mathur, A. Matusis, L. Rastelli, Correlation functions in the CFT(d) / AdS(d+1) correspondence. Nucl. Phys. **B546**, 96–118 (1999). arXiv:hep-th/9804058 [hep-th]
22. S. Lee, S. Minwalla, M. Rangamani, N. Seiberg, Three point functions of chiral operators in D = 4, N=4 SYM at large N. Adv. Theor. Math. Phys. **2**, 697–718 (1998). arXiv:hep-th/9806074 [hep-th]
23. H.J. Kim, L.J. Romans, P. van Nieuwenhuizen, The mass spectrum of chiral N=2 D=10 supergravity on S**5. Phys. Rev. **D32**, 389 (1985). https://doi.org/10.1103/PhysRevD.32.389
24. S.S. Gubser, I.R. Klebanov, A.M. Polyakov, Gauge theory correlators from noncritical string theory. Phys. Lett. **B428**, 105–114 (1998). arXiv:hep-th/9802109 [hep-th]
25. E. Witten, Anti-de Sitter space and holography. Adv. Theor. Math. Phys. **2**, 253–291 (1998). arXiv:hep-th/9802150 [hep-th]
26. S.S. Gubser, I.R. Klebanov, A.W. Peet, Entropy and temperature of black 3-branes. Phys. Rev. **D54**, 3915–3919 (1996). https://doi.org/10.1103/PhysRevD.54.3915. arXiv:hep-th/9602135 [hep-th]
27. I. Affleck, A.W.W. Ludwig, The Kondo effect, conformal field theory and fusion rules. Nucl. Phys. B **352**, 849–862 (1991)
28. N. Read, D.M. Newns, On the solution of the Coqblin-Schrieffer Hamiltonian by the large-N expansion technique. J. Phys. C: Solid State Phys. **16**(17), 3273 (1983)
29. P. Coleman, N. Andrei, Diagonalisation of the generalised Anderson model. J. Phys. C: Solid State Phys. **19**(17), 3211 (1986)
30. S. Harrison, S. Kachru, G. Torroba, A maximally supersymmetric Kondo model. Class. Quant. Grav. **29**, 194005 (2012). arXiv:1110.5325 [hep-th]
31. P. Benincasa, A.V. Ramallo, Holographic kondo model in various dimensions. JHEP **06**, 133 (2012). arXiv:1204.6290 [hep-th]
32. J. Erdmenger, C. Hoyos, A. O'Bannon, J. Wu, A holographic model of the kondo effect. JHEP **12**, 086 (2013). arXiv:1310.3271 [hep-th]
33. E.I. Buchbinder, J. Gomis, F. Passerini, Holographic gauge theories in background fields and surface operators. JHEP **12**, 101 (2007). arXiv:0710.5170 [hep-th]
34. J.A. Harvey, A.B. Royston, Gauge/gravity duality with a chiral N=(0,8) string defect. JHEP **08**, 006 (2008). arXiv:0804.2854 [hep-th]
35. P. Breitenlohner, D.Z. Freedman, Stability in gauged extended supergravity. Ann. Phys. **144**, 249 (1982). https://doi.org/10.1016/0003-4916(82)90116-6
36. E. Witten, Multitrace operators, boundary conditions, and AdS / CFT correspondence. arXiv:hep-th/0112258 [hep-th]
37. O. Aharony, M. Berkooz, E. Silverstein, Multiple trace operators and nonlocal string theories. JHEP **08**, 006 (2001). arXiv:hep-th/0105309 [hep-th]
38. J.M. Luttinger, An exactly soluble model of a many-fermion system. J. Math. Phys. **4**(9), 1154–1162 (1963)
39. S. Ryu, T. Takayanagi, Holographic derivation of entanglement entropy from AdS/CFT. Phys. Rev. Lett. **96**, 181602 (2006). arXiv:hep-th/0603001 [hep-th]

40. P. Calabrese, J.L. Cardy, Entanglement entropy and quantum field theory. J. Stat. Mech. **0406**, P06002 (2004). arXiv:hep-th/0405152 [hep-th]
41. J. Erdmenger, M. Flory, M.-N. Newrzella, Bending branes for DCFT in two dimensions. JHEP **01**, 058 (2015). arXiv:1410.7811 [hep-th]
42. J. Erdmenger, M. Flory, C. Hoyos, M.-N. Newrzella, J.M.S. Wu, Entanglement entropy in a holographic kondo model. Fortsch. Phys. **64**, 109–130 (2016). arXiv:1511.03666 [hep-th]
43. W. Israel, Singular hypersurfaces and thin shells in general relativity. Nuovo Cim. **B44S10**, 1 (1966). [Nuovo Cim. **B44**,1 (1966)]
44. E.S. Sørensen, M.-S. Chang, N. Laflorencie, I. Affleck, Quantum impurity entanglement. J. Stat. Mech.: Theory Exp. **2007**(08), P08003 (2007)
45. E. Eriksson, H. Johannesson, Impurity entanglement entropy in kondo systems from conformal field theory. Phys. Rev. B **84**, 041107 (2011)
46. J. Erdmenger, M. Flory, M.-N. Newrzella, M. Strydom, J.M.S. Wu, Quantum quenches in a holographic kondo model. arXiv:1612.06860 [hep-th]
47. D.T. Son, A.O. Starinets, Minkowski space correlators in AdS / CFT correspondence: recipe and applications. JHEP **09**, 042 (2002). arXiv:hep-th/0205051 [hep-th]
48. S. de Haro, S.N. Solodukhin, K. Skenderis, Holographic reconstruction of space-time and renormalization in the AdS / CFT correspondence. Commun. Math. Phys. **217**, 595–622 (2001). arXiv:hep-th/0002230 [hep-th]
49. J. Erdmenger, C. Hoyos, A. O'Bannon, I. Papadimitriou, J. Probst, J.M.S. Wu, Holographic kondo and fano resonances. Phys. Rev. D **96**(2), 021901 (2017). arXiv:1611.09368 [hep-th]
50. J. Erdmenger, C. Hoyos, A. O'Bannon, I. Papadimitriou, J. Probst, J.M.S. Wu, Two-point functions in a holographic kondo model. JHEP **03**, 039 (2017). arXiv:1612.02005 [hep-th]
51. U. Fano, Effects of configuration interaction on intensities and phase shifts. Phys. Rev. **124**, 1866–1878 (1961)
52. O. Parcollet, A. Georges, G. Kotliar, A. Sengupta, Overscreened multichannel SU(N) kondo model: large-N solution and conformal field theory. Phys. Rev. **B58**(7), 3794 (1998). arXiv:cond-mat/9711192 [cond-mat.str-el]
53. P. Coleman, *Introduction to Many-Body Physics* (Cambridge University Press, Cambridge, 2015)
54. S. Sachdev, J. Ye, Gapless spin fluid ground state in a random, quantum Heisenberg magnet. Phys. Rev. Lett. **70**, 3339 (1993). arXiv:cond-mat/9212030 [cond-mat]
55. J. Maldacena, D. Stanford, Remarks on the Sachdev-Ye-Kitaev model. Phys. Rev. D **94**(10), 106002 (2016). arXiv:1604.07818 [hep-th]
56. A. Georges, O. Parcollet, S. Sachdev, Quantum fluctuations of a nearly critical heisenberg spin glass. Phys. Rev. B **63**, 134406 (2001)
57. S. Sachdev, Bekenstein-Hawking entropy and strange metals. Phys. Rev. X **5**(4), 041025 (2015). arXiv:1506.05111 [hep-th]

Chapter 7
Local Probe of the Kondo Length at a Y-Junction of Critical Quantum Ising Chains

Domenico Giuliano and Pasquale Sodano

Abstract We estimate the screening length associated to the Kondo effect at a Y-junction of quantum Ising chains from the scaling behavior of the local magnetization at the junction. We propose a simple way to probe the Kondo screening length in a system whose parameters, including the effective length of the chains, can be in principle tuned at wish.

7.1 Introduction

The Kondo effect was discovered as an upturn, below a certain, nonuniversal temperature, of the resistivity in conducting metals containing magnetic impurities as, for instance, in Cu doped with Co atoms. It was explained as a consequence of the magnetic exchange interaction between the magnetic impurities and the conduction electrons [1–3]. It arises due to the cooperative effect of the conduction electrons surrounding the magnetic impurity, whose spins adjust themselves so as to dynamically screen the impurity magnetic moment [2].

In the energy domain, the above screening is determined by electrons with energy ranging all the way down to a characteristic dynamically generated energy scale D_K that is invariant along the renormalization group trajectories, dubbed the (Boltzmann constant times the) Kondo temperature [4, 5]. As the running energy scale D approaches D_K, a crossover takes place between the weakly coupled regime, in which the itinerant electrons perturbatively interact with the magnetic impurity with

D. Giuliano (✉)
Dipartimento di Fisica, Università della Calabria Arcavacata di Rende, 87036 Cosenza, Italy
e-mail: domenico.giuliano@fis.unical.it

I.N.F.N., Gruppo collegato di Cosenza, Arcavacata di Rende, 87036 Cosenza, Italy

P. Sodano
I.N.F.N., Sezione di Perugia, Via A. Pascoli, 06123 Perugia, Italy
e-mail: pasquale.sodano01@gmail.com

International Institute of Physics, Universidade Federal do Rio Grande do Norte, 59078-400 Natal-RN, Brazil

A. Ferraz et al. (eds.), *Strongly Coupled Field Theories for Condensed Matter and Quantum Information Theory*, Springer Proceedings in Physics 239,
https://doi.org/10.1007/978-3-030-35473-2_7

a small "bare" coupling constant, and the strongly coupled regime, in which cooperative, nonperturbative low-energy electron dynamics leads to the formation of the Nozières Fermi-liquid state, with the itinerant electrons cooperating to screen the impurity spin into a localized spin singlet [6]. Thus, the Kondo effect is regarded as the crossover between an ultraviolet, weakly coupled fixed point (the isolated impurity) and an infrared, strongly correlated fixed point (the Kondo correlated state).

While the crossover is typically driven by acting on a dimensionful scale such as the energy, or the temperature, using the Fermi velocity of fermions at the Fermi surface, v_f, it is possible to trade the energy scale D_K for a length scale $\lambda_K \sim v_f/D_K$. Having an emerging length scale suggests the possibility to regard the Kondo crossover as taking place in real space, as well. In fact, λ_K has been identified with the size of the electronic cloud screening the impurity spin and accordingly it has been dubbed the "Kondo screening length" [5, 7].

While there is a solid theoretical argument implying the existence of the Kondo screening length, any attempt to experimentally detect it in conducting metals has failed so far. Indeed, when considering magnetic impurities in conducting metals, there are several reasons for this failure, such as having a finite density of magnetic impurities combined with the large value of λ_K, with the correspondingly non-negligeable interference effects among impurities, spurious effects due to the electronic interactions, et cetera [8]. Given the difficulty of detecting the Kondo cloud in electronic systems, it becomes of the utmost importance to design protocols of a possible detection of the Kondo screening length in condensed matter systems.

The Kondo effect has become by now a paradigm in the physics of many-body correlated systems, providing a testing ground for a number of methods, including Wilson's numerical renormalization group technique [5], as well as Anderson's analytical scaling approach [4]. Indeed, a renewed interest in the Kondo effect has recently arisen, since it became possible to realize it in a controlled way in solid state systems, such as small quantum dots with metallic leads [9–11], or in magnetic impurities in gapless spin chains [12], or in "topological" devices [13–16], in which an effective magnetic impurity could be realized by means of Majorana modes emerging at the interface between topological superconductors in their topological phase, and normal conducting wires [17].

The Kondo effect emerges also in Y-junctions of XX and Ising chains [18–20] and of Tonks–Girardeau gases [21].

In the following, we see how the Kondo screening length may emerge at a Y-junction of quantum Ising chains and point out how to estimate it by means of a pertinent local measurement. While the onset of Kondo effect at a junction is by now a well-established feature [19, 22], the corresponding issue of the origin and the definition of the Kondo screening length in spin networks has so far never been addressed in detail. To this purpose, we argue here for the emergence of a length scale associated to the Kondo energy scale.

We shall employ the standard renormalization group approach to the Kondo problem [4], adapted to a symmetric, finite-size version of the junction (that is, with the three chains all assumed to have a finite length ℓ). We show that, at given values of all the other junction parameters, the system may, or may not, crossover to the Kondo regime depending on the value of ℓ. In terms of energies, this is readily related to the appearance of a finite energy gap in the excitation spectrum of the chains, $\Delta_{\rm FS} \propto \ell^{-1}$, which depletes the corresponding spectral density at low energy, that is, in the energy window that is relevant for the Kondo impurity screening [2]. Conversely, moving to real space, we propose to relate the finite-ℓ suppression of the effect to the need for the system to fit in it the whole Kondo screening cloud, in order for Kondo screening to be effective. Thus, we propose to naturally identify the Kondo screening length as the length scale ℓ_* such that the junction does not show/does show the onset of the Kondo effect for $\ell < \ell_*$ ($\ell > \ell_*$). Eventually, we show how our approach provides us with estimates for the Kondo screening length that are numerically consistent with the expected relation $\lambda_K \propto D_K^{-1}$.

The real-space picture of the Kondo cloud lies at the hearth of our proposal of a local probe of λ_K, and it can be most readily formulated in Kondo spin chains [12]. In these systems, it has been shown [23–25]. That the Kondo spin cloud is formed by all the spins fully entangled with the impurity. As a result, the impurity and the spins of the cloud should form an extended singlet [23], and, thus, one expects that the magnetization of the spins belonging to the cloud is equal to 0. Indeed, when fully formed within the spin system, the Kondo screening cloud corresponds to a fully entangled state of the magnetic impurity with the spins [23–25]. Here, we propose to probe the Kondo length in a Y-junction of quantum Ising chains by extracting λ_K from the scaling behavior (with the chain length ℓ) of the average value of a local spin operator at the impurity location [20, 26].

The paper is organized as follows:

- In Sect. 7.2 we review the model Hamiltonian for a Y-junction of quantum Ising chains and we show how to implement the Jordan–Wigner transformation to map it onto an effective Kondo Hamiltonian;
- In Sect. 7.3 we address the onset of the Kondo regime in the junction with chains of infinite length and with the parameters chosen so that their excitation spectrum is gapless ("critical chains"). We review then the approach of [20] to estimate the corresponding Kondo temperature;
- In Sect. 7.4 we analyze the junction made with finite-length chains. There, we shall focus on how the finite length of the chains—and the corresponding nonzero finite-size gap- affects the onset of Kondo regime;
- In Sect. 7.5 we show how to effectively vary the chain length by acting on local parameters only and we analyze how to probe the onset of Kondo regime by looking at the local magnetization at the junction;
- Section 7.6 is devoted to concluding remarks;
- In the appendix we review known facts about Jordan–Wigner approach to a single quantum Ising chain.

7.2 Y-Junction of Quantum Ising Chains: Jordan–Wigner Transformation and Effective Kondo-Like Hamiltonian

The Kondo effect at a Y-junction of quantum Ising chains (QIC's) was originally addressed in [19] where an appropriate version of the Jordan–Wigner mapping [27] from quantum spin operators onto lattice fermion operators was employed, leading to an effective overscreened Kondo Hamiltonian [6]. The model Hamiltonian proposed in [19] consists of three critical QIC's connected to each other at one of their endpoints via an additional Ising exchange interaction between the spin operators lying at the endpoints. Assuming for simplicity that the Hamiltonians of the three chains all have the same parameters, the "bulk" Hamiltonian of the system, $H_{\rm Bulk}$ for three critical QIC's of length ℓ is given by

$$H_{\rm Bulk} = \sum_{\lambda=1}^{3} \left\{ -2J \sum_{j=1}^{\ell-1} S_{j,\lambda}^x S_{j+1,\lambda}^x + H \sum_{j=1}^{\ell} S_{j,\lambda}^z \right\} \quad , \tag{7.1}$$

with $\mathbf{S}_{j,\lambda}$ being a quantum spin-1/2 variable lying at site-j of the chain λ, J being the Ising magnetic exchange strength, H being the applied, transverse magnetic field. As stated above, the three chains are connected to each other at their endpoints ($j = 1$) via the junction "boundary" Hamiltonian H_Δ, given by

$$H_\Delta = -2J_\Delta \sum_{\lambda=1}^{3} S_{1,\lambda}^x S_{1,\lambda+1}^x \quad , \tag{7.2}$$

with, in general, $J_\Delta / J < 1$. In the "critical" limit, $J = \pm H$, the total Hamiltonian $H_{3-\rm Chain} = H_{\rm Bulk} + H_\Delta$ is a paradigmatic example of a junction of three quantum spin chains mapping onto an effective two-channel Kondo Hamiltonian [19].

A multi-channel generalization has been proposed in [22] still using quantum spin chains, and the corresponding model Hamiltonians have proven to be exactly solvable in the continuum limit [28]. In addition, a generalization of $H_{3-\rm Chain}$ to a junction of three quantum XY-spin chains was proposed in [20]. It provides a means to continuously interpolate between the Y-junction of three QIC's, and the Y-junction of three XX-chains discussed in [18]. Finally, it should be mentioned that an effective Kondo-like Hamiltonian may arise in pertinent Josephson junction networks [26].

To resort to the effective Kondo-like Hamiltonian, one recasts $H_{3-\rm Chain}$ in fermionic coordinates by implementing a pertinent generalization of the Jordan–Wigner transformation, in which real-fermionic Klein factors $\{\sigma^1, \sigma^2, \sigma^3\}$ are introduced, one per each chain, to recover the correct commutation relations between spin operators lying at sites located on different chains. Specifically, one sets $\{\sigma^\lambda, \sigma^{\lambda'}\} = 2\delta_{\lambda,\lambda'}$. Accordingly, one introduces the generalized Jordan–Wigner transformations, given by [18]

$$
\begin{aligned}
S_{j,\lambda}^{+} &= i c_{j,\lambda}^{\dagger}\, e^{i\pi \sum_{r=1}^{j-1} c_{r,\lambda}^{\dagger} c_{r,\lambda}}\, \sigma^{\lambda} \\
S_{j,\lambda}^{-} &= i c_{j,\lambda}\, e^{i\pi \sum_{r=1}^{j-1} c_{r,\lambda}^{\dagger} c_{r,\lambda}}\, \sigma^{\lambda} \\
S_{j,\lambda}^{z} &= c_{j,\lambda}^{\dagger} c_{j,\lambda} - \frac{1}{2} \quad ,
\end{aligned} \tag{7.3}
$$

with the fermionic lattice operators satisfying the algebra

$$
\begin{aligned}
&\{ c_{j,\lambda} , c_{j',\lambda'} \} = \{ c_{j,\lambda}^{\dagger}, c_{j',\lambda'}^{\dagger} \} = 0 \\
&\{ c_{j,\lambda} , c_{j',\lambda'}^{\dagger} \} = \delta_{j,j'}\, \delta_{\lambda,\lambda'} \\
&\{ \sigma^{\lambda}, \sigma^{\lambda'} \} = \delta^{\lambda,\lambda'} \quad .
\end{aligned} \tag{7.4}
$$

Inserting (7.3) into (7.1), one obtains the fermionic version of the bulk Hamiltonian

$$
\begin{aligned}
H_{\mathrm{Bulk}} = \sum_{\lambda=1}^{3} \Bigg\{ & -\frac{J}{2} \sum_{j=1}^{\ell-1} \{ c_{j,\lambda}^{\dagger} c_{j+1,\lambda} + c_{j+1,\lambda}^{\dagger} c_{j,\lambda} \} + \frac{J}{2} \sum_{j=1}^{\ell-1} \{ c_{j,\lambda} c_{j+1,\lambda} + c_{j+1,\lambda}^{\dagger} c_{j,\lambda}^{\dagger} \} \\
& + H \sum_{j=1}^{\ell} c_{j,\lambda}^{\dagger} c_{j,\lambda} \Bigg\} \quad .
\end{aligned} \tag{7.5}
$$

From (7.5) one sees that the generalized Klein factors disappear from the bulk Hamiltonian. At variance, they do explicitly appear in the boundary junction Hamiltonian; indeed, the boundary is the only place where the three chains explicitly interact with each other. In particular, in terms of the Jordan–Wigner fermions, plus the generalized Klein factors, one obtains [19, 20]

$$
H_{\Delta} = 2 J_{\Delta} \Sigma_1 \cdot \mathscr{R}, \tag{7.6}
$$

with

$$
\Sigma_j = -\frac{i}{2} \begin{bmatrix} (c_{j,2}^{\dagger} + c_{j,2})(c_{j,3}^{\dagger} + c_{j,3}) \\ (c_{j,3}^{\dagger} + c_{j,3})(c_{j,1}^{\dagger} + c_{j,1}) \\ (c_{j,1}^{\dagger} + c_{j,1})(c_{j,2}^{\dagger} + c_{j,2}) \end{bmatrix} \quad ; \quad \mathscr{R} = -\frac{i}{2} \begin{bmatrix} \sigma^2 \sigma^3 \\ \sigma^3 \sigma^1 \\ \sigma^1 \sigma^2 \end{bmatrix} . \tag{7.7}
$$

It is straightforward to check that the components of Σ_j, as well as the ones of $\mathscr{R}$, close the algebra of the quantum angular momentum. Therefore, the Hamiltonian in (7.5), (7.6) describes a magnetic exchange interaction between the "effective" isolated magnetic impurity $\mathscr{R}$ and the angular momentum density of the network of the three chains, Σ_j, at the endpoint of the chains, $j = 1$. In particular, the boundary Kondo Hamiltonian in (7.6) corresponds to the peculiar realization of overscreened, 2-channel Kondo effect put forward in [29].

As a first observation we note that (see the appendix), even for critical chains, i.e., for $J = \pm H$, in a finite-length QIC of length ℓ, the excitation spectrum of the chain has a finite-size gap $\Delta_{\rm FS} \sim \pi J/\ell$. As $\ell \to \infty$, $\Delta_{\rm FS} \to 0$; accordingly, one finds that the low-energy excitation spectrum of $H_{\rm Bulk}$ contains gapless excitations carrying a nonzero Σ-spin-density. In perfect analogy with what happens with the Kondo effect in normal metals [2], we expect that those excitation provide an effective screening of $\mathscr{R}$, that is, the mechanism that lies at the hearth of the Kondo effect in the Y-junction of three QIC's. Accordingly, the onset of the Kondo effect at the Y-junction corresponds to a crossover in the running coupling associated to J_Δ towards the nonperturbative regime, $J_\Delta \sim J$.

In the following, we review the onset of the Kondo regime in the $\ell \to \infty$-limit by means of a pertinent application of Anderson's poor-man scaling approach to the specific problem we are investigating here.

Before concluding this section, it is worth mentioning that the Kondo effect may be also recovered at a Y-junction of off-critical QIC's, provided the Kondo temperature T_K is much larger than the bulk gap of the system, $\Delta_{\rm Bulk} \sim ||J| - |H||$. Specifically, one sees that the off-critical junction maps onto a remarkable two-channel version of the Kondo model for a magnetic impurity connected to two superconductors [30], which has been used to analyze the transport across a quantum dot at Coulomb blockade connected to two superconducting leads [31–34].

In the next section, we study the onset of the Kondo regime and we estimate the Kondo temperature in the critical system.

7.3 The Kondo Regime and the Kondo Temperature

For studying the onset of the Kondo effect at a Y-junction of critical QIC's, we resort to the imaginary-time formulation, in which the boundary dynamics is encoded in the imaginary time boundary action S_Δ, given by

$$S_\Delta = \int_0^\beta {\rm d}\tau \; H_\Delta(\tau) = 2J_\Delta \int_0^\beta {\rm d}\tau \; \Sigma_1(\tau) \cdot \mathscr{R}(\tau) \quad , \tag{7.8}$$

and $\beta = (k_B T)^{-1}$, where T is the temperature and k_B is Boltzmann's constant.

To resort to the standard poor man's analysis, one first rewrites (7.8) in (Matsubara) frequency space and in the $T \to 0$-limit. Sending the temperature to 0 requires the introduction of a cutoff D with the dimension of an energy (high-energy cutoff—typically identified with the excitation bandwidth). Accordingly, the imaginary-time action is rewritten as [20]

$$S_\Delta \to 2J_\Delta \int_{-D}^{D} \frac{{\rm d}\Omega}{2\pi} \sum_{\lambda=1}^{3} \Sigma_1^\lambda(-\Omega) \mathscr{R}^\lambda(\Omega) \quad . \tag{7.9}$$

The renormalization group (RG) approach leads to an appropriate scaling equation for the running coupling associated to J_Δ. For this purpose, one has to effectively reduce the cutoff D, by pertinently taking into account the corresponding corrections to J_Δ. We first split the integral in (7.9) into the sum of an integral over $[-D, -D/\kappa]$, an integral over $[D/\kappa, D]$, and an integral over $[-D/\kappa, D/\kappa]$, with $\kappa = 1+\epsilon$ and $0 < \epsilon \ll 1$; on rescaling D, one finally obtains

$$2J_\Delta \int_{-\frac{D}{\kappa}}^{\frac{D}{\kappa}} \frac{d\Omega}{2\pi} \sum_{\lambda=1}^{3} \Sigma_1^\lambda(-\Omega)\mathscr{R}^\lambda(\Omega) \to \frac{2J_\Delta}{\kappa} \int_{-D}^{D} \frac{d\Omega}{2\pi} \sum_{\lambda=1}^{3} \Sigma_1^\lambda\left(-\frac{\Omega}{\kappa}\right)\mathscr{R}^\lambda\left(\frac{\Omega}{\kappa}\right) \quad . \tag{7.10}$$

As a result [8], H_Δ corresponds to a marginal boundary interaction, that is, the right-hand side of (7.10) does not depend on κ. Therefore, the only renormalization of J_Δ comes from summing over the frequency windows $[-D, -D/\kappa]$ and $[D/\kappa, D]$.

In the sequel, one resorts to a method discussed in detail in [2] for the Kondo problem in a metal and in [20] for the specific case we are addressing here. As a result, one obtains an additional correction to the boundary action, $\delta S_\Delta^{(2)}$, arising to the second order in J_Δ and given by [20]

$$\delta S_\Delta^{(2)} = 2J_\Delta^2 \int_{-D}^{D} \frac{d\Omega}{2\pi} \sum_{\lambda=1}^{3} \Sigma_1^\lambda(-\Omega)\mathscr{R}^\lambda(\Omega)\left[\Gamma(\ell; D) + \Gamma(\ell; -D)\right] D\left(1-\kappa^{-1}\right) \quad , \tag{7.11}$$

with

$$\Gamma(\ell; \Omega) = \lim_{\beta\to\infty} \frac{1}{\beta} \sum_{\omega} \left[\frac{2}{i\Omega - i\omega}\right] G_{1,1}(i\omega) \quad ; \tag{7.12}$$

here the sum is taken over the fermionic Matsubara frequencies ω, with the Jordan–Wigner fermion Green's function $G_{1,1}(i\omega)$ defined in (7.59) of the appendix. In the $\ell \to \infty$-limit, $\Gamma(\Omega)$ has been derived in [20] along the whole critical line connecting the Y-junction of XX-spin chains [18] to the junction of QIC's. At finite-ℓ, a straightforward generalization of the metod of [20] yields

$$\Gamma(\ell; D) = \frac{1}{D}\frac{d}{dD}\left\{\frac{2D}{(\ell+1)}\sum_{n=0}^{\ell-1}\left\{\frac{2J\sin\left[\frac{\pi}{2}\frac{2n+1}{2\ell+1}\right]\cos^2\left[\frac{3\pi}{2}\frac{2n+1}{2\ell+1}\right]}{D^2+4J^2\sin^2\left[\frac{\pi}{2}\frac{2n+1}{2\ell+1}\right]}\right\}\right\} \quad . \tag{7.13}$$

Note that, differently from [20], here we explicitly keep the dependence of Γ on ℓ since, in the following, we shall use ℓ as a control parameter. Eventually, we check that the results of [20] are recovered in the thermodynamic limit, $\ell \to \infty$.

Considering (7.11), (7.13) alltogether, we see that summing over the frequency windows $[-D, -D/\kappa]$ and $[D/\kappa, D]$ provides a renormalization of the coupling strength J_Δ as $J_\Delta \to J_\Delta + \delta J_\Delta$, with

$$\delta J_\Delta = -J_\Delta^2\,\rho(\ell; D)\,\frac{\delta D}{D} \quad . \tag{7.14}$$

Note that, in (7.14), we have defined δD so that $\delta D/D = (1-\kappa^{-1})$ and we have set $\rho(\ell; D) \equiv D[\Gamma(\ell; D) + \Gamma(\ell; -D)]/2$, in order to recover the standard functional form of the RG equations for the Kondo problem [2].

Introducing a running coupling strength at a given ℓ, $J_\Delta(\ell; D)$, one may trade (7.14) for the differential equation given by

$$\frac{dJ_\Delta(\ell; D)}{dl} = J_\Delta^2(\ell; D)\rho(\ell; D) \quad , \tag{7.15}$$

with $dl = d\ln(D_0/D)$, D being the running energy scale, and D_0 being an high-energy cutoff $\sim J$. On integrating (7.15), one obtains for $J_\Delta(\ell; D)$ the expression [20, 35]

$$J_\Delta(\ell; D) = \frac{J_\Delta^{(0)}}{1 - J_\Delta^{(0)} \int_D^{D_0} \frac{\rho(\ell;x)}{x} dx} \quad . \tag{7.16}$$

The integrated scaling flow for $J_\Delta(\ell; D)$ allows to define a scale $D_*(\ell)$ as the scale at which the denominator at the right-hand side of (7.16) vanishes. As $\ell \to \infty$, $D_*(\ell \to \infty) \to D_K$, with D_K being the Kondo energy scale associated to the Y-junction of critical QIC's.

To estimate D_K, we note that, as $\ell \to \infty$, one gets

$$\int_D^{D'} \frac{\rho(\ell \to \infty; x)}{x} dx = \int_0^{\frac{\pi}{2}} dq \left[\frac{8J\sin(q)\cos^2(3q)}{4J^2\sin^2(q) + (D')^2} - \frac{8J\sin(q)\cos^2(3q)}{4J^2\sin^2(q) + (D)^2} \right] \quad . \tag{7.17}$$

As a result, since $J_\Delta(\ell \to \infty; D) \equiv J_\Delta(D)$, one eventually obtains

$$\frac{J_\Delta(D)}{D_0} \approx \frac{J_\Delta^{(0)}/D_0}{1 - \frac{J_\Delta^{(0)}}{D_0}\frac{4}{\pi}\ln\left(\frac{D_0}{D}\right)} \Rightarrow D_K = D_0\, e^{-\frac{\pi D_0}{4J_\Delta^{(0)}}} \quad , \tag{7.18}$$

with $J_\Delta^{(0)}$ being the (bare) running coupling at the reference scale D_0. Equation (7.18) contains the expected result for the Kondo scale of the Y-junction of QIC's in the $\ell \to \infty$-limit.

The key observation about D_K is that, in general, the running coupling (as well as the physical observables) must be a scaling function of D/D_K [4]. For instance, for the running coupling one expects a scaling relation of the form

$$\frac{J_\Delta(D)}{D_0} = \frac{J_\Delta^{(0)}}{D_0} \mathscr{F}\left(\frac{D}{D_K}\right) \quad , \tag{7.19}$$

with the perturbative (in J_Δ) asymptotic form of $\mathscr{F}\left(\frac{D}{D_K}\right)$ derived from (7.18) and given by

$$\mathscr{F}\left(\frac{D}{D_K}\right) \approx \frac{1}{\ln\left(\frac{D}{D_K}\right)} + \mathscr{O}\left(\frac{J_\Delta}{D_0}\right) \quad . \tag{7.20}$$

For finite-ℓ QIC's one expects the scaling relations to be violated by the presence of the additional nonuniversal energy scale $\Delta_{\rm FS}$. Roughly speaking, as long as $\Delta_{\rm FS} \ll D_K$, the net effect of the finite-size gap is just to provide a slight renormalization of D_K itself, in perfect analogy with what happens when an additional energy scale is introduced by means of a local magnetic field acting on the spin impurity in the Kondo effect in the XXZ-spin chain [36, 37]. Eventually, resorting to larger values of $\Delta_{\rm FS}/D_K$ can strongly affect Kondo effect till possibly leading to its suppression.

To evidence this last point, one has to generalize to finite-ℓ the definition of the "critical Kondo coupling" $J_{\Delta,c}(\ell)$. According to [35], one defines $J_{\Delta,c}(\ell)$ so that a necessary condition for the junction to crossover towards Kondo regime is that the bare coupling $J_\Delta^{(0)} > J_{\Delta,c}(\ell)$. Thus, one sets

$$\frac{1}{J_{\Delta,c}(\ell)} - \int_0^{D_0} \frac{\rho(\ell;x)}{x} {\rm d}x = 0 \quad , \tag{7.21}$$

which allows to recast (7.15) in the form

$$J_\Delta(\ell;D) = \frac{J_\Delta^{(0)} J_{\Delta,c}(\ell)}{J_{\Delta,c}(\ell) - J_\Delta^{(0)} + J_\Delta^{(0)} J_{\Delta,c}(\ell) \int_0^D \frac{\rho(\ell;x)}{x} {\rm d}x} \quad . \tag{7.22}$$

Equation (7.22)shows that the nonperturbative onset of the Kondo regime can only be recovered provided that $J_\Delta^{(0)} > J_{\Delta,c}(\ell)$. In [20], $J_{\Delta,c}$ was derived for a junction between infinite XY-chains, in which case the Ising case was recovered in the extremely asymmetric, $\gamma \to 0$ limit (see [20] for details). While we expect (7.21) to give us back the result for $\gamma = 0$ once $\ell \to \infty$, for the purpose of our following discussion it is important to discuss the explicit dependence on ℓ, which is induced in $J_{\Delta,c}$ by the dependence of ρ on each QIC length.

On qualitative grounds, the reason for this dependence can be understood from the fact that the onset to Kondo regime is basically due to impurity screening via low energy spinful excitations in the "bulk" of the system (for instance, by low-energy conduction electrons, close to the Fermi energy, in the "standard" electronic Kondo effect). In our specific case, the screening is dynamically induced by the presence of low-lying Jordan–Wigner fermions in the bulk spectrum of the chains, whose energy is cutoff at $\Delta_{\rm FS}$ in the finite-size system. Therefore, a finite value of ℓ implies a "depletion" of the low-energy part of the fermionic excitation spectrum, so as to make it harder for the system to crossover towards Kondo regime. Accordingly, the less effective is the dynamical screening from bulk fermions, the larger has to be the bare coupling for the junction to be able to crossover to Kondo regime. An explicit dependence of $J_{\Delta,c}$ on ℓ is, therefore, expected and, in particular, one may infer that, as $\ell \to \infty$, $J_{\Delta,c}$ tends to the value estimated in [20] in the Y-junction of QIC's, for $\ell \to \infty$. This trend is ultimately confirmed by the plot in Fig. 7.1, where we display the estimate of $J_{\Delta,c}/D_0$ as as a function of ℓ, for $1 \leq \ell \leq 300$ and with the band cutoff $D_0 = 2J$. While we see that, at large values of ℓ, $J_{\Delta,c}/D_0$ tends to the expected value of about 0.13 [20], it definitely increases, as ℓ gets lowered, as expected from the discussion above.

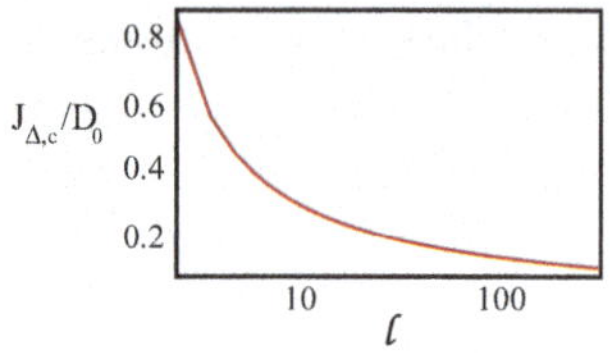

Fig. 7.1 Semilogarithmic plot of $J_{\Delta,c}/D_0$ as a function of ℓ for ℓ ranging from $\ell = 1$ to $\ell = 300$. At large values of ℓ, one obtains $J_{\Delta,c}/D_0 \approx 0.13$, which is the appropriate value for a Y-junction of infinite quantum Ising chains [20]. For small values of ℓ, $J_{\Delta,c}$ exhibits the expected increas, as ℓ gets smaller, due to the depletion of the low-energy part of the Jordan–Wigner fermion spectrum, because of the finite-size gap in their spectrum

In the following section, we analyze the effects of having a chain of finite size ℓ. We shall introduce a new scale ℓ_* such that the Kondo effect is suppressed in chains of length ℓ when $\ell < \ell_*$. In this approach ℓ_* is used to estimate the Kondo screening length [38].

7.4 Modified Scaling of the Kondo Coupling at Finite-ℓ

Having finite-length QIC's amounts to introduce an additional dimensionful scale, corresponding to the finite-size gap $\Delta_{\rm FS}$. Accordingly, the one-parameter scaling relation in (7.19) is modified into

$$\frac{J_\Delta(\ell; D)}{D_0} = \frac{J_\Delta^{(0)}}{D_0} \mathscr{G}\left[\frac{D}{D_K}; \frac{\Delta_{\rm FS}}{D_K}\right] \quad , \tag{7.23}$$

where the generalized scaling function $\mathscr{G}(x, y)$ must satisfy the condition $\mathscr{G}(x, 0) = \mathscr{F}(x)$. When $D_K/\Delta_{\rm FS} \ll 1$, one might trade (7.23) for another one in which $\Delta_{\rm FS}$ works as a reference scale, that is

$$\frac{J_\Delta(\ell; D)}{D_0} \approx \frac{J_\Delta^{(0)}}{D_0} \hat{\mathscr{G}}\left[\frac{D}{\Delta_{\rm FS}}\right] \quad . \tag{7.24}$$

A rigorous derivation of the scaling functions in (7.23), (7.24) would ultimately require a numerical derivation of the appropriate scaling functions from which to extract the running coupling at a given ℓ [39]. In the following, we shall provide an approximate asymptotic expression for the right-hand side of (7.23).

By resorting to the perturbative RG approach in the infinite-ℓ limit, one may exhibit an approximate expression for the function $\hat{\mathscr{G}}(x)$ by keeping ℓ finite and considering the approximate numerical fit given by

Table 7.1 Table of values of $\ell_*[J_\Delta^{(0)}]$ as a function of $J_{\Delta,0}$

$J_\Delta^{(0)}/D_0$	$\ell_*[J_\Delta^{(0)}]$ [Numerically estimated]	$\ell_*[J_\Delta^{(0)}]$ [From (7.27)]
0.40	5	5.96
0.35	7	7.70
0.30	12	10.82
0.25	16	17.41
0.20	36	35.57
0.15	130	117.10
0.10	$\to\infty$	1265.04

$$\frac{4}{(\ell+1)}\sum_{n=0}^{\ell-1}\left\{\frac{2J\sin\left[\frac{\pi}{2}\frac{2n+1}{2\ell+1}\right]\cos^2\left[\frac{3\pi}{2}\frac{2n+1}{2\ell+1}\right]}{D^2+4J^2\sin^2\left[\frac{\pi}{2}\frac{2n+1}{2\ell+1}\right]}\right\}-\frac{4}{(\ell+1)}\sum_{n=0}^{\ell-1}\left\{\frac{2J\sin\left[\frac{\pi}{2}\frac{2n+1}{2\ell+1}\right]\cos^2\left[\frac{3\pi}{2}\frac{2n+1}{2\ell+1}\right]}{4J^2+4J^2\sin^2\left[\frac{\pi}{2}\frac{2n+1}{2\ell+1}\right]}\right\}$$
$$\approx\frac{\alpha}{2J}\ln(\ell)-\frac{\gamma}{2J}\left(\frac{\ell D}{2J}\right)^2 , \tag{7.25}$$

with $\alpha\approx 2.8$, $\gamma\approx 8.7$. On inserting (7.25) into the (7.15) for the running coupling at a given ℓ, one obtains

$$J_\Delta(\ell;D)\approx\frac{J_\Delta^{(0)}}{1-\frac{\alpha J_\Delta^{(0)}}{D_0}\ln(\ell)+\frac{\gamma\pi^2 J_\Delta^{(0)}}{2D_0}\left(\frac{D}{\Delta_{\rm FS}}\right)^2} . \tag{7.26}$$

In (7.26), the left-over term $\propto\ln(\ell)$, which accounts for the explicit dependence of $J_{\Delta,c}$ on ℓ, has the expected scaling form, being an explicit function of the ratio $\frac{D}{\Delta_{\rm FS}}$.

The functional form of the right-hand side of (7.26) leads to the important observation that, in order for the running coupling to crossover towards the strongly coupled regime, one needs that $1-\frac{\alpha J_\Delta^{(0)}}{D_0}\ln(\ell)<0$. Thus, a necessary condition for the onset of the Kondo regime is given by

$$\ell>\ell_*[J_\Delta^{(0)}]\equiv e^{\frac{2D_0}{\alpha J_\Delta^{(0)}}} . \tag{7.27}$$

To double check the result in (7.27)—which depends on an approximate numerical fit to a not explicitly known functional relation involving the various dimensionful scales in the problem—in Table 7.1 we plotted some extrapolated values for $\ell_*[J_\Delta^{(0)}]$ as a function of $J_\Delta^{(0)}$. As expected, $\ell_*[J_\Delta^{(0)}]\to\infty$ as soon as $J_\Delta^{(0)}<J_{\Delta,c}$ taken in the $\ell\to\infty$ limit. On increasing $J_\Delta^{(0)}$, $\ell_*[J_\Delta^{(0)}]$ decreases, till it becomes 0 for $J_\Delta^{(0)}$ greater than the value taken by $J_{\Delta,c}$ for $\ell\to 0$.

As one readily checks, once the appropriate numbers are inserted in (7.27), one recovers the results for $\ell_*[J_\Delta^{(0)}]$, showing a good level of consistency between the numerical and the analytical results obtained from (7.27).

The emergence of the scale $\ell_*[J_\Delta^{(0)}]$ can be interpreted by stating that, in order for the junction to crossover towards the Kondo regime, one necessarely has to have $\ell > \ell_*(J_\Delta^{(0)})$; i.e., the junction must be "large enough" to make a block of size $\ell_*(J_\Delta^{(0)})$ fit into it. This is nothing but a way for defining the Kondo cloud, so that one naturally identifies $\ell_*[J_\Delta^{(0)}]$ with the corresponding value of the Kondo screening length (in units of the lattice step, which we have set to 1 from the very beginning) [8, 38, 39]. Incidentally, we note that the identification of $\ell_*[J_\Delta^{(0)}]$ with the Kondo screening length can be also inferred by comparing (7.27) with (7.18) for D_K and taking into account that 2/2.8(=0.714) $\approx \pi/4$(=0.7854) (of course, the mismatch can be readily attributed to different regularization schemes), which yields $\ell_*[J_\Delta^{(0)}] \propto D_K^{-1}$.

Our analysis suggests that one can directly probe the Kondo screening length at the Y-junction of quantum Ising chains, provided one has a way to actually rescale the length of the system ℓ, and a physical quantity $M(\ell)$ that is sensitive to the onset of the Kondo regime. Indeed, on increasing ℓ starting from rather small values, one expects that, as $\ell \sim \ell_*[J_\Delta^{(0)}]$, $M(\ell)$ crosses over towards the value expected when the system is within Kondo regime. In the following section, we address these two specific points.

7.5 Probing the Onset of Kondo Regime

In this section, we first illustrate how introducing an impurity, may lead to an effective reduction of the chain length. Then, we show how the scaling properties of a pertinently defined local impurity magnetization at fixed chain length-ℓ- point to the emergence of the Kondo effect in a Y-junction of QIC's.

To show how an impurity may effectively vary ℓ, one begins from the Bogoliubov–de Gennes equations for the lattice wavefunctions in (7.45) and slightly modifies them from the homogenous chain discussed in the appendix, by assuming a different exchange amplitude (J') between sites $\bar{j}-1$ and $\bar{j}$ and between sites $\bar{j}$ and $\bar{j}+1$. For the sake of generality, we assume that the magnetic field at site $\bar{j}$, H', is $\neq H$.

The Bogoliubov–de Gennes equations for the eigenfunctions corresponding to the eigenvalue ϵ at sites $\bar{j}-1, \bar{j}, \bar{j}+1$ are given by

$$\begin{aligned}
\epsilon u_{\bar{j}-1,\epsilon} &= -\frac{J}{2} u_{\bar{j}-2,\epsilon} - \frac{J'}{2} u_{\bar{j},\epsilon} + \frac{J}{2} v_{\bar{j}-2,\epsilon} - \frac{J'}{2} v_{\bar{j},\epsilon} + H u_{\bar{j}-1,\epsilon} \\
\epsilon v_{\bar{j}-1,\epsilon} &= \frac{J}{2} v_{\bar{j}-2,\epsilon} + \frac{J'}{2} v_{\bar{j},\epsilon} - \frac{J}{2} u_{\bar{j}-2,\epsilon} + \frac{J'}{2} u_{\bar{j},\epsilon} - H v_{\bar{j}-1,\epsilon} \quad ,
\end{aligned} \tag{7.28}$$

by

$$\begin{aligned}
\epsilon u_{\bar{j}+1,\epsilon} &= -\frac{J}{2} u_{\bar{j}+2,\epsilon} - \frac{J'}{2} u_{\bar{j},\epsilon} - \frac{J}{2} v_{\bar{j}+2,\epsilon} + \frac{J'}{2} v_{\bar{j},\epsilon} + H u_{\bar{j}+1,\epsilon} \\
\epsilon v_{\bar{j}+1,\epsilon} &= \frac{J}{2} v_{\bar{j}+2,\epsilon} + \frac{J'}{2} v_{\bar{j},\epsilon} + \frac{J}{2} u_{\bar{j}+2,\epsilon} - \frac{J'}{2} u_{\bar{j},\epsilon} - H v_{\bar{j}+1,\epsilon} \quad ,
\end{aligned} \tag{7.29}$$

and, finally, by

$$\begin{aligned}\epsilon u_{\bar{j},\epsilon} &= -\frac{J'}{2}u_{\bar{j}+1,\epsilon} - \frac{J'}{2}u_{\bar{j}-1,\epsilon} - \frac{J'}{2}v_{\bar{j}+1,\epsilon} + \frac{J'}{2}v_{\bar{j}-1,\epsilon} + H'u_{\bar{j},\epsilon} \\ \epsilon v_{\bar{j},\epsilon} &= \frac{J'}{2}v_{\bar{j}+1,\epsilon} + \frac{J'}{2}v_{\bar{j}-1,\epsilon} + \frac{J'}{2}u_{\bar{j}+1,\epsilon} - \frac{J'}{2}u_{\bar{j}-1,\epsilon} - H'v_{\bar{j},\epsilon} \quad . \end{aligned} \tag{7.30}$$

To simplify the derivation, we use different labels for the wavefunctions at the left- and at the right-hand side of site $\bar{j}$. Specifically, we set

$$\begin{bmatrix} u_{j,\epsilon} \\ v_{j,\epsilon} \end{bmatrix} \to \begin{bmatrix} u^<_{j,\epsilon} \\ v^<_{j,\epsilon} \end{bmatrix} \ (j < \bar{j}) \quad ; \quad \begin{bmatrix} u_{j,\epsilon} \\ v_{j,\epsilon} \end{bmatrix} \to \begin{bmatrix} u^>_{j,\epsilon} \\ v^>_{j,\epsilon} \end{bmatrix} \ (j > \bar{j}) \quad . \tag{7.31}$$

Assuming for both $\begin{bmatrix} u^<_{j,\epsilon} \\ v^<_{j,\epsilon} \end{bmatrix}$ and $\begin{bmatrix} u^>_{j,\epsilon} \\ v^>_{j,\epsilon} \end{bmatrix}$ the form given in (7.53), (7.28), (7.29) become

$$\begin{aligned} Ju^<_{\bar{j},\epsilon} - J'u_{\bar{j},\epsilon} + Jv^<_{\bar{j},\epsilon} - J'v_{\bar{j},\epsilon} &= 0 \\ Ju^>_{\bar{j},\epsilon} - J'u_{\bar{j},\epsilon} - Jv^>_{\bar{j},\epsilon} + J'v_{\bar{j},\epsilon} &= 0 \quad , \end{aligned} \tag{7.32}$$

while (7.30) imply

$$\begin{aligned} u_{\bar{j},\epsilon} &= -\frac{J'}{2[\epsilon - H']} \left\{ u^<_{\bar{j}-1,\epsilon} + u^>_{\bar{j}+1,\epsilon} - v^<_{\bar{j}-1,\epsilon} + v^>_{\bar{j}+1,\epsilon} \right\} \\ v_{\bar{j},\epsilon} &= \frac{J'}{2[\epsilon + H']} \left\{ v^<_{\bar{j}-1,\epsilon} + v^>_{\bar{j}+1,\epsilon} - u^<_{\bar{j}-1,\epsilon} + u^>_{\bar{j}+1,\epsilon} \right\} \quad . \end{aligned} \tag{7.33}$$

When $J' \to 0$, (7.28) imply

$$\begin{aligned} \epsilon u_{\bar{j}-1,\epsilon} &= -\frac{J}{2}u_{\bar{j}-2,\epsilon} + \frac{J}{2}v_{\bar{j}-2,\epsilon} + Hu_{\bar{j}-1,\epsilon} \\ \epsilon v_{\bar{j}-1,\epsilon} &= \frac{J}{2}v_{\bar{j}-2,\epsilon} - \frac{J}{2}u_{\bar{j}-2,\epsilon} - Hv_{\bar{j}-1,\epsilon} \quad , \end{aligned} \tag{7.34}$$

which leads to (see Appendix 7.6)

$$u_{\bar{j},\epsilon} + v_{\bar{j},\epsilon} = 0 \quad . \tag{7.35}$$

Equation (7.35), together with the first line of (7.46), corresponds to an effective shortening of the chain to $\bar{j} - 1$-sites. Clearly, the same result is obtained by sending to ∞ the field H' in (7.33). Thus, we find two, basically equivalent, ways to effectively shorten the spin chains by tuning the impurity. Being able to effectively tune the single-chain length provides us with a tool to probe the Kondo effect arising at a Y-junction of QIC's.

We now show how to probe the crossover towards the Kondo regime by measuring the local magnetization at the impurity. As stated in [20, 26], a suitable procedure to detect the onset of Kondo regime at a Y-junction of critical Ising chain consists in measuring the local magnetization along the z-axis in spin space, right at the junction. This, for a Y-junction of a QIC's, is defined as

$$m = \frac{1}{3}\langle \sum_{\lambda=1}^{3} S_{1,\lambda}^z \rangle \quad . \tag{7.36}$$

In the disconnected chain limit ($J_\Delta \to 0$), one may readily express m at finite ℓ, $m_0(\ell)$, in terms of the (imaginary-time representation of the) Green's function $G_{1,1}(\tau)$ we provided in (7.59). In particular, one obtains

$$\begin{aligned} m_0(\ell) &= \left\langle c_{1,1}^\dagger c_{1,1} - \frac{1}{2} \right\rangle = \lim_{\tau \to 0^+} \left\{ -\frac{1}{2} + G_{1,1}(\tau) \right\} \\ &= -\frac{1}{2} + \frac{2}{\ell+1} \sum_{n=0}^{\ell-1} \cos^2 \left[\frac{3\pi}{2} \frac{2n+1}{2\ell+1} \right] \quad . \end{aligned} \tag{7.37}$$

In Fig. 7.2a, $m_0(\ell)$ is plotted as a function of ℓ. Thus, by acting on ℓ, one has minor effects on the magnetization of the disconnected junction.

When $J_\Delta(D) \to \infty$ with ℓ, one has that, at the onset of the Kondo regime, J_Δ takes over, due to the running $J_\Delta(D)$ with ℓ. The value of m can be inferred by noting that, to minimize the boundary energy, the junction sets into a "locally ferromagnetic" state, involving the three spins $S_{j,\lambda}^x$ at $j = 1$. This is either one of the states labeled with $| \Rightarrow\rangle$ and $| \Leftarrow\rangle$ in [20], corresponding to all the spins $S_{1,\lambda}^x$ pointing upwards, or downwards (for a detailed discussions of the effects of the corresponding residual degeneracy in the system's groundstate see, for instance, [29, 40]). When averaging the total spin at $j = 1$ in the z-direction over either one of these states, one obtains $\langle \Rightarrow | \sum_{\lambda=1}^3 S_{1,\lambda}^z | \Rightarrow\rangle = \langle \Leftarrow | \sum_{\lambda=1}^3 |S_{1,\lambda}^z \Leftarrow\rangle = 0$, which eventually leads us to conclude that $m \to 0$ well within Kondo regime [20, 26].

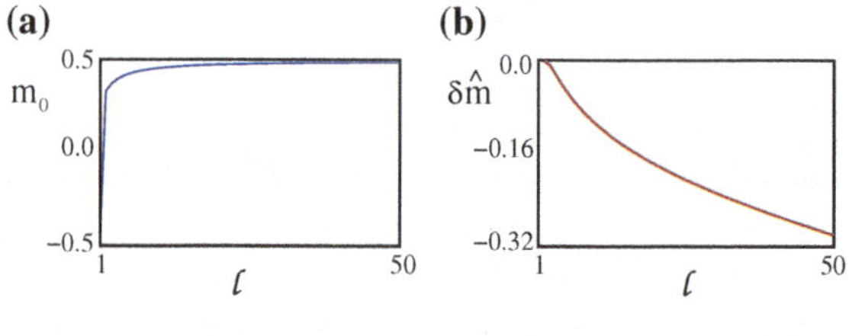

Fig. 7.2 **a**: Junction magnetization in the disconnected limit, $m_0(\ell)$, as a function of ℓ, for $1 \leq \ell \leq 50$. In a few lattice sites, $m_0(\ell)$ converges towards the expected asymptotic value $m_0(\ell \to \infty) = \frac{1}{2}$; **b**: First nontrivial correction to m, $\delta\hat{m}$, (see (7.38)) as a function of ℓ. For large enough ℓ, $\delta\hat{m}$ converges to its asymptotic value $\delta\hat{m}(\ell \to \infty) = -0.32$

To see how the Kondo regime is approached, one has to compute the leading (in J_Δ) corrections to m, δm. From [20], one obtains

$$\delta m(\ell) = -2J_\Delta^2 \sum_{n_1,n_2,n_3=0}^{\ell-1} \left\{ \frac{\mathscr{A}(n_1)\mathscr{A}(n_2)\mathscr{B}(n_2)\mathscr{A}^2(n_3)\epsilon_{n_2}}{(\epsilon_{n_1}+\epsilon_{n_2})(\epsilon_{n_2}+\epsilon_{n_3})(\epsilon_{n_3}+\epsilon_{n_1})} \right\} \equiv 2J_\Delta^2 \delta\hat{m}(\ell) \quad , \tag{7.38}$$

with

$$\begin{aligned} \mathscr{A}(n) &= \sqrt{\frac{2}{\ell+1}} \cos\left[\frac{3\pi}{2}\frac{2n+1}{2\ell+1}\right] \\ \mathscr{B}(n) &= \sqrt{\frac{2}{\ell+1}} \sin\left[\pi\frac{2n+1}{2\ell+1}\right] \\ \epsilon_n &= 2J\sin\left[\frac{\pi}{2}\frac{2n+1}{2\ell+1}\right] \quad . \end{aligned} \tag{7.39}$$

In Fig. 7.2b, $\delta\hat{m}(\ell)$ is plotted as a function of ℓ. One sees that it rapidly grows till, for large enough ℓ, it converges to the "asymptotic" value $\delta\hat{m}_\infty \approx -0.32$ [20]. As expected, turning on J_Δ works to effectively reduce m. To improve this result, nontrivial renormalization effects of the running coupling $J_\Delta(D)$ have to be included.

From the second-order renormalization group equations leading to poor man's scaling running parameter flow, one may safely assume that the only effect of rescaling the cutoff is to induce a renormalization of J_Δ, according to (7.22) for the associated running coupling $J_\Delta(\ell; D)$. As a result, the renormalization group flow of the local magnetization at a given ℓ, $m(\ell; D)$, is obtained by setting

$$m(\ell; D) = m_0(\ell) + 2J_\Delta^2(\ell; D)\delta\hat{m}(\ell) \quad , \tag{7.40}$$

with $m_0(\ell)$ defined in (7.37) and $\delta\hat{m}(\ell)$ in (7.38).

In Fig. 7.3 the curves obtained from the formula in (7.40) are plotted by drawing $m(\ell; D)$ as a function of D for $J_\Delta^{(0)}/D_0 = 0.3$ and for $J_\Delta^{(0)}/D_0 = 0.6$. Specifically, referring to Fig. 7.1, the curves corresponding to $\ell = 8$ and to $\ell = 30$ are drawn. Indeed, from Fig. 7.1, $J_{\Delta,c}(\ell = 8)/D_0 > 0.3$. Thus, for $J_\Delta^{(0)}/D_0 = 0.3$, one expects that, on lowering the reference scale D, $m(\ell = 8; D)$ renormalizes to a finite nonuniversal value for $D \to 0$; this shows how the junction never enters the Kondo regime (see the red curve of Fig. 7.3a). At variance, $J_{\Delta,c}(\ell = 30)/D_0 < 0.3$ and, accordingly, one expects that, on sending $D \to 0$, $m(\ell = 30; D \to 0)$ renormalizes all the way down to 0 as, indeed, it is apparent from the blue curve of Fig. 7.3a. From Fig. 7.1 one also sees that $J_{\Delta,c}^{(0)}(\ell = 30) < J_{\Delta,c}(\ell = 8)/D_0 < 0.6$. Thus, one expects that both $m(\ell = 8; D)$ and $m(\ell = 30; D)$ renormalize all the way down to 0, as $D \to 0$. Indeed, this appears from both plots in Fig. 7.3b, which are drawn assuming $J_\Delta^{(0)}/D_0 = 0.6$.

Ideally, one may construct a sequence of plots such as the ones drawn in Fig. 7.3, by holding $J_\Delta^{(0)}$ fixed and by drawing plots of $J_\Delta(\ell; D)$ at fixed ℓ, as a function of D.

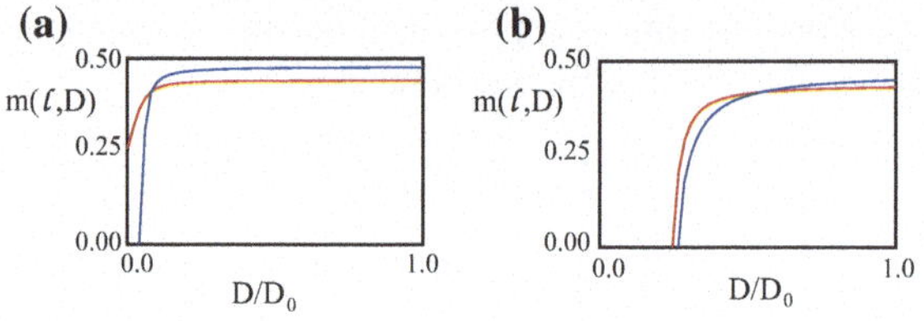

Fig. 7.3 **a** $m(\ell; D)$ as a function of D/D_0 for $J_\Delta/D_0 = 0.3$ and for $\ell = 8$ (red curve), and $\ell = 30$ (blue curve). For $\ell = 8$ $m(\ell = 8; D \to 0)$ renormalizes to a finite value, consistently with the plot in Fig. 7.1, that shows that $J_{\Delta,c}(\ell = 8)/D_0 > 0.3$. For $\ell = 30$, $m(\ell = 30; D \to 0)$ renormalizes to 0 (note that the points below $D = D_K$ have been removed from the plots, due to the expected unreliability of the perturbative renormalization group equation in this regime), consistently with the result that $J_{\Delta,c}(\ell = 30)/D_0 < 0.3$; **b** Same as in (**a**), but with $j_\Delta/D_0 = 6.0$. In this case both $m(\ell = 8; D \to 0)$ and $m(\ell = 30; D \to 0)$ renormalizes to 0, consistently with Fig. 7.1 that shows that $J_{\Delta,c}(\ell = 8)/D_0 < 0.6$

Beginning with very small values of ℓ (or, which is the same, with very large values of Δ_{FS}), since the finite-size gap implies a depletion of the low-energy window of excitation spectrum in the Y-junction, one expects a strong suppression of Kondo screening and, accordingly, a mild renormalization of $m(\ell; D)$ toward values smaller than $m_0(\ell)$, as D runs towards 0. On increasing ℓ, the renormalization of $m(\ell; D)$ becomes more effective, so that, the larger is ℓ, the stronger is the renormalization towards smaller values of $m(\ell; D)$ as $D \to 0$. By construction, assuming that $J_\Delta^{(0)} > J_{\Delta,c}$, there will be a value of ℓ, $\ell_*[J_\Delta^{(0)}]$, such that $m(\ell_*[J_\Delta^{(0)}]; D \to 0) \to 0$.

According to the discussion of Sect. 7.4, one is led to identify $\ell_*[J_\Delta^{(0)}]$ with the Kondo screening length for the Y-junction. As a result, one has, on one hand, identified how a Kondo screening length emerges at a Y-junction of QIC's, on the other hand one has proposed a way to probe it, by looking at how the local magnetization at the impurity, $m(\ell; D)$, scales. Indeed, tuning ℓ in each one of the three QIC's just means acting on appropriate local parameters in the chains, such as the magnetic exchange strength between nearest neighboring sites, or the local applied magnetic field.

These observations complete our proposal for a protocol suitable to estimate the Kondo screening length at a Y-junction of QIC's.

7.6 Concluding Remarks

We analyzed the emergence of the screening length $\lambda_K (= \ell_*[J_\Delta^{(0)}])$ associated to the Kondo effect at a Y-junction of quantum Ising chains. In particular, we showed that effectively varying the length ℓ of the QIC's allows for directly probing the Kondo screening length. Specifically, we proposed to look at the scaling behavior of the local magnetization at the impurity on lowering the reference energy scale D, at fixed ℓ.

When $\ell < \lambda_K$, the full Kondo screening cloud does not fit into the Y-junction. This suppresses the onset of the Kondo regime and, consequently, prevents the local magnetization from flowing all the way down to zero. When $\ell > \lambda_K$, instead, the Kondo cloud does fit into the Y-junction, the system enters the Kondo regime and, accordingly, the magnetization flows all the way down to 0.

We proposed to identify the Kondo length λ_K as the value of ℓ at which the junction crosses over between the two complementary behaviors. Thus, our method provides an effective mean to probe λ_K from the scaling behavior of the local magnetization at the junction at given ℓ, in analogy with the results provenin the infinite-chain limit [20, 26].

We performed our derivation by means of an adapted version of the perturbative RG approach to Kondo problem [4]. While our approach allows us for qualitatively recovering the main features associated to the onset of Kondo regime at the junction, it relies on a number of rough estimates of nonuniversal quantities (cutoff scales, etc.), so that it would be of the utmost importance to perform an indipendent numerical analysis of the problem.

For a single spin chain, DMRG has already proven to be able to provide the correct scaling between the two different regimes (weakly- and strongly-coupled fixed points) [37]. Efficient algorithms to extend DMRG to Y-junctions of spin chains have been by now well established [41] and, as a further development of our work, we plan to employ them to numerically derive the scaling properties of $m(\ell; D)$ (which, in our derivation, takes the place of the integrated real-space spin-spin correlations $\Sigma[x]$ of [37]), to eventually complement our analytical derivation with quantitatively rigorous numerical results.

Appendix

The one-dimensional quantum Ising model is exactly solvable via the Jordan–Wigner fermionization of the spin variables [27]. This method is especially suited for a chain with open boundary conditions. In particular, it allows for an exact solution of the model with open boundary conditions. In this appendix we set the parameters of the chain at their "critical" values. A QIC of finite length ℓ is defined by

$$H_{1-\text{Chain}} = -2J \sum_{j=1}^{\ell-1} S_j^x S_{j+1}^x + H \sum_{j=1}^{\ell} S_j^z \quad , \tag{7.41}$$

with $\mathbf{S}_j$ being a quantum spin-1/2 impurity residing at site j, J being the Ising magnetic exchange strength and H being the (transverse) applied magnetic field. In particular, as $\ell \to \infty$, the chain becomes critical for $H = J$, with a corresponding gapless excitation spectrum [42].

The Jordan–Wigner tranformation is based on the realization of the spin-1/2 operators in terms of spinless complex lattice fermions $\{c_j, c_j^\dagger\}$ as

$$
\begin{aligned}
S_j^+ &= c_j^\dagger e^{i\pi \sum_{r=1}^{j-1} c_r^\dagger c_r} \\
S_j^- &= c_j e^{i\pi \sum_{r=1}^{j-1} c_r^\dagger c_r} \\
S_j^z &= c_j^\dagger c_j - \frac{1}{2} \quad ,
\end{aligned} \tag{7.42}
$$

with the spin ladder operators given by $S_j^\pm = S_j^x \pm i S_j^y$. On inserting (7.42) into the lattice spin Hamiltonian in (7.41), one gets

$$
H_{1-\text{Chain}} \to H_F = -\frac{J}{2}\sum_{j=1}^{\ell-1}\{c_j^\dagger c_{j+1} + c_{j+1}^\dagger c_j\} + \frac{J}{2}\sum_{j=1}^{\ell-1}\{c_j c_{j+1} + c_{j+1}^\dagger c_j^\dagger\} + H\sum_{j=1}^{\ell} c_j^\dagger c_j + \text{const.} \tag{7.43}
$$

H_F in (7.43) is similar to the Kitaev Hamiltonian [17] for a spinless, p-wave one-dimensional superconductor. It is then diagonalized by defining the energy eigenmodes Γ_ϵ via the Bogoliubov–Valatin transformations as

$$
\Gamma_\epsilon = \sum_{j=1}^{\ell}\{u_{j,\epsilon} c_j + v_{j,\epsilon} c_j^\dagger\} \quad , \tag{7.44}
$$

and with the wavefunctions $(u_{j,\epsilon}, v_{j,\epsilon})$ satisfying pertinent Bogoliubov–de Gennes equations. These are obtained by requiring that $[\Gamma_\epsilon, H_F] = \epsilon \Gamma_\epsilon$, and are given by

$$
\begin{aligned}
\epsilon u_{j,\epsilon} &= -\frac{J}{2}\{u_{j+1,\epsilon} + u_{j-1,\epsilon}\} - \frac{J}{2}\{v_{j+1,\epsilon} - v_{j-1,\epsilon}\} + H u_{j,\epsilon} \\
\epsilon v_{j,\epsilon} &= \frac{J}{2}\{v_{j+1,\epsilon} + v_{j-1,\epsilon}\} + \frac{J}{2}\{u_{j+1,\epsilon} - u_{j-1,\epsilon}\} - H v_{j,\epsilon} \quad ,
\end{aligned} \tag{7.45}
$$

for $1 < j < \ell$.

Over the open chain, (7.45) must be complemented with the appropriate boundary conditions. On the fermionic wavefunctions, these are given by [20]

$$
\begin{aligned}
u_{0,\epsilon} - v_{0,\epsilon} &= 0 \\
u_{\ell+1,\epsilon} + v_{\ell+1,\epsilon} &= 0 \quad .
\end{aligned} \tag{7.46}
$$

To solve (7.45), one sets

$$
\begin{bmatrix} u_{j,\epsilon} \\ v_{j,\epsilon} \end{bmatrix} = \begin{bmatrix} u_k \\ v_k \end{bmatrix} e^{ikj} \quad , \tag{7.47}
$$

thus obtaining the secular equations for the amplitudes u_k, v_k given by

$$\begin{aligned} &[\epsilon - J(1-\cos(k))]u_k + iJ\sin(k)v_k = 0 \\ &-iJ\sin(k)u_k + [\epsilon + J(1-\cos(k))]v_k = 0 \quad . \end{aligned} \tag{7.48}$$

The amplitudes u_k, v_k satisfy

$$\begin{aligned} &[\epsilon - H + J\cos(k))]u_k + iJ\sin(k)v_k = 0 \\ &-iJ\sin(k)u_k + [\epsilon + H - J\cos(k))]v_k = 0 \quad . \end{aligned} \tag{7.49}$$

Solving (7.49), one finds the dispersion relation

$$\epsilon^2 = \epsilon_k^2 = \sqrt{(J\cos(k) - H)^2 + J^2\sin^2(k))} \quad , \tag{7.50}$$

which implies a gap in the excitation spectrum $\Delta_{\text{Bulk}} = ||J| - |H||$. Specifically, in the following one looks at the "critical" limit, defined by setting $H = J$. In this case, (7.50) implies

$$\epsilon = \pm\epsilon_k = \pm J\sin\left(\frac{k}{2}\right) \quad , \tag{7.51}$$

with the corresponding solutions given by

$$\begin{bmatrix} u_k \\ v_k \end{bmatrix} = \begin{bmatrix} \cos\left(\frac{\pi-k}{4}\right) \\ i\sin\left(\frac{\pi-k}{4}\right) \end{bmatrix} \quad . \tag{7.52}$$

To recover the boundary conditions in (7.47), one considers the linear combination between the wavefunction with momentum k and the one with momentum $2\pi - k$, that is

$$\begin{bmatrix} u_{j,k} \\ v_{j,k} \end{bmatrix} = \alpha \begin{bmatrix} \cos\left(\frac{\pi-k}{4}\right) \\ i\sin\left(\frac{\pi-k}{4}\right) \end{bmatrix} e^{ikj} + \beta \begin{bmatrix} \cos\left(\frac{\pi-k}{4}\right) \\ -i\sin\left(\frac{\pi-k}{4}\right) \end{bmatrix} e^{-ikj} \quad . \tag{7.53}$$

Imposing the boundary conditions, one obtains the secular equation for the allowed values of k

$$\sin\left[k(\ell+1) + \frac{\pi - k}{2}\right] = 0 \quad , \tag{7.54}$$

which implies $k = \pi\left[\frac{2n+1}{2\ell+1}\right]$, $n = 0, \ldots, \ell - 1$. As a result, from (7.51) one sees that, even at the critical value of the bulk parameters, due to the constraint on the values of k stemming from the secular equation for the finite-ℓ system in (7.54), at finite ℓ, the QIC exhibits a finite-size gap $\Delta_{\text{FS}}(\ell) = J\sin\left(\frac{\pi}{2\ell+1}\right) \sim \frac{J\pi}{2\ell}$. As expected, $\Delta_{\text{FS}}(\ell) \to 0$ as $\ell \to \infty$.

From the boundary conditions in (7.47), one sees that the wavefunction for the single Jordan–Wigner fermion excitation must be given by [20]

$$\begin{bmatrix} u_{j,k} \\ v_{j,k} \end{bmatrix} = \sqrt{\frac{2}{\ell+1}} \begin{bmatrix} \cos\left(\frac{\pi-k}{4}\right)\sin\left[kj - \frac{\pi-k}{4}\right] \\ -\sin\left(\frac{\pi-k}{4}\right)\cos\left[kj - \frac{\pi-k}{4}\right] \end{bmatrix} \quad . \tag{7.55}$$

From the results given above, one constructs the Jordan–Wigner fermion operators in the Heisenberg representation at imaginary time τ as:

$$\begin{aligned} c_j(\tau) &= \sum_k \sqrt{\frac{2}{\ell+1}} \left\{\cos\left(\frac{\pi-k}{4}\right)\sin\left[kj - \frac{\pi-k}{4}\right]\Gamma_k e^{-\epsilon_k \tau} - \sin\left(\frac{\pi-k}{4}\right)\cos\left[kj - \frac{\pi-k}{4}\right]\Gamma_k^\dagger e^{\epsilon_k \tau}\right\} \\ c_j^\dagger(\tau) &= \sum_k \sqrt{\frac{2}{\ell+1}} \left\{\cos\left(\frac{\pi-k}{4}\right)\sin\left[kj - \frac{\pi-k}{4}\right]\Gamma_k^\dagger e^{\epsilon_k \tau} - \sin\left(\frac{\pi-k}{4}\right)\cos\left[kj - \frac{\pi-k}{4}\right]\Gamma_k e^{-\epsilon_k \tau}\right\} , \end{aligned} \tag{7.56}$$

with the set of allowed values of k determined by $k = \pi\left(\frac{2n+1}{2\ell+1}\right)$, $n = 0, \ldots, \ell-1$.

For our purpose we need to compute the imaginary time-ordered Green's function $G_{j,j'}(\tau) = -\langle \mathbf{T}_\tau \xi_j(\tau)\xi_{j'}(0)\rangle$, with $\mathbf{T}_\tau$ being the imaginary time-ordering operator and $\xi_j(\tau) = c_j(\tau) + c_j^\dagger(\tau)$. This is computed as

$$G_{j,j'}(\tau) = -\epsilon(\tau)\frac{2}{\ell+1}\sum_k \cos\left[k\left(j+\frac{1}{2}\right)\right]\cos\left[k\left(j'+\frac{1}{2}\right)\right]\{e^{-\epsilon_k|\tau|}f(-\epsilon_k) + e^{\epsilon_k|\tau|}f(\epsilon_k)\}, \tag{7.57}$$

with $f(\epsilon)$ being Fermi distribution function. In the Fourier-Matsubara space one gets

$$G_{j,j'}(i\omega) = \frac{2}{\ell+1}\sum_k \cos\left[k\left(j+\frac{1}{2}\right)\right]\cos\left[k\left(j'+\frac{1}{2}\right)\right]\left\{\frac{1}{i\omega+\epsilon_k} + \frac{1}{i\omega-\epsilon_k}\right\} \quad . \tag{7.58}$$

For $j = j' = 1$ (which is the case of relevance for us), one obtains

$$G_{1,1}(i\omega) = \frac{2i\omega}{\ell+1}\sum_{n=0}^{\ell-1}\left\{\frac{\cos^2\left[\frac{3\pi}{2}\left(\frac{2n+1}{2\ell+1}\right)\right]}{\omega^2 + 4J^2\sin^2\left[\frac{\pi}{2}\left(\frac{2n+1}{2\ell+1}\right)\right]}\right\} \quad . \tag{7.59}$$

References

1. J. Kondo, Prog. Theor. Phys. **32**, 37 (1964)
2. A.C. Hewson, *The Kondo Effect to Heavy Fermions* (Cambridge University Press, Cambridge, 1993)
3. L. Kouwenhoven, L. Glazman, Phys. World **14** (2001)
4. P.W. Anderson, J. Phys. C Solid State Phys. **3**, 2436 (1970)
5. K.G. Wilson, Rev. Mod. Phys. **47**, 773 (1975)
6. Ph. Nozières, A. Blandin, J. Phys. France **41**, 193 (1980)
7. P. Nozières, J. Low Temp. Phys. **17**, 31 (1974)
8. I. Affleck, *Perspectives of Mesoscopic Physics* (World Scientific, Singapore, 2010), pp. 1–44

9. L.P. Kouwenhoven, C. Marcus, Phys. World **11**, 35 (1998)
10. S.M. Cronenwett, T.H. Oosterkamp, L.P. Kouwenhoven, Science **281**, 540 (1998)
11. D. Goldhaber-Gordon, H. Shtrikman, D. Mahalu, D. Abusch-Magder, U. Meirav, M.A. Kastner, Nature **391**, 156 (1998)
12. N. Laflorencie, E.S. Sørensen, I. Affleck, J. Stat. Mech. Theory Exp. **2008**, P02007 (2008)
13. B. Béri, N.R. Cooper, Phys. Rev. Lett. **109**, 156803 (2012)
14. A. Altland, B. Béri, R. Egger, A.M. Tsvelik, Phys. Rev. Lett. **113**, 076401 (2014)
15. F. Buccheri, H. Babujian, V.E. Korepin, P. Sodano, A. Trombettoni, Nucl. Phys. B **896**, 52 (2015)
16. E. Eriksson, A. Nava, C. Mora, R. Egger, Phys. Rev. B **90**, 245417 (2014)
17. A.Y. Kitaev, Physics-Uspekhi **44**, 131 (2001)
18. N. Crampé, A. Trombettoni, Nucl. Phys. B **871**, 526 (2013)
19. A.M. Tsvelik, Phys. Rev. Lett. **110**, 147202 (2013)
20. D. Giuliano, P. Sodano, A. Tagliacozzo, A. Trombettoni, Nucl. Phys. B **909**, 135 (2016)
21. F. Buccheri, G.D. Bruce, A. Trombettoni, D. Cassettari, H. Babujian, V.E. Korepin, P. Sodano, New J. Phys. **18**, 075012 (2016)
22. A.M. Tsvelik, W.-G. Yin, Phys. Rev. B **88**, 144401 (2013)
23. A. Bayat, S. Bose, P. Sodano, Phys. Rev. Lett. **105**, 187204 (2010)
24. P. Sodano, A. Bayat, S. Bose, Phys. Rev. B **81**, 100412 (2010)
25. A. Bayat, P. Sodano, S. Bose, Phys. Rev. B **81**, 064429 (2010)
26. D. Giuliano, P. Sodano, EPL (Europhysics Letters) **103**, 57006 (2013)
27. P. Jordan, E. Wigner, Zeitschrift für Physik **47**, 631 (1928)
28. A.M. Tsvelik, New J. Phys. **16**, 033003 (2014)
29. P. Coleman, L.B. Ioffe, A.M. Tsvelik, Phys. Rev. B **52**, 6611 (1995)
30. D. Giuliano, G. Campagnano, A. Tagliacozzo, Eur. Phys. J. B **89**, 251 (2016)
31. M.-S. Choi, M. Lee, K. Kang, W. Belzig, Phys. Rev. B **70**, 020502 (2004)
32. Y. Avishai, A. Golub, A.D. Zaikin, Phys. Rev. B **67**, 041301 (2003)
33. G. Campagnano, D. Giuliano, A. Naddeo, A. Tagliacozzo, Phys. C Supercond. **406**, 1 (2004)
34. A. Oguri, Y. Tanaka, J. Bauer, Phys. Rev. B **87**, 075432 (2013)
35. L. Dell'Anna, J. Stat. Mech. Theory Exp. **2010**, P01007 (2010)
36. D. Giuliano, P. Sodano, A. Trombettoni, Phys. Rev. A **96**, 033603 (2017)
37. D. Giuliano, D. Rossini, A. Trombettoni, Phys. Rev. B **98**, 235164 (2018)
38. E.S. Sørensen, I. Affleck, Phys. Rev. B **53**, 9153 (1996)
39. V. Barzykin, I. Affleck, Phys. Rev. Lett. **76**, 4959 (1996)
40. I. Affleck, A.W. Ludwig, Nuclear Physics B **360**, 641 (1991)
41. H. Guo, S.R. White, Phys. Rev. B **74**, 060401 (2006)
42. S. Sachdev, *Quantum Phase Transitions*, 2nd edn. (Cambridge University Press, Cambridge, 2011)

Chapter 8
Gauge Theories with Ultracold Atoms

João C. Pinto Barros, Michele Burrello and Andrea Trombettoni

Abstract We discuss and review, in this chapter, the developing field of research of quantum simulation of gauge theories with ultracold atoms.

8.1 Introduction

During the School on "Strongly Coupled Field Theories for Condensed Matter and Quantum Information Theory", held in Natal (Brazil) in the days 2–14 August 2015, one of the authors gave a course on "Quantum Simulations of Gauge Fields with Ultracold Atoms". The course was meant to be informal, at the blackboard, with time for discussions and to interact with the younger part of the audience. Subsequently, in 2017 J. C. Pinto Barros obtained the Ph.D. in SISSA (Trieste) defending a Thesis on "Field and Gauge Theories with Ultracold Atoms", under the supervision of Andrea Trombettoni and Marcello Dalmonte and with Michele Burrello and Enrique Rico Ortega acting as external referees. The present chapter is based on the Natal's course and on the above mentioned Ph.D. Thesis of Pinto Barros [1]. The latter Thesis is available at http://www.statphys.sissa.it/wordpress/?page_id=1095.

J. C. Pinto Barros (✉)
Albert Einstein Center for Fundamental Physics, Institute for Theoretical Physics,
University of Bern, Sidlerstrasse 5, CH-3012 Bern, Switzerland
e-mail: jpintobarros@itp.unibe.ch

M. Burrello
Center for Quantum Devices, Niels Bohr Institute, Niels Bohr International Academy,
University of Copenhagen, 2100 Copenhagen, Denmark
e-mail: michele.burrello@nbi.ku.dk

A. Trombettoni
CNR-IOM DEMOCRITOS Simulation Center and SISSA, Via Bonomea 265,
34136 Trieste, Italy
e-mail: andreatr@sissa.it

A. Ferraz et al. (eds.), *Strongly Coupled Field Theories for Condensed Matter and Quantum Information Theory*, Springer Proceedings in Physics 239,
https://doi.org/10.1007/978-3-030-35473-2_8

The course gave an introductory discussion on lattice gauge theories, and then moved to explain how to simulate gauge potentials and gauge fields. A prior knowledge of ultracold atomic systems was assumed, even though during the lectures the corresponding concepts and notions were briefly reminded.

8.2 Gauge Theories

A gauge theory is a model that has a gauge symmetry. Such symmetry can be seen as a redundancy in the description of the degrees of freedom. In other words, this means that one can have two mathematically distinct solutions of the equations describing the system and nonetheless they describe the same physical situation. The paradigmatic example is classical electrodynamics. It describes the behavior of the electric field $\mathbf{E}(t, \mathbf{x})$ and the magnetic field $\mathbf{B}(t, \mathbf{x})$ in the presence of an electric charge density $\rho(t, \mathbf{x})$ and the current density $\mathbf{j}(t, \mathbf{x})$. The system is governed by the Maxwell equations:

$$\begin{aligned} &\nabla \cdot \mathbf{E}(t, \mathbf{x}) = \rho(t, \mathbf{x}) \; ; && \nabla \times \mathbf{B}(t, \mathbf{x}) - \partial_t \mathbf{E}(t, \mathbf{x}) = \mathbf{j}(t, \mathbf{x}) \; ; \\ &\nabla \cdot \mathbf{B}(t, \mathbf{x}) = 0 ; && \nabla \times \mathbf{E}(t, \mathbf{x}) + \partial_t \mathbf{B}(t, \mathbf{x}) = 0 . \end{aligned} \tag{8.1}$$

In the above equations and in the rest of this chapter, natural units shall be adopted. The homogeneous equations, which are independent of charges and currents, can be straightforwardly solved by introducing a scalar potential $\phi(t, \mathbf{x})$ and a vector potential $\mathbf{A}(t, \mathbf{x})$:

$$\mathbf{E}(t, \mathbf{x}) = -\nabla\phi(t, \mathbf{x}) - \partial_t \mathbf{A}(t, \mathbf{x}) \; , \qquad \mathbf{B}(t, \mathbf{x}) = \nabla \times \mathbf{A}(t, \mathbf{x}) \; . \tag{8.2}$$

Using these two relations the last two equations in (8.1) are fulfilled and the ones from the first row can be written in terms of $\phi(t, \mathbf{x})$ and $\mathbf{A}(t, \mathbf{x})$. After a solution is found, it can be plugged in (8.2) in order to obtain the electric and magnetic fields. However, not all different $\phi(t, \mathbf{x})$ and $\mathbf{A}(t, \mathbf{x})$ will give different electric and magnetic fields. In fact if two fields $\phi(t, \mathbf{x})'$ and $\mathbf{A}(t, \mathbf{x})'$ are related to another solution $\phi(t, \mathbf{x})$ and $\mathbf{A}(t, \mathbf{x})$ by:

$$\phi(t, \mathbf{x})' = \phi(t, \mathbf{x}) + \partial_t \alpha(t, \mathbf{x}) \; , \qquad \mathbf{A}(t, \mathbf{x})' = \mathbf{A}(t, \mathbf{x}) - \nabla\alpha(t, \mathbf{x}) \; , \tag{8.3}$$

for some regular function $\alpha(t, \mathbf{x})$, then the electric and magnetic fields, given by (8.2), remain unchanged. This means that the solutions ϕ, $\mathbf{A}$ and ϕ', $\mathbf{A}'$ correspond to the same physical situation and therefore they are just redundant descriptions of the same physics. The transformations of (8.3) are called gauge transformations.

The existence of a gauge symmetry does not require the field to be dynamical. Consider a charged quantum particle in a background of a classical electromagnetic field. The Schrödinger equation for this system can be written as the equation in the absence of any field and "correcting" the canonical momentum $\mathbf{p} \to \mathbf{p} - e\mathbf{A}$. In the presence of an electromagnetic field the mechanical momentum, associated with the kinetic energy of the particle and denoted here by $\boldsymbol{\pi}$, is no longer the canonical momentum given by $\mathbf{p}$. The relation between them is $\boldsymbol{\pi} = \mathbf{p} - e\mathbf{A}$ which is at the core of this substitution. The same happens for the time derivative with the scalar potential $i\partial_t \to i\partial_t - e\phi$. Therefore, the Schrödinger equation reads, in the absence of any other interactions:

$$(i\partial_t - e\phi)\,\psi(t,\mathbf{x}) = (-i\nabla - e\mathbf{A})^2\,\psi(t,\mathbf{x})\ . \tag{8.4}$$

Also this equation is invariant under the transformation (8.3) provided that the wave function is transformed by a position-dependent phase:

$$\psi(t,\mathbf{x}) = e^{-ie\alpha(t,\mathbf{x})}\psi(t,\mathbf{x})\ . \tag{8.5}$$

Given the space and time dependence of this transformation, it is denoted as a local gauge symmetry.

In quantum field theory an illustrative example is provided by QED. The Lagrangian is given by:

$$\mathcal{L} = \bar{\psi}\left(\gamma^\mu\left(i\partial_\mu - eA_\mu\right) - m\right)\psi - \frac{1}{4}F_{\mu\nu}F^{\mu\nu}\ . \tag{8.6}$$

Implicit sum over repeated indices is assumed. γ^μ are the gamma matrices satisfying the Clifford algebra $\{\gamma^\mu,\gamma^\nu\} = 2\eta^{\mu\nu}$, $\eta^{\mu\nu}$ is the Minkowski metric $\eta = \mathrm{Diag}\,(1,-1,-1,-1)$, ψ the Dirac spinor and $\bar{\psi} = \psi^\dagger\gamma^0$. The indices μ run from 0 to 3 where 0 corresponds to the time index. A_μ is called gauge field and the last term of the Lagrangian corresponds to its kinetic term where $F_{\mu\nu} = \partial_\mu A_\nu - \partial_\nu A_\mu$. Also in this case there is a local set of transformations that leave this Lagrangian invariant. Explicitly:

$$A_\mu(x) \to A_\mu(x) - \frac{1}{e}\partial_\mu\alpha(x)\ , \qquad \psi(x) \to e^{i\alpha(x)}\psi(x)\ , \tag{8.7}$$

One can define the covariant derivative $D_\mu = \partial_\mu + ieA^a_\mu$ such that, under a gauge transformation, $D_\mu\psi \to e^{i\alpha(x)}D_\mu\psi$. In this way, the local gauge symmetry becomes apparent.

This is an example of a $U(1)$ gauge theory: a gauge transformation is defined, at each point, by phases $\alpha \in [0, 2\pi[$ which combine according to the group $U(1)$. This construction can be generalized to other gauge groups, like $\mathbb{Z}_N$, or even non-Abelian, like $SU(N)$, for N an integer number. For example, the Kitaev toric code is a $\mathbb{Z}_2$ (Abelian) gauge theory [2] whereas Quantum Chromodynamics (QCD), the theory that describes strong interactions in particle physics, is a $SU(3)$ (non-Abelian) gauge theory [3–5]. In the following a brief description of non-Abelian $SU(N)$ gauge invariance in quantum field theory is provided. For more details see, for example, [6].

In order to explore these other symmetries, extra indices must be inserted (in the paradigmatic example of QCD these are the color indices). To simplify the notation, whenever ψ it is used it is meant:

$$\psi \equiv \begin{pmatrix} \psi_1 \\ \psi_2 \\ \vdots \\ \psi_n \end{pmatrix} \tag{8.8}$$

where each one of the ψ_i corresponds to a (four-component in $3+1$ dimensions) Dirac spinor. Consider then a general symmetry group and a respective set of generators represented by Hermitian $n \times n$ matrices t^a. The goal is to build a Lagrangian which is invariant under the set of local transformations

$$\psi(x) \to e^{i\alpha^a(x)t^a}\psi(x)\ . \tag{8.9}$$

This is a unitary transformation that mixes the n components of the vector (8.8) following a n-dimensional representation of the gauge group element $e^{i\alpha^a(x)t^a}$. The gauge field becomes, in turn, a matrix which can be parametrized as $A^a_\mu t^a$. Under a gauge transformation the field transforms as

$$A^a_\mu t^a \to e^{i\alpha^a(x)t^a}\left(A^a_\mu t^a + \frac{i}{g}\partial_\mu\right)e^{-i\alpha^a(x)t^a}. \tag{8.10}$$

Writing the covariant derivative as $D_\mu = \partial_\mu - igA^a_\mu t^a$ one finds $D_\mu\psi(x) \to e^{i\alpha^a(x)t^a} D_\mu\psi(x)$. In this way $\bar{\psi}\left(\gamma^\mu D_\mu - m\right)\psi$ is a gauge invariant operator which includes the fermionic kinetic term and the matter-gauge coupling. Note that $\bar{\psi}$ is to be interpreted as line vector with components $\bar{\psi}_i$ and γ^μ are diagonal on the color indices, i.e. act the same for every color by standard matrix multiplication $\gamma^\mu\psi_i$. In order to define the gauge field dynamics, its gauge invariant kinetic term must be inserted. A possible way to derive its form is by considering the commutator $\left[D_\mu, D_\nu\right] = it^a F^a_{\mu\nu}$. Putting it differently, a general form for $F^a_{\mu\nu}$ can be obtained from this formula. Explicit computation yields $F^a_{\mu\nu} = \partial_\mu A^a_\nu - \partial_\nu A^a_\mu + gf^{abc}A^b_\mu A^c_\nu$ where the structure constants f^{abc} are given by $\left[t^a, t^b\right] = it^c f^{abc}$ and depend only on the symmetry group. From the transformation law for the covariant derivatives, one can see that $F^a_{\mu\nu}F^{a\mu\nu}$ is gauge invariant. The full Lagrangian can then be written as

$$\mathcal{L} = \bar{\psi}\left(\gamma^\mu D_\mu - m\right)\psi - \frac{1}{4}F^a_{\mu\nu}F^{a\mu\nu}. \tag{8.11}$$

The perspective of implementing these kind of models in table top experiments is very appealing. First of all, it could give answers to very fundamental questions in physics like, for example, the exploration of the phase diagram of QCD. That is certainly a long term challenge and the path envisioned towards it involves the implementation of simpler intermediate steps. While QCD has a $SU(3)$ gauge symmetry and involves $3+1$ dimensions, this does not need to be the main target. A much simpler case of a $U(1)$ gauge symmetry in $1+1$ dimensions is already of great interest. In fact, this was the target of the first experimental implementation of a lattice gauge theory [7](to be discussed in Sect. 8.4.3). Step by step one may think to be able to realize more and more complex models. It is clear that if these models are realized they become interesting on their own both theoretically and experimentally. In particular, for example, it may also be advantageous to have situations where only certain degrees of freedom live in higher dimensions keeping others in lower dimensionality [8–10] which could be used to simulate systems with long-range interactions. This has been the subject of an intense investigation in the last years [11–20].

8.2.1 Gauge Symmetry on the Lattice

8.2.1.1 Static Fields

Following the discussion in the previous section, a many-body Hamiltonian in the presence of a magnetic field can be obtained by replacing the momentum components for each particle by $p_i \to p_i - eA_i$. On the lattice, instead, this can be approximated by the Peierls substitution where the hopping parameters become complex. This is valid in a tight-binding regime and for a slow varying magnetic field. Explicitly the kinetic term is modified according to

$$K = \sum_{\mathbf{r},j} t_j \hat{a}^\dagger_{\mathbf{r}+\hat{j}} \hat{a}_\mathbf{r} + \text{h.c.} \to \sum_{\mathbf{r},j} t_j \hat{a}^\dagger_{\mathbf{r}+\hat{j}} e^{i\theta_j(\mathbf{r})} \hat{a}_\mathbf{r} + \text{h.c.} \tag{8.12}$$

In the previous equation the sum of $\mathbf{r}$ is taken over the lattice sites and the sum of j is taken over all d directions corresponding to the dimensionality of the system. The angles $\theta_j(\mathbf{r})$ are just phases that can depend, on general grounds, on both the direction of the hopping and the position. The key difference is that this phase here is non dynamical, so there is no kinetic term for it. This simply corresponds to allow the hopping parameter of the particles, on the lattice, to be complex. Similarly to the models in continuum space, not all complex hoppings represent different physical scenarios as there is gauge invariance. In Sect. 8.3, several examples of techniques to engineer complex phases on the hopping parameters are discussed. Reviews can be found in [21–24].

8.2.1.2 Dynamical Fields

In order to study a dynamical quantum (lattice) gauge theory, the lattice system under analysis must include also the degrees of freedom for the gauge fields and the complex hopping parameters are therefore promoted to operators acting on these degrees of freedom. Such degrees of freedom are usually associated to the lattice edges and their kinetic term must be supplied.

A constructive way to define such a system consists on taking the Lagrangian (8.11), write an Hamiltonian and perform a naive discretization. This offers a recipe to engineer possible quantum simulations of these systems: a straightforward way of proceeding is indeed to create a system implementing the specific Hamiltonian of the lattice gauge theory. As a result it is useful to consider such theories in their Hamiltonian formulation [25].

To describe the $U(1)$ case one can introduce the following link operators, acting on the gauge degrees of freedom:

$$U_{\mathbf{r},j} = \exp\left(ie \int_{a\mathbf{r}}^{a\left(\mathbf{r}+\hat{j}\right)} dx A_j(x) \right), \qquad L_{\mathbf{r},j} = \frac{E_{\mathbf{r},j}}{e}. \tag{8.13}$$

U and L are operators corresponding respectively to the connection and electric field of the theory (see, for example, [26] for more details). Based on these operators, we can define the Hamiltonian

$$H = -\frac{i}{2a}\sum_{\mathbf{r},j}\left(\psi_{\mathbf{r}}^{\dagger} U_{\mathbf{r},j}\psi_{\mathbf{r}+\hat{j}} - \text{h.c.}\right) + m\sum_{n}\psi_{\mathbf{r}}^{\dagger}\psi_{\mathbf{r}} + \frac{ae^2}{2}\sum_{n} L_{\mathbf{r},j}^2, \tag{8.14}$$

which reproduces the correct continuum theory when the *naive* continuum limit is taken. In the expressions above a is the lattice spacing, $\mathbf{r}$ are the lattice points and j labels the links connected to it. $E_{\mathbf{r}j}$ is the discretized version of the electric field which is the conjugate momentum of A_j in the continuum version. The commutation relations between the link operators are

$$\left[L_{\mathbf{r},i}, U_{\mathbf{r}',j}\right] = \delta_{\mathbf{r}\mathbf{r}'}\delta_{ij}U_{\mathbf{r},i}, \quad \left[L_{\mathbf{r},i}, U_{\mathbf{r}',j}^{\dagger}\right] = -\delta_{\mathbf{r}\mathbf{r}'}\delta_{ij}U_{\mathbf{r},i}^{\dagger}, \quad \left[U_{\mathbf{r},i}, U_{\mathbf{r}',j}\right] = \left[U_{\mathbf{r},i}^{\dagger}, U_{\mathbf{r}',j}\right] = 0. \tag{8.15}$$

We pause here to point out a couple of subtleties. The first concerns the so-called *naive* continuum limit, obtained by simply sending $a \to 0$. While this works well for bosons, fermions suffer from the so-called "fermion doubling problem". When this limit is taken with more care each fermion flavor on the lattice gives rise to 2^d fermion flavors on the continuum, being d the number of discretized dimensions. The Nielsen–Ninomiya Theorem [27–29] states that this is always the case when the fermion action is real, local and invariant under lattice translations and chiral transformations. There are alternative approaches to evade the Nielsen–Ninomiya Theorem which have their own advantages and disadvantages. A possible choice, popular among the quantum simulation community, is provided by staggered fermions [30] (also known as Kogut–Susskind fermions). The idea consists on distributing the spinor components among different lattice sites. In this way, instead of a spinor per site, one has only one fermion. Only for the Hamiltonian formulation of the $1 + 1D$ theory the fermion doubling problem can be completely solved in this way. In this case the Hamiltonian becomes

$$H = -\frac{i}{2a}\sum_n \left(c_n^\dagger U_n c_{n+1} - \text{h.c.}\right) + m\sum_n (-1)^n c_n^\dagger c_n + \frac{ae^2}{2}\sum_n L_n^2. \tag{8.16}$$

Spinors can be reconstructed from $\psi_n = (c_{2n}, c_{2n+1})^T / \sqrt{a_{st}}$.

In higher dimensions the most non-trivial step consists on the existence of plaquette terms or, in other words, an energy cost for magnetic fields. These are gauge-invariant terms which must be present in order to fully represent the gauge theory. The absence of these terms is related to the strong coupling limit of the theory. On a 2D square lattice, the plaquette term originating at the point $\mathbf{r}$ is $U_\square = U_{\mathbf{r},x} U_{\mathbf{r}+\hat{x},y} U^\dagger_{\mathbf{r}+\hat{y},x} U^\dagger_{\mathbf{r},y}$ consisting on the smallest loops possible to draw on the lattice. The Hamiltonian for d spatial dimensions takes the form:

$$\begin{aligned} H =& -\frac{i}{2a}\sum_{\mathbf{r},i} (-1)^{r_1+\ldots+r_{i-1}} \left(c_\mathbf{r}^\dagger U_{\mathbf{r},i} c_{\mathbf{r}+\hat{i}} - \text{h.c.}\right) + m\sum_\mathbf{r} (-1)^{r_1+\ldots+r_d} c_\mathbf{r}^\dagger c_\mathbf{r} \\ &+ \frac{a^{2-d}e^2}{2}\sum_{\mathbf{r},i} L_{\mathbf{r},i}^2 - \frac{a^{d-4}}{4e^2}\sum_\square \left(U_\square + U_\square^\dagger\right). \end{aligned} \tag{8.17}$$

The extra alternating signs on the first term are required to obtain the correct Dirac Hamiltonian in the continuum limit with staggered fermions [30].

Another fundamental point, associated with the Hamiltonian formulation, consists on the restriction of the Hilbert space to physical states. This can be derived from the Lagrangian formulation by noting that the component A_0 is non-dynamical (there is no term $\partial_0 A_0$). As a consequence it acts as a Lagrange multiplier enforcing the Gauss' law as a constraint. Therefore the physical states are defined by the relation

$$G_\mathbf{r} |\Psi\rangle = 0 \tag{8.18}$$

for each lattice site $\mathbf{r}$, where

$$G_\mathbf{r} = \sum_i \left(L_{\mathbf{r},j} - L_{\mathbf{r}-\hat{j},j}\right) - Q_\mathbf{r}, \tag{8.19}$$

where $Q_\mathbf{r}$ is the dynamical matter charge. For the $1+1D$ case, for example, $Q_n = c_n^\dagger c_n + \frac{1-(-1)^n}{2}$. The alternating term, which may look odd, is related to the staggered formulation. Considering a state with no electric field. The Gauss law demands that fermions populate the odd sites while leaving the even empty. This is because the spinor degrees of freedom are distributed along the lattice. Occupied odd sites have the interpretation of a filled Dirac sea. When a fermion hops from an odd to an even site it creates a hole in the Dirac sea while creating a particle above it. This is interpreted as the creation of particle/anti-particle pair where the hole plays the role of an anti-particle. In the presence of gauge fields, the hopping described above must be accompanied by a change on the electric field preserving Gauss' law, as described by the connection operator in the first term of the Hamiltonian (8.17).

The $G_{\mathbf{r}}$ are also generators of the gauge transformation and can be extended for the $U(N)$ and $SU(N)$ gauge theories. To this purpose, one can consider matter fields $\psi_{\mathbf{r}}$ that transform under the gauge symmetries under a suitable representation of dimension n of the gauge group. The generators of the gauge symmetries must therefore satisfy the relation $\left[G^a_{\mathbf{r}}, \psi_{\mathbf{r}}\right] = t^a \psi_{\mathbf{r}}$ where t^a are n-dimensional representations of the (left) group generators.

In order to preserve the gauge-invariance of the Hamiltonian, the connection operators must transform like tensors under the gauge transformations and they must follow the same representation of the matter fields: $U_{\mathbf{r},j} \rightarrow e^{i\alpha^a_{\mathbf{r}} t^a} U_{\mathbf{r},j} e^{-i\alpha^a_{\mathbf{r}+\hat{j}} t^a}$. In particular the connection is multiplied on the left side by the transformation inherited from the lattice site on its left and on the right side by the inverse of the transformation inherited from the lattice site on its right. When we deal with a non-Abelian group, it is thus useful to distinguish left and right generators for the group transformations [25], labelled by $L_{\mathbf{r},i}$ and $R_{\mathbf{r},i}$ respectively (see, for example, [26]). The local generators of the gauge transformation can therefore be defined as:

$$G^a_{\mathbf{r}} = \sum_i \left(L^a_{\mathbf{r},i} + R^a_{\mathbf{r}-\hat{i},i}\right) + \psi^\dagger_{\mathbf{r}} t^a \psi_{\mathbf{r}}. \tag{8.20}$$

Finally, the lattice Hamiltonian for the non-Abelian theory will be:

$$\begin{aligned} H = & -\frac{i}{2a}\sum_{\mathbf{r},i} (-1)^{r_1+\ldots+r_{i-1}} \left(\psi^\dagger_{\mathbf{r}} U_{\mathbf{r}i} \psi_{\mathbf{r}+\hat{i}} - \text{h.c.}\right) + m \sum_{\mathbf{r}} (-1)^{r_1+\ldots+r_d} \psi^\dagger_{\mathbf{r}} \psi_{\mathbf{r}} \\ & + \frac{a^{2-d} g^2}{2} \sum_{\mathbf{r},i,a} \left(\left(L^a_{\mathbf{r}i}\right)^2 + \left(R^a_{\mathbf{r}i}\right)^2\right) - \frac{a^{d-4}}{4g^2} \sum_{\Box} \mathrm{Tr}\left(U_{\Box} + U^\dagger_{\Box}\right). \end{aligned} \tag{8.21}$$

Again, the Gauss law should be imposed on physical states $G^a_{\mathbf{r}} |\Psi\rangle = 0$.

Often times, in the proceeding sections, the matter-gauge correlated hopping will be written as $\psi^\dagger_{\mathbf{r}} U_{\mathbf{r}i} \psi_{\mathbf{r}+\hat{i}} + \text{h.c.}$ rather than $i\left(\psi^\dagger_{\mathbf{r}} U_{\mathbf{r}i} \psi_{\mathbf{r}+\hat{i}} - \text{h.c.}\right)$ as above. While the latter reproduces the familiar continuum Hamiltonian in the naive continuum limit, both are related by a gauge transformation.

8.2.1.3 Challenges, Limitations and Quantum Link Models

Cold atom systems offer the possibility to construct Hubbard-like Hamiltonians with tunable hopping parameters and on-site interactions. However, gauge potentials and gauge fields demand more than that.

When the field is static, as described in Sect. 8.2.1.1, the hopping parameters become complex. This is not readily available in simple optical lattices, but, thanks to recent experimental developments, it is nowadays possible to engineer static gauge fields, as we will discuss in Sect. 8.3.

For dynamical gauge fields, as discussed in Sect. 8.2.1.2, the matter hopping and link operators must be correlated in such a way to guarantee the existence of the local gauge symmetry (at each lattice site). Such kind of hopping is not natural in a cold atomic system and a discussion on how to implement is presented in Sect. 8.4.

There is, yet, a further difficulty for dynamical gauge fields. Take, for example, the commutation relations in (8.15) pertaining a certain link. The operator $U_{\mathbf{r}}$ acts as a raising operator of the electric field (or equivalently of $L_{\mathbf{r}}$). But, for a $U(1)$ theory, this corresponds to an infinite Hilbert space per link. Constructing such links is certainly an experimental challenge, even for small lattice sizes. A solution of this problem is provided by quantum link models which are characterized by a finite Hilbert space per link, without violating the required gauge symmetry. These models were introduced by Horn in 1981 [31] and were further studied in [32–36]. Proposed as an alternative formulation to Wilson gauge field theories on the lattice, they became an attractive realization of gauge symmetries for quantum simulation purposes.

In quantum link models the link degrees of freedom are replaced by quantum spins, such that the algebra in (8.15) is replaced by the algebra of angular momentum. In particular this correspond to consider alternative link operators:

$$L_{+\mathbf{r},i} = S^x_{\mathbf{r},i} + i S^y_{\mathbf{r},i}\,, \quad L_{\mathbf{r},i} = S^z_{\mathbf{r},i} \tag{8.22}$$

where the raising and lowering operators $L_{\pm\mathbf{r},i}$ replace $U_{\mathbf{r}}$ and its conjugate. With this construction, the first two relations of (8.15) are still satisfied. However, the last no longer holds because U and $U^\dagger$ do not commute any longer. In particular, $L_{\pm\mathbf{r}i}$ are not unitary whereas $U_{\mathbf{r}}$ was.

Even though the algebra itself is different, the angular momentum operators can be equally used to construct a gauge theory without compromising the gauge symmetry. In particular, we can choose the dimension of the Hilbert space in each of the links to be $2S+1$ with S a positive half integer (corresponding, in the spin language, to the total spin). It is expected that in the limit of large S the Wilson formulation should be recovered. Explicitly one can use the following link variables $U_{\mathbf{r}i} \to L_{+\mathbf{r}i}/\sqrt{S\,(S+1)}$. The new non-zero commutation relation is $\left[U_{\mathbf{r}i}, U^\dagger_{\mathbf{r}i}\right] = 2L_{\mathbf{r}i}/S\,(S+1)$. In the limit of $S \to +\infty$ the right hand side goes to zero and the initial algebra is recovered.

There is an analogous construction for $U\,(N)$ non-Abelian symmetries. One can see that the symmetry can be realized using an $SU\,(2N)$ algebra (note that for $N=1$ this gives, correctly, $SU\,(2)$). It is possible to construct the new algebra using the so-called "rishon fermions" [36]. They are written in terms of pairs of fermionic operators $l^m_{\mathbf{r},j}$ and $r^m_{\mathbf{r}+\hat{j},j}$ for each link between the sites $\mathbf{r}$ and $\mathbf{r}+\hat{j}$. These operators define additional left and right gauge modes laying on the lattice sites $\mathbf{r}$ and $\mathbf{r}+\hat{j}$, with the aim of describing the link degrees of freedom. m labels their color index. We can write:

$$L^a_{\mathbf{r},j} = \frac{1}{2} l^{m\dagger}_{\mathbf{r},j} t^a_{mn} l^n_{\mathbf{r},j}\,, \qquad R^a_{\mathbf{r},j} = \frac{1}{2} r^{m\dagger}_{\mathbf{r}+\hat{j},j} t^a_{mn} r^n_{\mathbf{r}+\hat{j},j}\,, \tag{8.23}$$

$$E_{\mathbf{r},j} = \frac{1}{2}\left(r^{m\dagger}_{\mathbf{r}+\hat{j},j} r^m_{\mathbf{r}+\hat{j},j} - l^{m\dagger}_{\mathbf{r},j} l^m_{\mathbf{r},j} \right)\,, \tag{8.24}$$

$$U^{mn}_{\mathbf{r},j} = l^m_{\mathbf{r},j} r^{n\dagger}_{\mathbf{r}+\hat{j},j}\,. \tag{8.25}$$

The finiteness of the Hilbert space is a feature desirable for future quantum simulation schemes. Even though not a primary concern at this stage, it is reassuring that the effective continuum limit can be achieved even if one uses quantum link models [34].

8.3 Simulation of Gauge Potentials

In accordance with the previous discussion, the goal of this section is to show specific examples on how a complex hopping parameter can be engineered. The two main strategies described will be two contrasting situations. In one external parameters are varied adiabatically (Sect. 8.3.1), while in the other fast modes are integrated out (Sect. 8.3.2).

8.3.1 Adiabatic Change of External Parameters

The idea of this approach has, in its core, the tight relation between the Aharonov–Bohm phase [37] and the Berry phase which was a concept introduced by Berry in [38]. The first is the phase acquired by a particle traveling around a closed contour. At the end of the path, when it is back to the initial position, the wave function acquires a new phase which is independent of the details of how the path was done and only depends on the total magnetic flux through the contour. On the other side, the Berry phase corresponds to the phase acquired when some external parameters of the system are varied in time, "slowly", coming back again to their initial value for a non-degenerate state. In a more precise way, the starting point is an Hamiltonian $H(q^a, \lambda_i)$ where q^a are degrees of freedom and λ_i are a set of external parameters. If these parameters are varied sufficiently slow returning, in the end, to their initial value, and if the initial state is an non-degenerate eigenstate, then the system will be back to its initial state. During the process, however, it will acquire a phase:

$$|\psi\rangle \underset{\text{adiabatic change}}{\longrightarrow} e^{i\gamma}|\psi\rangle\,. \tag{8.26}$$

The phase γ can be derived by computing the time evolution operator and subtract the "trivial" dynamical phase acquired simply due to the time evolution. Let us consider the adiabatic evolution of a system such that each energy eigenstates remain non-degenerate during the whole process. In this case, starting from one of the eigenstates of the initial Hamiltonian, the system will continuously evolve remaining in the corresponding eigenstate, with energy $E(t)$, at each time. Therefore the dynamical phase is $e^{-i\int E(t)dt}$. The additional Berry phase reads instead:

$$\gamma = \oint_C \tilde{A}_i(\lambda)\, d\lambda_i \,, \tag{8.27}$$

where C is the closed path in the space of the parameters λ_i and $\tilde{A}_i$ are given by:

$$\tilde{A}_i(\lambda) = i\,\langle\phi(\lambda)|\,\frac{\partial}{\partial\lambda_i}\,|\phi(\lambda)\rangle \,, \tag{8.28}$$

where $|\phi(\lambda)\rangle$ are reference eigenstates taken with an arbitrary choice of their overall phases. $\tilde{A}(\lambda)$ is the Berry connection. Different choices of the reference eigenstates with different phases, for example $|\phi'(\lambda)\rangle = e^{i\alpha(\lambda)}\,|\phi(\lambda)\rangle$, would just reproduce a gauge transformation on $\tilde{A}$:

$$\tilde{A}_i \to \tilde{A}'_i = \tilde{A}_i - \frac{\partial\alpha}{\partial\lambda_i} \,. \tag{8.29}$$

This principle can be applied in multi-level atomic systems in order to reproduce artificial gauge fields in an ultracold atomic setting. As an example, the computation can be done for a two level atom, where it is shown how this vector potential appears explicitly at the Hamiltonian level. These two levels correspond to two internal states of the atom, the ground state $|g\rangle$ and an excited state $|e\rangle$. The center of mass Hamiltonian, assumed diagonal on the internal states, is taken to be simply the free particle Hamiltonian. The total Hamiltonian is $H = H_0 + U$. By a suitable shift of the energy spectrum we can assume that the ground and excited state energies are related by $E_g = -E_e$. Then U can be written as

$$U = \frac{\Omega}{2}\begin{pmatrix} \cos\theta & e^{i\phi}\sin\theta \\ e^{i\phi}\sin\theta & -\cos\theta \end{pmatrix} \tag{8.30}$$

where θ and ϕ may depend on the position. The frequency Ω characterizes the strength of the coupling between the two states and it is assumed to be position independent. The eigenstates of this operator, denoted as "dressed states", are given by:

$$|\chi_1\rangle = \begin{pmatrix} \cos\frac{\theta}{2} \\ e^{i\phi}\sin\frac{\theta}{2} \end{pmatrix}, \qquad |\chi_2\rangle = \begin{pmatrix} -e^{-i\phi}\sin\frac{\theta}{2} \\ \cos\frac{\theta}{2} \end{pmatrix} \tag{8.31}$$

with eigenvalues $\pm\hbar\Omega/2$ respectively. We assume that the initial internal state is $|\chi_1\rangle$ and that the evolution is adiabatic, such that the system remains in the state $|\chi_1\rangle$ at all times. Hence the state of the system can be described by a wave function $|\psi(t, \mathbf{r})\rangle = \varphi(t, \mathbf{r})\,|\chi_1(\mathbf{r})\rangle$ where $\varphi(t, \mathbf{r})$ will obey a modified Schrödinger equation due to the dependence of $|\chi_1(\mathbf{r})\rangle$ on the position. Plugging this into the Schrödinger equation and projecting on $|\chi_1(\mathbf{r})\rangle$, we find the effective Hamiltonian governing φ:

$$H_{\text{eff}} = \frac{\left(p_i - i\left\langle\chi_1(\mathbf{r})\right|\frac{\partial}{\partial x_i}\left|\chi_1(\mathbf{r})\right\rangle\right)^2}{2m} + \frac{\left|\left\langle\chi_2(\mathbf{r})\right|\frac{\partial}{\partial x_i}\left|\chi_1(\mathbf{r})\right\rangle\right|^2}{2m} + \frac{\Omega}{2}. \tag{8.32}$$

As expected, a vector potential $\tilde{A}_i(\mathbf{r})$ corresponding to the Berry connection is found. Additionally a potential $\tilde{V}(\mathbf{r})$ is also created and related to virtual transitions to the other state $|\chi_2(\mathbf{r})\rangle$. In this two level approximation, these two quantities are given by $\tilde{A}_i(\mathbf{r}) = \frac{\cos\theta-1}{2}\frac{\partial\phi}{\partial x_i}$ and $\tilde{V}(\mathbf{r}) = \frac{(\nabla\theta)^2+\sin^2\theta(\nabla\phi)^2}{8m}$. Discussions about the practical implementation on optical lattices can be found in [21, 22, 39, 40]. First experimental evidence of scalar potentials in quantum optics was found in [41] and the first observation of geometric magnetic fields in cold atomic physics was done in [42]. By considering a set of degenerate or quasi-degenerate dressed states it is possible to achieve non-Abelian gauge potentials as well [21, 22].

8.3.2 Effective Hamiltonian in Periodic Driven System

In contrast to the approach of the previous subsection, where the creation of the magnetic field relied on a slow change in time (i.e. the particle moves slowly enough such that the position dependent internal state is followed adiabatically), the following technique relies on fast oscillations. The basic principle consists on having two very distinct timescales. A fast oscillating time dependent potential will give rise to an effective time independent Hamiltonian which will present the desired complex hopping term. A general technique was proposed in [43] and it is based on a generic time-dependent periodic Hamiltonian:

$$H = H_0 + V(t) \tag{8.33}$$

where all the the time dependence is relegated to $V(t) = V(t+\tau)$ where τ is the time period. $V(t)$ can be decomposed as:

$$V(t) = \sum_n \left(V_{n+}e^{in\omega t} + V_{n-}e^{in\omega t}\right) \tag{8.34}$$

where $V_{n\pm}$ are operators and $\omega = 2\pi/\tau$. The condition $V_{n+} = V_{n-}^\dagger$ guarantees the Hermiticity of the Hamiltonian.

A unitary transformation $e^{iK(t)}$ generates an effective Hamiltonian given by:

$$H_{\text{eff}} = e^{iK(t)}He^{-iK(t)} + i\left(\frac{\partial}{\partial t}e^{iK(t)}\right)e^{-iK(t)}. \tag{8.35}$$

We choose a periodic operator $K(t)$ such that the effective Hamiltonian is time independent. Under this requirement, the time evolution operator can be represented as:

$$U\left(t_i \to t_f\right) = e^{iK(t_f)} e^{-iH_{\text{eff}}(t_f - t_i)} e^{-iK(t_i)}, \tag{8.36}$$

and it can be shown that, at lowest order, the effective Hamiltonian can be written as [43]:

$$H_{\text{eff}} = H_0 + \tau \sum_n \frac{1}{n} \left[V_{n+}, V_{n-}\right] + \mathcal{O}\left(\tau^2\right). \tag{8.37}$$

This expansion relies on the small parameter $\tilde{V}\tau$ where $\tilde{V}$ is the typical energy scale of $V(t)$. This expansion turns out to be very useful in the effective description of ultracold atomic systems though care should be taken, in a case by case scenario, in order to be sure about the convergence of the series.

8.3.2.1 Lattice Shaking

The lattice shaking approach consists on having an external time dependent optical potential that is changing in time in accordance to the previous description. Then a change of basis is performed for a co-moving frame that, along with a time average, will create an effective Hamiltonian with the desired complex hopping. As an example, a brief prescription is presented along the lines of the realization in a Rb Bose–Einstein condensate [44]. The Hamiltonian considered is the usual tight-biding Hamiltonian in 2D with the usual hopping and on-site part H_{os} (by on-site it is intended one body potential and scattering terms that act in single sites). In addition, there is an extra time dependent potential:

$$H = -\sum_{\mathbf{r},j} t_{\mathbf{r}j} \hat{a}^{\dagger}_{\mathbf{r}+j} \hat{a}_{\mathbf{r}} + H_{\text{os}} + \sum_{\mathbf{r}} v_{\mathbf{r}}(t)\, \hat{a}^{\dagger}_{\mathbf{r}} \hat{a}_{\mathbf{r}}. \tag{8.38}$$

The function $v_{\mathbf{r}}(t)$ is periodic on time with period τ: $v_{\mathbf{r}}(t) = v_{\mathbf{r}}(t+\tau)$. A unitary transformation on the states is performed and plugged in on the Schrödinger equation, thus defining new states $\left|\psi'\right\rangle$ such that $\left|\psi(t)\right\rangle = U(t)\left|\psi'(t)\right\rangle$. The Hamiltonian becomes $H'(t) = U(t)^{\dagger} HU(t) - iU(t)^{\dagger} \dot{U}(t)$ (where the dot stands for time derivative). The transformation is given by

$$U(t) = e^{-i\int_0^t dt' \sum_{\mathbf{r}} v_{\mathbf{r}}(t') \hat{a}^{\dagger}_{\mathbf{r}} \hat{a}_{\mathbf{r}}}. \tag{8.39}$$

It is straightforward to see that this transformation cancels the part of H (which will be present also on $U^{\dagger}HU$) corresponding to $v_i(t)\, \hat{a}^{\dagger}_{\mathbf{r}} \hat{a}_{\mathbf{r}}$. On the other side, since this does not commute with the kinetic term, a time dependence will be inherited by the hopping terms. For a set of rapidly oscillating function $v_i(t)$ the Hamiltonian can be replaced by an effective one, resulting from time averaging over a period. The new hopping parameters will read:

$$t_{\mathbf{r}j} \to t_{\mathbf{r}j} \left\langle e^{i\Delta v_{\mathbf{r}j}} \right\rangle_{\mathcal{T}} \tag{8.40}$$

where $\langle\rangle_{\mathcal{T}}$ stands for the average over a period: $\mathcal{T}^{-1} \int_0^{\mathcal{T}} dt$ and $\Delta v_{\mathbf{r}j} = v_{\mathbf{r}}(t) - v_{\mathbf{r}+\hat{j}}(t) - \left\langle v_{\mathbf{r}}(t) - v_{\mathbf{r}+\hat{j}}(t) \right\rangle_{\mathcal{T}}$.

8.3.2.2 Laser-Assisted Hopping

In this case the effective dynamics is induced by the coupling of the atoms on the optical lattice with a pair of Raman lasers. A fundamental ingredient consists on introducing an energy offset Δ on neighboring sites. It is enough to consider such scenario along a single direction. Considering a $2D$ lattice:

$$H = -t \sum_{\mathbf{r},j} \left(\hat{c}^{\dagger}_{\mathbf{r}+\hat{j}} \hat{c}_{\mathbf{r}} + \text{h.c.} \right) + \frac{\Delta}{2} \sum_{\mathbf{r}} (-1)^x c^{\dagger}_{\mathbf{r}} \hat{c}_{\mathbf{r}} + V(t) \tag{8.41}$$

where $\mathbf{r} = (x, y)$ runs through the lattice sites. The offset term, characterized by Δ, can be obtained by tilting the lattice, introducing magnetic gradients or through superlattices. The potential $V(t)$ is the result of the two external lasers that induce an electric field $E_1 \cos(\mathbf{k}_1 \cdot \mathbf{r}_1 - \omega_1 t) + E_2 \cos(\mathbf{k}_2 \cdot \mathbf{r}_2 - \omega_2 t)$. It is assumed that the frequencies are fine-tuned such that they match the offset $\omega_1 - \omega_2 = \Delta$. Neglecting fast oscillating terms the potential is written as:

$$V(t) = 2E_1 E_2 \sum_{\mathbf{r}} e^{i(\mathbf{k}_R \cdot \mathbf{r} - \Delta t)} c^{\dagger}_{\mathbf{r}} c_{\mathbf{r}} + \text{h.c.} \tag{8.42}$$

with $\mathbf{k}_R = \mathbf{k}_1 - \mathbf{k}_2$. Then one can get the effective Hamiltonian in two steps. First performing an unitary transformation $\exp[-it\frac{\Delta}{2} \sum_{\mathbf{r}} (-1)^x c^{\dagger}_{\mathbf{r}} \hat{c}_{\mathbf{r}}]$ will create oscillatory hopping terms (with $\exp(\pm i\Delta t)$ in front). Then one may apply the previous formalism building an effective Hamiltonian using (8.37):

$$\begin{aligned} H = &- t \sum_{x,y} \left(\hat{c}^{\dagger}_{x,y} \hat{c}_{x,y+1} + \text{h.c.} \right) \\ &- \frac{2tE_1E_2}{\Delta} \sum_{x \text{ even}, y} \left[\left(e^{i\mathbf{k}_R \cdot \mathbf{r}} - 1 \right) \left(e^{i\mathbf{k}_R \cdot \mathbf{r}} \hat{c}^{\dagger}_{x,y} \hat{c}_{x+1,y} + e^{-i\mathbf{k}_R \cdot \mathbf{r}} \hat{c}^{\dagger}_{x-1,y} \hat{c}_{x,y} \right) + \text{h.c.} \right] + \mathcal{O}\left(\Delta^{-2}\right). \end{aligned} \tag{8.43}$$

It is clear that this generates complex hopping and one finds that the lattice has staggered flux. With a choice $\mathbf{k}_R = (\Phi, \Phi)$ (as also made in the experiment [45]) one can write upon a gauge transformation:

$$H = -t\sum_{x,y}\left(\hat{c}^{\dagger}_{x,y}\hat{c}_{x,y+1} + \text{h.c.}\right)$$
$$-\frac{2tE_1E_2\sin\Phi/2}{\Delta}\sum_{x\text{ even},y}\left[\left(e^{i\Phi y}\hat{c}^{\dagger}_{x,y}\hat{c}_{x+1,y} + e^{-i\Phi y}\hat{c}^{\dagger}_{x-1,y}\hat{c}_{x,y}\right) + \text{h.c.}\right] + \mathcal{O}\left(\Delta^{-2}\right) \tag{8.44}$$

where it is clear that a sequence of fluxes $\pm\Phi$ alternates in the plaquettes along the x direction. More refined techniques allow for the realization of systems with uniform fluxes [46]. In such systems the Chern number of the Hofstadter bands was measured in [47]. It is worth noting that other kind of one-body terms, beyond the staggered term, can be used as it was done in the first quantum simulations of this model with ultracold atoms [48, 49]. In that case a linear potential is used. These kind of approaches can be adapted to more general scenarios including different geometries and multi-component species. The latter, for example, can be achieved by introducing spin dependent potentials as done in [48].

8.4 Simulation of Gauge Fields

In the context of Abelian gauge theories, the goal of simulating gauge fields consists in attributing dynamics to the complex phases on the hopping parameters that were identified in the previous section. In order to construct such dynamics one should identify degrees of freedom that will play the role of the gauge field. Several proposals have been put forward which map the gauge degrees of freedom into some other controllable variables. The platforms used include ultracold atoms, trapped ions and superconducting qubits. They may be analogue or digital quantum simulators and include Abelian or non-Abelian symmetries [50–76]. A more detailed description of two particular approaches in analogue cold atomic simulators will follow: the gauge invariance will be obtained by either penalizing with a large energy cost the non-physical states or by exploiting microscopic symmetries.

There are other symmetries which have been explored, namely $\mathbb{Z}_n$ [54, 65] which, in particular, can provide an alternative route towards $U(1)$ symmetry in the large n limit [54] and can be addressed with similar approaches. Proposal for the realization of $\mathbb{C}P(N-1)$ [77, 78] models have been put forward in [71, 72]. These models can serve as toy models for QCD and are also relevant in studying the approach to the continuum limit, in the context of D-theories, where the continuum limit is taken via dimensional reduction [35, 36]. Furthermore other formulations are possible for specific groups [26, 79–82]. Gauge theories with Higgs fields have also been the target of quantum simulation proposals [83–86].

Another relevant approach is the so-called quantum Zeno dynamics which takes inspiration on the quantum Zeno effect, stating that a system being continuously observed does not evolve on time. Furthermore, if the measurement commutes with a certain part of the Hamiltonian, then it can freeze a certain part of the Hilbert space but still enables the dynamics in another subspace [87]. This feature can be used in order to freeze gauge dependent quantities and let the system evolve in the gauge invariant subspace. The Hamiltonian to be implemented has the form $H_{\text{noise}} = H_0 + H_1 + \sqrt{2\kappa}\sum_{x,a} G_x^a$ where H_0 and H_1 are time independent and are, respectively, gauge invariant and gauge variant parts of the Hamiltonian. The operators G_x^a are associated to the constraint one wishes to impose $G_x^a\,|\psi\rangle = 0$. In the case of gauge theories G_x^a are the generators of gauge transformations. An advantage of this approach, with respect to the energy punishment approach of the next section, is that only linear terms on the generators must be imposed on the Hamiltonian (energy punishment requires quadratic terms). By other side leakage from the gauge invariant subspace of the Hilbert space happens as a function of time, which does not happen in the energy penalty approach. This approach was developed in [63].

Another approach, that was successfully implemented in the first quantum simulator of a gauge theory using trapped ions [7], is the digital quantum simulator [88]. The key idea consists on in dividing the full time evolution operator e^{-iHt} into smaller pieces of sizes $\tau = t/N$ and apply time evolution of smaller parts of the Hamiltonian at a time. Consider for example an Hamiltonian which is a sum of M contributions: $H = \sum_\alpha^M H_\alpha$. Each part H_α can represent, for example, a nearest neighbor spin interaction in which case only two spins are coupled on each H_α. For large enough N one can write:

$$e^{-iHt} = \left(e^{-iH\tau}\right)^N \simeq \left(\prod_{\alpha=1}^{M} e^{-iH_\alpha \tau}\right)^N . \tag{8.45}$$

Each time step can now be interpreted as an individual gate. While in the analogue simulation the great difficult lies on building the appropriate gauge invariant Hamiltonian, in digital quantum simulations that is not a problem. The difficulty lies, however, in building an efficient sequence of gates. Other then the scheme used in the first experimental realization [89], other proposals towards digital quantum simulations of lattice gauge theories have been put forward [50, 52, 61, 69, 73, 74].

8.4.1 Gauge Invariance from Energy Punishment

The energy punishment approach is a quite general approach which allows for the theoretical construction of models that will exhibit a given symmetry in its low energy sector. It consists on building a Hamiltonian which does not prohibit the symmetry violation to occur but instead punishes it with a large energy. In a more concrete way, let suppose one wants to implement a set of symmetries corresponding to a set of generators $\{G_x\}$ commuting with each other $\left[G_x, G_y\right] = 0$. Furthermore consider a typical Hamiltonian H_0 which does not respect these symmetries. Then one constructs the following Hamiltonian:

$$H = H_0 + \Gamma \sum_x G_x^2 \tag{8.46}$$

where Γ is a large energy scale, meaning much larger than the energy scales involved in H_0. Since G_x are Hermitian G_x^2 have non-negative eigenvalues. One can choose the lowest eigenvalue to be zero by an appropriate definition of G_x. Then, at low energy ($\ll\Gamma$), the states will respect approximately the condition $G_x |\psi\rangle \simeq 0$. If not, this would give a state automatically in an energy scale $\sim\Gamma$. It is then possible to construct an effective Hamiltonian, valid in low energy, which will respect the symmetries generated by $\{G_x\}$. Let G be the projector operator on the subspace of the total Hilbert space obeying $G_x |\psi\rangle = 0$ and let $P = 1 - G$. Then the low energy Hamiltonian can be written as:

$$H_{\text{eff}} = GH_0G - \frac{1}{\Gamma} GH_0P \frac{1}{\sum_x G_x^2} PH_0G + \mathcal{O}\left(\Gamma^{-2}\right) \tag{8.47}$$

which fulfills the symmetries. Within this framework an effective Abelian gauge theory can be constructed. In non-Abelian theories the generators of the gauge transformation do not commute and this construction fails. There are, of course, several possible drawbacks even on the theoretical level. For example the Hamiltonian (8.47), even though gauge invariant, may contain unwanted interactions or miss some particular terms which are present on the target system.

In order to construct a quantum simulator the first task is naturally to map the degrees of freedom of the target theory into the laboratory controlled ones, in this case the atomic variables. The matter fields, which are fermionic, will naturally be described by fermionic atomic species. Regarding gauge fields, the target will be the quantum links formulation discussed in Sect. 8.2.1.3. Therefore the goal consists on building the quantum links satisfying the algebra $\left[L_{\mathbf{r},i}, U_{\mathbf{r}',j}\right] = \delta_{ij}\delta_{\mathbf{rr}'}U_{\mathbf{r}',j}$ and $\left[U_{\mathbf{r},i}, U^{\dagger}_{\mathbf{r}',j}\right] = \delta_{ij}\delta_{\mathbf{rr}'}2L_{\mathbf{r},i}/S\,(S+1)$.

This can be achieved using the Schwinger representation. Given two bosonic species $b^{(\sigma)}$ with $\sigma = 1, 2$ which are associated to each link, one can write

$$U_{\mathbf{r}i} = \frac{1}{\sqrt{S\,(S+1)}} b^{(2)\dagger}_{\mathbf{r}i} b^{(1)}_{\mathbf{r}i}, \; L_{\mathbf{r}i} = \frac{1}{2}\left(b^{(2)\dagger}_{\mathbf{r}i} b^{(2)}_{\mathbf{r}i} - b^{(1)\dagger}_{\mathbf{r}i} b^{(1)}_{\mathbf{r}i}\right). \tag{8.48}$$

Each link is loaded with a total of $2S$ bosons where S is an half integer. Then one has the desired representation for the quantum links in terms of atomic variables. Now the variables are identified. One then can then build a d dimensional optical lattice where fermions are allowed to hop among lattice points and in each links there are a total of $2S$ bosons. For $1D$, the target Hamiltonian is of the form:

$$H = -t\sum_n \left(c_n^{\dagger} U_n c_{n+1} + \text{h.c.}\right) + m\sum_n (-1)^n c_n^{\dagger} c_n + \frac{g^2}{2}\sum_n L_n^2. \tag{8.49}$$

When comparing to the general structure of (8.21) there are two differences: the plaquette term and the position-dependent coefficient of the kinetic term. The plaquettes are naturally absent in $1D$, whereas the tunneling amplitude can be fixed by a gauge transformation $c_n \rightarrow (-i)^n c_n$. The Hamilton (8.49) has, therefore, the required structure and can be targeted with the Schwinger boson approach. It then becomes

$$H = -\frac{t}{\sqrt{S(S+1)}} \sum_n \left(c_n^\dagger b_n^{(\bar{\sigma})\dagger} b_n^{(\sigma)} c_n + \text{h.c.} \right) + m \sum_n (-1)^n c_n^\dagger c_n + \frac{g^2}{8} \sum_n \left(b_n^{(2)\dagger} b_n^{(2)} - b_n^{(1)\dagger} b_n^{(1)} \right)^2 . \tag{8.50}$$

The two last terms can be, in principle, implemented directly using a proper tuning of the interactions between the bosons and the potential for the fermions. The first term, instead, is a correlated hopping between bosons and fermions which is obtained less easily. Furthermore the additional terms that are not gauge invariant, like $b_n^{(\bar{\sigma})\dagger} b_n^{(\sigma)}$ and $c_n^\dagger c_{n+\hat{i}}$, must be suppressed. This is solved by the energy punishment approach. In general the non-gauge invariant Hamiltonian with the ingredients described has the form:

$$\begin{aligned} H_0 = &- \sum_{n,i} \left[t_F \left(c_n^\dagger c_{n+1} + \text{h.c.} \right) - t_B \left(b_n^{(2)\dagger} b_n^{(1)} + \text{h.c.} \right) \right] \\ &+ \sum_n \left(v_n^F c_n^\dagger c_n + \sum_\sigma v_n^{B\sigma} b_{ni}^{(\sigma)\dagger} b_{ni}^{(\sigma)} \right) + U \sum_n \left(b_n^{(2)\dagger} b_n^{(2)} - b_n^{(1)\dagger} b_n^{(1)} \right)^2 . \end{aligned} \tag{8.51}$$

Using the generators for the $U(1)$ gauge symmetry in (8.19) one considers the full Hamiltonian:

$$H = H_0 + \Gamma \sum_n \left(L_n - L_{n-1} - c_n^\dagger c_n + \frac{1-(-1)^n}{2} \right)^2 . \tag{8.52}$$

It is crucial that one has access to the interactions that are introduced on the last term corresponding to the energy punishment. To see that this is the case it useful to be more specific about the labels σ. One can take, as in [53], the labels $\sigma = 1, 2$ meaning respectively left and right part of the link, which can be thought to coincide with the lattice site. In this way $b_n^{(2)\dagger} b_n^{(1)}$ are just regular hopping terms. Furthermore it is recalled that the total number of bosons associated to each link is conserved. Therefore one can write: $L_n = -S + b_n^{(2)\dagger} b_n^{(2)} = S - b_n^{(1)\dagger} b_n^{(1)}$. This means that terms like L_n^2 and and $L_n L_{n-1}$ can be written as a density-density interaction. Regarding the last case, recall that $b_n^{(1)}$ and $b_{n-1}^{(2)}$ are effectively in the same site, see Fig. 8.1. Now (8.47) can be applied. The number of particles in each site is a good quantum number to describe the eigenstates of G_x. The number of particles in the site j are denoted by $n_j^F = c_j^\dagger c_j$, $n_j^1 = b_j^{(1)\dagger} b_j^{(1)}$ and $n_j^2 = b_{j-1}^{(2)\dagger} b_{j-1}^{(2)}$. The subspace of gauge invariant states is then characterized by:

$$n_j^F + n_j^1 + n_j^2 = 2S + \frac{1-(-1)^j}{2} . \tag{8.53}$$

At the lowest order only the two last terms of (8.51) survive as any single hopping destroys the above relation. At the next order there are three possible virtual processes

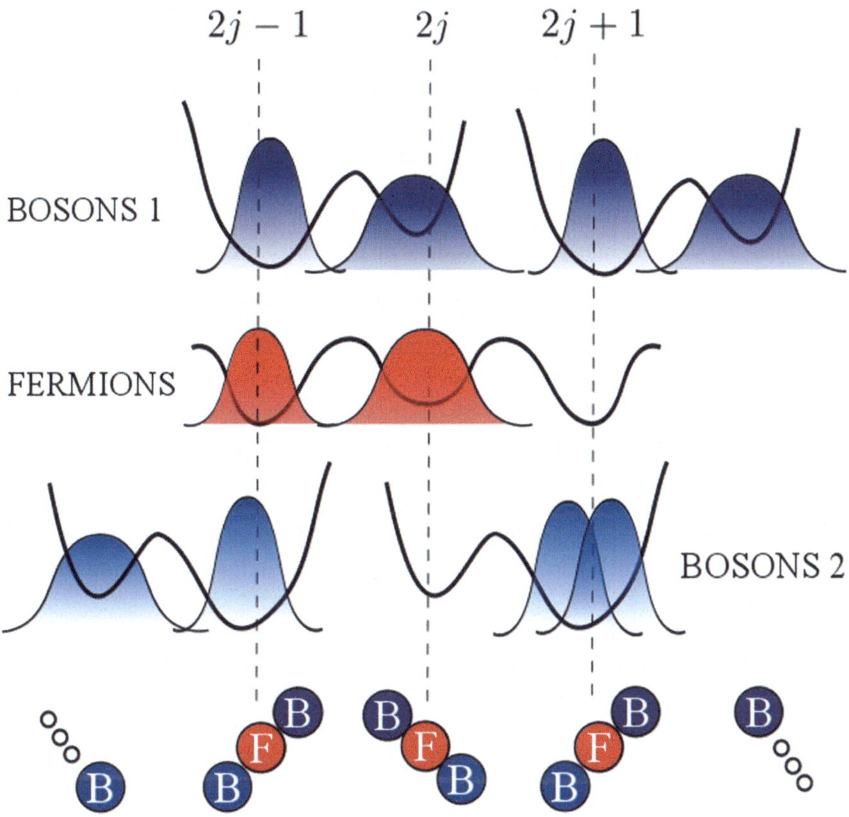

Fig. 8.1 Superlattice configurations for the two boson species and the fermionic one. Bosons of the species 1 at an even site $2j$ can only hop to $2j-1$ while a boson of species 2 has only access to the site $2j+1$. The Figure presents an example of a gauge invariant state configuration (on these three sites) where $G_x\,|\psi\rangle = 0$

that preserve this condition. Up to some linear terms on the particle density operator, they are:

i Boson-boson hopping: a boson hops to the neighboring site on the same link and another boson hops back. It gives rise to a boson density-density interaction.
ii Fermion-Fermion hopping: a fermion hops to a neighboring site and then hops back. Only possible if neighboring site is unoccupied and gives rise to a nearest neighbor fermion density-density interaction.
iii Boson–Fermion hopping: a fermion hops to a neighboring site and a boson belonging to the link that connects the two sites does the opposite path. It gives rise to a correlated hopping.

The terms coming from (i) should be joined with the last term of (8.51) in order to form the correct kinetic term for the gauge fields. The terms in (ii) are somehow unwanted and correspond to a repulsion between neighbor fermions $n_j^F n_{j+1}^F$. Naturally, they do not spoil gauge invariance and their inclusion should not be a problem [53]. Finally the terms originating from (iii) give rise to the correlated hopping responsible for the matter-gauge coupling as written on the first term of (8.48). There is another issue which should be addressed. From the beginning it was assumed that the the number of bosons in each link is conserved. In particular this means that bosons are not allowed to pass to a neighboring link. In order to guarantee this condition in an experiment one should introduce an extra bosonic species and this is the reason that bosons in neighboring links were represented with different colors on Fig. 8.1. Then one bosonic species is trapped on the even links and the other in the odd links. This will prevent bosonic hopping between links. A numerical study of real dynamics of the the model as well as accuracy of the effective gauge invariance obtained was also done in [53].

Finally, in a possible experimental realization, the first fundamental step is to guarantee that the system is initialized on a gauge invariant state. This can be done by loading the atoms in a deep lattice such that they are in Mott phase. Afterwards the system should evolve according to the fine tuned Hamiltonian described above (after lowering the lattice barriers). Finally measures of relevant quantities can be performed.

This principle is valid in higher dimensionality where one has to face the difficulty of generating plaquette terms. This was done for the pure gauge in [59, 90] by suitably allowing hopping between links. In the first case each link has an infinite dimensional Hilbert space that is represented by a Bose–Einstein condensate. In the second the proposal is simplified by considering a quantum link model.

8.4.2 Gauge Invariance from Many Body Interaction Symmetries

This approach consists on building a lattice which will have the necessary local gauge invariance arising from microscopic symmetries. Specific proposals may vary significantly even though the same principle is used. For example in [76] the simulation is built upon the global symmetry conserving the total number of excitations and is achieved via a state-dependent hopping. In turn, see for example [65, 91], are built upon conservation of angular momentum. For concreteness the later approach will be described in more detail below. In the case of [62] $SU(N)$ symmetries of the ground state manifold of alkaline-earth-like atoms could be exploited in order to built non-Abelian gauge theories.

Symmetries only allow for certain type of processes to occur and, by exploiting these constraints, one can build a gauge symmetry. This can be done, as said before, considering angular momentum conservation. The Schwinger model is taken as an illustrative example. Bosons, that will make up the gauge fields, are placed at the two boundaries of the links. Because the goal consists, partially, in forbidding gauge dependent terms like simple boson or fermion hopping, the lattice should be spin dependent. In this way a single hopping is forbidden as it does not conserve angular momentum. By other side one should guarantee that correlated spin between bosons and fermions is allowed. This can be achieved by a judicious choice of respective hyperfine angular momentum in each lattice site. For concreteness, consider a single link connecting two sites and a total of two bosonic ($b^{(1)}$,$b^{(2)}$) and two fermionic species (c,d). The site at the left of the link can only be populated by c while the right side by d. Analogously the left end of the link can only be populated by $b^{(1)}$ while the right end can only be populated by $b^{(2)}$. Then the conditions described above for allowed/forbidden hopping are automatically satisfied if one chooses the hyperfine angular momentum of each atomic species to satisfy:

$$m_F\,(d) - m_F\,(c) = m_F\left(b^{(1)}\right) - m_F\left(b^{(2)}\right). \tag{8.54}$$

It is intended that the lattice is, indeed, spin dependent so that $m_F\,(d) \neq m_F\,(c)$ and $m_F\left(b^{(1)}\right) \neq m_F\left(b^{(2)}\right)$. In other words, what this means is that the difference of angular momentum caused by a fermion hop can be exactly compensated by a bosonic hop in the opposite direction. This leads directly to the correlated hopping desired which, in fact, comes from the scattering terms between bosons and fermions. The only other allowed scattering term between fermions and bosons correspond to density-density interactions like $c^\dagger c\left(b^{(2)\dagger}b^{(2)} + b^{(1)\dagger}b^{(1)}\right)$. These are just linear terms on the fermionic number operator due to the conservation of the total number of bosons per link. Summing over all lattice sites will give just a constant shift of the energy. The scattering terms between bosons give rise to the gauge kinetic term as before (in $1+1$ dimensions).

Again, for higher dimensionality, there is a non-trivial extra step consisting on building plaquette interactions. If plaquettes are ignored and the model described above is loaded on an higher dimensional lattice the result corresponds to the strong coupling limit of the gauge theory.

The plaquette terms can be achieved by the so-called loop method. It uses perturbation theory in a similar way that was used in the energy penalty approach. In order to discuss the essence of the construction of the plaquette terms, one can consider just the pure gauge theory. The target Hamiltonian is

$$H_{\text{target}} = \frac{g^2}{2}\sum_{\mathbf{r},i} L_{\mathbf{r}i}^2 - \frac{1}{4g^2}\sum_{\Box}\left(U_\Box + U_\Box^\dagger\right). \tag{8.55}$$

The description will be specialized for $2+1$ dimensions but the theoretical construction for higher dimensions is analogous. The construction of the plaquette term relies on a perturbative expansion similar to (8.46) but, in this case, H_0 is already a gauge invariant Hamiltonian. For reasons that will be explained below one should have two fermionic species, say χ and ψ, and build the trivial part of the generalization of the $1+1$ process:

$$H_0 = -t\sum_{\mathbf{r},i}\left(\psi_\mathbf{r}^\dagger U_{\mathbf{r}i}\psi_{\mathbf{r}+\hat{i}} + \chi_\mathbf{r}^\dagger U_{\mathbf{r}i}\chi_{\mathbf{r}+\hat{i}} + \text{h.c.}\right) + \frac{g^2}{2}\sum_{\mathbf{r},i} L_{\mathbf{r}i}^2\,. \tag{8.56}$$

The fermionic species are auxiliary and in the effective model they will be integrated out. There should be no interacting term between them. Here the energy penalty must enforce the following conditions at each site $\mathbf{r} = (r_1, r_2)$:

- there is a fermion ψ if both r_1 and r_2 are even
- there is a fermion χ if both r_1 and r_2 are odd
- no fermion otherwise

The positions of these fermions is represented on Fig. 8.2a. This kind of constraint can be obtained, for large Γ, with a Hamiltonian of the form:

$$H_{\text{penalty}} = -\Gamma \sum_{\mathbf{r}} \left[\frac{\left(1+(-1)^{r_1}\right)\left(1+(-1)^{r_2}\right)}{4} \psi_{\mathbf{r}}^{\dagger} \psi_{\mathbf{r}} + \frac{\left(1-(-1)^{r_1}\right)\left(1-(-1)^{r_2}\right)}{4} \chi_{\mathbf{r}}^{\dagger} \chi_{\mathbf{r}} \right]. \tag{8.57}$$

Through perturbation theory, according to (8.47), one gets the plaquette terms at fourth order. This process is "cleaner" if the $U_{\mathbf{r}}$ in (8.56) are considered unitary. In particular we may consider a unitary limit, in which the total spin of the quantum link goes to infinity: $S \rightarrow = +\infty$. Order by order:

1. Only the pure gauge part of (8.56) contributes, no fermionic term occurs.
2. Trivial constant contribution assuming that U_n are unitary. The virtual process giving rise to this contribution is a single link interaction where a fermionic-bosonic correlated hopping occurs back and forth restoring the initial state. There are never fermions on the neighbor lattice site. In the unitary limit there is an infinite number of bosons such that $\left[U, U^{\dagger}\right] \rightarrow 0$. In the case of finite bosonic number, extra contribution corresponding to a renormalization of the pure gauge term of (8.56) will appear, together with another term which can be discarded by application of the Gauss law.
3. Trivial constant contribution assuming that U_n are unitary. Virtual contributions evolving links constitute again back and forth hopping plus a pure gauge term at any stage of the process. The extra contributions coming from considering a finite number of boson per link cannot be disregarded trivially as at second order.
4. It gives the desired plaquette term plus renormalization of the pure gauge term of (8.56) assuming that U_n are unitary. The last case corresponds to the virtual process where a fermion goes around a plaquette and returns to the initial place. This virtual process is represented on Fig. 8.2b. Naturally, in the non-unitary case, more terms appear.

Plaquette terms only appear at fourth order. However, in the unitary limit, most contributions are trivial. One can then see that it is effectively a second order contribution [92].

When one considers a finite number of bosons in the links there are extra contributions appearing which cannot be disregarded. As in the case of the energy penalty, these contributions, even though unwanted, can be tolerated as they are naturally gauge invariant. However one should guarantee that these extra contributions are not more important than the plaquette term which is the target term. That can be achieved

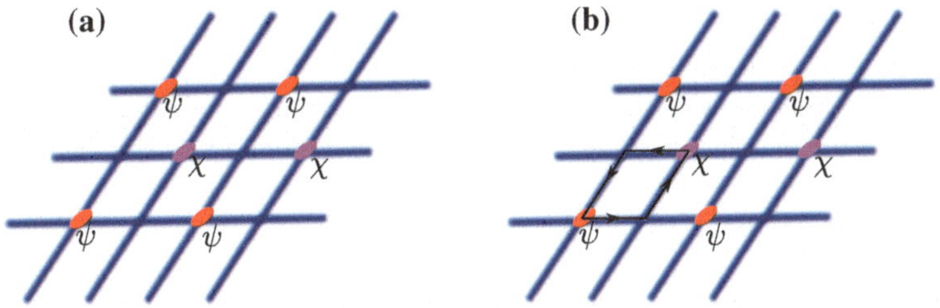

Fig. 8.2 Loop method for obtaining the plaquette terms. In the panel **a** it is depicted the positions of the auxiliary fermions that are used to construct the plaquette term using gauge invariant building blocks. One of the species, say ψ, is represented in red and placed on sites with both coordinates even. In turn χ, in pink, is placed on sites with both coordinates odd. This correspond to the ground-state of (8.57). In the panel **b** it is represented a virtual process that gives rise to a plaquette term

if the coupling term, parameterized by g^2, is taken to be small in units of t. By taking $g^2 \sim t^2/\Gamma$ one makes the unwanted terms at third order effectively of the same order as the plaquettes and unwanted terms of the fourth order effectively of higher order than the plaquettes.

On top of these, an extra species of fermions can be introduced to play the role of matter fields. They will consist, in the initial Hamiltonian, to the usual correlated hopping with the bosons. Furthermore the staggered mass term (of (8.17)) should also be introduced. In the unitary case this extra piece commutes with the interacting part of (8.56) and no further contribution is obtained in perturbation theory. In the truncated case there is an extra (gauge invariant) correlated hopping coming at third order. Another different aspect of the introduction of dynamical fermions is that the Gauss law ($\sum_i L_{\mathbf{r}i} - L_{\mathbf{r}-\hat{i},i} = \text{const}$) can no longer be used to trivialize terms. The divergence of the electric field gets a contribution from the charge density of the dynamical fermions. Nonetheless it can still be employed and the extra charge density terms can be compensated on the initial Hamiltonian if proper fine tuning is available experimentally.

In [91] it was proposed a realization of the Schwinger $(1+1)$ model using a mixture of ^{23}Na for the bosons and ^{6}Li for the fermions as well as an extensive study on the influence of the finiteness of the number of bosons per link in that case.

8.4.3 Encoding in $\mathbf{1+1}$ *Fermions*

The case of the Schwinger model, $1+1$ Dirac fermions coupled to a gauge field, is an interesting experimental and theoretical playground. It shares some non-trivial features with QCD like confinement, chiral symmetry breaking and a topological theta vacuum [93]. However, due to its simplicity, it allows for analytical and numerical studies which may become significantly harder in more complicated theories. Furthermore it was the target of the first experimental implementation of a lattice gauge theory [7]. In the context of quantum simulations it may not only provide the entrance door towards more complicated experimental realizations but also a way of benchmarking experimental techniques.

One of the reasons why this model bares an intrinsic simplicity, as mentioned previously, is the fact that the gauge fields are non-dynamical. This is reflected on the absence of plaquette terms in the Hamiltonian formulation. Furthermore the Gauss law fixes the gauge field and can be used to integrate out its degrees of freedom. This results in a long-range interacting model which will be addressed next. In the following the lattice Hamiltonian formulation is considered for N lattice sites:

$$H = -it\sum_{n=1}^{N-1}\left[c_n^\dagger U_n c_{n+1} - \text{h.c}\right] + m\sum_{n=1}^{N}(-1)^n c_n^\dagger c_n + \frac{g^2}{2}\sum_{n=1}^{N-1} L_n^2. \tag{8.58}$$

Here an infinite dimensional Hilbert space per link is considered, therefore the operators U_n are unitary and the non-trivial commutation relations on the links are given by $[L_m, U_n] = U_n\delta_{mn}$. Equivalently the link can be written as $U_n = e^{i\theta_n}$. The Gauss law is imposed in accordance with the relations (8.18) and (8.19). This model can be formulated in terms of Pauli spin operators [94] through the Jordan–Wigner transformation:

$$\begin{cases} c_n = \prod_{l<n}\left(i\sigma_z(l)\right)\sigma^-(n) \\ c_n^\dagger = \prod_{l<n}\left(-i\sigma_z(l)\right)\sigma^+(n) \end{cases} \tag{8.59}$$

where $\sigma_i(l)$ represent the Pauli matrices in the site l and $\sigma^\pm(n) = \sigma_x(n) \pm i\sigma_y(n)$. In terms of the spins the Gauss law is determined by:

$$G_n = L_n - L_{n-1} - \frac{1}{2}\left(\sigma_z(n) + (-1)^n\right). \tag{8.60}$$

By restricting ourselves to the physical space, through the Gauss law $G_n|\psi\rangle = 0$, the link variables can be almost completely eliminated. Using periodic boundary conditions ($L_0 = L_N$) one finds:

$$L_n = L_0 + \frac{1}{2}\sum_{l=1}^{n}\left(\sigma_z(l) + (-1)^n\right). \tag{8.61}$$

The value of L_0 is a parameter of the theory and corresponds to a background field. For simplicity it will be taken to zero at the present discussion. By using the above relations the Hamiltonian (8.58) can be rewritten as:

$$H = t\sum_{n=1}^{N}\left[\sigma^+(n)e^{i\theta_n}\sigma^-(n+1) + \text{h.c}\right] + \frac{m}{2}\sum_{n=1}^{N}(-1)^n\sigma_z(n) + \frac{g^2}{8}\sum_{n=1}^{N}\left[\sum_{l=1}^{n}\left(\sigma_z(l) + (-1)^n\right)\right]^2 \tag{8.62}$$

where a trivial constant term was dropped. The remaining gauge field variable θ_n can be eliminated by a residual gauge transformation [95]:

$$\sigma^{\pm}(n) \rightarrow \sigma^{\pm}(n) \prod_{j<n} e^{\pm i\theta_j}. \tag{8.63}$$

This is a non-trivial transformation as θ's are operators. More precisely, the above relation should be seen as defining a new set of operators $\bar{\sigma}^{\pm}(n) = \sigma^{\pm}(n) \prod_{j<n} e^{\pm i\theta_j}$. The $\bar{\sigma}$ still respect the angular momentum algebra between each other. Therefore they are still spin operators on the sites of the lattice, despite acting non-trivially on the links. Since the links degrees of freedom are being traced out using the Gauss law, one can arrive at an effective spin model for the sites. Plugging this transformation and expanding the interaction term, the resulting model is a long-range interacting spin model:

$$H = t\sum_{n=1}^{N} \left[\sigma^{+}(n)\sigma^{-}(n+1) + \text{h.c}\right] + \sum_{n=1}^{N} \left(\frac{m}{2}(-1)^n - \frac{g^2}{8}\left(1-(-1)^n\right)\right)\sigma_z(n)$$
$$+\frac{g^2}{4}\sum_{n=1}^{N-2}\sum_{l=1}^{N-1}(N-l)\,\sigma_z(n)\,\sigma_z(l). \tag{8.64}$$

This is a useful formulation for quantum simulations since the total of N particles and $N-1$ gauge fields are simulated by just N spins (with exotic long-range interactions), thanks to the gauge invariance. The difficulty was moved towards an efficient way of implementing the long-range asymmetric interaction between spins. This Hamiltonian was implemented as a digital quantum simulator in [7] using trapped ions ($^{40}Ca^{+}$). The system was composed of four qubits. The Schwinger mechanism of pair creation of particle-antiparticle was explored, as well as real time evolution of entanglement in the system. Based on the staggering prescription in Sect. 8.2.1.2, a particle on an odd site corresponds to the vacuum and a hole as an antiparticle (the contrary holds for particles in the even sites). Following this picture the number of particles at the site n is given by $\nu_n = \left(1-(-1)^n\right)/2 + (-1)^n c_n^{\dagger} c_n$ and therefore a relevant observable is the particle density $\nu(t) = (2N)^{-1}\sum_n \left\langle 1 + (-1)^n \sigma_z(n)\right\rangle$. Starting from a bare vacuum ($\nu(0) = 0$) it is observed a rapid increase of the particle density followed by a decrease which is due to particle/anti-particle recombination. Also the vacuum persistence $G(t) = \langle 0| e^{-iHt} |0\rangle$ and entanglement were evaluated. The latter is done by reconstructing the density matrix and evaluating the entanglement in one half of the system with the other half through logarithmic negativity. Entanglement is produced through particle creation that get distributed across the two halves. More detail on the simulation and experimental results can be found in [7, 89]. Future challenges include the simulation of larger systems as well higher dimensionality and non-Abelian symmetries.

Acknowledgements The authors want to thank the organizers of the Natal's school: Pasquale Sodano, Alvaro Ferraz, Kumar S. Gupta and Gordon Semenoff. They are also pleased to thank the participants to the course, in particular T. J. G. Apollaro, V. E. Korepin, T. Macrì, G. Mussardo, E.

Tonni and J. Viti, for useful discussions and to have contributed to set a stimulating and pleasant atmosphere. Finally, special thanks go to Marcello Dalmonte for discussions and common work on the topics discussed in this chapter.

References

1. J.C. Pinto Barros, Field and Gauge theories with ultracold gauge potentials and fields. Ph.D. Thesis, SISSA, 2017
2. A.Y. Kitaev, Fault-tolerant quantum computation by anyons. Ann. Phys. **303**(1), 2–30 (2003)
3. D.J. Gross, F. Wilczek, Ultraviolet behavior of non-abelian gauge theories. Phys. Rev. Lett. **30**(26), 1343 (1973)
4. H. Fritzsch, M. Gell-Mann, H. Leutwyler, Advantages of the color octet gluon picture. Phys. Lett. B **47**(4), 365–368 (1973)
5. H.D. Politzer, Reliable perturbative results for strong interactions? Phys. Rev. Lett. **30**(26), 1346 (1973)
6. M.E. Peskin, D.V. Schroeder, *An Introduction to Quantum Field Theory* (Westview press, Boulder, 1995)
7. E.A. Martinez, C.A. Muschik, P. Schindler, D. Nigg, A. Erhard, M. Heyl, P. Hauke, M. Dalmonte, T. Monz, P. Zoller et al., Real-time dynamics of lattice gauge theories with a few-qubit quantum computer. Nature **534**(7608), 516–519 (2016)
8. R.L.P.G. Do Amaral, E.C. Marino, Canonical quantization of theories containing fractional powers of the d'alembertian operator. J. Phys. A Math. Gen. **25**(19), 5183 (1992)
9. J.C. Pinto Barros, M. Dalmonte, A. Trombettoni, Long-range interactions from u (1) gauge fields via dimensional mismatch. J. Stat. Mech. Theory Exp. **2018**(10), 103103 (2018)
10. J.C. Pinto Barros, M. Dalmonte, A. Trombettoni, Robustness of confinement properties of schwinger-thirring models. To be published (2017)
11. M.C. Angelini, G. Parisi, F. Ricci-Tersenghi, Relations between short-range and long-range ising models. Phys. Rev. E **89**, 062120 (2014)
12. E. Brezin, G. Parisi, F. Ricci-Tersenghi, The crossover region between long-range and short-range interactions for the critical exponents. J. Stat. Phys. **157**(4), 855–868 (2014)
13. N. Defenu, A. Trombettoni, A. Codello, Fixed-point structure and effective fractional dimensionality for o(n) models with long-range interactions. Phys. Rev. E **92**, 052113 (2015)
14. N. Defenu, A. Trombettoni, S. Ruffo, Anisotropic long-range spin systems. Phys. Rev. B **94**, 224411 (2016)
15. M.F. Paulos, S. Rychkov, B.C. van Rees, B. Zan, Conformal invariance in the long-range ising model. Nucl. Phys. B **902**, 246–291 (2016)
16. T. Horita, H. Suwa, S. Todo, Upper and lower critical decay exponents of ising ferromagnets with long-range interaction. Phys. Rev. E **95**, 012143 (2017)
17. G. Gori, M. Michelangeli, N. Defenu, A. Trombettoni, One-dimensional long-range percolation: a numerical study. Phys. Rev. E **96**, 012108 (2017)
18. C. Behan, L. Rastelli, S. Rychkov, B. Zan, Long-range critical exponents near the short-range crossover. Phys. Rev. Lett. **118**, 241601 (2017)
19. C. Behan, L. Rastelli, S. Rychkov, B. Zan, A scaling theory for the long-range to short-range crossover and an infrared duality. J. Phys. A Math. Theor. **50**(35), 354002 (2017)
20. N. Defenu, A. Trombettoni, S. Ruffo, Criticality and phase diagram of quantum long-range o(n) models. Phys. Rev. B **96**, 104432 (2017)
21. J. Dalibard, F. Gerbier, G. Juzeliūnas, P. Öhberg, Colloquium: artificial gauge potentials for neutral atoms. Rev. Mod. Phys. **83**(4), 1523 (2011)
22. N. Goldman, G. Juzeliūnas, P. Öhberg, I.B. Spielman, Light-induced gauge fields for ultracold atoms. Rep. Prog. Phys. **77**(12), 126401 (2014)

23. J. Dalibard, Introduction to the physics of artificial gauge fields, in *Quantum Matter at Ultralow Temperatures* (2015)
24. M. Burrello, L. Lepori, S. Paganelli, A. Trombettoni, Abelian gauge potentials on cubic lattices, in *Advances in Quantum Mechanics* (Springer, Berlin, 2017), pp. 47–70
25. J. Kogut, L. Susskind, Hamiltonian formulation of wilson's lattice gauge theories. Phys. Rev. D **11**(2), 395 (1975)
26. E. Zohar, M. Burrello, Formulation of lattice gauge theories for quantum simulations. Phys. Rev. D **91**(5) (2015)
27. H.B. Nielsen, M. Ninomiya, A no-go theorem for regularizing chiral fermions. Phys. Lett. B **105**(2–3), 219–223 (1981)
28. H.B. Nielsen, M. Ninomiya, Absence of neutrinos on a lattice. Nucl. Phys. B **185**(1), 20–40 (1981)
29. H.B. Nielsen, M. Ninomiya, Absence of neutrinos on a lattice. Nucl. Phys. B **193**(1), 173–194 (1981)
30. L. Susskind, Lattice fermions. Phys. Rev. D **16**, 3031–3039 (1977)
31. D. Horn, Finite matrix models with continuous local gauge invariance. Phys. Lett. B **100**(2), 149–151 (1981)
32. P. Orland, D. Rohrlich, Lattice gauge magnets: local isospin from spin. Nucl. Phys. B **338**(3), 647–672 (1990)
33. P. Orland, Exact solution of a quantum gauge magnet in 2+ 1 dimensions. Nucl. Phys. B **372**(3), 635–653 (1992)
34. S. Chandrasekharan, U.-J. Wiese, Quantum link models: a discrete approach to gauge theories. Nucl. Phys. B **492**(1–2), 455–471 (1997)
35. R. Brower, S. Chandrasekharan, U.-J. Wiese, Qcd as a quantum link model. Phys. Rev. D **60**(9), 094502 (1999)
36. R. Brower, S. Chandrasekharan, S. Riederer, U.-J. Wiese, D-theory: field quantization by dimensional reduction of discrete variables. Nucl. Phys. B **693**(1), 149–175 (2004)
37. Y. Aharonov, D. Bohm, Significance of electromagnetic potentials in the quantum theory. Phys. Rev. **115**(3), 485 (1959)
38. M.V. Berry, Quantal phase factors accompanying adiabatic changes, in *Proceedings of the Royal Society of London A: Mathematical, Physical and Engineering Sciences*, vol. 392 (The Royal Society, 1984), pp. 45–57
39. R. Dum, M. Olshanii, Gauge structures in atom-laser interaction: Bloch oscillations in a dark lattice. Phys. Rev. Lett. **76**(11), 1788 (1996)
40. P.M. Visser, G. Nienhuis, Geometric potentials for subrecoil dynamics. Phys. Rev. A **57**(6), 4581 (1998)
41. S.K. Dutta, B.K. Teo, G. Raithel, Tunneling dynamics and gauge potentials in optical lattices. Phys. Rev. Lett. **83**(10), 1934–1937 (1999)
42. Y.-J. Lin, R.L. Compton, K. Jiménez-García, J.V. Porto, I.B. Spielman, Synthetic magnetic fields for ultracold neutral atoms. Nature **462**(7273), 628–632 (2009)
43. N. Goldman, J. Dalibard, Periodically driven quantum systems: effective hamiltonians and engineered gauge fields. Phys. Rev. X **4**(3), 031027 (2014)
44. J. Struck, C. Ölschläger, M. Weinberg, P. Hauke, J. Simonet, A. Eckardt, M. Lewenstein, K. Sengstock, P. Windpassinger, Tunable gauge potential for neutral and spinless particles in driven optical lattices. Phys. Rev. Lett. **108**(22) (2012)
45. M. Aidelsburger, M. Atala, S. Nascimbène, S. Trotzky, Y.-A. Chen, I. Bloch, Experimental realization of strong effective magnetic fields in an optical lattice. Phys. Rev. Lett. **107**(25) (2011)
46. N. Goldman, J. Dalibard, M. Aidelsburger, N.R. Cooper, Periodically driven quantum matter: the case of resonant modulations. Phys. Rev. A **91**(3) (2015)
47. M. Aidelsburger, M. Lohse, C. Schweizer, M. Atala, J.T. Barreiro, S. Nascimbène, N.R. Cooper, I. Bloch, N. Goldman, Measuring the chern number of hofstadter bands with ultracold bosonic atoms. Nat. Phys. **11**(2), 162–166 (2014)

48. M. Aidelsburger, M. Atala, M. Lohse, J.T. Barreiro, B. Paredes, I. Bloch, Realization of the hofstadter hamiltonian with ultracold atoms in optical lattices. Phys. Rev. Lett. **111**(18) (2013)
49. H. Miyake, G.A. Siviloglou, C.J. Kennedy, W.C. Burton, W. Ketterle, Realizing the harper hamiltonian with laser-assisted tunneling in optical lattices. Phys. Rev. Lett. **111**(18) (2013)
50. H. Weimer, M. Müller, I. Lesanovsky, P. Zoller, H.P. Büchler, A rydberg quantum simulator. Nat. Phys. **6**(5), 382–388 (2010)
51. A.W. Glaetzle, M. Dalmonte, R. Nath, I. Rousochatzakis, R. Moessner, P. Zoller, Quantum spin-ice and dimer models with rydberg atoms. Phys. Rev. X **4**(4), 041037 (2014)
52. L. Tagliacozzo, A. Celi, A. Zamora, M. Lewenstein, Optical abelian lattice gauge theories. Ann. Phys. **330**, 160–191 (2013)
53. D. Banerjee, M. Dalmonte, M. Müller, E. Rico, P. Stebler, U.-J. Wiese, P. Zoller, Atomic quantum simulation of dynamical gauge fields coupled to fermionic matter: from string breaking to evolution after a quench. Phys. Rev. Lett. **109**(17) (2012)
54. S. Notarnicola, E. Ercolessi, P. Facchi, G. Marmo, S. Pascazio, F.V. Pepe, Discrete abelian gauge theories for quantum simulations of qed. J. Phys. A Math. Theor. **48**(30), 30FT01 (2015)
55. V. Kasper, F. Hebenstreit, M.K. Oberthaler, J. Berges, Schwinger pair production with ultracold atoms. Phys. Lett. B **760**, 742–746 (2016)
56. E. Kapit, E. Mueller, Optical-lattice hamiltonians for relativistic quantum electrodynamics. Phys. Rev. A **83**(3), 033625 (2011)
57. E. Zohar, J.I. Cirac, B. Reznik, Simulating (2+ 1)-dimensional lattice qed with dynamical matter using ultracold atoms. Phys. Rev. Lett. **110**(5), 055302 (2013)
58. A. Bazavov, Y. Meurice, S.-W. Tsai, J. Unmuth-Yockey, J. Zhang, Gauge-invariant implementation of the abelian-higgs model on optical lattices. Phys. Rev. D **92**(7), 076003 (2015)
59. E. Zohar, J.I. Cirac, B. Reznik, Simulating compact quantum electrodynamics with ultracold atoms: probing confinement and nonperturbative effects. Phys. Rev. Lett. **109**, 125302 (2012)
60. S. Tewari, V.W. Scarola, T. Senthil, S.D. Sarma, Emergence of artificial photons in an optical lattice. Phys. Rev. Lett. **97**(20), 200401 (2006)
61. L. Tagliacozzo, A. Celi, P. Orland, M.W. Mitchell, M. Lewenstein, Simulation of non-abelian gauge theories with optical lattices. Nat. Commun. **4**, 2615 (2013)
62. D. Banerjee, M. Bögli, M. Dalmonte, E. Rico, P. Stebler, U.-J. Wiese, P. Zoller, Atomic quantum simulation of $u(n)$ and $su(n)$ non-abelian lattice gauge theories. Phys. Rev. Lett. **110**(12), 125303 (2013)
63. K. Stannigel, P. Hauke, D. Marcos, M. Hafezi, S. Diehl, M. Dalmonte, P. Zoller, Constrained dynamics via the zeno effect in quantum simulation: implementing non-abelian lattice gauge theories with cold atoms. Phys. Rev. Lett. **112**(12), 120406 (2014)
64. E. Zohar, J.I. Cirac, B. Reznik, Cold-atom quantum simulator for $su(2)$ yang-mills lattice gauge theory. Phys. Rev. Lett. **110**(12) (2013)
65. E. Zohar, J.I. Cirac, B. Reznik, Quantum simulations of gauge theories with ultracold atoms: local gauge invariance from angular-momentum conservation. Phys. Rev. A **88**(2) (2013)
66. P. Hauke, D. Marcos, M. Dalmonte, P. Zoller, Quantum simulation of a lattice schwinger model in a chain of trapped ions. Phys. Rev. X **3**(4), 041018 (2013)
67. D. Marcos, P. Rabl, E. Rico, P. Zoller, Superconducting circuits for quantum simulation of dynamical gauge fields. Phys. Rev. Lett. **111**(11), 110504 (2013)
68. D. Marcos, P. Widmer, E. Rico, M. Hafezi, P. Rabl, U.-J. Wiese, P. Zoller, Two-dimensional lattice gauge theories with superconducting quantum circuits. Ann. Phys. **351**, 634–654 (2014)
69. A. Mezzacapo, E. Rico, C. Sabín, I.L. Egusquiza, L. Lamata, E. Solano, Non-abelian su (2) lattice gauge theories in superconducting circuits. Phys. Rev. Lett. **115**(24), 240502 (2015)
70. B. Douçot, L.B. Ioffe, J. Vidal, Discrete non-abelian gauge theories in josephson-junction arrays and quantum computation. Phys. Rev. B **69**, 214501 (2004)
71. C. Laflamme, W. Evans, M. Dalmonte, U. Gerber, H. Mejía-Díaz, W. Bietenholz, U.-J. Wiese, P. Zoller, Proposal for the quantum simulation of the $cp(2)$ model on optical lattices (2015). arXiv:1510.08492
72. C. Laflamme, W. Evans, M. Dalmonte, U. Gerber, H. Mejía-Díaz, W. Bietenholz, U.-J. Wiese, P. Zoller, cp quantum field theories with alkaline-earth atoms in optical lattices. Ann. Phys. **370**, 117–127 (2016)

73. E. Zohar, A. Farace, B. Reznik, J.I. Cirac, Digital quantum simulation of $\mathbb{Z}_2$ lattice gauge theories with dynamical fermionic matter. Phys. Rev. Lett. **118**(7) (2017)
74. E. Zohar, A. Farace, B. Reznik, J.I. Cirac, Digital lattice gauge theories. Phys. Rev. A **95**(2) (2017)
75. G.K. Brennen, G. Pupillo, E. Rico, T.M. Stace, D. Vodola, Loops and strings in a superconducting lattice gauge simulator. Phys. Rev. Lett. **117**(24) (2016)
76. A.S. Dehkharghani, E. Rico, N.T. Zinner, A. Negretti, Quantum simulation of abelian lattice gauge theories via state-dependent hopping. Phys. Rev. A **96**, 043611 (2017)
77. A. d'Adda, M. Lüscher, P. Di Vecchia, A 1n expandable series of non-linear σ models with instantons. Nucl. Phys. B **146**(1), 63–76 (1978)
78. H. Eichenherr, $su(n)$ invariant non-linear σ models. Nucl. Phys. B **146**(1), 215–223 (1978)
79. L. Tagliacozzo, A. Celi, M. Lewenstein, Tensor networks for lattice gauge theories with continuous groups. Phys. Rev. X **4**(4) (2014)
80. M. Mathur, Harmonic oscillator pre-potentials in $su(2)$ lattice gauge theory. J. Phys. A Math. Gen. **38**(46), 10015 (2005)
81. R. Anishetty, M. Mathur, I. Raychowdhury, Prepotential formulation of $su(3)$ lattice gauge theory. J. Phys. A Math. Theor. **43**(3), 035403 (2009)
82. E. Zohar, M. Burrello, Building projected entangled pair states with a local gauge symmetry. New J. Phys. **18**(4), 043008 (2016)
83. Y. Kuno, S. Sakane, K. Kasamatsu, I. Ichinose, T. Matsui, Atomic quantum simulation of a three-dimensional $u(1)$ gauge-higgs model. Phys. Rev. A **94**(6) (2016)
84. K. Kasamatsu, I. Ichinose, T. Matsui, Atomic quantum simulation of the lattice gauge-higgs model: Higgs couplings and emergence of exact local gauge symmetry. Phys. Rev. Lett. **111**(11) (2013)
85. Y. Kuno, S. Sakane, K. Kasamatsu, I. Ichinose, T. Matsui, Quantum simulation of $(1 + 1)$-dimensional $u(1)$ gauge-higgs model on a lattice by cold bose gases. Phys. Rev. D **95**(9) (2017)
86. D. González-Cuadra, E. Zohar, J.I. Cirac, Quantum simulation of the abelian-higgs lattice gauge theory with ultracold atoms. New J. Phys. **19**(6), 063038 (2017)
87. P. Facchi, S. Pascazio, Quantum zeno subspaces. Phys. Rev. Lett. **89**(8), 080401 (2002)
88. S. Lloyd, Universal quantum simulators. Science, 1073–1078 (1996)
89. C. Muschik, M. Heyl, E.A. Martinez, T. Monz, P. Schindler, B. Vogell, M. Dalmonte, P. Hauke, R. Blatt, P. Zoller, $u(1)$ wilson lattice gauge theories in digital quantum simulators. New J. Phys. (2017)
90. E. Zohar, B. Reznik, Confinement and lattice quantum-electrodynamic electric flux tubes simulated with ultracold atoms. Phys. Rev. Lett. **107**(27) (2011)
91. V. Kasper, F. Hebenstreit, F. Jendrzejewski, M.K. Oberthaler, J. Berges, Implementing quantum electrodynamics with ultracold atomic systems. New J. Phys. **19**(2), 023030 (2017)
92. E. Zohar, J.I. Cirac, B. Reznik, Quantum simulations of lattice gauge theories using ultracold atoms in optical lattices. Rep. Prog. Phys. **79**(1), 014401 (2015)
93. U.-J. Wiese, Ultracold quantum gases and lattice systems: quantum simulation of lattice gauge theories. Annalen der Physik **525**(10–11), 777–796 (2013)
94. T. Banks, L. Susskind, J. Kogut, Strong-coupling calculations of lattice gauge theories: $(1 + 1)$-dimensional exercises. Phys. Rev. D **13**(4), 1043–1053 (1976)
95. C.J. Hamer, Z. Weihong, J. Oitmaa, Series expansions for the massive schwinger model in hamiltonian lattice theory. Phys. Rev. D **56**(1), 55–67 (1997)

Chapter 9
The Remarkable BEC Dimer

David K. Campbell

Abstract We provide a short review (to appear in the proceedings of 'Strongly Coupled Field Theories for Condensed Matter and Quantum Information Theory', held in Natal 2015) of several articles by the author and his collaborators on aspects of the 'Bose–Hubbard" dimer (also known as the "Bose Josephson junction") as realized by a Bose Einstein Condensate (BEC) in a double-well optical trapping potential. We discuss the semi-classical model for this system (the Gross-Pitaevski (GP) equation) which is valid for a large number of atoms in the condensate and show that it is equivalent to an integrable classical dynamical system with multiple fixed points whose locations depend on the parameters of the system. We then discuss the full quantum mechanical model (the Bose–Hubbard Hamiltonian) both with and without dissipation. We demonstrate the surprising result that dissipation can actually enhance the condensate fraction and the EPR entanglement by driving the system nearer to the (classical) fixed points. We also show that the full quantum model predicts a tunneling phenomenon that is absent in the GP equation and that, in principle, should be observable in near-term experiments.

9.1 Introduction

The experimental realizations of Bose–Einstein Condensates (BECs) and degenerate Fermi gases (DFGs) revolutionized the field of atomic, molecular, and optical (AMO) physics, and when combined with the remarkable precision of magnetic and optical trapping—in particular, optical lattices—have allowed these ultra cold atomic systems to serve as highly tunable analogues of solid state systems, enabling the study of otherwise unobtainable exotic phases [1]. In this short contribution, we discuss several examples of the interesting physics that can be observed in one of the simplest BEC systems in an optical lattice: a condensate trapped in a double well

D. K. Campbell (✉)
Department of Physics, Boston University, 590 Commonwealth Avenue,
Boston, MA 02215, USA
e-mail: dkcampbe@bu.edu

A. Ferraz et al. (eds.), *Strongly Coupled Field Theories for Condensed Matter and Quantum Information Theory*, Springer Proceedings in Physics 239,
https://doi.org/10.1007/978-3-030-35473-2_9

optical potential, or a "BEC dimer." The results we discuss are taken primarily from several recent papers by the author and his collaborators [2, 4, 5], to which the reader is referred for the full details.

9.2 Modeling Bose Einstein Condensates (BEC)

A Bose Einstein Condensate (BEC) is typically described by a single quantum wave function $\Phi(x,t)$. In the context of cold atomic vapors trapped in an external potential (typically combined magnetic confinement and an optical potential) in the mean field/semi-classical approximation [1], $\Phi(x,t)$ obeys the "Gross-Pitaevskii" (GP) equation

$$i\hbar\frac{\partial\Phi(x,t)}{\partial t} = -\frac{\hbar^2}{2m}(\nabla)^2\Phi + [V_{\text{ext}} + g_0|\Phi|^2]\Phi. \tag{9.1}$$

Nonlinear dynamicists will recognize this immediately as a variant of the nonlinear Schrödinger (NLS) equation, here in three dimensional form and with an external potential. The interaction term describes s-wave scattering of the atoms [1]. In the one-dimensional case (which is realizable depending on the external potential) we expect (continuum) NLS solitons among the excitations.

9.2.1 *BECs in Optical Lattices*

When a BEC is placed in a one-dimensional optical lattice, the counter-propagating laser pulses create a standing wave that interacts via the AC stark effect with the neutral atoms in the condensate so that the BEC experiences a periodic potential of the form [1]

$$V_{ext}(\mathbf{r}) = U_L(x,y)\sin^2(2\pi z/\lambda) \tag{9.2}$$

Here $U_L(x,y)$ is the transverse confining potential, λ is the laser wavelength (typically of order 850 nm), and "z" is the direction of motion. Equation 9.2 describes a 1D periodic potential in which the BEC resides. Just as in a periodic solid state system, we can describe this system in terms of (localized) Wannier wave functions centered on the wells of the potential (but allowing tunneling between the wells). Expanding the GPE equation in (9.1) in terms of Wannier functions and focusing on (1D) motion in the z-direction leads to the Discrete Nonlinear Schrödinger (DNLS) equation for the Wannier functions. In the normalization we shall later use, one finds

$$i\frac{\partial\psi_n}{\partial t} = \Lambda|\psi_n|^2\psi_n - \frac{1}{2}[\psi_{n-1} + \psi_{n+1}]. \tag{9.3}$$

which comes from the Hamiltonian

$$\mathscr{H} = \sum_{n=1}^{M} [U|\psi_n|^4 + \mu_n|\psi_n|^2] - \frac{J}{2}\sum_{n=1}^{M-1}(\psi_n^*\psi_{n+1} + c.c.) \tag{9.4}$$

where $n = 1, \ldots, M$ is the index of the lattice site, $|\psi_n(t)|^2 \equiv N_n(t)$ is the mean number of bosons at site n (also referred to as the norm $N_n(t)$), $U = 4\pi\hbar^2 g_0 V_{\text{eff}}/m$ describes the interaction between two atoms at a single site (V_{eff} is the effective mode volume of each site, m is the atomic mass, and g_0 is the s-wave atomic scattering length), μ_n is the on-site chemical potential, and J is the tunneling amplitude. In (9.3), the "nonlinearity" parameter $\Lambda = 2UN/J$.

The above mean-field/semi-classical approach applies in the limit of a large number of atoms N in the total system. As N is reduced, quantum effects can become significant. In the case of a sufficiently deep optical lattice so that the lowest band is widely separated from the first excited band and the Wannier functions decay sufficiently rapidly that only nearest neighbor tunneling is allowed (already assumed in the above GPE equation), the BEC in the optical lattice is described by the (quantum) Bose–Hubbard Hamiltonian [1], which for a general lattice takes the form

$$H = -J\sum_{n=1}^{M-1}(\hat{a}_n^{\dagger}\hat{a}_{n+1} + \hat{a}_{n+1}^{\dagger}\hat{a}_n) + U\sum_{n=1}^{M}(\hat{a}_n^{\dagger 2}\hat{a}_n^2) \tag{9.5}$$

9.2.2 *The BEC Dimer or "Bose Josephson Junction"*

In the present article, we shall focus on the problem of a single coherent BEC in a "double well" optical potential. This is analogous to a two-site "dimer" system in a solid, and can therefore be called a "BEC dimer". However, it is also analogous to a Josephson junction with tunneling between two superconducting states (which can have different phases), and is therefore also known as a "Bosonic Josephson Junction" (BJJ) [2]. In this case, the parameters of the system are the total number of particles in the BEC (N) and the number in each of the two wells N_1 and $N_2 = N - N_1$, the phase difference between the condensate fractions in each of the wells ($\phi = \phi_1 - \phi_2$), the tunneling/resonant coupling between the two wells (J) and the interaction of the bosons within the condensate, U (Fig. 9.1).

Our study will focus on the differences between the mean field approximation (defined by the appropriate two-site DNLS/GP equation) and the fully quantum BH dimer. We shall focus on two key observables and follow them as functions of time. The first is the "purity" or condensate fraction, $c_t(\phi, z)$, which is defined as the maximum eigenvalue of the reduced single-particle density matrix

$$\rho = \frac{1}{N}\begin{bmatrix}\langle a_1^{\dagger}a_1\rangle & \langle a_1^{\dagger}a_2\rangle \\ \langle a_2^{\dagger}a_1\rangle & \langle a_2^{\dagger}a_2\rangle\end{bmatrix} \tag{9.6}$$

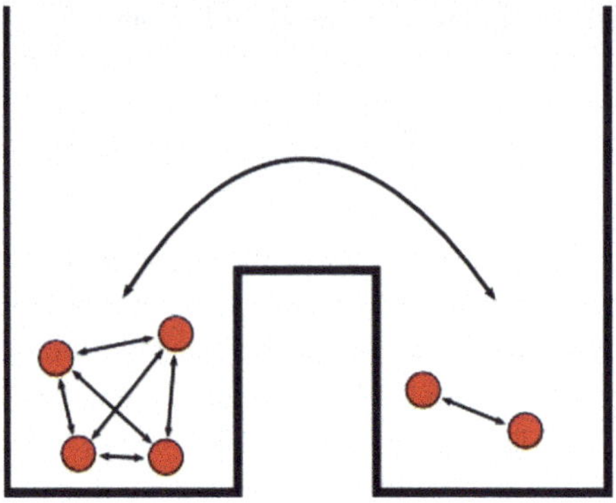

Fig. 9.1 Schematic of Bose–Hubbard Dimer showing the two separate parts of the condensate trapped in the two wells

The second is the Einstein–Podolsky–Rosen (EPR) entanglement, which is given by

$$E = |\langle a_1^\dagger a_2 \rangle|^2 - \langle a_1^\dagger a_1 a_2^\dagger a_2 \rangle . \tag{9.7}$$

For $E > 0$, the state is EPR entangled.

Our discussion is a much-reduced version of [2], to which the reader is referred for critical details and the complete set of references. We begin by studying the mean-field/ semiclassical GPE in what we term "the global phase space approach".

9.3 The Global Phase Space Approach

Introducing $z = N_1 - N_2$, one can show [2, 3], that the two-site DNLS/GP equation can be rewritten as an *integrable* classical Hamiltonian system with the Hamiltonian

$$H_{cl}(z, \phi) = \frac{\Lambda z^2}{2} - \sqrt{1 - z^2}\, \cos(\phi) \,, \tag{9.8}$$

with z and ϕ as conjugate variables. Recognizing that z plays the role of momentum, we see that the Hamiltonian in (9.8) looks like a pendulum equation but with a momentum-dependent length. The corresponding equations of motion are

$$\dot{z} = -\sqrt{1 - z^2} \sin \phi \tag{9.9}$$

and

$$\dot{\phi} = \Lambda z + \frac{z \cos \phi}{\sqrt{1 - z^2}} \tag{9.10}$$

These equations have fixed points that depend non-trivially on the nonlinearity parameter Λ. In particular, for $\Lambda < 1$, there are two stable fixed points at

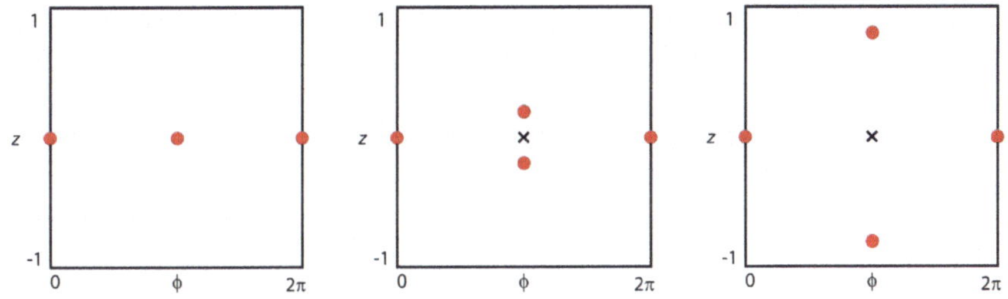

Fig. 9.2 The fixed points for $\Lambda < 1$, $\Lambda > 1$, and $\Lambda >> 1$

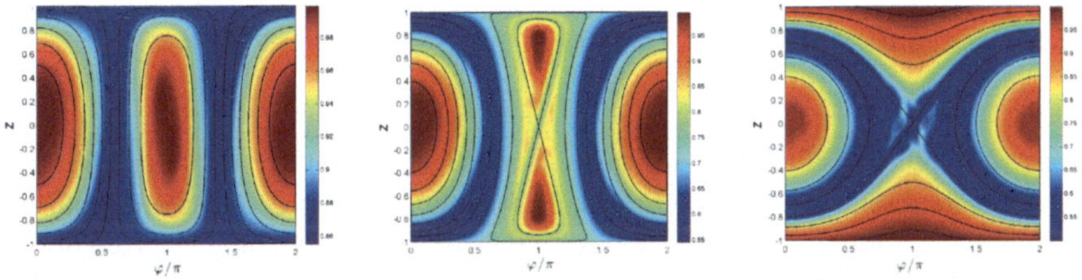

Fig. 9.3 The global phase space for (left panel) $\Lambda < 0.5$; (center panel) $\Lambda = 1.5$; (right panel) $\Lambda = 5$. The figures show the value of $c_t(\phi, z)$ at a given time after starting the system in a pure condensate (solid dark red in [2] for all values of (ϕ, z) at $t = 0$. (Color available in [2]: The regions around the classical fixed points are still dark red, indicating high condensate purity; they are separated by regions of intermediate purity from the low purity regions, shown in blue in [2]

$(z, \phi = (0, 0)$ and $(0, \pi))$, whereas for $\Lambda > 1$ there are three stable fixed points at $(z, \phi = (0, 0), (z_+, \pi)$, and $(z_-, \pi))$ where $z_+ = \frac{\sqrt{\Lambda^2-1}}{\Lambda}$ and $z_- = -\frac{\sqrt{\Lambda^2-1}}{\Lambda}$. Note that as $\Lambda \to \infty$, $z_+ \to 1$ and $z_- \to -1$. These fixed points are shown schematically in Fig. 9.2.

The classical trajectories follow the lines of constant energy $H_{cl}(z, \phi) = E$ and are determined by the initial position (ϕ_0, z_0) in global phase space and the dimensionless 'nonlinearity' parameter $\Lambda = 2UN/J$. The nature of the trajectories and the "global phase space" are shown for three distinct values of Λ in Fig. 9.3 above. In this figure, the classical orbits are shown by thin black lines. The actual figures, with the colored/distinct regions, reflect the results of a calculation that includes quantum features beyond the classical approximation, which we will describe below (see also [2]). First, we note that the semiclassical model provides a remarkably good fit to experimental data. This is shown in Fig. 9.4, which is taken from [6] and shows that experiments started with different values of (z_0, ϕ_0) "track" the semi-classical trajectories quite well for fairly long times. For the precise details, the reader is referred to [6].

Nonetheless, there are important quantum effects beyond the mean-field/semi-classical GPE model that affect experimental observables such as the condensate fraction and EPR entanglement. In particular the GPE implicitly assumes a pure BEC at all times. This would correspond to a single, solid (red) color in Fig. 9.3 at

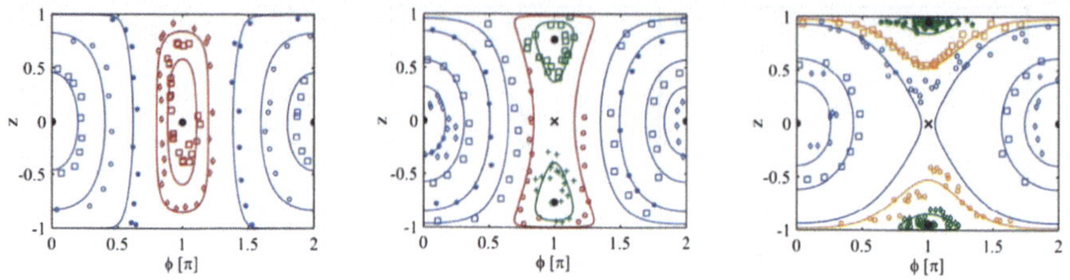

Fig. 9.4 Experimental results showing how the data follow very closely the predictions of the GPE in the global phase space. From left to right for $\Lambda = 0.78$, $\Lambda = 1.551$, and $\Lambda = 3.1$. The symbols show the experimental results at different times: each symbol is a different experiment. The ability to prepare an experiment at a given (ϕ_0, z_0) is remarkable. Data and figure from Zibold et al., PRL 105 204101 (2010)

time $t = 0$. What this figure actually represents is the value of the condensate fraction $c_t(z, \phi)$ at a subsequent time, where the c_t has remained high in certain regions but has decreased substantially in other regions. In the color versions of Fig. 9.3 (see online version of [4]), the red areas indicate high condensate fraction regions whereas blue is low condensate fraction, with green/yellow regions being intermediate condensate fraction. The reader should observe that the regions of highest condensate fraction, for all values of Λ shown, are around the classical fixed points, *including the self-trapped fixed points for* $\Lambda > 1$. The full explanation of this is beyond the scope of this short summary, but readers can find it in [4]. Here we content ourselves with a brief explanation of how the images in Fig. 9.3 are created (for full details, the reader should consult [2]). To capture quantum features beyond GPE approximation, including quantum mechanical spreading over time, one represents a quantum state $|\Psi\rangle$ by the Husimi function $Q(\phi, z) = |\langle\phi, z|\phi_0, z_0\rangle|^2$ (which is a (quantum) phase-space density) instead of a single classical trajectory. Here, $|\phi, z\rangle$ denotes an *atomic coherent state* which is nothing but a pure BEC. As discussed in [2], the dynamics of the Husimi function follows a classical Liouville equation with the Hamiltonian (9.8) plus quantum correction terms vanishing as $1/N$.

As shown by the color scale on the right side of Fig. 9.3, regions of red indicate high purity regions whereas blue is low purity, with green/yellow being intermediate.

9.4 Full Quantum Dynamics in the BJJ: The Bose Hubbard Hamitonian

For the two-well system, (9.5) reduces to the quantum Bose–Hubbard dimer equation

$$H = -J(\hat{a}_1^\dagger \hat{a}_2 + \hat{a}_2^\dagger \hat{a}_1) + U\left(\hat{a}_1^{\dagger 2}\hat{a}_1^2 + \hat{a}_2^{\dagger 2}\hat{a}_2^2\right). \tag{9.11}$$

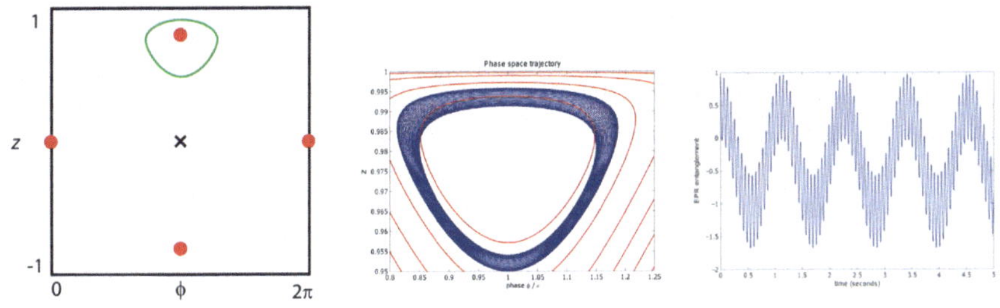

Fig. 9.5 The nature of quantum "orbits" near a classical fixed point in the self-trapped region of the BJJ. Left panel: A schematic graph of the original classical orbit in the GPS. Center panel: The corresponding quantum orbit plotted in GPS showing the "thickening" of the original classical orbit due to quantum effects described in the text. Right Panel: The EPR entanglement as a function of time showing clearly the existence of two frequencies. Note the systems is EPR entangled only for $E > 0$

For N bosons on the dimer, there are $N + 1$ states in the Hilbert space (think of the Fock space representation) rendering the solution to the quantum dynamics a straightforward matrix diagonalization problem. Importantly, as shown in [4] (to which readers are referred for full details), this dynamics can still be visualized in the GPS approach discussed above: namely, one starts with an initial pure BEC wave function of the form

$$|(z,\phi)\rangle = \frac{1}{\sqrt{N}}\left(\sqrt{(1+z)/2}\,\hat{a}_1^\dagger + \sqrt{(1-z)/2}\,\exp^{i\phi}\,\hat{a}_2^\dagger\right)^N |(0)\rangle \qquad (9.12)$$

and studies its time evolution as a function of the initial starting position in the GPS, (z_0, ϕ_0). The full quantum wave function at any given time is projected back onto the GPS, revealing the key features of the quantum dynamics. One finds ([4]) that instead of the classical line trajectories, "thick" trajectories, as shown in the center panel of Fig. 9.5, where the "thickness" results from a "quantum drift" motion perpendicular to the classical trajectory. In fact, the overall motion exhibits two clear frequencies: a "high frequency" motion corresponding to the classical trajectory and a "low frequency" motion, which can, at least in the vicinity of the classical fixed points, be understood in terms of beating among the three highest eigenstates of the Hamiltonian [4].

9.4.1 *Quantum "Orbits" Near Fixed Points: Dissipationless Case*

The GPS approach of [2] enables us to gain insight into the full quantum dynamics of entanglement and coherence in the BEC dimer. In [4], we found that the wells are entangled in the Einstein-Podolski-Rosen (EPR) sense *only* in the neighborhood of the classical fixed points and that the time dependence of the entanglement can be understood completely in terms of beats among three eigenstates of the BH Hamiltonian for the system. In [4] we found these eigenstates numerically (essentially to exact precision) but our analytical perturbative results in the limit of weak coupling between the wells were in excellent agreement with the (numerically) exact results. We found similar results for the dynamics of other observables, such as the well population imbalance and condensate fraction. We summarize these results briefly in this section. In Fig. 9.5, the left panel shows a schematic of the classical orbit around the fixed point z_+. The center panel shows an actual calculation of the quantum orbit (as projected onto the GPS), clearly indicating a "thickening" of the classical trajectory. The right panel demonstrates the existence of two distinct frequencies in this motion, reflected in the behavior of the EPR entanglement.

As discussed in [4], the higher/faster frequency is expected on the basis of the mean-field/GPE model. If one linearizes the equations of motion obtained from the Hamiltonian of (9.5) about the fixed point, one finds

$$f_{\mathrm{MF}} = \frac{\sqrt{\Lambda^2 - 1}}{\pi} \frac{J}{\hbar}. \tag{9.13}$$

As shown in the center panel of Fig. 9.6, this fit works remarkably well over a large range of Λ. The lower/slow frequency motion, which leads to the "thickening" of the orbit, cannot be captured within the mean-field/GPE approach, but we showed in [4] that this f_{SLOW} can effectively be interpreted in terms of a quantum revival (which in the limit $J \to 0$ would yield $f_{\mathrm{SLOW}} = \frac{U}{\pi\hbar}$). The fit for $J \neq 0$ is shown in the right panel of Fig. 9.6 and is quantitatively accurate again over a range of Λ such that the motion is near the classical fixed point.

9.4.2 *Quantum "Orbits" Near Fixed Point: The Dissipative Case*

As discussed in detail in [4], atoms can be removed for the double-well optical potential with single site precision, using a focused electron beam or strong resonant laser blasts. This removal adds (controlled) dissipation to the problem and drives the system closer to the classical fixed point. Based on our observation that the purity is highest near the classical fixed points, we expect that this (temporary) dissipation will lead to greater entanglement in the system: in short, the counterintuitive result of "dissipation-induced coherence." Indeed, this is the result we found in [4] and summarized in Fig. 9.7. The left panel again illustrates the schematic motion of the classical trajectory when the system is subjected to a burst of dissipation which is then removed: the classical orbit moves closer to the fixed point. The central panel shows the two corresponding "thickened" quantum orbits. Finally, the right panel shows that the EPR entanglement is driven to values $E > 0$ by the burst of dissipation and remains there in the evolution afterwards.

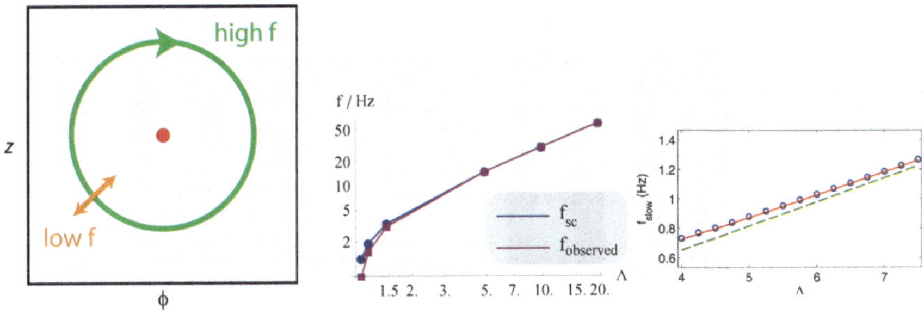

Fig. 9.6 Comparison of numerical results versus theoretical predictions for the high and low frequency oscillations observed in the quantum motion near a classical fixed point in the BJJ. Left panel: A schematic illustration showing the high frequency arising from the classical motion and the low frequency originating from quantum revivals. Center panel: Comparison of the semiclassical prediction of the high frequency oscillations observed in the EPR entanglement with the exact numerical results of the full BHH calculation; even down to small Λ, the agreement is remarkably good; Right panel: The low frequency oscillations near the fixed point as a function of Λ. The numerically exact results (circles) are well described by a second-order perturbation results (see [4] for full details) whereas the zeroth-order perturbation theory (dashed line) slightly underestimates the frequency. Adapted from [4] to which readers are referred for full details

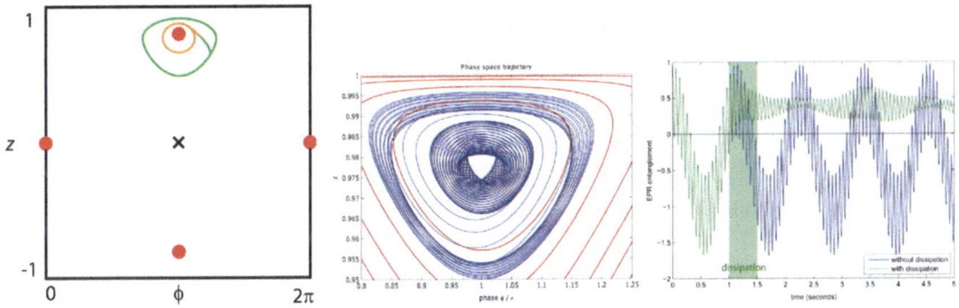

Fig. 9.7 Dissipation induced coherence: Left panel: Schematic depiction showing that a short pulse of dissipation drives the classical (and quantum) motion closer to the classical fixed points where the condensate fraction and EPR entanglement are expected to be highest; Center panel: The actual quantum orbit over the time period of the calculation. The motion remains in the over oval until the dissipation is turned on, and then moves to the inner oval under the dissipation, remaining there after the dissipation is turned off. Right panel: the EPR entanglement versus time. The shaded vertical region between times 1.0 and 1.5 s indicates that time interval in which the dissipation is on. Notice the greatly enhanced EPR entanglement after the dissipation. Adapted from [4] to which the reader is referred for full details

9.4.3 Quantum Tunneling Between Self-trapped Fixed Points

In this section we explore another quantum phenomenon in the BEC dimer that goes beyond the GPE model: namely, quantum tunneling between the two "self-trapped" regions—that is, the regions that contain z_+ and z_-—that exist within the mean-field/GPE model, as depicted in the center and right panels of Fig. 9.3. Within the mean field approach, these regions are divided by a classical separatrix and

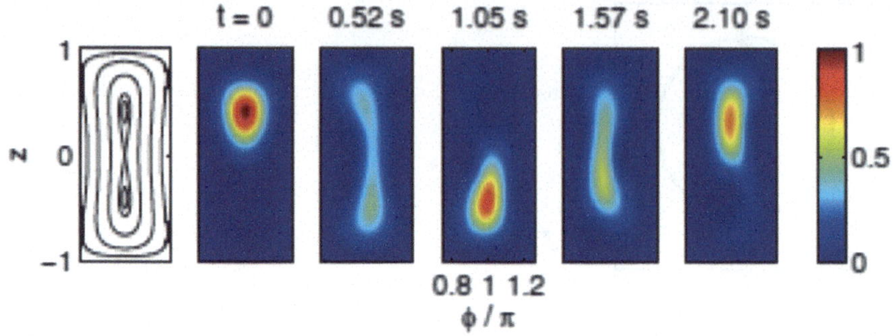

Fig. 9.8 (Color available in online version of [5]). Tunneling between the two fixed points as a function of time as illustrated by the Husimi function depicted at five times spaced by one quarter of the tunneling period expected from the two-state model. The relevant parameters are $N = 500$ atoms, $\Lambda = 1.025$, and $U = 2\pi(x)0.063$ Hz. Adapted from [5]

motions within them are distinct and uncoupled. Does this remain true in the full quantum description given by the Hamiltonian Equation 9.5? In [5] we investigated and answered this question and showed that in the full quantum model, for values of Λ slightly greater than 1 and moderate values of N, it is possible for the BEC to tunnel back and forth between the two self-trapped regions, just as in elementary quantum mechanics a particle in a double well potential whose wave function is initially localized in one well tunnels back and forth between the two wells. While this seems intuitively clear, the case of the BEC dimer is much more complicated than this simple quantum analogy suggests, and the semi-classical calculations to establish this tunneling are quite complicated [5]. For this reason, we refer readers interested in the details to [5] and to the video showing the dynamics of the tunneling in the supplemental material to [5]. Here we simply display two figures that illustrate the key results. The first figure (Fig. 9.8) depicts the Husimi function, $Q(\phi, z) = |\langle \phi, z|\phi_0, z_0\rangle|^2$, at several times during the tunneling process for $\Lambda = 1.1$ and the other parameters described by in the figure caption. In Fig. 9.9 the left panel shows the tunneling frequency as a function of Λ for different numbers of atoms (N) in the condensate. As expected, as N is decreased, the quantum effects are increased and tunneling becomes more prominent. The right panel In Fig. 9.9 shows the expected tunneling frequency as a function of Λ for the experimental parameters of [6]. While this suggests that one should be able to observe the tunneling in current or near-term experiments, the level of control on the parameter Λ and the length of time that one can control the condensate pose substantial challenges to the experimental observations. Nonetheless, given the ingenuity of our experimental colleagues, we are hopeful that this theoretical prediction will be confirmed in the future.

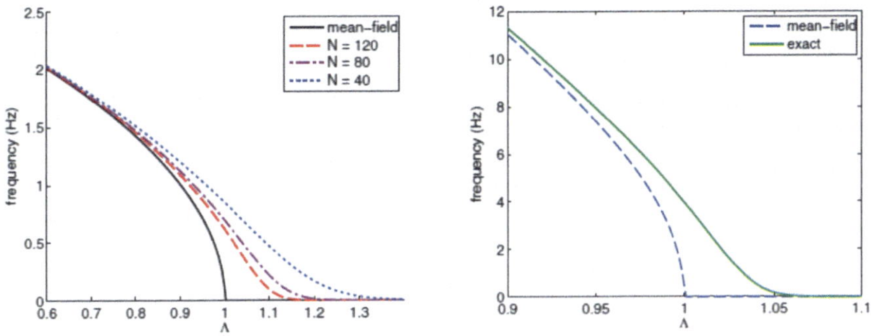

Fig. 9.9 Left Panel: The tunneling frequency between the two classical self-trapped regions of a BJJ as a function of Λ for different numbers (N) of atoms in the condensate. The mean field result of no tunneling is approached as $N \to \infty$ but for any finite N there is a non-zero tunneling frequency for any Λ and N. In all plots, $J = 10\,\text{Hz}$. Right Panel: The frequency of tunneling between the classical BJJ fixed points versus Λ for $U = 2\pi(x)0.063\,\text{Hz}$ and $N = 500$, which correspond to the experimental parameters of [6]. Both figures adapted from [5]

9.5 Conclusions

In conclusion, we believe that this short summary of our results has justified our claim that the BEC dimer is indeed "remarkable" and worthy of continued theoretical and experimental scrutiny. Importantly, the quantum effects we have found in the dimer's behavior are not only interesting in their own right but also suggest a new approach to understanding the less numerically tractable behavior of BECs in multi-well optical lattices. The self-trapping fixed points of the dimer are analogous to discrete breathers observed in larger systems in the GPE approach (see, e.g., [7, 8]). It is likely that these entities will serve as regions in the high dimensional phase space where entanglement and condensate purity are enhanced and that targeted dissipation can further enhance their coherence. There remain many exciting questions to be studied in the future.

Acknowledgements The author deeply indebted to his colleagues who worked with him on this research: Jerome Dorignac, Holger Hennig, Ted Pudlik, and Dirk Witthaut. None of these results would have been achieved without their contributions. He is also grateful to Pasquale Sodano for inviting him to the October 2015 conference on "Strongly Coupled Field Theories for Condensed Matter and Quantum Information Theory", held at the International Institute for Physics in Natal, Brazil.

References

1. For a thorough review of the field of ultra-cold atoms, see Bloch, I, Dalibard, J., Zwenger, W.: Many-body physics with ultracold gases. Rev. Mod. Phys. **80**, 885–964 (2008)
2. H. Hennig, D. Witthaut, D.K. Campbell, Global phase space of coherence and entanglement in a double-well Bose-Einstein condensate. Phys. Rev. A **86**, 051604(R) (2012)

3. S. Raghavan, A. Smerzi, S. Fantoni, S.R. Shenoy, Coherent oscillations between two weakly coupled Bose-Einstein condensates: Josephson effects, π oscillations, and macroscopic quantum self-trapping. Phys. Rev. A **59**, 620 (1999)
4. T. Pudlik, H. Hennig, D. WItthaut, G.K. Campbell, Dynamics of entanglement in a dissipative Bose-Hubbard dimer. Phys. Rev. A **88**, 063606 (2013)
5. T. Pudlik, H. Hennig, D. Witthaut, D.K. Campbell, Tunneling in the self-trapped regime of a two-well Bose-Einstein condensate. Phys. Rev. A **90**, 053610 (2014)
6. T. Zibold, C. Nicklas, C. Gross, M.K. Oberthaler, Classical bifurcation at the transition from Rabi to Josephson dynamics. Phys. Rev. Lett. **105**, 204101 (2010)
7. S. Mossmann, C. Jung, Semiclassical approach to Bose-Einstein condensates in a triple well potential. Phys. Rev. A **74**, 033601 (2006)
8. H. Hennig, J. Dorignac, D.K. Campbell, Transfer of Bose-Einstein condensates through discrete breathers in an optical lattice. Phys. Rev. A **82**, 053604 (2010)

Chapter 10
Quantized Vortex Lines in BECs with a Generalized Equation of State

Tommaso Macrì

Abstract Vortices are of fundamental importance for the understanding dynamical and thermodynamical properties of quantum fluids. In this chapter we focus on the effects of the corrections to the equation of state of a BEC on a vortex state. We consider the case of a condensate with short-range interactions in the absence and in the presence of an external harmonic confinement. We derive the equations obeyed by a static vortex line in both configurations and solve them both numerically and by a convenient ansatz for the wavefunction of the vortex state originally introduced by Fetter. Interesting generalization of this work include the study of the static properties of vortex lines in mixtures of BECs and condensates with non-local interactions, such as dipolar atoms.

10.1 Introduction

Quantized vortices have been observed in a variety of superfluid systems, from ^{4}He to condensates of alkali-metal bosons and ultracold Fermi gases along the BEC-BCS crossover. Vortices are a direct manifestation of the genuine quantum behavior of superfluid systems. A well known example is superfluid helium, which has been extensively studied over the last century. Ultracold quantum gases offer the interesting possibility to investigate vortex properties in a very different regime in terms of particle numbers, interaction strength, and range [1, 2] with either bosonic [3] or fermionic [4] atoms. Systems with long-range interactions, such as dipolar condensates may also display vortices with unique properties as a consequence of the long range and angular dependence of the interactions [5–8]. Excitations of vortex states are also important for the study of the stability of such configurations. As an example, for a three-dimensional condensate with a vortex line and in the presence of a periodic potential, the spectrum of transverse modes may display a rotonlike

T. Macrì (✉)
Departamento de Física Teórica e Experimental and International Institute of Physics, Universidade Federal do Rio Grande do Norte, Natal, RN 59072-970, Brazil
e-mail: macri@fisica.ufrn.br

A. Ferraz et al. (eds.), *Strongly Coupled Field Theories for Condensed Matter and Quantum Information Theory*, Springer Proceedings in Physics 239,
https://doi.org/10.1007/978-3-030-35473-2_10

minimum [9, 10], which destabilizes the straight vortex leading to a helical or a snakelike configuration [11, 12].

Modeling of superfluids of ultracold atomic gases at zero temperature are generally based on mean-field approximations which describe experiments [13] accurately. They span from analytic treatments and variational approaches to full numerical solutions. Terms in the equation of state beyond such mean-field description have been measured for strongly interacting Bose gases [14] and for ultracold fermions along the BCS-BEC crossover [15] and compared with ab initio quantum Monte Carlo calculations. The observation of self-bound droplets both in dipolar condensates [16–22], and in Bose-Bose mixtures [23, 24] established the importance of the fundamental role of quantum fluctuations in ultracold atomic systems [25–31].

In this chapter we will focus on the effect of beyond-mean field corrections to the equation of state for a single vortex line both in the absence and in the presence of an external harmonic trapping potential. We employ the local density approximation to analyze the effect of the beyond-mean field term of a single component Bose–Einstein condensate with contact interactions. After deriving the equation obeyed by a vortex with unitary circulation, we solve it numerically and by the use of a variational ansatz first proposed by Fetter [32, 33]. We find that the beyond-mean field term affects the width of the vortex core even for relatively small interactions. In the case of a trapped BEC, we will emphasize the regime of the so-called Thomas–Fermi limit [2, 34], where the contribution from the kinetic energy term becomes negligible compared to trapping and interaction energy. We derive analytical results for the energy density as well as for the critical angular velocity for the formation of a vortex in an axisymmetric trap.

This chapter is organized as follows: in Sect. 10.2 we derive the energy density of a condensate with contact interactions at zero temperature including beyond mean field corrections in the equation of state in the presence of a vortex line with unit circulation. Then, in Sect. 10.3 we make use of the results of Sect. 10.2 to extend them to the case of a trapped three-dimensional BEC within the so-called Thomas–Fermi approximation. This result is useful to obtain the critical angular velocity for the formation of a vortex. Finally, in Sect. 10.4, we draw some conclusions and list possible extensions of this work.

10.2 Vortex Line in a Three Dimensional BEC

In this section we derive the energy density of a vortex line in a three-dimensional BEC in a uniform background. We consider a weakly interacting condensate with average density n. The interactions we consider are short range, and the strength is parametrized by the scattering length a. We begin by writing the mean-field and the first beyond mean-field contribution, the well known Lee-Huang-Yang (LHY) correction [35], to the energy of a homogeneous weakly repulsive Bose gas reads in the local density approximation

$$E = \int \mathrm{d}\mathbf{r} \left[\frac{\hbar^2}{2m} |\nabla \psi|^2 + V(\mathbf{r}) |\psi|^2 + \frac{g}{2} |\psi|^4 \left(1 + \frac{128}{15\pi^{1/2}} |\psi| a_s^{3/2} \right) \right]. \quad (10.1)$$

The last term in the energy functional (10.1) is the celebrated Lee-Huang-Yang correction to the equation of state of a weakly interacting Bose gas [35]. The term $V(\mathbf{r})$ is an external (trapping) potential.

A generalized time-independent Gross-Pitaevskii equation obeyed by $\psi(\mathbf{r})$ can be obtained by the stationarity condition of the energy functional, under the constraint of the conservation of the total number of particles

$$\frac{\partial (E - \mu N)}{\partial \psi^*} = 0, \quad (10.2)$$

where μ is the chemical potential, resulting in

$$\left(-\frac{\hbar^2}{2m} \nabla^2 + V(\mathbf{r}) + g|\psi|^2 + \frac{32}{3\pi^{1/2}} |\psi|^3 a_s^{3/2} \right) \psi = \mu \psi. \quad (10.3)$$

For a uniform system the chemical potential reads

$$\mu = \mu_0 \left(1 + \frac{32}{3\pi^{1/2}} (n\, a_s^3)^{1/2} \right), \quad (10.4)$$

where $\mu_0 = g\, n$, and $n = |\psi|^2$ is the density of the system.

The presence of a vortex line centered at the origin, allows us to write the wave function in the form of a phase with circulation ℓ and an amplitude $f(\rho)$, $\psi(\mathbf{r}) = f(\rho) e^{i\ell\phi}$ [33, 36], where we employed cylindrical coordinates (ρ, z, ϕ). In the following we will focus on vortices with unit circulation $\ell = 1$. Substituting the expression for $\psi(\mathbf{r})$ into (10.3) in cylindrical coordinates we get

$$\left(-\frac{\hbar^2}{2m} \frac{1}{\rho} \frac{\mathrm{d}}{\mathrm{d}\rho} \left(\rho \frac{\mathrm{d}}{\mathrm{d}\rho} \right) + \frac{\hbar^2}{2m} \frac{1}{\rho^2} - \frac{\hbar^2}{2m} \frac{\mathrm{d}^2}{\mathrm{d}z^2} + V(\rho, z) + g\, f^2 + \frac{32}{3\pi^{1/2}} f^3 a_s^{3/2} \right) f = \mu f. \quad (10.5)$$

For the rest of this section we will assume that the external potential is absent, $V(\mathbf{r}) = 0$. Harmonic trapping potential will be reintroduced in Sect. 10.3 to discuss the energy of a vortex state inside a confined BEC. Equation (10.5) can be conveniently rewritten in terms of a rescaled units. To do so, we use the healing length ξ as a unit of length, defined by

$$\frac{\hbar^2}{2m\xi^2} = g\, n, \quad (10.6)$$

and the radial wavefunction $f(\rho)$ is rescaled by the asymptotic value f_0, $\chi = \frac{f}{f_0}$.

In rescaled units GPE reads

$$-\frac{1}{x}\frac{\mathrm{d}}{\mathrm{d}x}\left(x\frac{\mathrm{d}\chi}{\mathrm{d}x}\right)+\frac{\chi}{x^2}+\chi^3-\chi+\alpha(\chi^3-1)=0. \tag{10.7}$$

For clarity of notation we redefine the prefactor of the beyond mean-field term $\alpha = \frac{32}{3\sqrt{\pi}}\left(f_0^2 a^3\right)^{1/2} = \frac{32}{3\sqrt{\pi}}\left(na^3\right)^{1/2}$. Upon inspection of (10.7) it is clear that $\chi(\infty) = 1$, i.e. for large distances one recovers the density of the uniform state.

The energy per unit length on a disc of radius D becomes

$$\varepsilon = \frac{\pi\hbar^2 n}{m}\int_0^D \mathrm{d}x\, 2\pi x\left[\left(\frac{\mathrm{d}\chi}{\mathrm{d}x}\right)^2+\frac{\chi^2}{x^2}+\frac{\chi^4}{2}+\frac{2\alpha}{5}\chi^5\right]. \tag{10.8}$$

To compute the energy associated to a vortex state we subtract the energy of a uniform system with the same particle number $\nu = \bar{n}\pi D^2$, where $\bar{n}$ is the density of the uniform system [33]. Writing

$$\nu = \int_0^D \mathrm{d}\rho\rho f^2 = \pi D^2 f_0^2 - \int_0^D \mathrm{d}\rho\rho 2\pi(f_0^2 - f^2), \tag{10.9}$$

and substituting this expression into the equation for the energy density of the uniform system we obtain

$$\begin{aligned}\varepsilon_0 &= \bar{n}^2\tfrac{g}{2}\left(1+\tfrac{128}{15\pi^{1/2}}(\bar{n}a^3)^{1/2}\right)\pi D^2\\ &= \pi D^2\tfrac{g}{2}f_0^4\left(1-\tfrac{1}{\pi D^2 f_0^2}\textstyle\int_0^D \mathrm{d}\rho\rho 2\pi(f_0^2-f^2)\right)^2\times\\ &\left[1+\tfrac{128}{15\pi^{1/2}}a^{3/2}\left(\tfrac{1}{\pi D^2 f_0^2}\textstyle\int_0^D \mathrm{d}\rho\rho 2\pi(f_0^2-f^2)\right)^{1/2}\right]\end{aligned} \tag{10.10}$$

We notice that the integrals of the difference of the radial densities are proportional to the square of the healing length ξ

$$\int_0^D \mathrm{d}\rho\rho 2\pi(f_0^2 - f^2) \approx \xi^2 f_0^2. \tag{10.11}$$

Then the integrals of the type

$$\begin{aligned}\left(1-\tfrac{1}{\pi D^2 f_0^2}\textstyle\int_0^D \mathrm{d}\rho\rho 2\pi(f_0^2-f^2)\right)^\beta &= 1-\tfrac{\beta}{\pi D^2 f_0^2}\textstyle\int_0^D \mathrm{d}\rho\rho 2\pi(f_0^2-f^2)\\ &+\mathcal{O}((\xi/D)^4),\end{aligned} \tag{10.12}$$

where we neglected the term proportional to $(\xi/D)^4$).

Equation (10.10) can then be rewritten as

$$
\begin{aligned}
\varepsilon_0 &= \tfrac{1}{2} g\pi D^2 f_0^4 \Big[1 + \tfrac{128}{15\pi^{1/2}} (f_0 a^{3/2}) \Big(1 - \tfrac{1}{2\pi D^2 f_0^2} \int_0^D \mathrm{d}\rho\rho 2\pi (f_0^2 - f^2) \Big) \\
&\quad - g f_0^2 \Big(\int_0^D \mathrm{d}\rho\rho 2\pi (f_0^2 - f^2) \Big) \Big(1 + \tfrac{128}{15\pi^{1/2}} (f_0 a^{3/2}) \Big) \Big] \\
&= \mu_0 n \Big[\tfrac{\pi D^2}{2} \Big(1 + \tfrac{4\alpha}{5} \Big(1 - \tfrac{\xi^2}{2\pi D^2} \int_0^{D/\xi} x 2\pi (1 - \chi^2) \Big) \Big) \\
&\quad - \xi^2 \big(1 + \tfrac{4\alpha}{5} \big) \int_0^{D/\xi} x 2\pi (1 - \chi^2) \Big] .
\end{aligned}
\tag{10.13}
$$

To compute the energy of the vortex ε_V we substract ε_0 from the bare energy ε in (10.8)

$$
\begin{aligned}
\varepsilon_V &= \varepsilon - \varepsilon_0 \\
&= 2\pi \tfrac{\hbar^2 n}{2m} \int_0^{D/\xi} \mathrm{d}x x \Big[\Big(\tfrac{\mathrm{d}\chi}{\mathrm{d}x} \Big)^2 + \tfrac{\chi^2}{x^2} + \tfrac{\chi^4}{2} + \tfrac{2\alpha}{5} \chi^5 \\
&\quad + \big(\tfrac{1}{2} - \chi^2 \big) - \tfrac{2}{5}\alpha + \alpha \left(1 - \chi^2 \right) \Big]
\end{aligned}
\tag{10.14}
$$

After elementary manipulations one obtains

$$
\varepsilon_V = \tfrac{\pi\hbar^2 n}{m} \int_0^{D/\xi} \mathrm{d}x\, x \Big[\Big(\tfrac{\mathrm{d}\chi}{\mathrm{d}x} \Big)^2 + \tfrac{\chi^2}{x^2} + \tfrac{1}{2}(1-\chi^2)^2 + \tfrac{\alpha}{5}(2\chi^5 - 5\chi^2 + 3) \Big] .
\tag{10.15}
$$

We note that for vanishing $\alpha = 0$ we recover the well-known energy density by Ginzburg and Pitaevskii [37]. Numerical solution of the minimum energy density from equation (10.15) for the case $\alpha = 0$ predicts $\varepsilon_v = \frac{\pi\hbar^2 n}{m} \log \left(1.464 \frac{D}{\xi} \right)$.

We now provide a variational estimate, using the same wavefunction introduced by Fetter [32], of the LHY correction to (10.15), and we compare it with the full numerical solution of the GPE in (10.15). Then

$$
\chi(x) = \frac{x}{(x^2 + \delta)^{1/2}} .
\tag{10.16}
$$

Minimizing (10.15) with respect to δ one obtains

$$
\delta = \frac{2}{(1 + 2\alpha(31 - 30\log(2))/15)} .
\tag{10.17}
$$

Again, for $\alpha = 0$ one recovers the result $\delta = 2$ [32, 33]. Inserting the wavefunction into (10.15) with the optimal value above one gets

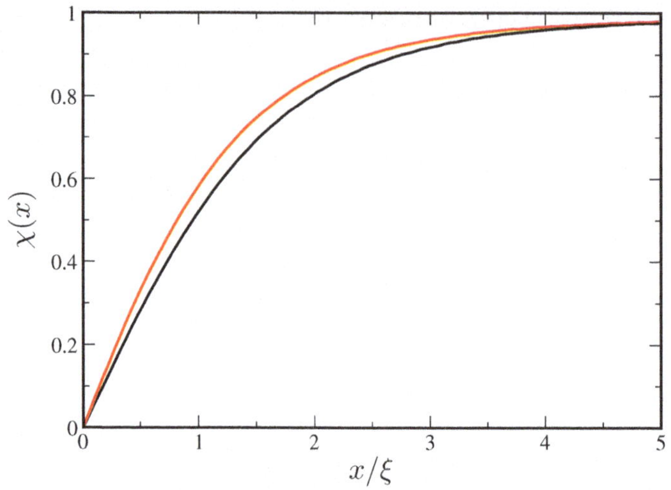

Fig. 10.1 Beyond mean-field correction to the density profile of a vortex configuration in uniform BEC. Density profile for $n\,a^3 = 0$ (black) and for $n\,a^3 = 10^{-3}$ (red)

$$\varepsilon_V = \frac{\pi\hbar^2 n}{m}\log\left(1.497\frac{D}{\xi}\left(1+\frac{2\alpha}{15}(31-30\log(2))\right)^{1/2}\right). \tag{10.18}$$

A much better approximation to the numerical result is obtained by substituting the $\delta = 0$ prefactor (1.497) with the one obtained by the numerical solution of (10.15) with $\delta = 0$ (1.464) leaving the beyond mean-field contribution depending explicitly on α unaltered:

$$\varepsilon_V = \frac{\pi\hbar^2 n}{m}\log\left(1.464\frac{D}{\xi}\left(1+\frac{2\alpha}{15}(31-30\log(2))\right)^{1/2}\right). \tag{10.19}$$

Upon separating the terms of the logarithm, it is interesting to notice that the beyond mean-field correction does not depend explicitly on the ratio D/ξ.

In Fig. 10.1 we plot the rescaled radial wavefunction for vanishing beyond mean-field correction, and for diluteness parameter $n\,a^3 = 10^{-3}$. The size of the vortex core shrinks because the effective chemical potential increases, see (10.4), then the healing length diminishes. Numerically, the size of the core is computed at the point $x_{|\chi=0.8}$ where the wavefunction reaches 80% of its asymptotic value. For the case shown in the figure the size of the vortex decreases by ≈12% with respect to the case without the inclusion of quantum fluctuations of a 3D BEC in the presence of a vortex line.

In Fig. 10.2 we plot the energy density in units of $\pi\hbar^2 n/m$ as a function of the diluteness parameter $n\,a^3$. The agreement between the full numerical solution of the minimization of (10.15) and the mixed numerical-variational result (10.19) is excel-

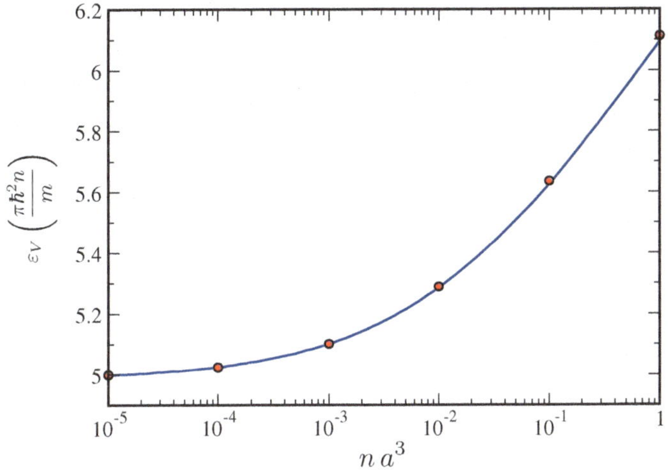

Fig. 10.2 Continuous line: Beyond mean-field correction to the energy density (10.19) of a vortex configuration in uniform BEC computed at fixed $D/\xi = 100$ as a function of the diluteness parameter $y = (na^3)$. Full dots: numerical minimization of (10.15) for $n\,a^3 = 10^{-5}$, 10^{-4}, 10^{-3}, 10^{-2}, 10^{-1}, 1

lent. We point out that the beyond mean-field term is crucial for proper calculation of the energy density, even for very large distances ($\sim 100\xi$). It is then impressive that the very same variational ansatz captures extremely well the contribution even for very large values of the diluteness parameter. We notice, however, that the validity of this result is based on two crucial assumptions, namely that perturbation theory providing the LHY contribution holds in this regime, and the local density approximation. Both these assumptions ought to be compared with full ab initio numerical approaches such as Quantum Monte Carlo calculations.

10.3 Vortex Line in a Trapped Three Dimensional BEC

We now compute the energy of a vortex in a BEC in a trap, when the conditions of the Thomas–Fermi (TF) approximation are met [33, 34]. To compute the energy of a vortex in a trapped configuration one can proceed by noting that $\xi_0/R \approx \hbar\omega_\perp/2\mu$ [33], where R is the characteristic Thomas–Fermi radius of the BEC. When the radial trapping frequency is much smaller than the chemical potential, the radial extension of the BEC is much larger than the size of the vortex. One can then proceed by separately adding vortex contribution in a uniform medium and the hydrodynamic kinetic energy of the bulk. Following [33], outside the vortex core, up to a distance ρ_1, much smaller than the TF radius R and greater than the healing length ξ, we can employ (10.19) for the energy of the vortex inside the BEC. At larger distances from the vortex, the density is essentially the density of the condensate in the absence of the vortex, with the difference that the velocity field is that imposed by the vorticity of the wavefunction. In this region one can use standard hydrodynamics to derive the energy contribution of the system.

First, we consider the 2D problem in which we neglect the trapping along the direction of the vortex line. According to the considerations above, the energy per unit length ε_V^{TF} is given by

$$\begin{aligned}\varepsilon_V^{TF} &= \pi n(0)\tfrac{\hbar^2}{m}\log\left(1.464\tfrac{\rho_1}{\xi}\left(1+\tfrac{2\alpha}{15}(31-30\log(2))\right)^{1/2}\right)+\\ &+\tfrac{1}{2}\int_{\rho_1}^{\rho_2} mn(\rho)v^2(\rho)2\pi\rho\,\mathrm{d}\rho.\end{aligned} \tag{10.20}$$

To find a closed expression for the energy density we use the fact that the velocity field is $v(\rho)=\hbar/m\rho$, and the density in the $2D$ harmonic trap varies as

$$n(\rho)=n(0)\left(1-\frac{\rho^2}{\rho_2^2}\right), \tag{10.21}$$

where $n(0)$ is the peak density at the origin. In (10.21) we neglect corrections to the TF density given by the beyond mean field terms. Substituting (10.21) into (10.20) and by standard integration, we obtain

$$\begin{aligned}\varepsilon_V^{TF} &= \pi n(0)\tfrac{\hbar^2}{m}\log\left(1.464\tfrac{\rho_1}{\xi}\left(1+\tfrac{2\alpha}{15}(31-30\log(2))\right)^{1/2}\right)+\\ &+\pi n(0)\tfrac{\hbar^2}{m}\left(\log\left(\tfrac{\rho_2}{\rho_1}\right)-\tfrac{1}{2}\right)+\mathcal{O}((\rho_1/\rho_2)^2)\end{aligned} \tag{10.22}$$

Neglecting terms $\mathcal{O}((\rho_1/\rho_2)^2)$ and recasting the terms in (10.22) we have

$$\varepsilon_V^{TF}=\pi n(0)\frac{\hbar^2}{m}\log\left(0.888\frac{\rho_2}{\xi}\left(1+\frac{2\alpha}{15}(31-30\log(2))\right)^{1/2}\right). \tag{10.23}$$

We now extend our derivation to the three-dimensional case with density $n(\rho,z)$. If the extension of the cloud is $2Z$ along the direction of the vortex line. Under the assumption that $\xi << Z$, we can compute the total energy by summing the energy of the BEC by integrating (10.23) along the vertical direction. Therefore we obtain

$$\mathcal{E}_V^{TF}=\pi n(0)\frac{\hbar^2}{m}\int_{-Z}^{Z}\mathrm{d}z\,n(0,z)\log\left(0.888\frac{\rho_2(z)}{\xi}\left(1+\frac{2\alpha}{15}(31-30\log(2))\right)^{1/2}\right). \tag{10.24}$$

The TF density $n(\rho,z)$ and the radius $\rho_2(z)$ are given by

$$
\begin{aligned}
n(0,z) &= n(0,0)\left(1-\tfrac{z^2}{Z^2}\right) \\
\rho_2(z) &= R\left(1-\tfrac{z^2}{Z^2}\right)^{\frac{1}{2}}.
\end{aligned}
\tag{10.25}
$$

where $n(0,0)$ is the peak density and R the TF radial dimension. All such quantities can be computed by making use of the normalization condition on $n(\rho,z) = n(0,0)\left(1-\frac{1}{2}m\omega_\perp^2\rho^2-\frac{1}{2}m\omega_z^2z^2\right)$, where $\omega_{\perp,z}$ are the radial and longitudinal trapping frequencies respectively.

Upon substitution into (10.24), and making use of elementary integrations, one then obtains

$$
\mathscr{E}_V^{TF} = \frac{4\pi}{3}n(0,0)\frac{\hbar^2}{m}Z\log\left(0.671\frac{R}{\xi}\left(1+\frac{2\alpha}{15}(31-30\log(2))\right)^{1/2}\right), \tag{10.26}
$$

where Z is the linear extension of the BEC along the vortex line, R the radial dimension. Again, as in the previous section, when $\alpha = 0$ one recovers the known expression for the TF energy of a BEC in the presence of a vortex line. Finally, we would like to point the attention to the fact that, within the approximation above that does not include beyond mean field interactions into the density, the total angular momentum is not affected by the beyond mean-field correction dependent on α. Such observation is helpful for the calculation of the critical velocity for the formation of a vortex line in an axisymmetric trap, which can be readily derived and it reads [33]

$$
\Omega_c = \frac{5}{2}\frac{\hbar}{mR^2}\log\left(0.671\frac{R}{\xi}\left(1+\frac{2\alpha}{15}(31-30\log(2))\right)^{1/2}\right). \tag{10.27}
$$

10.4 Conclusions

In this chapter we studied the effects of the beyond mean-field corrections to the equation of state of a BEC in the presence of a vortex line. We considered the case of a condensate with short-range interactions in the absence and in the presence of an external harmonic confinement. We derived the equations obeyed by a static vortex line in both configurations and solved them both numerically and by a convenient ansatz for the wavefunction of the vortex state. Before concluding we would like to mention that the use of the local density approximation for the equation of state derives from the assumption that the spatial variations of the wavefunction take place on distances greater than the healing length of the condensate. However, this might not be the case for the vortex state, especially close to the origin. Therefore, it would be helpful to have ab initio verifications of the results derived in this work, as e.g. Quantum Monte Carlo calculations of the energy density of a BEC in the presence of a vortex line [38, 39]. Interesting generalization of this work include the study of the static properties of vortex lines in mixtures of BECs [10, 33, 40, 41] and condensates with non-local interactions, such as dipolar atoms, or Rydberg-dressed systems [42–45]. An intriguing question may be especially related to the relevance of the results obtained here for condensates in self-bound states in three dimensions, observed recently in ^{39}K mixtures and Dy and Er dipolar condensates [46].

Acknowledgements T.M. thanks the organizers of the event on "Strongly Coupled Field Theories for Condensed Matter and Quantum Information Theory", and particularly Pasquale Sodano, for discussions and support during and after the event. T.M. acknowledges for support CAPES through the Project CAPES/Nuffic n. 88887.156521/2017-00 and CNPq through Bolsa de produtividade em Pesquisa n. 311079/2015-6.

References

1. F. Dalfovo, S. Stringari, Phys. Rev. A **53**, 2477 (1996)
2. A. Fetter, A. Svidzinsky, J. Phys.: Condens. Matter **13**, R135–R194 (2001)
3. K.W. Madison, F. Chevy, W. Wohlleben, J. Dalibard, Phys. Rev. Lett. **84**, 806 (2000)
4. M.W. Zwierlein, J.R. Abo-Shaeer, A. Schirotzek, C.H. Schunck, W. Ketterle, Nature (London) **435**, 1047–1051 (2005)
5. T. Lahaye, C. Menotti, L. Santos, M. Lewenstein, T. Pfau, Rep. Prog. Phys. **72**, 126401 (2009)
6. A.M Martin, N.G Marchant, D.H.J. ODell, N.G. Parker, J. Phys.: Condens. Matter **29**, 103004 (2017)
7. M. Abad, M. Guilleumas, R. Mayol, M. Pi, D.M. Jezek, Phys. Rev. A **79**, 063622 (2009)
8. I. Tikhonenkov, B.A. Malomed, A. Vardi, Phys. Rev. A **78**, 043614 (2008)
9. R.M. Wilson, S. Ronen, J.L. Bohn, H. Pu, Phys. Rev. Lett. **100**, 245302 (2008)
10. W.E. Shirley, B.M. Anderson, C.W. Clark, R.M. Wilson, Phys. Rev. Lett. **113**, 165301 (2014)
11. M. Klawunn, R. Nath, P. Pedri, L. Santos, Phys. Rev. Lett. **100**, 240403 (2008)
12. M. Klawunn, L. Santos, New J. Phys. **11**, 055012 (2009)
13. F. Dalfovo, S. Giorgini, L.P. Pitaevskii, S. Stringari, Rev. Mod. Phys. **71**, 463 (1999)
14. N. Navon, S. Piatecki, K. Günter, B. Rem, T.C. Nguyen, F. Chevy, W. Krauth, C. Salomon, Phys. Rev. Lett. **107**, 135301 (2011)
15. N. Navon, S. Nascimbène, F. Chevy, C. Salomon, Science **328**, 729 (2010)
16. H. Kadau, M. Schmitt, M. Wenzel, C. Wink, T. Maier, I. Ferrier-Barbut, T. Pfau, Nature (London) **530**, 194 (2016)
17. I. Ferrier-Barbut, H. Kadau, M. Schmitt, M. Wenzel, T. Pfau, Phys. Rev. Lett. **116**, 215301 (2016)
18. L. Chomaz, S. Baier, D. Petter, M.J. Mark, F. Wächtler, L. Santos, F. Ferlaino, Phys. Rev. X **6**, 041039 (2016)
19. I. Ferrier-Barbut, M. Schmitt, M. Wenzel, H. Kadau, T. Pfau, J. Phys. B **49**, 214004 (2016)
20. M. Schmitt, M. Wenzel, B. Boöttcher, I. Ferrier-Barbut, T. Pfau, Nature **539**, 259 (2016)
21. M. Wenzel, F. Böttcher, T. Langen, I. Ferrier-Barbut, T. Pfau, Phys. Rev. A **96**, 053630 (2017)
22. L. Chomaz et al., Nat. Phys. **14**, 442–446 (2018)
23. C.R. Cabrera, L. Tanzi, J. Sanz, B. Naylor, P. Thomas, P. Cheiney, L. Tarruell, Science **359**, 301 (2018)
24. G. Semeghini, G. Ferioli, L. Masi, C. Mazzinghi, L. Wolswijk, F. Minardi, M. Modugno, G. Modugno, M. Inguscio, M. Fattori, Phys. Rev. Lett. **120**, 235301 (2018)
25. D.S. Petrov, Phys. Rev. Lett. **115**, 155302 (2015)
26. D.S. Petrov, G.E. Astrakharchik, Phys. Rev. Lett. **117**, 100401 (2016)
27. F. Cinti, M. Boninsegni, Phys. Rev. A **96**, 013627 (2017)
28. F. Cinti, A. Cappellaro, L. Salasnich, T. Macrì, Phys. Rev. Lett. **119**, 215302 (2017)

29. D. Laghi, T. Macrì, A. Trombettoni, Phys. Rev. A. **96**, 043605 (2017)
30. A. Cappellaro, T. Macrì, G.F. Bertacco, L. Salasnich, Sci. Rep. **7**, 13358 (2017)
31. A. Cappellaro, T. Macrì, L. Salasnich, Phys. Rev. A **97**, 053623 (2018)
32. A.L. Fetter, in *Lectures in Theoretical Physics*, vol. XIB, ed. K.T. Mahanthappa, W.E. Brittin (New York, Gordon and Breach, 1969), p. 351
33. C.J. Pethick, H. Smith, *Bose Einstein Condensation in Dilute Gases*, 2nd edn. (Cambridge University Press, Cambridge, 2011)
34. E. Lundh, C.J. Pethick, H. Smith, Phys. Rev. A **55**, 2126 (1997)
35. T.D. Lee, C.N. Yang, Phys. Rev. **105**, 1119 (1957). See also T.D. Lee, K. Huang, C.N. Yang, Phys. Rev. **106**, 1135 (1957)
36. A.L. Fetter, Rev. Mod. Phys. **81**, 647 (2009)
37. V.L. Ginzburg, L.P. Pitaevskii, Zh. Eksp. Teor. Fiz.**34**, 1240 (1958) [Sov. Phys. JETP **7**, 858 (1958)]
38. D.E. Galli, L. Reatto, M. Rossi, Phys. Rev. B **89**, 224516 (2014)
39. S.A. Vitiello, L. Reatto, G.V. Chester, M.H. Kalos, Phys. Rev. B **54**, 1205 (1996)
40. E.J. Mueller, T.-L. Ho, Phys. Rev. Lett. **88**, 180403 (2002)
41. K. Kasamatsu, M. Tsubota, M. Ueda, Phys. Rev. Lett. **91**, 150406 (2003)
42. N. Henkel, F. Cinti, P. Jain, G. Pupillo, T. Pohl, Phys. Rev. Lett. **108**, 265301 (2012)
43. F. Cinti, T. Macrì, W. Lechner, G. Pupillo, T. Pohl, Nat. Commun. **5**, 3235 (2014)
44. T. Macrì, F. Maucher, F. Cinti, T. Pohl, Phys. Rev. A **87**, 061602 (2013)
45. T. Macrì, S. Saccani, F. Cinti, J. Low Temp. Phys. **177**, 59–71 (2014)
46. A. Cidrim, F.E.A. dos Santos, E.A.L. Henn, T. Macrì, Phys. Rev. A **98**, 023618 (2018)

Chapter 11
Topological View on Entanglement and Complexity

Dmitry Melnikov

Abstract Topological Quantum Field Theories are examples of quantum field theories with a discrete and even finite-dimensional Hilbert space. In this respect they are an intermediate step between quantum mechanics and quantum field theory. Such a special position allows one to study some non-trivial aspects of quantum field theories in more accessible and familiar quantum-mechanical terms. In particular, since the topological theories do not possess local dynamical degrees of freedom, one can study, in such theories, various interesting non-local correlations. Quantum entanglement is an example of non-classical, non-local correlation between parts of a quantum system. Description of entanglement in topological terms is a familiar idea in quantum information theory. In this note I will review some recent developments of this idea bearing explicit connections of entanglement with knots and their topological invariants. The new formulation implies a somewhat updated view on quantum computation. I will illustrate some new aspects by reviewing the notion of complexity.

11.1 Introduction

One idea that quantum entanglement can be interpreted in terms of topological relations between objects representing parts of a quantum system is known in the theory of quantum information as the Aravind's conjecture. There are certain links in knot theory that have a property that removing one component unlinks all the remaining ones. Such links are generally called Brunnian and the simplest example are the so-called Borromean rings

---

D. Melnikov (✉)
International Institute of Physics, Federal University of Rio Grande do Norte, Campus Universitário, Lagoa Nova, Natal 59078-970, Brazil
e-mail: dmitry@iip.ufrn.br

A. Ferraz et al. (eds.), *Strongly Coupled Field Theories for Condensed Matter and Quantum Information Theory*, Springer Proceedings in Physics 239,
https://doi.org/10.1007/978-3-030-35473-2_11

(11.1)

a three-component link with this exact property. Aravind [1] notices that this is reminiscent to a class of quantum entangled states with more than two subsystems with a property that projecting on (making measurement in) one subsystem makes the other subsystems unentangled. Such states are called Greenberger–Horne–Zeilinger (GHZ) states and the simplest example thereof is the three qubit state

$$|\mathrm{GHZ}\rangle = \frac{|000\rangle + |111\rangle}{\sqrt{2}} \,. \tag{11.2}$$

In case of tripartite systems GHZ states furnish one of the two classes of non-biseparable states (the other type is called W), which cannot be mixed by local quantum operations [2]. Hence GHZ and W states characterize specific patterns of quantum entanglement. Understanding and classifying different entanglement patterns is an interesting question in the theory of quantum information.

The analogy of Aravind is in fact basis dependent. Several more recent works make this proposal more specific. Kauffman and Lomonaco [3] (see also [4] and references therein), introduce "quantum knots" to make knots more on par with quantum mechanics. In the meantime the proposal of Balasubramanian et al. [5] uses classical knots and associates basis-independent quantum states to links using the approach of topological quantum field theories (TQFT).

In these notes I will review a connection between knots and entanglement more in the spirit of the proposal of [5]. The idea naturally follows from a definition of TQFT given by Witten and Atiyah [6–9], which associates a quantum mechanical interpretation to topological spaces. This helps to give a topological meaning to the entanglement of quantum states [10].

One of the standard tools to quantify entanglement is the entanglement entropy. Computing the entanglement entropy in TQFT is much simpler, than in regular quantum field theory [11, 12]. In particular, tools like replica trick express the entropy in terms of topological invariants of knots [13]. In this sense the relation between quantum entanglement and knots might be even deeper than originally expected.

Topological interpretation of basic notions reviewed in this text prepares the stage for extending the technology further, to the subject of quantum computation. The relation between knots and quantum computers is known for a while [14], but I hope that some ideas discussed in these notes can enrich this relation. Thus, following the discussion of entanglement in TQFT, I describe a topological invitation to the notion of complexity of quantum states.

11.2 Topological Quantum Field Theory

Let us define a Topological Quantum Field Theory (TQFT). I will start from the most familiar example of the Chern–Simons theory and then give a formal definition.

11.2.1 Chern–Simons Theory. A Definition

The prime example of a TQFT is the Chern–Simons (CS) theory. It is a three-spacetime-dimensional gauge theory with action

$$S = \frac{k}{4\pi}\int_{\mathcal{M}_3} d^3x\, \varepsilon^{ijk}\,\mathrm{Tr}\left(A_i\partial_j A_k + \frac{2}{3}A_iA_jA_k\right)\,, \tag{11.3}$$

where A_i is a three-dimensional gauge field, in general non-Abelian (assume group $SU(N)$). Parameter k is the level of the CS theory. It is an analog of the coupling constant. Gauge invariance requires k to be an integer in the quantum theory.

The integral in (11.3) is taken over a three-dimensional manifold, which will most of the time be assumed closed, e.g. $\mathcal{M}_3 = S^3$, or $\mathcal{M}_3 = S^2 \times S^1$. This theory is often referred as $SU(N)_k$ CS theory on $\mathcal{M}_3$.

We note that the CS action depends on $\mathcal{M}_3$, but not on the metric g_{ij} on it. Hence, one expects that observables in the theory are metric-independent. Such theories are called topological, because they only depend on the topology, but not geometry of the spacetime. One consequence is triviality of the TQFT Hamiltonian. Indeed, the Hamiltonian density is the component T^{00} of the stress energy tensor, but

$$T^{ij} \propto \frac{\delta S}{\delta g_{ij}} = 0\,. \tag{11.4}$$

Therefore, the theory does not have a non-trivial dynamics. The solutions to equations of motion

$$F_{ij} = \partial_i A_j - \partial_j A_i + [A_i, A_j] = 0\,, \tag{11.5}$$

are pure gauge fields $A_i = g\partial_i g^{-1}$. The only non-trivial degrees of freedom are possible singularities of A_i. They can be thought as of heavy non-dynamical particles coupled to the gauge fields. Nothing depends on local perturbations of the particles, but they possess non-trivial non-local correlations, expressed in topological terms.

The correlations can be seen in the quantum theory. Define a functional integral

$$Z[\mathcal{M}_3] = \int \mathcal{D}A\, e^{i\,S[A;\mathcal{M}_3]}\,. \tag{11.6}$$

In this formula, the dependence of the partition function on the manifold $\mathcal{M}$ is made explicit, but functionally the result of the integration depends on two parameters: the level k and the rank of the gauge group N.

For a probe particle an observable sensitive to the spacetime topology is the holonomy

$$W(\gamma) = \mathscr{P} \exp \oint_{\gamma} A_i \, \mathrm{d}x^i \,, \tag{11.7}$$

where the exponential should be path ordered if A_i is a non-Abelian field. Holonomy transforms under the action of the gauge group. To have a gauge-invariant quantity one takes the trace. The trace can be taken in different representations of the group. Consequently, interesting observables in the quantum CS theory are correlation functions

$$Z(\mathscr{M}_3; \gamma_1, \ldots, \gamma_n) = \langle \mathrm{Tr}_{R_1} W(\gamma_1) \cdots \mathrm{Tr}_{R_n} W(\gamma_n) \rangle \,, \tag{11.8}$$

where traces of holonomies are called Wilson loop operators. Implicitly, functional integral Z depends on the representations R_i associated to closed paths γ_i, k and N. Witten [6] and others [15] showed that correlators (11.8) are topological invariants of knots associated with γ_i. After an appropriate redefinition of parameters k and N, they can be explicitly mapped to invariants known in knot theory (linking number, Jones polynomials, HOMFLY-PT polynomials etc.)

To actually compute quantum partition functions and explicitly describe the Hilbert space of the quantum CS theory canonical quantization is more appropriate [6]. In mathematical terms, the problem on $\mathscr{M} = \Sigma \times R$ can be formulated as finding the space of holomorphic determinant line bundles on Σ. The corresponding spaces also appear in the study of conformal fields theories (CFTs), in particular $U(N)$ level k Wess–Zumino–Witten (WZW) theories on Σ, where this space is called the space of conformal blocks. Conformal blocks are solutions of certain differential equations, which provide a basis for the CFT correlation functions. In this sense they can be understood as a basis of wavefunctions in some Hilbert space. I will not explain the quantization procedure here, but instead introduce the Hilbert spaces and states rather formally, associating them to the data defining functional integral (11.6).

11.2.2 Axiomatic Definition

Let us now review a formal axiomatic definition of a TQFT in terms of the category theory. The reader may skip this definition and go directly to the summary, where the meaning of this definition is explained.

Following the definition of Atiyah and others [8, 9], TQFT is a functor Z between the cobordism category Cob and the category of linear spaces Lin, $Z : Cob \to Lin$. The Cob_d category has as its objects d dimensional oriented topological spaces, while morphisms are $d+1$-dimensional manifolds, whose boundaries are the d-dimensional spaces (objects). One can think of the morphisms (cobordisms) as of evolution histories of the d-dimensional spaces. Lin is the standard collection of linear space with morphisms—linear maps between linear spaces. As usual, the functor maps objects and morphisms of one category, to objects and morphisms of another category. In particular,

- to any d-dimensional manifold Σ in Cob_d functor Z associates a vector space V in Lin, $Z(\Sigma) = V$;
- to any $d+1$-dimensional manifold $\mathscr{M}$, such that $\partial\mathscr{M} = \Sigma$ functor Z associates a vector v in V, $Z(\mathscr{M}) = v$;
- to any $d+1$-dimensional manifold $\mathscr{M}$, whose boundary is a disjoint union $\Sigma_1 \cup \Sigma_2 = \partial\mathscr{M}$ functor Z associates a linear map M between $Z(\Sigma_1)$ and $Z(\Sigma_2)$, $Z(\mathscr{M}) = M : Z(\Sigma_1) \to Z(\Sigma_2)$.

The TQFT functor satisfies a number of axioms:

1. If $Z(\Sigma) = V$ then $Z(\bar{\Sigma}) = V^{\dagger}$, where $\overline{\Sigma}$ denotes space Σ with a reversed orientation and $V^{\dagger}$ is a linear space dual to V.
2. For a disjoint union $\Sigma_1 \cup \Sigma_2$ the functor associates a tensor product of vector spaces, $Z(\Sigma_1 \cup \Sigma_2) = Z(\Sigma_1) \otimes Z(\Sigma_2)$.
3. Consider $\mathscr{M}_1 : \Sigma_1 \to \Sigma_2$ and $\mathscr{M}_2 : \overline{\Sigma}_2 \to \Sigma_3$ then one can form a new morphism $\mathscr{M}$, which is a composition $\mathscr{M} = \mathscr{M}_2 \circ \mathscr{M}_1 : \Sigma_1 \to \Sigma_3$. Functor Z maps the composition of the morphisms to the composition of the linear maps, $Z(\mathscr{M}_2 \circ \mathscr{M}_1) = Z(\mathscr{M}_2) \circ Z(\mathscr{M}_1)$.
4. Cob_d is also assumed to contain empty manifold $\Sigma = \emptyset$, which can also be introduced through a morphism $\mathscr{M} = \partial\Sigma'$, such that $\mathscr{M} : \Sigma' \to \emptyset$, or through a morphism $\mathscr{M} : \emptyset \to \emptyset$ with no boundaries. Functor Z associates to $\emptyset$ complex numbers $\mathbf{C}$, $Z(\emptyset) = \mathbf{C}$.
5. There is a trivial morphism in Cob_d which maps any Σ to itself. Such a map is represented by a cylinder $\Sigma \times I$, where I is a unit interval. This is mapped to the identity map of the linear spaces, $Z(\Sigma \times I) = \mathbf{1}$.

The general idea of the category theory definition is to introduce a quantum mechanical, or Hilbert space structure associated with certain spaces Σ, for example two-dimensional Riemann manifolds. Quantum mechanical structure means states (vectors in the Hilbert space) $|\psi\rangle$, scalar products $\langle\psi|\psi'\rangle$ and (unitary) operators $U : |\psi\rangle \to |\psi'\rangle$. Provided that the map (functor) is constructed we can endow the following heuristic diagrams with a quantum-mechanical meaning.

A diagram of a manifold $\mathscr{M}$ with a boundary Σ can be viewed as a state:

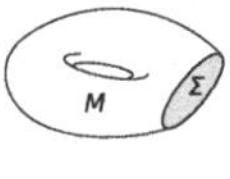

$$\equiv |\psi\rangle\,. \tag{11.9}$$

Scalar product of two states in the same Hilbert space $\mathscr{H}$ is "gluing" of two spaces $\mathscr{M}$ and $\mathscr{M}'$ along the same boundary Σ:

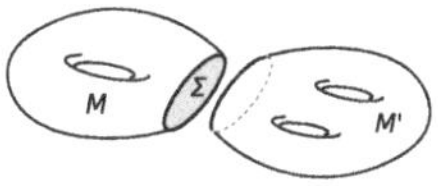

$$\longrightarrow$$

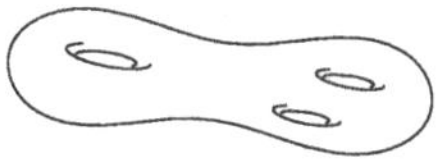

$$\equiv \langle\psi|\psi'\rangle \tag{11.10}$$

The result is a manifold without boundaries. Since these are topological theories, the shape of the manifolds on the diagrams do not matter, but only topology.

Generalizing the above diagrams one can consider manifolds $\mathcal{M}$ with multiple boundaries $\Sigma_1 \cup \Sigma_2 \cdots \cup \Sigma_n$

$$\equiv T \in \mathcal{H}_1 \otimes \mathcal{H}_2 \cdots \otimes \mathcal{H}_n\,. \tag{11.11}$$

In the simplest case with two boundaries T can be viewed either as an operator $T : \mathcal{H}_1 \to \mathcal{H}_2$, or as a state in the tensor product of the two Hilbert spaces. Formally, in the first case $T \in \mathcal{H}_1^\dagger \otimes \mathcal{H}_2$, since the operator should have a reversed orientation to be able to "act" on $\mathcal{H}_1$.

One can notice that the idea of associating spaces to the vectors and operators in some Hilbert spaces is similar to the idea of tensor networks. Manifolds $\mathcal{M}$, or T are analogous to tensors, whose legs (Σ) belong to the corresponding Hilbert spaces (see also [16]). So far, the TQFT tensors were "black boxes". However, once map (functor) Z is specified, the tensors-manifolds acquire a specific meaning.

11.2.3 Connecting Two Definitions

In some formal way one can propose that the functional integral of the CS theory is an explicit realization of the TQFT functor Z. The key observation is that given a "time" slice $t = 0$, one can define a wave function as a functional integral,

$$|\psi(t=0)\rangle = \frac{1}{\mathcal{N}} \int \mathcal{D}A\Big|_{A|_\Sigma = A_\Sigma} e^{i\,S[A;\mathcal{M},\Sigma]} \equiv \frac{1}{Z[\mathcal{M} \cup \overline{\mathcal{M}}]^{1/2}}\, Z[\mathcal{M}, \Sigma]\,, \tag{11.12}$$

where $\mathcal{N}$ is the normalization defined through the square of the norm $Z[\mathcal{M} \cup \overline{\mathcal{M}}]$. The integral is computed over the gauge field configurations with prescribed boundary conditions on Σ.

We will assume that Σ might have a set $\{P_i\}$ of points removed, so that we actually associate the Hilbert space to $\Sigma\backslash\{P_i\}$. Quantum mechanics on such spaces should then be equivalent to cobordisms of spaces with punctures in such a way that punctures should be cobordant to other punctures and no punctures can disappear in the bulk of the corresponding $\mathcal{M}$.

In CS theory punctures correspond to non-dynamical particles sourcing the gauge field. Their worldlines correspond to open or closed Wilson lines in CS theory, which are cobordisms of the punctures. Each worldline is "colored" by an irreducible representation of the gauge group, as in correlation function (11.8).

Calculation of scalar products in the Hilbert space requires changing the orientation of the boundary Σ. For consistency, this should imply replacing all the representations of lines ending on $\overline{\Sigma}$, and inside a corresponding $\mathcal{M}$, by their conjugate. This rule is consistent with assigning orientations to the worldlines, so that when Σ and $\overline{\Sigma}$ are glued together, the glued worldlines have a continuous flow of the orientation.

11.3 Examples of TQFT

Let us consider some explicit examples of Hilbert spaces $\mathscr{H}_\Sigma$ that can be associated to two-dimensional Riemann surfaces Σ and to $\Sigma\backslash\{P_i\}$, with removed points [6].

11.3.1 Hilbert Spaces of Punctured Spheres

To a two-sphere S^2 with no marked points the $SU(N)_k$ CS TQFT associates a one-dimensional Hilbert space. From the WZW CFT point of view there is only one conformal block for the descendants of the identity operator.

From the analysis of CFT on S^2 we know that any one-point function vanishes unless it is a one-point function of an identity operator. Consequently, the Hilbert space associated to $S^2\backslash P$ is zero-dimensional unless the representation corresponding to P is trivial $R = 0$, in which case $\dim \mathscr{H} = 1$. The latter fact can also be related to the observation that it is not possible to accommodate a non-trivial "charge" in a closed space. The total charge must vanish.

For two punctures vanishing of total charge tells that the Hilbert space has dimension only if the corresponding representations are conjugated $R_1 = \overline{R}_2$, in which case $\dim \mathscr{H} = 1$. There are more possibilities for three "charges". In general there can be more than one way for two representations to "fuse" producing a third one. The corresponding number is called multiplicity of the third representation. As follows from WZW CFT the dimensions of the Hilbert space are smaller or equal than multiplicities, because of additional CFT restrictions. The corresponding numbers are called "fusion" numbers. They are denoted $N_{R_1R_2}^{\overline{R}_3}$.

Equipped with this basic information one can compute basic examples of topological invariants on $S^2 \times S^1$, with periodic Euclidean time. Since the Hamiltonian is zero, the partition function is just the trace of the identity operator over the Hilbert space. For a sphere with one point marked by representation R the result is $Z(S^2 \times S^1) = \delta_{R0}$, which includes the case of no marked points for trivial R.

The following two results are also straightforward:

$$Z(S^2 \times S^1; R_1, R_2) = \delta_{R_1}^{\overline{R}_2} \tag{11.13}$$

and

$$Z(S^2 \times S^1; R_1, R_2, R_3) = N_{R_1R_2}^{\overline{R}_3} \,. \tag{11.14}$$

What one computes in these equations are correlations functions (11.8) of two and three colored oriented Wilson lines winding around the S^1 of $S^2 \times S^1$.

11.3.2 Hilbert Space of T^2

It is quite instructive to quantize CS theory on $T^2 \times S^1$ (see [17], for example). The Hilbert space $\mathscr{H}_{T^2}$ of the quantum theory is finite and has dimension $\begin{pmatrix} k+N-1 \\ N-1 \end{pmatrix}$ in the case of $SU(N)_k$ CS. The states of this theory are cobordisms of T^2, of which the simplest are the solid tori.

There is a canonical choice of basis on $\mathscr{H}_{T^2}$. To define the basis vectors we consider solid tori with Wilson lines inserted along the "longitude" of the torus:

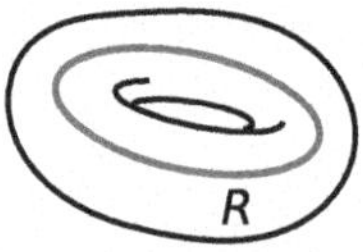

$$\equiv |R\rangle \,. \tag{11.15}$$

The vectors are labeled by the representations of the inserted Wilson lines. In particular, an empty solid torus is understood as having a Wilson line in the trivial representation. In $SU(N)_k$ CS the number of admissible representations is finite, equal to dim $\mathscr{H}_{T^2}$ (cf. "integrable" representations in the corresponding WZW CFT).

States (11.15) are orthonormal with respect to the scalar product defined in Sect. 11.2.2. To calculate $\langle R_1|R_2\rangle$ one constructs an $\mathscr{M}_3$ by gluing two solid tori along Σ with an appropriate orientation. Since a solid torus is $D_2 \times S^1$, a disk crossed with a circle, then gluing corresponds to identifying two disks along the boundary (producing a 2-sphere) and transporting the result along S^1. Hence, $\mathscr{M}_3 = S^2 \times S^1$ and

$$\langle R_1|R_2\rangle = Z(S^2 \times S^1; \overline{R}_1, R_2) = \delta_{R_1}^{R_2} \,, \tag{11.16}$$

where we applied (11.13). Indeed, when we glue the solid tori in the most obvious way, the Wilson lines remain parallel in $S^2 \times S^1$. In particular, they never braid, and the partition function is the one of two unlinked circles winding S^1 of $S^2 \times S^1$.

One can now ask about a general state with an arbitrary knot or link placed inside the solid torus, for example,

$$|\Psi_R\rangle \equiv$$

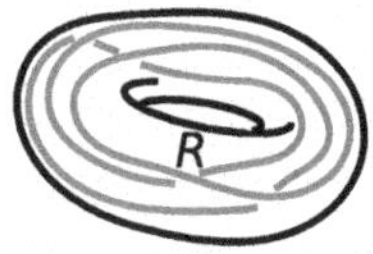

$$= \sum_Q c_{QR}|Q\rangle \,. \tag{11.17}$$

Indeed, any such state can be expanded in the basis (11.15) labeled by integrable representations with coefficients c_{QR}. This coefficients have a special meaning, which can be understood if one computes the scalar product in TQFT

$$c_{QR} = \langle Q|\Psi_R\rangle = Z(S^2 \times S^1; \mathscr{L}_{QR}) \,. \tag{11.18}$$

The expansion coefficients are the partition functions of a link $\mathscr{L}_{QR}$ composed of a circle and the original knot both wrapping the S^1 of $S^2 \times S^1$.

We note that the basis of eigenvectors (11.15) has an explicit representation in terms of differential operators acting on the torus, e.g. [18].

11.4 Entanglement in TQFT

With the definition of TQFT in Sect. 11.2.2, which reinterprets quantum mechanics in terms of topological spaces, it is straightforward to discuss entanglement in topological theories. Let us first discuss a naive interpretation of entanglement in TQFT.

11.4.1 General Idea

In quantum theory a state consisting of subsystems A and B is entangled if the wavefunction (density matrix) is not separable, that is it cannot be presented as a direct product of the wavefunctions (density matrices) of the subsystems, $\Psi_{A\cup B} \neq \Psi_A \otimes \Psi_B$.

To discuss the analogous situation in a TQFT we start with a Hilbert space $\mathscr{H} = \mathscr{H}_A \otimes \mathscr{H}_B$ where we assume all $\Psi_{A\cup B}$ belong [10]. By one of the above axioms, this should be represented by a disjoint union of spaces $\Sigma_A \cup \Sigma_B$. Then there are two obvious candidates for a separable and an entangled states in $\mathscr{H}$:

$|\Psi_1\rangle =$ 

and $|\Psi_2\rangle =$ 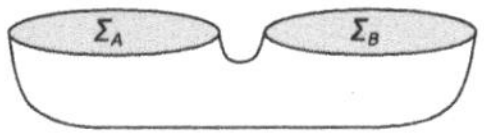

(11.19)

To check this proposal one can compute the von Neumann entropy of subsystems A or B, given by formula $S_{\mathrm{E}}(A) = -\operatorname{Tr}_A \rho_A \log \rho_A$ in terms of the reduced density matrix of subsystem A, $\rho_A = \operatorname{Tr}_B \rho_{A\cup B}$. Given the path integral, or formal TQFT definition it is not obvious whether $S_E(A)$ can be computed for states in (11.19). A useful tool is the replica trick, which produces a clear graphical interpretation in TQFT [10, 13]. It is useful to keep in mind the analogy with tensors—a contraction of legs of two tensors corresponds to gluing two $\mathscr{M}_3$ along a common instance of Σ. A network of tensors (density matrices) becomes a three-manifold with "handles".

We start the calculation by introducing the density matrices, which are defined in general as $\rho = |\Psi\rangle\langle\Psi|$. For states in (11.19), we have to consider

$\widehat{\rho}_1 =$ 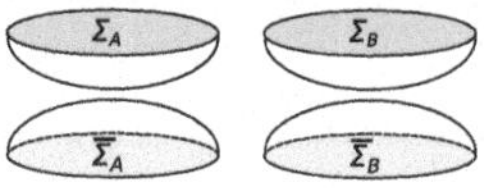

and $\widehat{\rho}_2 =$ 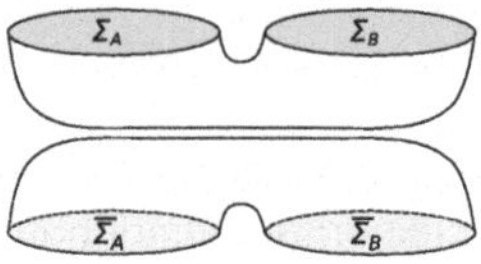

, (11.20)

which represent the tensors to be contracted. Note that these matrices are not normalized, so we have to divide them by their traces. Since trace is a contraction of a pair of chosen indices in a tensor, we have to glue Σ_A with $\overline{\Sigma}_A$ and Σ_B with $\overline{\Sigma}_B$ in diagrams of (11.20). The normalization factors will be closed three-manifolds, which in TQFT correspond to numbers, and we define the normalized density matrices as

$\rho_1 =$ []$^{-2}$ $\widehat{\rho}_1$ and $\rho_2 =$ [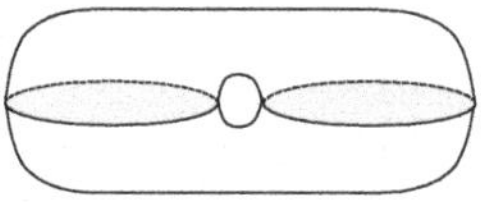]$^{-1}$ $\widehat{\rho}_2$. (11.21)

We stress that the manifolds on the above diagrams are schematic representations. To understand what they actually are, one has to specify Σ_A and Σ_B.

It is straightforward to find the reduced density matrices. For subsystem A,

$\rho_1(A) =$ []$^{-1}$ 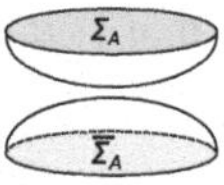

and $\rho_2(A) =$ [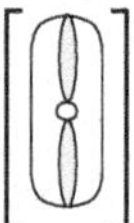]$^{-1}$ 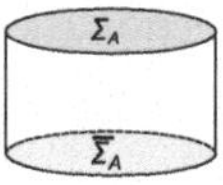

. (11.22)

One can already guess, that the second reduced density matrix is that of a mixed state. One can now apply the replica trick computing

$$S_{\rm E} = -\lim_{n\to 1}\frac{d}{dn}\,{\rm Tr}_A\,\rho^n(A)\,. \tag{11.23}$$

To construct powers of $\rho(A)$ with tensors (11.22) one glues Σ_A with $\overline{\Sigma}_B$ $n-1$ times. Assuming an analytic continuation in n one computes the derivative. The result is

$S_{\rm E1} = 0$ and $S_{\rm E2} = \log$ [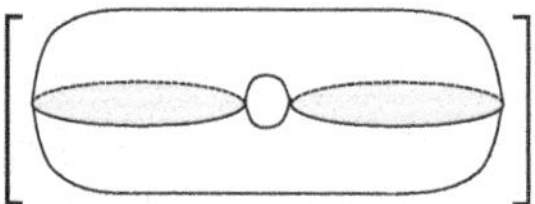] , (11.24)

confirming our expectations that the left state in (11.19) is separable, while the right one is entangled. Without Wilson lines the object appearing in the second expression is easy to compute. As in Sect. 11.3.1 it is nothing but the logarithm of the dimension of the Hilbert space $\mathscr{H}_{\Sigma_A}$. More examples appear below.

11.4.2 Back to the Aravind's Conjecture

An interesting proposal to interpret entangled states in terms of links was made in [5]. Consider a state in a TQFT, obtained by cutting tubular neighborhoods (cylinders) around a link inside a three-dimensional sphere S^3. The complement of an n-component link is a space $\mathcal{M}_3$, whose boundary is a disjoint union of n tori.

A simple example to consider is a thick circle (a solid torus) inside S^3. Cutting the torus out produces a space, which is topologically also a solid torus. However such an inverted torus is not fully equivalent to the original one: consider a circle drawn on its boundary common to both the cut and inverted tori. If the circle is contractible in the bulk of one of the tori it cannot be contracted inside the other one. This shows a subtlety related to defining a TQFT state: given a T^2 we can consider two cobordisms, which fill the space either inside or outside, producing two inequivalent states. More precisely, since the orientation of T^2 is opposite for the two cobordisms, one state is not the dual of the other one.

The relation between different choices of contractible and non-contractible cycles on tori is well-known. It is realized by the modular group $PSL(2, Z) = SL(2, Z)/Z_2$. The element of the modular group, exchanging the longitude and latitude of a torus is denoted S. Two states in a TQFT, with Wilson lines drawn along the longitude and the latitude and colored by representations R_i and R_j respectively, are related by a matrix $S_{R_i R_j}$ in an appropriate representation of $SL(2, Z)$. For an $SU(N)_k$ CS the representation is defined by k and N.

From the point of view of the TQFT, the complement of a thick link on S^3 is a state in the tensor product of Hilbert spaces $\mathcal{H}_{T^2}$, $|\Psi\rangle \in \otimes_{i=1}^{n} \mathcal{H}_{T^2}^{(i)}$. Such a state can be decomposed in the basis of that Hilbert space obtained by tensoring (11.15):

$$|\Psi\rangle = \sum c_{R_1 \cdots R_n} |R_1\rangle \otimes \cdots \otimes |R_n\rangle \,. \tag{11.25}$$

Again, the coefficients have a special meaning. In [5] they are matrix elements

$$c_{R_1 \cdots R_n} = \langle R_1, \ldots, R_n | \Psi \rangle = Z(S^3; R_1, \ldots R_n) \,, \tag{11.26}$$

which correspond to an opposite operation of gluing tori to $\mathcal{M}_3$, in such a way that one recovers the original link in S^3, now colored by the given set of representations. The coefficients are then the corresponding topological link invariants on S^3.

We consider an example of the Hopf link, the simplest linking of two circles. It is not hard to see that it corresponds to an entangled state. A direct way is to find coefficients $c_{R_A R_B}$, which are matrix elements of the modular group element S

$$c_{R_A R_B} = \langle R_A | S | R_B \rangle = S_{\overline{R}_A R_B} \,. \tag{11.27}$$

One does not need to know their explicit form. As a modular group element, S is subject to constraint $S^2 = 1$, so that the reduced density matrix of state (11.25) is

$$\rho_A = \frac{S_{\overline{R}_A R_B} S^{\overline{R}'_B R'_A}}{\text{Tr}\, S^2} = \frac{\delta_{\overline{R}_A}^{R'_A}}{\dim \mathcal{H}_{T^2}} \,. \tag{11.28}$$

In the numerator of the last expression appears the identity matrix of the space $\mathscr{H}_{T^2}$. Hence, the von Neumann entropy is the logarithm of the dimension of that space,

$$S_{\mathrm{E}}(\text{Hopf link}) = \log \dim \mathscr{H}_{T^2} = \log \begin{pmatrix} k+N-1 \\ N-1 \end{pmatrix} . \tag{11.29}$$

The calculation can also be done following the approach of Sect. 11.4.1. State (11.25) is heuristically encoded by the right example in (11.19) with $\Sigma_A = T^2$ and $\Sigma_B = T^2$. Consider such a state obtained by cutting a longitudinal tube inside a solid torus

$$|\Psi\rangle =$$

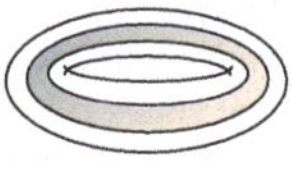

. (11.30)

Section 11.4.1 applies directly to (11.30) and predicts that the entropy is given by (11.29).

There is a subtlety in the last argument. State (11.30) only represents state (11.25) up to the action of S on one of the boundaries. However, from the point of view of entanglement entropy, this would be a local unitary transformation, which leaves the entropy invariant. More precisely, state (11.30) has expansion

$$|\Psi\rangle = \sum \delta_{R_1 R_2} |R_1\rangle \otimes |R_2\rangle \, . \tag{11.31}$$

This can be seen evaluating the scalar products of $|\Psi\rangle$ with the basis vectors and observing that in such a way one finds partition functions on $S^2 \times S^1$, namely (11.13). Consequently, $|\Psi\rangle$ in (11.30) is a simple analog of a Bell pair.

Authors of [5, 19] considered different links in S^3 and computed the associated entanglement entropy. Some of the conclusions of those studies are

- The link is trivial (all components of it are unlinked) if and only if the associated von Neumann entropy vanishes. This follows from the property of unlinked components that their partition functions factorize. Associated states (11.25) are separable.
- The Hopf link is maximally entangled. It can be viewed as an analog of a Bell pair.
- The set of torus links (links that can be drawn on the two-dimensional surface of a torus without self-intersections) with three or more components, correspond to the GHZ-type states in quantum information theory.
- Considered hyperbolic links are consistent with the W-type entanglement.

This is similar to the Aravind's proposal in the sense that entanglement implies linking. However, the correspondence does not work exactly as expected: nor Borromean rings, nor Brunnian links in general, give a GHZ-type of states in the sense of [5]. Such multipartite states, are realized by torus links.

11.4.3 Reinforcing Entanglement with Wilson Lines

As one could notice from Sect. 11.3.1, the case of $\Sigma = S^2$ is a bit different. In particular, to have a non-trivial Hilbert space one should have at least three punctures on the sphere. Therefore to discuss entanglement, one should always include Wilson lines piercing the spheres [10]. Otherwise, following (11.24) the entanglement entropy for the case (11.19) with $\Sigma_A = S^2$ and $\Sigma_B = S^2$ is $\log\dim \mathscr{H}_{S^2} = \log 1 = 0$.

With three or four punctures the simplest example to consider is the state

$|\Psi\rangle =$ 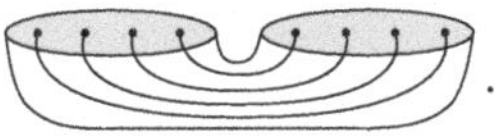 . (11.32)

It is in general an entangled state. Its entropy can be calculated following the steps of Sect. 11.4.1 giving

$S_{\mathrm{E}}(A) =$ 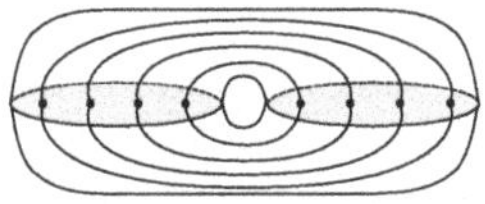 $= \log\dim \mathscr{H}_A\,,$ (11.33)

that is logarithm of the dimension of the Hilbert space of a sphere with four punctures with fixed representations satisfying the zero "net charge" condition $0 \subset R_1 \otimes R_2 \otimes R_3 \otimes R_4$. Hence (11.32) is maximally entangled.

One can ask what happens, if the parallel Wilson lines in (11.32) are replaced by a braid. A braid is produced by "local" permutation of punctures on either Σ_A, or Σ_B. From the steps of the calculation in Sect. 11.4.1 it is clear that any such braid will be unbraided in the reduced density matrix. Result (11.33) will persist as a consequence of independence of the von Neumann entropy from local unitary operations, where "local" means applied to either of subsystems, here A or B. A non-trivial example would be considering a non-local braiding, which transforms (11.32) to either "tangles"

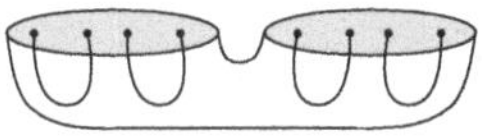 or 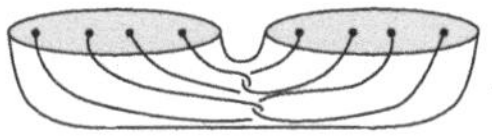 . (11.34)

One can show that in the first case there is no entanglement, since the corresponding partition functions will only compute topological invariants of disjoint circles in $S^2 \times S^1$. As long as the circles do not wind the S^1, which they do not do in the first example of (11.34), such partition functions factorize giving zero von Neumann entropy. The second example is less trivial an involve calculation of "chains" in $S^2 \times S^1$, as in the following diagram

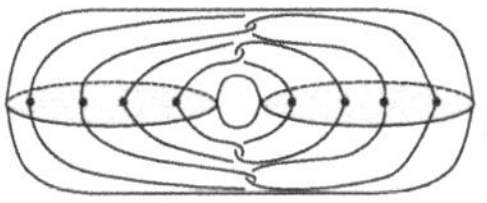 , (11.35)

but with n sections, analytically continued to real n. Evaluation of such invariants is a subject of current research [20].

In case of two-spheres chosen as Σ it is not enough to consider simple cobordisms as in the right example of (11.19) to guarantee entanglement. The cobordisms should be endowed with Wilson lines in order to support the correlations between the subsystems. For maximal entanglement the Wilson lines should have a braid structure, as in state (11.32) rather than tangle structure, as in (11.34).

11.5 Complexity

Classical and quantum computing operates with a notion of complexity of a given computational task. Complexity is supposed to tell how hard is the task in terms of time and resource cost. Different quantifications of complexity were being discussed in computer science and related matters, but more recently complexity was proposed to have interesting (geometrical) meaning in studies of quantum gravity and quantum field theory (see [21] and references therein). While it is rather straightforward to import the notion of complexity to quantum mechanics and quantum computing, it is not easy to generalize these ideas to quantum field theory, where locality and causality are important. It is proposed here to approach this problem in terms of TQFT, which are intermediate theories between quantum mechanics and quantum field theory. I will discuss how a standard notion of complexity can be applied to TQFT, further details will appear in [22].

11.5.1 Complexity in Quantum Mechanics

In quantum mechanics, one can ask how difficult it is to prepare a "target" state $|\Psi_T\rangle$ given a "reference" state $|\Psi_R\rangle$. Such a task is fulfilled by a unitary operator,

$$|\Psi_T\rangle = U|\Psi_R\rangle \ . \tag{11.36}$$

Hence, U is a quantum algorithm or "circuit", solving the problem. Assume that we have available a number of discrete operations U_a, whose composition can approximate U with a given accuracy. In principle, there can be more than one way to approximate U with a composition of U_a. The number of U_a in a given circuit is called the "depth" of the circuit $d(\{U_a\})$, while the complexity is the minimum depth of the circuit necessary to solve the problem

$$\mathscr{C}(\Psi_R, \Psi_T; \{U_a\}) = \min_{\{U_a\}|\prod_i U_{ai}\simeq U} d(\{U_a\}) \ . \tag{11.37}$$

Circuit U can always be viewed as an evolution operator with a time dependent Hamiltonian. Since the values of the Hamiltonian at different times a priori do not commute, the evolution is a path-ordered exponential of a such Hamiltonian

$$U(t) = \mathscr{P} \exp\left(i\int_0^t H(t)\right). \tag{11.38}$$

Note that practically, the path integral is a sequence of discrete time evolutions.

An interesting possibility is $H(t)$ taking values in some Lie algebra, such that U_a are exponentials of a minimal set of generators, $H(t) = H_a(t)T^a$. The path ordered exponential in (11.38) can be cast in terms of paths on the group space, that is the space of available unitary operators. Complexity $\mathscr{C}$ in such a case can be defined as the minimum length, or the geodesic distance, of the path in the group space connecting the identity element and the operator $U(t)$ [23].

Integral in (11.38) can be written in terms of a "Berry connection", $H(t) = \mathscr{A}_i(x)\dot{x}^i$, endowing $U(t)$ with a Wilson line interpretation. When the connection is flat, the integral does not depend on the particular path taken, but rather on the topology of the space of operators. This makes the determination of complexity a discrete problem, as in the following example [22].

11.5.2 Complexity of Torus Knot States

Complexity is also an intuitive notion in knot theory. The standard classification of knots is by the minimum number of crossings in the diagram of a given knot. While the knot is defined as an embedding of a circle in the three-dimensional space up to an isotopy, the diagram of the knot is a projection of the embedding onto an arbitrary two-dimensional plane. Smooth deformation of the embedding do not change the isotopy class and one can find a projection with a minimal number of self-intersections (crossings). Thus, crossings provide a natural definition of the complexity of the knot.

The initial idea about knot complexity may be found a bit naive, because studies of the topological invariants of knots show that some knots, or links, are "simpler" than others, even if their number of crossings is larger. Torus knots give an example of a simple family. Such knots can be drawn on the surface of a torus without self-intersections—all of them are classified by a pair of coprime integers (m, n). Torus knots are simpler for few reasons. One of them is that an explicit formula for most general topological invariants is known for any pair (m, n) [24]. The results are generalized to the case of torus links: a ℓ-component torus link can be characterized by a pair $(\ell m, \ell n)$, where m and n are coprime.

I would like to review some basic ideas about the definition of complexity for torus knots. Further details will appear in the future work [22]. The ideas are based on the following construction of a torus knot.

Given the Hilbert space $\mathscr{H}_{T^2}$ described in Sect. 11.3.2 one can discuss particular states in this Hilbert space corresponding to a solid torus with a torus knot (m, n) inside. In general, such a state can be expanded in basis (11.15), as in (11.17). On the other hand, torus knot state can be obtained by the action of $SL(2, Z)$ modular transformations on the boundary torus with a circle along the longitude (unknot). We already know the action of the element S, which takes the knot parallel to the longitude to the knot parallel to the latitude of the torus. $SL(2, Z)$ can be generated by two elements: S and T with relations $S^2 = 1$ and $(ST)^3 = 1$. In particular, for the circle around the longitude, generator T adds a loop around the latitude, which in general is contractible in the bulk.

One can understand the two generators as the following matrices modulo sign (the actual modular group is $PSL(2, Z)$, which identifies g and $-g$ for $g \in SL(2, Z)$)

$$S = \begin{pmatrix} 0 & 1 \\ -1 & 0 \end{pmatrix}, \qquad T = \begin{pmatrix} 1 & 0 \\ 1 & 1 \end{pmatrix}. \tag{11.39}$$

An $SL(2, Z)$ element of the form

$$W_{m,n} = \begin{pmatrix} m & a \\ n & b \end{pmatrix}, \qquad bm - an = 1 . \tag{11.40}$$

will take the circle along the longitude to the torus knot (m, n).

Let us call reference state the vector $|R\rangle_{(1,0)}$, which is a solid torus with a longitude circle colored with R. We would like to define complexity of the target state $|R\rangle_{(m,n)}$, in which the Wilson line is the (m, n) knot. From what we said above,

$$|R\rangle_{(m,n)} = W_{m,n}|R\rangle_{(1,0)} . \tag{11.41}$$

The element $W_{m,n}$ of $SL(2, Z)$ can be presented as a word of generators S and T: $ST^{a_1}ST^{a_2}S\cdots ST^{a_r}$. Due to the relations for S and T, there are many words of such form that give the same element $W_{m,n}$. Hence, we would like to consider the shortest word of the generators and call complexity the total number of generators in the shortest word.

The shortest word representing $W_{m,n} \in SL(2, Z)$ is known to be related to the following continued fraction presentation of m/n (assuming $m > n$):

$$\frac{m}{n} = a_1 - \cfrac{1}{a_2 - \cfrac{1}{\ddots - \cfrac{1}{a_r}}} , \tag{11.42}$$

where all $|a_i| > 1$. The complexity of torus knot (m, n) can be defined as sum

$$\mathscr{C} = \sum_{i=1}^{r} |a_i| + r . \tag{11.43}$$

In [22] further details and interpretations of this definition of complexity are considered. One of the conclusions of that work in progress is that quantity (11.43) grows slower than the number of crossings. Moreover, since $SL(2, Z)$ also realizes unitary representations of the three-element braid group B_3, the above discussion can be extended to the class of two-bridge knots and links.

11.6 Conclusions

In these notes I reviewed different interpretations of quantum entanglement in terms of topological quantum field theories and knots. It was explained how intuitive ideas of quantum entanglement as a physical "linking" of subsystems can be explicitly realized in topological theories. In particular, it was explained, how the fundamental examples of entangled states, such as Bell pair and GHZ states can be constructed and how the entanglement entropy can be expressed in terms of the topological invariants of knots and links. An interesting further direction outlined in my joint work [25] is to understand the role of knots in quantum computation and quantum algorithms. The material of these notes contain a step in that direction. One example considered in the notes is the notion of complexity of algorithms, which can be connected to the natural notion of complexity of knots. Some interesting related material have been left behind. The reader is referred to [4, 16, 26, 27] for a subjective choice of relevant works.

Acknowledgements Most of the original ideas and results described in this note were developed in a collaboration project with Andrei Mironov, Sergei Mironov, Alexei Morozov and Andrey Morozov dedicated to applications of knot theory to quantum information. For discussions of complexity and other aspects of quantum information theory I am thankful to the members of the string theory group at the IIP in Natal, especially to Gian Camilo, and to my students João Aires, Victoria Leite and Alberto Palhares. Some essential steps in my understanding would not be possible without communication with Pawel Kaputa, Fabio Novaes and Francesco Toppan. This work was supported by the Russian Science Foundation grant No 18-71-10073.

References

1. P.K. Aravind, in *Potentiality, Entanglement and Passion-at-a-Distance*, ed. by R.S. Cohen et al (Kluwer, 1997), pp. 53–59
2. W. Dur, G. Vidal, J.I. Cirac, Phys. Rev. A **62**, 062314 (2000)
3. S.J. Lomonaco, L.H. Kauffman, in *Quantum Information and Computation II*, ed. by Donkoe, Pirich and Brandt. Proceedinbgs of Spie, 12–14 April 2004 (2004), p. 268
4. L.H. Kauffman, in *Knot Logic and Topological Quantum Computing with Majorana Fermions*, arXiv:1301.6214 [quant-ph], https://arxiv.org/abs/1301.6214. Cited 26 Jan 2013
5. V. Balasubramanian, J.R. Fliss, R.G. Leigh, O. Parrikar, JHEP **1704**, 061 (2017)
6. E. Witten, Commun. Math. Phys. **121**, 351 (1989)
7. E. Witten, Commun. Math. Phys. **117**, 353 (1988)
8. M. Atiyah, Inst. Hautes Etudes Sci. Publ. Math. **68**, 175 (1989)

9. M.F. Atiyah, *The Geometry and Physics of Knots*, 1st edn. (Cambridge University Press, Cambridge, 1990), pp. 1–108
10. D. Melnikov, A. Mironov, S. Mironov, A. Morozov, A. Morozov, JHEP **1905**, 116 (2019)
11. A. Kitaev, J. Preskill, Phys. Rev. Lett. **96**, 110404 (2006)
12. M. Levin, X.G. Wen, Phys. Rev. Lett. **96**, 110405 (2006)
13. S. Dong, E. Fradkin, R.G. Leigh, S. Nowling, JHEP **0805**, 016 (2008)
14. M.H. Freedman, A. Kitaev, Z. Wang, Commun. Math. Phys. **227**, 587 (2002)
15. A.M. Polyakov, Mod. Phys. Lett. A **3**, 325 (1988)
16. G. Salton, B. Swingle, M. Walter, Phys. Rev. D **95**(10), 105007 (2017)
17. M. Bos, V.P. Nair, Phys. Lett. B **223**, 61 (1989)
18. A. Brini, B. Eynard, M. Marino, Annales Henri Poincare **13**, 1873 (2012)
19. V. Balasubramanian, M. DeCross, J. Fliss, A. Kar, R.G. Leigh, O. Parrikar, Physics **2018**, 38 (2018)
20. A. Mironov, A. Morozov, A. Morozov, JHEP **1809**, 128 (2018)
21. D.W.F. Alves, G. Camilo, JHEP **1806**, 029 (2018)
22. G. Camilo, D. Melnikov, F. Novaes, A. Prudenziati, JHEP **1907**, 163 (2019)
23. M.A. Nielsen, M.R. Dowling, M. Gu, A.C. Doherty, Science **311**, 1133 (2006)
24. M. Rosso, V. Jones, J. Knot Theor. Ramifications **2**, 97 (1993)
25. D. Melnikov, A. Mironov, S. Mironov, A. Morozov, A. Morozov, Nucl. Phys. B **926**, 491 (2018)
26. A.I. Solomon, C.L. Ho, in Links and quantum entanglement, in *Proceedings of the Conference in Honour of Murray Gell-Mann's 80th Birthday* (NTU, Singapore, 2010), p. 646
27. L.H. Kauffman, E. Mehrotra, in *Topological Aspects of Quantum Entanglement*, arXiv:1611.08047 [math.GT], https://arxiv.org/abs/1611.08047. Cited 24 Nov 2016

Chapter 12
Finite Size Effects in Topological Quantum Phase Transitions

Mucio A. Continentino, Sabrina Rufo and Griffith M. Rufo

Abstract The interest in the topological properties of materials brings into question the problem of topological phase transitions. As a control parameter is varied, one may drive a system through phases with different topological properties. What is the nature of these transitions and how can we characterize them? The usual Landau approach, with the concept of an order parameter that is finite in a symmetry broken phase is not useful in this context. Topological transitions do not imply a change of symmetry and there is no obvious order parameter. A crucial observation is that they are associated with a diverging length that allows a scaling approach and to introduce critical exponents which define their universality classes. At zero temperature the critical exponents obey a quantum hyperscaling relation. We study finite size effects at topological transitions and show they exhibit universal behavior due to scaling. We discuss the possibility that they become discontinuous as a consequence of these effects and point out the relevance of our study for real systems.

12.1 Topological Phase Transitions

Topology studies the stability of forms, shapes under different operations. These may occur in abstract spaces as in momentum space reciprocal to crystalline structures [1–3]. If certain symmetries are present, they give rise to invariants that are robust under different operations. In many cases, in condensed matter systems, these topological invariants are directly related to physical observables [1–3]. The existence of non-trivial topological phases derives from their symmetry properties, but may occur only for restricted regions of the parameter space characterizing the system. As a consequence, if these parameters are changed, the system may transit from one non-trivial topological phase to another or even to a trivial topological phase. Here we will be interested in topological transitions that occur at zero temperature ($T = 0$) [4], as a

M. A. Continentino (✉) · S. Rufo · G. M. Rufo
Centro Brasileiro de Pesquisa Físicas, Rua Dr. Xavier Sigaud, 150,
Rio de Janeiro, RJ 22290-180, Brazil
e-mail: mucio@cbpf.br

A. Ferraz et al. (eds.), *Strongly Coupled Field Theories for Condensed Matter and Quantum Information Theory*, Springer Proceedings in Physics 239,
https://doi.org/10.1007/978-3-030-35473-2_12

physical parameter like the chemical potential is varied. The critical fluctuations in this case are purely quantum mechanical and the topological transition is a quantum phase transition [5]. These phase transitions that are of great interest nowadays differ [6], but also share many features with conventional ones. A significant difference is the lack of an order parameter since in general there is no symmetry breaking at a topological transition. The use of a topological invariant as an order parameter is not a valid option as it changes abruptly. This may wrongly suggest that the phase transition is discontinuous and does not fully develops. The main consequence of the absence of an order parameter is that a Landau expansion [7] of the ground state energy in terms of a small quantity near the transition is not possible.

The most important feature that characterizes a topological transition as a genuine critical phenomenon is the existence of a characteristic length ξ that diverges at this transition. If g is a control parameter, such that, the transition occurs at $g = 0$, we can write

$$\xi = \xi_0 |g|^{-\nu}, \tag{12.1}$$

where we will refer to ν as the *correlation length exponent* and ξ_0 is a natural length of the system, as the lattice spacing.

The identification of this characteristic length is guided by a unique attribute of non-trivial topological phases, namely, the existence of surface states that decay as they penetrate the bulk of the material [8]. This *penetration length* diverges at the topological transition and can be identified as the characteristic length scale associated with this critical phenomenon [4, 9–13].

The existence of this diverging length allows to develop a scaling theory for topological transitions [5]. The singular part of the temperature dependent free energy as a function of the distance g to the transition can be written as [5],

$$f_s \propto |g|^{\nu(d+z)} F\left[\frac{T}{|g|^{\nu z}}\right]. \tag{12.2}$$

If hyperscaling holds, the quantum hyperscaling relation implies

$$2 - \alpha = \nu(d + z), \tag{12.3}$$

where we introduced two new quantum critical exponents, α and z. Since the scaling function $F[0] =$ constant, the former characterizes the singular behavior of the ground state energy density [14]. The latter is the dynamic critical exponent and d is the dimension of the system.

The dynamic critical exponent z plays a fundamental role in quantum critical phenomena [5]. Here, it is defined by the form of the dispersion relation of the excitations at the QCP, $g = 0$, i.e., $\omega(g = 0) \propto k^z$. In general for isotropic systems close to the topological transition, the spectrum of excitations can be written as, $\omega = \sqrt{|g|^{2\nu z} + k^{2z}}$ [15]. The *wavevector* k is that for which the gap $\Delta = |g|^{\nu z}$ closes at the transition. In the cases of interest here the dynamic exponent z takes the Lorentz invariant value $z = 1$, as a consequence of the Dirac-like nature of the dispersion relation *at* the transition [15].

It is important to mention that the quantum hyperscaling relation, (12.3), that relates the quantum critical exponents to the dimension of the system can be violated in several ways [5]. For example, when the critical exponent α determined by this relation becomes negative. For the systems studied here with $z = 1$ and $\nu = 1$, as obtained below, this occurs for $d > 1$. In this case there may be analytic contributions to the free energy, like $f \propto |g|^2$ that for $\alpha < 0$ will vanish more slowly close to the QCP than the scaling contribution [16]. This implies that the exponent α remains fixed at $\alpha = 0$ for all $d \geq 1$. For $d = 1$, the marginal dimension, there may be also logarithmic corrections for the ground state energy (see below). Hyperscaling may also breakdown if the dispersion relation of the system is highly anisotropic, such that, the correlation length exponent is not uniquely defined but depends on a given direction [17].

Notice that in conventional quantum phase transitions the algebraic decay of correlations of the order parameter at the QCP requires introducing a critical exponent η [5]. This is related to the exponent β of the order parameter through another hyperscaling relation $2\beta = \nu(d + z - 2 + \eta)$ [5]. The exponents η and β play no role in the characterization of topological quantum phase transitions as discussed here.

In the next sections, we study two models exhibiting topological transitions and determine their universality classes, essentially the critical exponents ν, z and α. We start with the one-dimensional (1*d*) Su–Schrieffer–Heeger (SSH) [3] model for a dimerized tight-binding chain, which is one of the simplest model to exhibit a quantum topological phase transition. We also consider the two-dimensional (2*d*) Bernevig–Hugues–Zhang model [3] and obtain the correlation length exponents ν for both models. Finally, we discuss a 3*d* model of a topological insulator and the possible occurrence of a discontinuous transition in this system.

12.2 The Su–Schrieffer–Heeger Model

The Su–Schrieffer–Heeger (SSH) model [3] has been proposed to study the electronic properties of the polymer composed of repeating units of polyacetylene organic molecules $(C_2H_2)_n$. The Hamiltonian in real space can be written as

$$\mathcal{H} = \sum_n \psi_n^\dagger A \psi_n + \psi_n^\dagger B \psi_{n-1} + \psi_n^\dagger B^\dagger \psi_{n+1}, \tag{12.4}$$

where $\psi_n = (\psi_n^a, \psi_n^b)^T$ is the wave function vector of a unit cell n with wave function components ψ^a and ψ^b from a and b sublattices, respectively. The intra and inter cell hoppings are given by 2×2 matrices $(A)_{i,j} = t_1^* \delta_{i,j-1} + t_1 \delta_{i-1,j}$ and $(B)_{i,j} = t_2^* \delta_{i,j-1}$, respectively, where t_1 and t_2 are real numbers that represent the intra and inter cell hopping terms. After a Fourier transformation of the Hamiltonian, (12.4), we get

$$\mathcal{H} = \sum_k \psi_k^\dagger H(k) \psi_k, \tag{12.5}$$

such that, $\psi_k = (\psi_k^a, \psi_k^b)^T$ and $(H(k))_{i,j} = t(k)\delta_{i,j-1} + t^*(k)\delta_{i-1,j}$ with $t(k) = t_1 + t_2 e^{ika}$. A diagonalization process allows to obtain the energies of the electronic states of the model as

$$E(k) = \pm |t(k)| = \pm \sqrt{t_1^2 + t_2^2 + 2t_1 t_2 \cos k}, \tag{12.6}$$

where the lattice spacing was taken equal to unity. Notice that, for $|t_1| \neq |t_2|$, this energy dispersion presents a gap around zero energy. Therefore, if the Fermi level μ is taken at zero energy, the ground state describes an insulating phase. On the other hand, this model undergoes a topological phase transition at the quantum critical point, $g = t_1 - t_2 = 0$, with a gap closing at $k = \pi$.

The insulating phase that arises when $|t_1| > |t_2|$ is a trivial topological phase, since the topological invariant winding number W is equal to zero. For $|t_1| < |t_2|$, the insulating phase is topologically non-trivial with winding number equal to one. In the topological non-trivial phase, there are edge states with zero energy, $(\psi_n^a(E=0), \psi_n^b(E=0))$, that are protected by the topology of the Bloch bulk electronic states.

Solving recursively for the zero-energy eigenstates of the Hamiltonian (12.4), we find for the ratio of the wave functions at sites n and 1 at the edge of the a sub-lattice,

$$\delta\psi_n^a = \frac{\psi_n^a(E=0)}{\psi_1^a(E=0)} = \left(-\frac{t_1}{t_2}\right)^{n-1}. \tag{12.7}$$

These edge states are mostly located at the edges of the chain, more precisely in the unit cells 1 and N of the SSH model. Their existence is guaranteed by the condition $E = 0$ in (12.6) that leads to $t(\tilde{k}_0) = 0$ or $e^{i\tilde{k}_0} = -(t_1/t_2)$. Notice that in the case of edges states with zero energy, $\tilde{k}_0$ is a complex number. Substituting, $(t_1/t_2) = -e^{i\tilde{k}_0}$ in (12.7), the ratio of wave functions for the a sub-lattice can be written as

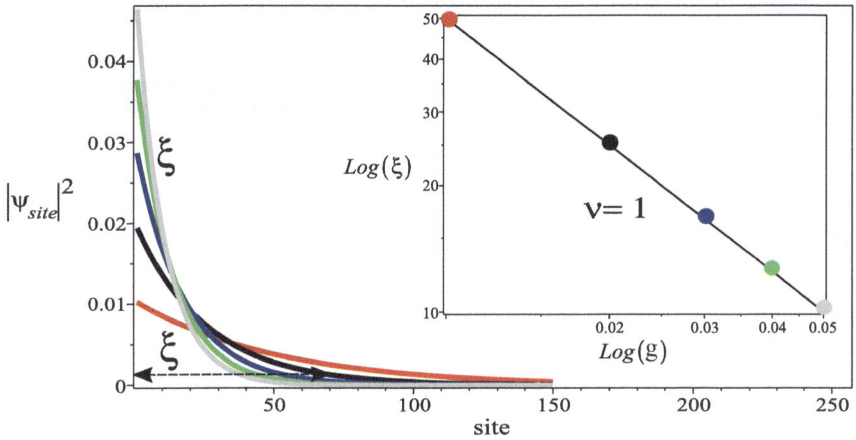

Fig. 12.1 The square of the wave function (Edge states) as a function of the sites. Solid lines are solutions for the a sublattice. The penetration depth ξ is shown for $g = 0.02$ (black curve). At ξ the wave function satisfies the condition $\Psi_n(\xi) = \Psi_1/e$. The inset shows the penetration depth versus g. Different colors represent different values of g as depicted in the inset. The angular coefficient of the straight line is formally the critical exponent, $\nu = 1$

$$\delta\psi_n^a = e^{i\tilde{k}_0(n-1)}. \tag{12.8}$$

The value of $\tilde{k}_0$ as a function of the distance from the critical point can be obtained from the following equation

$$E(k) \sim \sqrt{g^2 + t_1 t_2 k^2}, \tag{12.9}$$

which is a series expansion of (12.6) near the QCP. We have introduced $g = t_1 - t_2$ to represent the distance from this QCP. For an edge state $E(\tilde{k}_0) = 0$ and therefore (12.9) yields $\tilde{k}_0 = i(g/\sqrt{t_1 t_2})$.

Finally, substituting $\tilde{k}_0$ in (12.8), we obtain for the wave functions ratio

$$\delta\psi_n^a = e^{-(n-1)/\xi}, \tag{12.10}$$

where $\xi = \sqrt{t_1 t_2}|g|^{-1}$. The normalized wave function decays exponentially with n within the bulk with a penetration depth ξ that diverges with critical exponent $\nu = 1$. Notice that this result can also be obtained directly from (12.7).

Figure 12.1 shows the square of the wave functions ($|\psi_n^a(E = 0)|^2$) of the edge states obtained numerically from (12.4). The solid lines in the figure are the solutions for a sub-lattice as a function of the sites. There are similar solutions for the b sub-lattice (not shown) that in this case are localized near the last site.

We have defined the penetration depth ξ as the distance, relative from the initial site, for which $\Psi_n(x_0 + \xi) = \Psi_1(x_0)/e$. Now, by considering several values of g, varying from $g = 0.01$ to 0.05, we obtain $\xi(g)$ and determine the critical exponent

ν, as shown in the inset of Fig. 12.1. We get, $\nu = 1$ in perfect agreement with the analytic result of (12.10), showing that the numerical method is very reliable. In the next section, we use the numerical approach to obtain the critical exponent ν of the 2*d* BHZ model.

12.3 The Bernevig–Hugues–Zhang (BHZ) Model

The first experimental observation of a 2*d* topological quantum phase transition was in a CdTe/HgTe/CdTe heterostructure. This consisted of a layer of HgTe sandwiched between CdTe yielding a semiconductor quantum well [18]. At some critical thickness value of these quantum wells, the topological quantum phase transition takes place, from a conventional insulating phase to a quantum Hall effect phase with helical edge states protected by the non-trivial topology of the bulk. This topological quantum phase transition can be described by the BHZ model [19] associated with the following Hamiltonian

$$H(k_x, k_y) = \boldsymbol{\sigma} \cdot \mathbf{h}(k), \tag{12.11}$$

where $\mathbf{h}(k)$ takes values on the two-dimensional Brillouin zone (k_x, k_y) and $\boldsymbol{\sigma} = \{\sigma_x, \sigma_y, \sigma_z\}$ are the Pauli matrices. Specifically, $h_x = t_{sp} \sin k_x$, $h_y = t_{sp} \sin k_y$ and $h_z = 2t_1(\cos k_x + \cos k_y) + t_2 - 4t_1$. In this Hamiltonian, the sub-lattice space represents the orbitals *s* and *p* for each atom. In order to describe the quantum wells in HgTe/CdTe layers, the simplified spinless BHZ model introduces the hopping terms t_{sp} and t_1, as well as, a *mass* term t_2. The antisymmetric hybridization between the orbitals of different parities, *s* and *p* has an amplitude given by t_{sp}, and the hopping between the same orbitals *s* or *p* of nearest neighbors atoms has an amplitude t_1.

A topological phase can be identified by some proper topological invariant. For the 2*d* BHZ model, we can consider the Chern number invariant $\mathscr{C}$ [20, 21] obtained at the high-symmetry points $[k_x, k_y] = \{[0, 0], [0, \pi], [\pi, 0], [\pi, \pi]\}$. It predicts a non-trivial topological phase for the intervals $0 < t_2 < 4t_1$ with $\mathscr{C} = 1$ and $4t_1 < t_2 < 8t_1$ with $\mathscr{C} = -1$. A trivial phase with $\mathscr{C} = 0$ occurs for $t_2 > 8t_1$. The Chern number signs $\mathscr{C} = \pm 1$ are related to edge states with propagation in opposite directions.

Here we are interested in determining numerically the correlation length critical exponent for the two-dimensional BHZ model. For this purpose, we study the penetration of the edge states, which requires one of the dimensions of the lattice to be finite. Since these edge states are indeed connected to the real terminations of the system, for a square lattice to keep one of the dimensions finite means to deform the lattice into a cylinder. The finite axis takes the direction of the main axis of the cylinder and the other dimension with periodic boundary conditions is represented by the body of the cylinder.

The correlation length critical exponent as before characterizes the decay of the edge states into the bulk close to the topological transition. Let us consider the edge states of the BHZ model in one dimension. One way to get one of the dimensions finite is to perform a Fourier transformation as

$$H_{I,J}(k_y) = \frac{1}{N_x}\sum_{k_x} e^{ik_x(m-m')} H_{I,J}(k_x, k_y), \tag{12.12}$$

where N_x is the number of sites along the finite x-axis and $\{I, J\}$ indexes run over the matrix elements of (12.11). The positions of the atoms along the finite x-axis are denoted by m and vary from 0 to N_x. For example, considering the element $H_{1,1}(k_x, k_y) = h_z$ we have

$$\begin{aligned} H_{1,1}(k_y) &= \frac{1}{N_x}\sum_{k_x} e^{ik_x(m-m')} H_{1,1}(k_x, k_y) \\ &= \frac{1}{N_x}\sum_{k_x} e^{ik_x(m-m')} [2t_1 \cos k_x + C] \\ &= \frac{1}{N_x}\sum_{k_x} e^{ik_x(m-m')} \left[t_1(e^{ik_x} + e^{-ik_x}) + C\right] \\ &= \frac{1}{N_x}\sum_{k_x} \left[t_1 \left(e^{ik_x(m-m'+1)} + e^{ik_x(m-m'-1)}\right) + Ce^{ik_x(m-m')}\right] \\ &= t_1[\delta_{m,m'+1} + \delta_{m,m'-1}] + C\delta_{m,m'}, \end{aligned} \tag{12.13}$$

where $C = 2t_1 \cos k_y + (t_2 - 4t_1)$ is independent of k_x and the same procedure should be applied to all the other matrix elements.

The sum over k_x allows to work in real space along the x-axis. Notice that we chose the k_x to be in the finite direction, but since we consider a square lattice the choice between k_x or k_y is irrelevant due the symmetry of the lattice. For the diagonal directions of the square lattice [22], or for more complex lattices this is not necessarily true. For instance, for the honeycomb lattice, the choice of the finite axis along one or other direction means different edge arrangements [23].

Accordingly, after Fourier transforming (12.12), from momentum to real space along the x-axis, we have

$$H(k_y) = \begin{pmatrix} H_{11} & H_{12} \\ H_{21} & H_{22} \end{pmatrix}. \tag{12.14}$$

Following the procedure of (12.13) yields $H_{11} = [2t_1 \cos k_y + (t_2 - 4t_1)]\delta_{m,m'} + t_1[\delta_{m,m'+1} + \delta_{m,m'-1}]$ that stands for sub-lattice a and $H_{22} = -H_{11}$ for sub-lattice b. Here, the sub-lattices indexes a and b represent the subspace of the orbitals s and p, respectively. The matrix elements responsible for the mixing of the different orbitals or sub-lattices are given by $H_{12} = -it_{sp} \sin k_y \delta_{m,m'} - \frac{it_{sp}}{2}[\delta_{m,m'-1} - \delta_{m,m'+1}]$ and $H_{21} = H_{12}^{\dagger}$. The m index counts the unit cells or atoms along the finite x-axis and in the same way m' can be interpreted as a neighbor site in the real space Hamiltonian, (12.14). Besides, the order of each matrix element $H_{I,J}$ is increased to $N_x \times N_x$, which means that the order of the final matrix becomes $2N_x \times 2N_x$.

For the purpose of obtaining the energy dispersion in real space, a numerical study of the $2d$ BHZ model was developed to diagonalize the Hamiltonian, (12.14). We fix

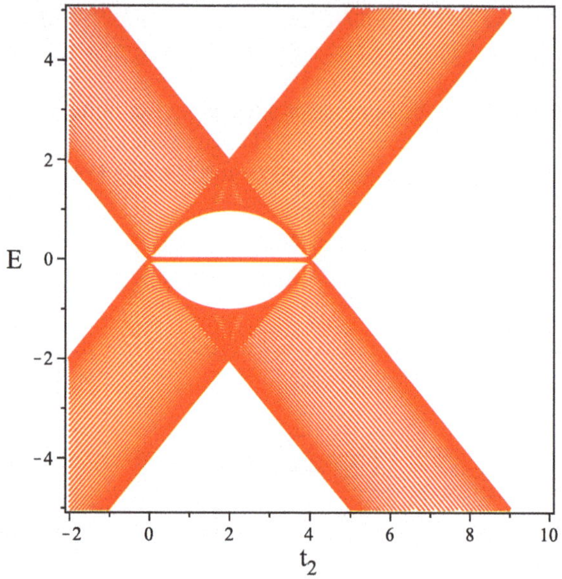

Fig. 12.2 (Color online) Energy dispersion of the BHZ model as a function of t_2, for a fixed $t_1 = 1$. The red lines are obtained for $k_y = 0$ and for $N_x = 50$ sites. Topological quantum phase transition takes place for $t_2 = 0$ and $t_2 = 4$. Along the line E = 0, we highlighted the presence of the two edge states with thick lines

the energy scale as $t_1 = 1$ and take $N_x = 50$ sites. In Figs. 12.2 and 12.3, respectively, we present the energy E as a function of the topological transition control parameter (mass) t_2 at the high symmetry points, $k_y = 0$ and $k_y = \pi$. In the first case, $k_y = 0$, topological quantum phase transitions take place for $t_2 = 0$ and $t_2 = 4$. The thick lines in the figures show the presence of the edge states with zero energy. The same is observed for $k_y = \pi$, but the transition points are now given by $t_2 = 4$ and $t_2 = 8$.

For the study of the penetration of the edge states, we identify the eigenvectors responsible for the zero energy dispersions in Figs. 12.2 and 12.3. For $N_x = 500$, we show in Fig. 12.4 the square of the wave function of the edge states in the vicinity of the critical points. Actually, just one half of the lattice is presented, since the behavior is the same on both sides. In addition, the results for sub-lattice a and b coincide. The edge states are obtained for distances to the critical point ranging from $g = 0.01$ to $g = 0.05$. As g increases, the edge states become more localized at the edges of the lattice. The inset presents the characteristic length ξ as a function of g and the points are obtained from the numerical study of the model. From the linear fitting of these points, we can conclude with accuracy that the correlation length critical exponent for the $2d$ BHZ model is $\nu = 1$. As mentioned before, in real space the lattice is a cylinder and Fig. 12.4 presents a pictorial view of the penetration of the edge states from the perspective of this cylinder. The top cylinder represents the case where the edge states penetration decays very fast. The color gradient follows the penetration intensity of the edge state. In the same way, the bottom cylinder shows a case where the edge state extends almost along the entire lattice. The color gradient here holds inside the cylinder body. These results reflect strictly the behavior obtained for all critical points $t_2 = 0$, $t_2 = 4$ and $t_2 = 8$.

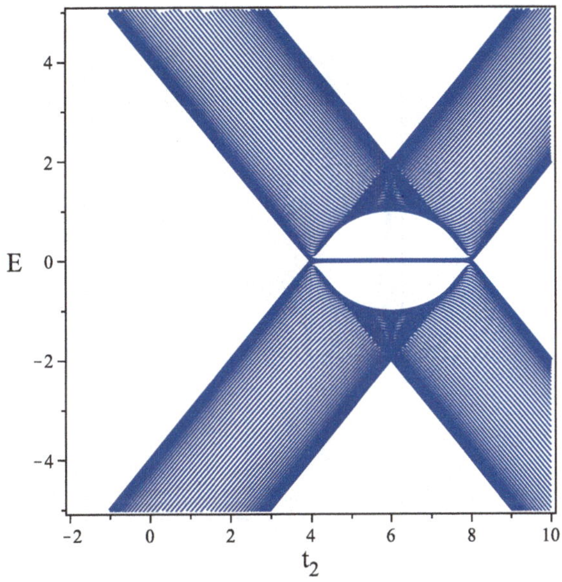

Fig. 12.3 (Color online) The same of Fig. 12.2, but the blue lines are obtained for $k_y = \pi$. In this case, the topological quantum phase transition takes place for $t_2 = 4$ and $t_2 = 8$. Again, we highlighted the presence of the two edge states with thick lines along E = 0

In the process of varying the distance to the quantum critical point, we notice that as the system moves away from the QCP, the behavior of the penetration length for the orbitals s and p (sub-lattices a and b) becomes distinct at the different edges. Figure 12.5 shows that for $g \geq 0.068$, the wave function of the left edge state is nearly localized and has mostly s-character, while that of the right edge has mostly p-character. We also observe that the amplitude of the wave functions at the edges and consequently their localization at these sites becomes larger as g increases. The cylinders here indicate the correspondence between the edge states of the subspaces and the termination of the lattice for each case. Finally, close to the QCP the wave functions of the edge states have a mixed character, as shown in Fig. 12.4, due to their strong hybridization.

In summary our numerical study of the $2d$ BHZ model shows that the critical exponent for the penetration depth takes the value $\nu = 1$, the same we have obtained for the $1d$ SSH and for a 1d sp-chain [5]. We have also pointed out a qualitative change in the nature of the edge states for the 2d BHZ model as the distance to the QCP of the topological transition changes.

12.4 Finite Size Effects at Topological Transitions

Quantum topological transitions as conventional phase transitions also exhibit finite size scaling properties [15]. For a finite system close to quantum criticality, the characteristic length ξ and the finite size L are the relevant length scales. The singular

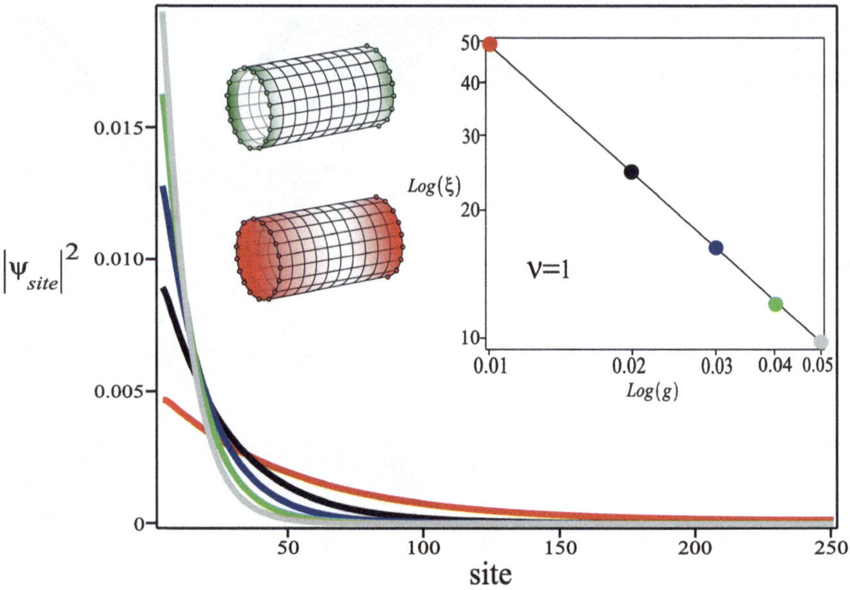

Fig. 12.4 The square of the wave function (edge states) as a function of sites ($N_x = 500$) sites close to the quantum critical point, $t_2 = 0$. We show just half of the lattice ($N_x = 250$) for one sub-lattice, since the behavior is the same on both sides for the two sub-lattices a and b. The color scheme and the penetration depth follows that of Fig. 12.1. The inset shows the penetration depth versus g and yields with high accuracy the value $\nu = 1$ for the critical exponent of the penetration depth. For all the QCPs as g increases and the system moves away from these QCPs, the edge states become more localized at the ends. The cylinders represent the finite BHZ lattice along the x-axis direction (main axis of the cylinder) with periodic boundary conditions in y-axis (body of the cylinder). The top cylinder shows the penetration of the wave-function of the edge states for $g = 0.04$ according to the color gradient (green). Similarly, the bottom cylinder shows the same (red), but for $g = 0.01$. In this case the wave function of the edge state penetrates almost the entire lattice

part of its free energy $\delta\mathscr{F}_C(g, L)$ is expected to have a finite size contribution that can be written as [15],

$$\delta\mathscr{F}_C^{\text{sing}}(g, L) = \Delta_C L^{-(d+z-1)} f(L/\xi). \tag{12.15}$$

This follows from dimensional analysis and a finite size scaling assumption. It is a natural generalization of the classical result for the quantum case [24] and for topological transitions [15]. In (12.15), the dimension d of the classical system is replaced by the effective dimension $d + z$ as in the quantum hyperscaling relation [25]. At the QCP of the bulk system, the characteristic length is infinite, and the scaling function $f(L/\xi = 0) = 1$. For $d + z = 1 + 1$, conformal invariance implies that the amplitudes Δ_C are universal quantities [26].

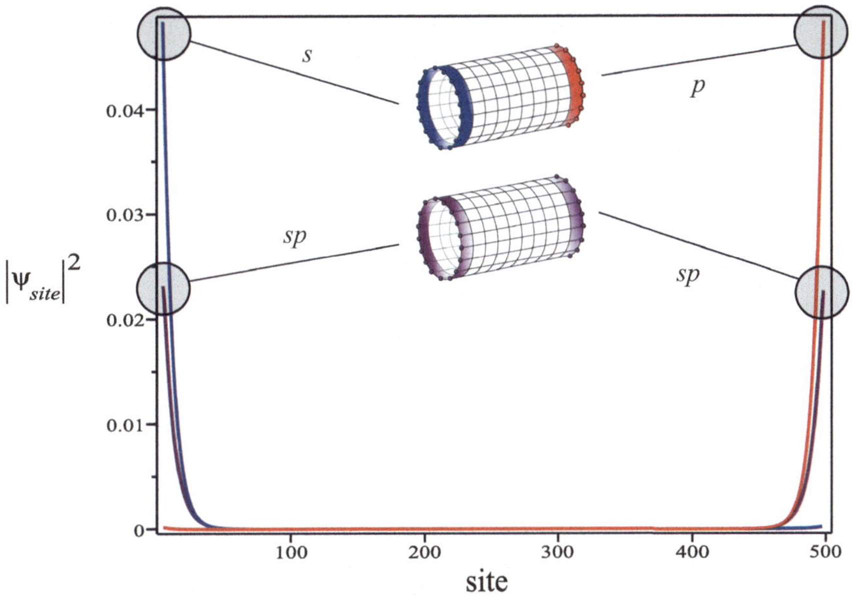

Fig. 12.5 (Color online) The same of Fig. 12.4, but exploring the effect of the proximity to criticality on the nature of the edge states. In the close vicinity of the topological quantum phase transition the edge states of the subspace of the orbitals s and p, in purple color, coincide and exhibit the same amplitudes in both sides of the lattice. This is represented by the bottom cylinder filled with the same purple color in each edge. However, far away from the critical point ($g \geq 0.068$) the edge states on different sides begin to present mostly s (left) or p (right) character. For instance, the wave function of the edge state nearly localized at the left end has a very strong s character. The opposite for the right end with a wave function with mostly p-character. This is indicated by the top cylinder where the blue and red colors represent the s and p orbitals, respectively

The scaling form, (12.15), of the finite size contribution to the free energy has been successfully verified for several systems exhibiting topological quantum phase transitions [15], as the $1d$ p-wave superconductor model of Kitaev, the $3d$ SSH model and a $3d$ model for topological insulators [15]. In all these cases, the dynamic exponent is given by $z = 1$ and the correlation length exponent turns out to be $\nu = 1$.

For the purpose of calculating the finite size properties of a system, it is useful to consider it as confined within two parallel planes of area S separated by a distance L. The free energy per unit area of this slab can be written as [24, 25, 27]

$$\lim_{S\to\infty} \frac{\mathscr{F}(g, L)}{JS} = L F_{\text{bulk}}(g) + F_{\text{surface}}(g) + \delta\mathscr{F}_C(g, L), \tag{12.16}$$

where $F_{\text{bulk}}(g)$ is the dimensionless bulk free energy per unit volume of the unconfined system, F_{surface} is the sum of the free energies of the surfaces, per unit surface area, due to the confining planes and J is a natural energy scale of the bulk material. If one uses periodic boundary conditions, the surface terms do not arise in the expression above [15]. The last term represents the finite size contribution to the free energy per unit area from the slab of width L. It can be imagined as mediating interface-interface interactions at a distance L. Close to criticality its singular part has the scaling form given by (12.15) above. In the next sections, we use (12.16) to obtain this term and explore the physics it contains.

12.4.1 Multi-band Topological Insulator

In this section we focus on a 3d multi-band topological insulator. Recently [28], a theory has been formulated that points out the existence of a time-reversal invariant in these systems with $\Theta^2 = -1$. This occurs whenever a band of conduction electrons hybridizes with the $m_J = \pm 1/2$ doublet arising from the f-multiplet of a rare-earth system in a crystalline environment for which this doublet is the ground state. The theory considers an effective four-band model of dispersive quasi-particles, with different effective masses. The parity of the orbitals forming these bands is such that the k-dependent hybridization between them is antisymmetric [28]. The Hamiltonian belongs to class AII, and is characterized by a $\mathbb{Z}_2$ invariant. The dispersion relations of the hybridized bands [28] of the model are given by,

$$\omega_{1/2} = \frac{1}{2}\left[(\varepsilon_k^a + \varepsilon_k^b) \pm \sqrt{(\varepsilon_k^a - \varepsilon_k^b)^2 + 4|V(k)|^2}\right], \tag{12.17}$$

where

$$\varepsilon_k^a = -\varepsilon_0^a + 2t(\cos k_x a + \cos k_y a + \cos k_z a),$$

$$\varepsilon_k^b = \varepsilon_0^b + 2\tilde{\alpha}t(\cos k_x a + \cos k_y a + \cos k_z a)$$

are the dispersions of the originals non-hybridized bands.

The quantity $\tilde{\alpha}$ multiplying the hopping term above accounts for the different *effective masses* of the quasi-particles and $\varepsilon_0^{a,b}$ are the centers of the bands. The k dependent hybridization is given by

$$|V(k)|^2 = V_0^2(\sin^2 k_x a + \sin^2 k_y a + \sin^2 k_z a),$$

where V_0 measures the intensity of the (antisymmetric) effective hybridization.

We consider here the simplest case of $\tilde{\alpha} = 1$ and *inverted bands*, i.e, $\varepsilon_k^b = -\varepsilon_k^a = -\varepsilon_k$, such that, $\varepsilon_0^a = \varepsilon_0^b = \varepsilon_0$. This preserves the topological properties of the original model. In this case we get,

$$\omega_{1/2} = \pm\sqrt{\varepsilon_k^2 + |V(k)|^2}. \tag{12.18}$$

In the continuum limit and for $k \to 0$, we obtain

$$\varepsilon_k = -\varepsilon_0 + 6t - t(ak)^2 = g - t(ak)^2$$

and

$$|V(k)|^2 = V_0^2(ak)^2,$$

with $k^2 = k_x^2 + k_y^2 + k_z^2$ and $g = 6t - \varepsilon_0$. This model has a topological transition at $g = 0$ from a non-trivial topological insulator for $g < 0$ to a trivial one for $g > 0$ [28]. The dispersion relations close to the transition can be cast in the general form [29],

$$\omega_{1/2}/V_0 = \pm\sqrt{M^2 + (1 - 2MB)(ak)^2 + B^2(ak)^4}, \tag{12.19}$$

where $M = g/V_0$ and $B = t/V_0$. Notice that at the QCP, $M = 0$ and for $k \to 0$, $\omega \propto k^z$ with the dynamic exponent $z = 1$. Alternatively, at $k = 0$ there is a gap in the spectrum, $\omega \propto |g|$ that vanishes at the QCP with the *gap exponent* $\nu z = 1$. The dispersion relations, (12.19), describe a large variety of topological insulators [29]. The ground state energy density associated with these dispersions is given by

$$f_s = \frac{E_{GS}}{V_0 V} = \frac{1}{(2\pi)^3}\int d^3k\sqrt{M^2 + (1 - 2MB)(ak)^2 + B^2(ak)^4}, \tag{12.20}$$

where V is the volume of the system. Close to the topological transition, we introduce a characteristic length $\xi \propto M^{-1} \propto g^{-1}$, such that, the ground state energy density can be written in the scaling form,

$$f_s \propto \xi^{-4}\int_0^{\Lambda\xi} 4\pi d(k\xi)(k\xi)^2\sqrt{1 + (k\xi)^2}, \tag{12.21}$$

where Λ is a cut-off and we considered only the most singular terms close to the QCP. This equation can be cast in the scaling form,

$$f_s \propto |g|^{\nu(d+z)} F[\Lambda\xi], \tag{12.22}$$

where $\nu = 1$, $z = 1$, as identified previously and $d = 3$.

Performing the integration of (12.21) and taking the limit $\Lambda\xi \to \infty$, one obtains different contributions for the free energy,

- a cut-off independent term that corresponds to the scaling contribution, $f_S \propto |g|^{\nu(d+z)} = |g|^4$.
- a cut-off independent term, $f_s \propto |g|^4 \log|g|$ that violates hyperscaling [29].

Cut-off dependent contributions including,

- a constant term, i.e., independent of g, that represents to the most singular cut-off dependent term.
- a term of order $|g|^2$ with a cut-off dependent coefficient. This appears for all $d \geq 1$.

For the one, two and three dimensional systems studied here, the correlation length exponents take the value $\nu = 1$, the dynamical exponents $z = 1$ and consequently the *gap exponents* $\nu z = 1$. When these are substituted in the quantum hyperscaling relation, (12.3), we obtain that $\alpha < 0$ for $d > 1$. In this case the non-universal, cut-off dependent $|g|^2$ term in the free energy, present for all $d \geq 1$, dominates its behavior as $g \to 0$. Since this is the leading term for $d > 1$, then $d = 1$ plays the role of an upper critical dimension for these topological transitions. According to this interpretation, we expect the critical exponents to be fixed at their $1d$ values for all $d > 1$. The presence of a logarithmic correction to the ground state energy in $d = 1$ is consistent with its role as a marginal dimension.

If one considers an expansion of the more general expression for the free energy, (12.20), in powers of B ($B = (t/V_0) < 1$), we find that the contribution proportional to $|g|^4 \log |g|$ remains and acquires a B dependent coefficient [29]. Subtracting the diverging, cut-off dependent terms in this expansion, this simple type of renormalization leads to a free energy $f_s(M, B)$ that exhibits a discontinuous transition between the trivial, $M < 0$ and topological insulator, $M > 0$ as a function of B [29]. This possibility of a first order topological transition associated with a gap that never closes [30] is very interesting and we wish to examine it using a type of renormalization different from that of [29].

12.4.2 Casimir Effects in Topological Insulators

The first order topological transition found in [29] at $B = B_c$ relies on the renormalization procedure to deal with the cut-off in (12.20). We explore here the possibility of a discontinuous topological transition using a new scenario and a different renormalization procedure. For this purpose we consider, as in Sect. 12.4, that the system with the spectrum of excitations corresponding to (12.17) is confined within two parallel plates of area S separated by a distance L. The free energy per unit area of this system is given by (12.16). Here we present calculations of the quantity $\delta\mathscr{F}_C^{\text{sing}}(g, L)$ for a slab of a multi-band topological insulator using a method similar to that for obtaining the Casimir force between parallel plates in the theory of electromagnetism [31, 32]. Since Casimir's calculation is also a renormalization procedure, we investigate the possibility of a discontinuous topological transition in the multi-band topological insulator using this approach. The boundary conditions in the slab are that the wave functions assume the same constant value in both planes, at $z = 0$ and $z = L$. The energy of the insulating slab can be written as,

$$\frac{E_S}{V_0}=\frac{4\pi^2 Sa}{L^3}\int_0^\infty \mathrm{d}yy\sum_{n=-\infty}^{\infty}\sqrt{M_L^2+(1-2M_L B_L)(y^2+n^2)+B_L^2(y^2+n^2)^2}. \tag{12.23}$$

where $y=k_\perp L/2\pi$, with $k_\perp^2=k_x^2+k_y^2$, $M_L=M(L/2\pi a)$ and $B_L=B(2\pi a/L)=L_0/L$ where we introduced a new length scale $L_0=2\pi aB=2\pi at/V_0$ associated with the hybridization (a is the lattice spacing). The energy of the insulator occupying the whole space is given by,

$$\frac{E_B}{V_0}=\frac{4\pi^2 Sa}{L^3}\int_0^\infty \mathrm{d}yy\int_{-\infty}^{\infty}\mathrm{d}t\sqrt{M_L^2+(1-2M_L B_L)(y^2+t^2)+B_L^2(y^2+t^2)^2}, \tag{12.24}$$

with $t=k_z L/2\pi$.

The calculation of the energy difference, $\Delta E=E_S-E_B$ yields the scaling contribution according to (12.16). It is carried out in [15] using the techniques to obtain the Casimir force in critical slabs. We obtain for this energy difference at $M=0$, or $\xi=\infty$, i.e., at the topological transition

$$\frac{\Delta E}{SV_0}=\frac{-\pi^2 a}{15}L^{-3}, \tag{12.25}$$

which obeys the finite size scaling form

$$\frac{\Delta E}{SV_0}=\Delta_C L^{-(d+z-1)}, \tag{12.26}$$

with $d=3, z=1$ and the Casimir amplitude $\Delta_C=-\pi^2 a/15$. Away from criticality, since $M_L=L/\xi$, we can write

$$\frac{\Delta E}{SV_0}=-16\pi^2 aL^{-3}f(L/\xi). \tag{12.27}$$

For $L/\xi\gg 1$, the scaling function $f(L/\xi)\propto\exp(-2\pi L/\xi)$ and the finite size contribution vanishes exponentially for $L\gg\xi$.

The full expression for the energy difference is given by [15],

$$\frac{\Delta E}{SV_0}=\frac{-32\pi^3 B}{L^4}\left(\int_{x_2}^{x_1}\mathrm{d}t\frac{f_1(t)}{e^{2\pi t}-1}+\int_{x_1}^{\infty}\mathrm{d}t\frac{f_2(t)}{e^{2\pi t}-1}\right), \tag{12.28}$$

which is a function of M, B and L. The quantities $x_{1,2}$ are given by

$$x_{1,2}^2=\frac{1}{2B_L^2}\left[(1-2MB)\pm\sqrt{1-4MB}\right]. \tag{12.29}$$

The functions in the integrand are

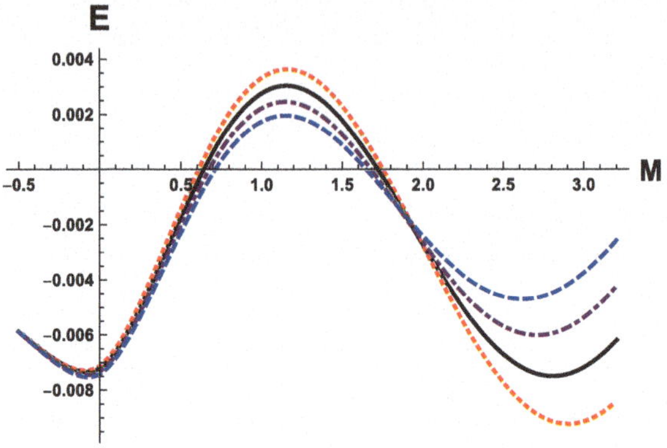

Fig. 12.6 (Color online) The ground state energy $E(M, B, L)$ as a function of M for $L = 3$ fixed and different values of B. Dotted line (red) $B = 0.41$, Full line (black) $B = B_c = 0.39$, Dot-dashed line (purple) $B = 0.37$ and dashed line (blue) $B = 0.35$. For $B = B_c \approx 0.39$, the minimum at small, negative M exchanges stability with the one at positive M

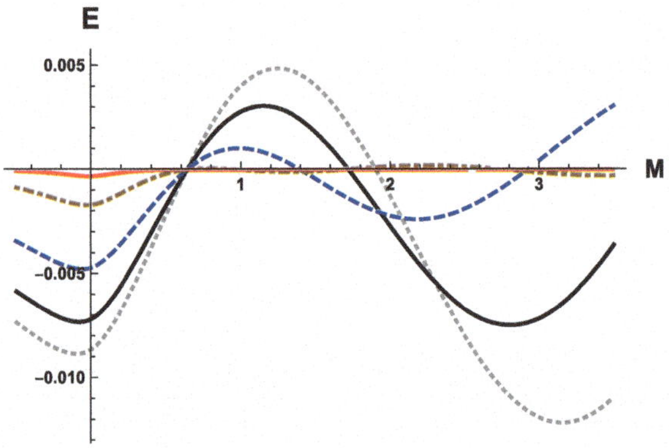

Fig. 12.7 The ground state energy $E(M, B, L)$ as a function of M for different values of L and $B = B_c = 0.39$ fixed. From bottom to top (at $M = 0$): dotted line (grey) $L = 2.8$, full line (black) $L = 3$, dashed line (blue) $L = 3.5$, dot-dashed line (brown) $L = 5$ full line (red) $L = 9$. For $L \approx 3$ there is a first order topological transition (full line) (see text)

$$f_1(t) = \frac{1}{16}x_1^4\left[\frac{\pi}{2}(1-\alpha^2)^2 - 2\sqrt{(\eta^2-\alpha^2)(1-\eta^2)}(1-2\eta^2+\alpha^2) - (1-\alpha^2)^2\tan^{-1}\frac{1-2\eta^2+\alpha^2}{2\sqrt{(\eta^2-\alpha^2)(1-\eta^2)}}\right], \tag{12.30}$$

for $x_2 < t < x_1$, where $\eta = t/x_1$, $\alpha^2 = (x_2^2/x_1^2)$ and

$$f_2(t) = \frac{\pi}{16}x_1^4(1-\alpha^2)^2, \tag{12.31}$$

that is independent of t ($t > x_1$).

Finally, the expression for the free energy difference $E(M, B, L) = \Delta E/SV_0$, (12.28), can be integrated numerically and the results are shown in Figs. 12.6 and 12.7. In Fig. 12.6, $E(M, B, L)$ is plotted as a function of M for a fixed separation $L = 3$ between the plates and different values of the parameter B. One notices the presence of two minima, one at small negative values of M and another for positive M. These minima exchange stability at a critical value of $B = B_c \approx 0.39$. The quantity M plays the role of an order parameter being negative in the trivial phase and positive in the topologically non-trivial phase [29]. For $B > B_c$ the stable minimum occurs for positive M and the system is in the topological phase. For $B < B_c$ the minimum at small negative M is the more stable and the system is in the trivial insulating phase. They exchange stability at $B = B_c$ where a first order transition occurs.

In Fig. 12.7, $E(M, B, L)$ is plotted as a function of M, now for a fixed value of $B = B_c$ and increasing separations L between the plates. For $L \approx 3$ there is a first order topological phase transition (full black line), such that, for systems with $L > 3$, the stable phase is the topologically trivial with $M < 0$. As L increases the minimum at negative M moves to zero and the curve $E(M, B_c, L \to \infty)$ has a single minimum at this value of M. The amplitude of the minimum at $M = 0$ decreases according to the finite size scaling law, (12.25) and the curve for $E(M, B, L)$ becomes progressively flat and small as a function of M.

A phenomenon similar to the one we have obtained, i.e., a first order transition in finite slabs that eventually evolves to a continuous one for large separations between plates has also been shown to occur in a strongly interacting system [33] exhibiting a fermionic condensate. In both cases the discontinuous character of the transition is due to finite size effects. Ultrathin films of topological insulators can provide ideal platforms to investigate these finite size effects [34].

12.5 Conclusions

In this work we discussed how to describe and characterize topological quantum phase transitions. We identified a characteristic length in this problem, namely the penetration length of the surface modes in the non-trivial topological phase of the system. It diverges as $\xi \propto |g|^{\nu}$ where ν is the correlation length exponent and g the distance to the transition. For simplicity, we neglected interactions, to put in evidence the purely topological aspects of the phenomenon and avoid the interference of any competing long range ordering. The role of interactions in topological systems is an active area of investigation [35] and these may give rise to new universality classes.

We have obtained numerically the critical exponent $\nu = 1$ for two well known systems exhibiting topological transitions, the SSH model in one dimension and the two dimensional BHZ model. Besides ν, two other critical exponents, z and α determine the universality class of the topological transition. The former is the dynamic critical exponent that for the systems studied here assumes the value $z = 1$ implying their Lorentz invariance. This value of z is also connected with the Dirac-like spectrum of excitations at the QCP. The exponent α determines the singular behavior of the free energy at zero temperature. These exponents are not independent but related through the quantum hyperscaling relation [5]. We have however pointed out that hyperscaling can break down and indicated how this may occur for non-interacting systems. We discussed the existence of an upper critical dimension d_C for the Lorentz invariant systems treated here and argued that it takes the value $d_C = 1$. We expect that for all $d > d_C$, the critical exponents remain fixed at their values for $d = d_C$.

Finally, we have studied the possibility of discontinuous topological transitions where the gap in the spectrum never closes. Our approach is inspired on that used to study the Casimir effect, It turns out to be an efficient method of renormalization that allows to get rid of infinities. We have shown that for a $3d$ slab with one finite dimension, finite size effects can give rise to an exchange of stability between the trivial and topological phases in a discontinuous transition. However, as the distance between the plates of the slab increases, these effects disappear.

Acknowledgements We would like to thank the Brazilian agencies, CNPq, CAPES and FAPERJ for partial financial support.

References

1. J. Alicea, Rep. Prog. Phys. **75**, 076501 (2012)
2. M.Z. Hasan, C.L. Kane, Rev. Mod. Phys. **82**, 3045 (2010)
3. S.-Q. Shen, Topological insulators, in *Dirac Equation in Condensed Matter*. Springer Series in Solid-State Sciences, 2nd edn., vol. 187 (Springer, Berlin, 2017)
4. M.A. Continentino, Phys. B: Condens. Matter **505**, A1–A2 (2017)
5. M.A. Continentino, *Quantum Scaling in Many-Body Systems: an Approach to Quantum Phase Transitions*, 2nd edn. (Cambridge University Press, Cambridge, 2017)
6. J.M. Kosterlitz, Rev. Mod. Phys. **89**, 040501 (2017)
7. L.D. Landau, Zh. Eksp. Teor. Fiz. **7**, 19 (1937); Ukr. J. Phys. **53**, 25 (2008)
8. A.Y. Kitaev, Physics-Uspekhi **44**, 131 (2001); A. Kitaev, Ann. Phys. **303**, 2 (2003)
9. M.A. Continentino, F. Deus, H. Caldas, Phys. Lett. **A378**, 1561 (2014)
10. S.N. Kempkes, A. Quelle, C. Morais Smith, Sci. Rep. **6**, 38530 (2016); A. Quelle, E. Cobanera, C. Morais Smith, Phys. Rev. **B94**, 075133 (2016)
11. M.A. Continentino, H. Caldas, D. Nozadze, N. Trivedi, Phys. Lett. A **378**, 3340 (2014)
12. W. Chen, M. Legner, A. Ruegg, M. Sigrist, Phys. Rev. B **95**, 075116 (2017)
13. E.P.L. van Nieuwenburg, A.P. Schnyder, W. Chen, Phys. Rev. **B97**, 155151 (2018)
14. M.A. Continentino, G.M. Japiassu, A. Troper, Phys. Rev. **39**, 9734 (1989)
15. M.A. Griffith, M.A. Continentino, Phys. Rev. E **97**, 012107 (2018)
16. F. Sun, J. Ye, Phys. Rev. B **96**, 035113 (2017)
17. B. Roy, P. Goswami, V. Juričić, Phys. Rev. **B95**, 201102(R) (2017); B. Roy, M.S. Foster, Phys. Rev. **X8**, 011049 (2018)
18. M. König, S. Wiedmann, C. Bröne, A. Roth, H. Buhmann, L.W. Molenkamp, X.-L. Qi, S.-C. Zhang, Science **318**(5851), 766 (2007)
19. B.A. Bernevig, T.L. Hughes, S.-C. Zhang, Quantum spin hall effect and topological phase transition in HgTe quantum wells. Science **314**, 1757 (2006)
20. B.A. Bernevig, T. Hughes, *Topological Insulators and Topological Superconductors* (Princeton University Press, Princeton, 2013)
21. M.Z. Hasan, C.L. Kane, Colloquium: topological insulators. Rev. Mod. Phys. **82**, 3045 (2010)
22. K.-I. Imura, A. Yamakage, S. Mao, A. Hotta, Y. Kuramoto, Phys. Rev. B **82**, 085118 (2010)
23. K.-I. Imura, S. Mao, A. Yamakage, Y. Kuramoto, Nanoscale Res. Lett. **6**, 358 (2011)
24. M. Krech, *The Casimir Effect in Critical Systems* (World Scientific Publishing Co. Pte. Ltd., Singapore, 1994) and references within
25. J.G. Brankov, D.M. Danchev, N.S. Tonchev, *Theory of Critical Phenomena in Finite-Size Systems: Scaling and Quantum Effects* (World Scientific Publishing Co. Pte. Ltd., Singapore, 2000)
26. T. Gulden, M. Janas, Y. Wang, A. Kamenev, Phys. Rev. Lett. **116**, 026402 (2016)

27. A. Gambassi, J. Phys.: Conf. Ser. **161**, 012037 (2009)
28. M.A. Griffith, M.A. Continentino, T.O. Puel, Phys. Rev. **B99**, 075109 (2019)
29. V. Juričić, D.S.L. Abergel, A.V. Balatsky, Phys. Rev. **B95**, 161403(R) (2017)
30. G. Krizman, B.A. Assaf, M. Orlita, T. Phuphachong, G. Bauer, G. Springholz, G. Bastard, R. Ferreira, L.A. de Vaulchier, Y. Guldner, Phys. Rev.**B98**, 161202(R) (2018); G. Krizman, B.A. Assaf, T. Phuphachong, G. Bauer, G. Springholz, L.A. de Vaulchier, Y. Guldner, Phys. Rev. **B98**, 245202 (2018)
31. M. Bordag, G.L. Klimchitskaya, U. Mohideen, V.M. Mostepanenko, *Advances in the Casimir Effect* (Oxford University Press, 2009), p. 22
32. J. Schiefele, C. Henkel, J. Phys. A: Math. Theor. **42**, 045401 (2009)
33. A. Flachi, M. Nitta, S. Takada, R. Yoshii, Phys. Rev. Lett. **119**, 031601 (2017); A. Flachi, Phys. Rev. **D86**, 104047 (2012)
34. J.L. Collins, A. Tadich, W. Wu, L.C. Gomes, J.N.B. Rodrigues, C. Liu, J. Hellerstedt, H. Ryu, S. Tang, S.-K. Mo, S. Adam, S.A. Yang, M.S. Fuhrer, M.T. Edmonds, Nature, **564**, 390 (2018)
35. B.-J. Yang, E.-G. Moon, H. Isobe, N. Nagaosa, Nat. Phys. **10** 774 (2014)

Chapter 13
From Quantum Spin Chains to Chiral Spin Liquids

Rodrigo G. Pereira

Abstract Chiral spin liquids are highly entangled phases of matter in which interacting spins break time reversal and reflection symmetries, but do not develop magnetic order even at zero temperature. The conventional analytical approach to describe quantum spin liquids employs parton representations for the spin operator, but the resulting gauge theories are hard to handle beyond mean-field approximations. In this chapter, we review an alternative approach that starts from the conformal field theory for weakly coupled Heisenberg spin chains. We provide two examples of such coupled-chain constructions. The first one is the Kalmeyer–Laughlin chiral spin liquid, a gapped topological phase that can be obtained by coupling parallel spin chains with three-spin interactions that favor uniform spin chirality. The second example is a gapless chiral spin liquid on a geometry of crossed chains with a staggered chirality pattern. Using a renormalize group analysis, we identify the conditions necessary to stabilize these nontrivial phases and discuss how to calculate their properties explicitly at the low-energy fixed points.

13.1 Introduction

The spin-1/2 Heisenberg chain was the first quantum many-body Hamiltonian to be solved exactly by H. Bethe in 1931 [6]. This pioneer work led to the development of the Bethe ansatz for integrable one-dimensional models, of which the spin-$1/2$ Heisenberg model and its anisotropic extension, the XXZ chain, are notable examples [20]. In addition to the Bethe ansatz solution, analytical field theory methods [1] and sophisticated numerical techniques [38] have been successfully applied over several decades to reveal some remarkable properties of the antiferromagnetic Heisenberg chain. For instance, it is known that the exact ground state for a chain with an even number of spins is a singlet, i.e. a nondegenerate state invariant under global SU(2)

R. G. Pereira (✉)
International Institute of Physics and Departamento de Física Teórica e Experimental, Universidade Federal do Rio Grande do Norte, Campus Lagoa Nova, Natal, RN 59078-970, Brazil
e-mail: rpereira@iip.ufrn.br

A. Ferraz et al. (eds.), *Strongly Coupled Field Theories for Condensed Matter and Quantum Information Theory*, Springer Proceedings in Physics 239,
https://doi.org/10.1007/978-3-030-35473-2_13

spin rotations. While this behavior is familiar from the ground state of the two-spin problem with antiferromagnetic exchange interaction, it contrasts starkly with the classical picture of Néel order later proposed to describe the ground state of antiferromagnets in two- or three-dimensional bipartite lattices [26]. Moreover, the excitation spectrum on top of this ground state is organized in terms of elementary excitations that carry spin-1/2, as opposed to the spin 1 of a magnon in ordered magnets [10]. Selection rules then imply that these excitations, nowadays called spinons, must be created in pairs. Such fractionalization of the spin quantum number in spin chain materials has been confirmed experimentally through the observation of a two-spinon continuum in the spectrum probed by inelastic neutron scattering [24].

Inspired by the physics of the antiferromagnetic spin-1/2 chain and by Pauling's idea of valence bonds as a theory for unconventional metals, in 1973 P. Anderson proposed the concept of a quantum spin liquid [3]. While a precise definition is still lacking [31], the term has been used to refer to phases of interacting spin systems which evade long-range magnetic order even at zero temperature, due to the strong quantum fluctuations in the ground state. Hallmarks of quantum spin liquids are a high degree of many-body entanglement and an excitation spectrum characterized by deconfined fractional excitations analogous to the spinons in the Heisenberg spin chain. In contrast with the exact methods available in one dimension, to describe fractionalization in two and three dimensions one usually resorts to approximate solutions of effective theories where the elementary excitations are slave fermions or bosons (generically called partons) coupled to emergent gauge fields. Essentially, a spin liquid phase arises when the fluctuations of the gauge field are harmless enough to allow for parton deconfinement. In this case, mean-field approximations manage to capture the essential features of Anderson's resonating valence bond state [36]. Beyond the mean-field level, there are a few examples of spin models, such as Kitaev's honeycomb model [19], which are exactly solved by parton representations involving Majorana fermions and static Z_2 gauge fields.

Kalmeyer and Laughlin put forward one of the earliest proposals of wave functions that can describe a quantum spin liquid ground state [16]. Starting from a Holstein-Primakoff representation for spin operators, they argued that the antiferromagnetic Heisenberg model on the triangular lattice was equivalent to a model of strongly interacting bosons in the presence of a fictitious magnetic field at filling factor $\nu = 1/2$. This observation motivated the proposal of a gapped ground state corresponding to a fractional quantum Hall effect of bosonic spin excitations. As a result, the Kalmeyer–Laughlin state shares some properties with the electronic fractional quantum Hall effect, such as topological order, anyonic quasiparticles and gapless chiral edge states. A crucial difference is that the quasiparticles can carry spin and energy, but no electric charge. Furthermore, rather than an external magnetic field, the mechanism for breaking time reversal and mirror symmetries in the Kalmeyer–Laughlin state is a nonzero expectation value of the scalar spin chirality operator $\hat{\chi}_{ijk} = \mathbf{S}_i \cdot (\mathbf{S}_j \times \mathbf{S}_k)$, where i, j, k are any three sites that form a triangular plaquette on the lattice [4, 37]. In fact, a finite value of $\langle \hat{\chi}_{ijk} \rangle$ in the absence of long-range magnetic order is the defining property of a *chiral* spin liquid. A direct

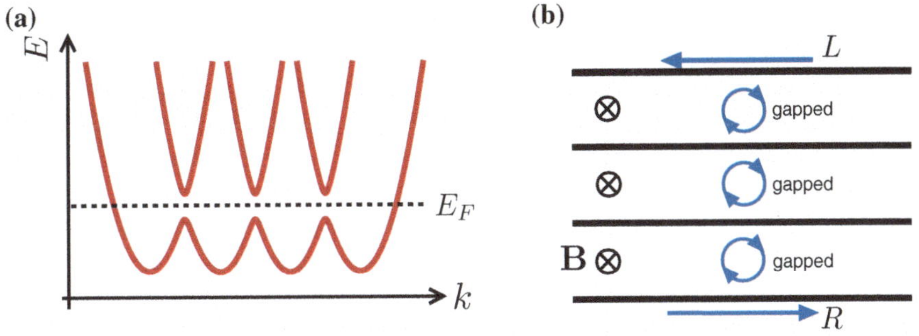

Fig. 13.1 Schematic representation of a coupled-wire construction of the integer quantum hall effect. **a** Dispersion relation for tunnel-coupled wires, showing the avoided level crossings at the Fermi level E_F. **b** In real space, the gapped bulk states correspond to pairs of right and left movers living in different adjacent wires, while the edge harbors gapless chiral states

way to drive the system into a chiral spin liquid phase is to add to the Hamiltonian the three-spin interaction $J_\chi \hat{\chi}_{ijk}$, which favors either sign of the chirality for each triangle depending on the sign of the coupling constant J_χ. Such term breaks time reversal symmetry explicitly while preserving the spin SU(2) symmetry. This route was used to stabilize the Kalmeyer–Laughlin state on the kagome lattice with uniform three-spin interactions [5]. Other examples of gapped chiral spin liquids include non-Abelian phases in spin-1 systems [13] and in the Kitaev model in a magnetic field [19]. On the experimental side, perhaps the best evidence for a chiral spin liquid state comes from a recent measurement of quantized thermal Hall conductance associated with a Majorana edge state in a Mott insulating material with strong spin-orbit coupling [18].

As examples of strongly correlated topological phases, chiral spin liquids present a challenge for analytical approaches which do not rely on mean-field approximations. The question then is: can we go back to the beginning and use our knowledge about the Heisenberg spin chain to construct and analyze chiral spin liquid states? In principle, all we need to do is to couple spin chains in such a way that the spinons living in each chain are set free to propagate in two dimensions. This is more easily said than done, because interchain exchange interactions usually give rise to magnetically ordered or dimerized phases [33]. On the other hand, a precedent for how to obtain microscopic Hamiltonians for topological phases starting from one-dimensional critical systems is the so-called coupled-wire approach [17, 32, 35]. Figure 13.1 illustrates the idea for the simple example of an integer quantum Hall phase [32]. Consider an array of noninteracting quantum wires contained in the xy plane which is threaded by an external magnetic field $\mathbf{B} = B\hat{\mathbf{z}}$. If we work in the Landau gauge $\mathbf{A} = -By\hat{\mathbf{x}}$, where $\hat{\mathbf{x}}$ is the direction parallel to the wires, the dispersion relations associated with different wires are shifted in momentum space as shown in Fig. 13.1a. Turning on a finite tunnelling amplitude between neighboring wires gaps out the crossing points in the spectrum. If the Fermi level lies exactly at the energy of the crossing points, in the low-energy limit we are left with a single pair of gapless chiral fermionic modes: a left-moving mode associated with the upper wire and a right-moving one associated with the lower wire; see Fig. 13.1b. This describes an incompressible phase with chiral edge states which is adiabatically connected with the integer quantum Hall effect (in this example, $\nu = 1$) in a spatially isotropic two-dimensional system. The advantage of the coupled-wire approach is that we can adapt it to include electron-electron interactions nonperturbatively in the wires by treating them as Luttinger liquids. The result is a coupled-wire construction of fractional quantum Hall states [17]. More generally, the idea is to start with a set of critical one-dimensional systems described by the appropriate conformal field theory, and gap out pairs of modes with right and left movers that live in different wires, until the total central charge is reduced to that of a particular chiral edge state [35].

The purpose of this chapter is to review the coupled-chain construction of chiral spin liquids that use Heisenberg chains as building blocks [12, 23, 29]. The key step is to couple the chains with suitable time-reversal-symmetry-breaking perturbations that involve three-spin interactions on triangular geometries. We start in Sect. 13.2 by discussing the low-energy effective field theory for a single Heisenberg chain. In Sect. 13.3, we explain the coupled-chain construction of the Kalmeyer–Laughlin chiral spin liquid using arrays of parallel spin chains. Section 13.4 deals with the construction of a gapless chiral spin liquid on a lattice constructed from weakly-coupled crossing chains. Finally, in Sect. 13.5 we present some concluding remarks and discuss some open questions.

13.2 Effective Field Theory for the Spin-1/2 Heisenberg Chain

The Heisenberg chain is described by the Hamiltonian

$$H_{1D} = J \sum_{j=1}^{N} \mathbf{S}_j \cdot \mathbf{S}_{j+1}, \tag{13.1}$$

where $\mathbf{S}_j$ are spin-1/2 operators acting the local Hilbert space at site j, $J > 0$ is the antiferromagnetic exchange coupling, and N is the number of sites with periodic boundary conditions. The low-energy physics of the spin-1/2 chain is described by the $SU(2)_1$ Wess-Zumino-Novikov-Witten model, which is a conformal field theory with central charge $c = 1$ [2]. To derive bosonization formulas for the local operators in the theory, one can start from the one-dimensional Hubbard model at half-filling and gap out the charge degrees of freedom in the Mott insulating phase. The spin operator is represented in the continuum limit by

$$\mathbf{S}_j \mapsto \mathbf{S}(x) \sim \mathbf{J}_L(x) + \mathbf{J}_R(x) + (-1)^j \mathbf{n}(x), \tag{13.2}$$

where $\mathbf{J}_\nu$, with $\nu = R, L = \pm$, are chiral SU(2) currents associated with right- or left-moving modes, respectively, and $\mathbf{n}$ is the staggered magnetization field. The components of the chiral currents can be written in terms of chiral bosons in the form

$$J^z_\nu(x) = \frac{\nu}{\sqrt{4\pi}}\partial_x\varphi_\nu(x), \qquad J^\pm_\nu(x) = \frac{1}{2\pi}\,e^{\pm i\sqrt{4\pi}\varphi_\nu(x)}. \tag{13.3}$$

where the bosonic fields obey the algebra $[\varphi_\nu(x), \partial_{x'}\varphi_{\nu'}(x')] = i\nu\delta_{\nu\nu'}\delta(x-x')$. The components of the staggered magnetization are

$$n^\pm(x) = \mathscr{A}\,e^{\pm i\sqrt{\pi}(\varphi_L+\varphi_R)}, \qquad n^z(x) = \mathscr{A}\sin\left[\sqrt{\pi}\,(\varphi_L-\varphi_R)\right], \tag{13.4}$$

where $\mathscr{A}$ is a non-universal constant of order 1. Another local operator in the theory is the dimerization operator $\varepsilon(x)$, which appears in the expansion of the scalar $\mathbf{S}_j\cdot\mathbf{S}_{j+1}\sim(-1)^j\varepsilon(x)$ in the continuum limit. Its bosonized form reads

$$\varepsilon(x) = \mathscr{A}'\cos\left[\sqrt{\pi}\,(\varphi_L-\varphi_R)\right], \tag{13.5}$$

where $\mathscr{A}'$ is another non-universal constant.

The symmetry properties of the local operators are important to analyze the perturbations of the conformal field theory. Under translation by one site, $x\mapsto x+1$, the fields transform in the form

$$\mathscr{L}:\mathbf{J}_\nu\mapsto\mathbf{J}_\nu, \quad \mathbf{n}\mapsto-\mathbf{n}, \quad \varepsilon\mapsto-\varepsilon. \tag{13.6}$$

Time reversal $\mathscr{T}$ involves complex conjugation and acts on the fields as follows:

$$\mathscr{T}:\mathbf{J}_R\leftrightarrow-\mathbf{J}_L, \quad \mathbf{n}\mapsto-\mathbf{n}, \quad \varepsilon_l\mapsto\varepsilon_l. \tag{13.7}$$

Site parity, defined as the reflection $\mathscr{P}_s$ about an axis passing through a given site, takes $x\to-x$ and

$$\mathscr{P}_s:\mathbf{J}_R\leftrightarrow\mathbf{J}_L, \quad \mathbf{n}\mapsto\mathbf{n}, \quad \varepsilon\mapsto-\varepsilon. \tag{13.8}$$

Site parity, defined as the reflection $\mathscr{P}_l$ about an axis passing through the middle of link between two sites, takes $x\to-x$ and

$$\mathscr{P}_l:\mathbf{J}_R\leftrightarrow\mathbf{J}_L, \quad \mathbf{n}\mapsto-\mathbf{n}, \quad \varepsilon\mapsto\varepsilon. \tag{13.9}$$

The low-energy effective Hamiltonian for a single Heisenberg chain is written in terms of the chiral currents as

$$H_{1D}\sim\int\mathrm{d}x\left[\frac{2\pi v}{3}\left(\mathbf{J}_R^2+\mathbf{J}_L^2\right)+2\pi v g_{bs}\mathbf{J}_L\cdot\mathbf{J}_R\right]. \tag{13.10}$$

Fig. 13.2 Two parallel Heisenberg chains coupled by exchange and three-spin interactions in a zigzag geometry

Here $v \propto J$ is the spin velocity and $g_{bs} < 0$ is the dimensionless coupling constant for the marginally irrelevant backscattering operator [1]. As the effective g_{bs} vanishes (logarithmically) in the low-energy limit, the theory becomes equivalent to free bosons and the asymptotic decay of correlation functions can be calculated exactly.

13.3 Coupling Parallel Chains: The Kalmeyer–Laughlin State

Let us now couple spin-1/2 chains to build towards the two-dimensional limit. We start with two parallel chains with a zigzag geometry as illustrated in Fig. 13.2 [12]. We consider the interchain exchange coupling J' together with a three-spin interaction J_χ,

$$H = \sum_j \left[J \sum_{l=1,2} \mathbf{S}_{j,l} \cdot \mathbf{S}_{j+1,l} + J'\mathbf{S}_{j,1} \cdot (\mathbf{S}_{j,2} + \mathbf{S}_{j+1,2}) \right] + J_\chi \sum_{ijk \in \triangle} \hat{\chi}_{ijk}, \quad (13.11)$$

in the regime $|J'|, |J_\chi| \ll J$. The ordering of the spins i, j, k in the triple product $\hat{\chi}_{ijk}$ is set to correspond to a configuration with uniform chirality, as indicated by all the arrows in Fig. 13.2 pointing either clockwise or counterclockwise. Note that the operator $\hat{\chi}_{ijk}$ is odd under time reversal, which takes $\mathbf{S}_i \mapsto -\mathbf{S}_i$, and under a reflection that exchanges two of the three sites, $\hat{\chi}_{ijk} = -\hat{\chi}_{jik} = -\hat{\chi}_{ikj}$.

The unperturbed effective Hamiltonian for $J' = J_\chi = 0$ is given by two copies of the chain Hamiltonian in (13.10), with operators in each chain labeled by the index $l = 1, 2$. For small J' and J_χ, the leading interchain couplings in the effective field theory are all marginal operators. There is a twist operator [27] that depends on the staggered magnetization fields,

$$\delta H_{tw} = 2\pi v g_{tw} \int \mathrm{d}x \, \mathbf{n}_1 \cdot \partial_x \mathbf{n}_2, \quad (13.12)$$

and two operators that couple the chiral currents in different chains

$$\delta H_{RL} = 2\pi v \int \mathrm{d}x \, (g_+ \mathbf{J}_{1,R} \cdot \mathbf{J}_{2,L} + g_- \mathbf{J}_{1,L} \cdot \mathbf{J}_{2,R}), \quad (13.13)$$

where $g_\pm = g \pm g_\chi$ with $g \propto J'$ and $g_\chi \propto J_\chi$ to first order. For $J' > 0$ and $J_\chi = 0$, this problem reduces to the familiar zigzag chain, whose ground state in the regime $J/J' \gtrsim 0.241$ is fully gapped and breaks translational invariance due to dimerization $\langle \varepsilon(x) \rangle \neq 0$ (as in the Majumdar–Ghosh model [22]). Within the effective field theory, the dimerized phase is understood as the regime where both g_+ and g_- flow to strong coupling under the renormalization group, eventually pinning the bosonic fields to the minima of a sine-Gordon potential [14]. The new interesting possibility that appears when we add the time-reversal-symmetry-breaking interaction J_χ is that g_+ and g_- can have opposite signs if J_χ is large enough, in the regime $|g_\chi| > |g|$. In this case, one of them flows to strong coupling while the other flows to zero. As a result, we obtain a gapless quasi-one-dimensional phase with one pair of gapless chiral modes, either $J_{1,R}$ and $J_{2,L}$ for $J_\chi > 0$, or $J_{1,L}$ and $J_{2,R}$ for $J_\chi < 0$. Such phase of the two-leg system can actually be verified exactly for a special value of J_χ where the model is integrable [12].

The next step is to generalize this approach to an array of N_c spin chains with the same coupling between neighboring chains, forming an anisotropic triangular lattice. From the effective field theory perspective, we have to write down all the possible local interchain interactions allowed by symmetry. Here it is important to analyze how the local operators of the SU(2)$_1$ Wess-Zumino-Novikov-Witten model transform under the various symmetry operations. Particularly important are the roles of time reversal $\mathscr{T}$ and the reflection $\mathscr{P}$ that acts as site parity for chains labeled by odd values of l and as link parity for chains with even values of l. While both $\mathscr{P}$ and $\mathscr{T}$ are broken by the three-spin interaction, the Hamiltonian and the chiral spin liquid ground state we are looking for preserve the product $\mathscr{P}\mathscr{T}$. The symmetries of the lattice model are enough to rule out relevant operators in the effective field theory for two chains, but not for $N_c > 2$. In fact, if we allow for coupling between next-nearest-neighbor chains, the effective Hamiltonian for the two-dimensional array contains two relevant operators which involve the staggered magnetization and the dimerization fields:

$$\delta H_{\text{rel}} = v\Lambda \sum_l \int \mathrm{d}x \, (g_n \mathbf{n}_l \cdot \mathbf{n}_{l+2} + g_\varepsilon \varepsilon_l \varepsilon_{l+2}) \,, \tag{13.14}$$

where Λ is an ultraviolet cutoff with units of momentum and g_n and g_ε are dimensionless coupling constants. The g_n interaction favors a transition to a collinear antiferromagnetic phase in which the $\mathbf{n}_l(x)$ fields are pinned, while g_ε favors a dimerized or valence bond crystal phase where $\varepsilon_l(x)$ acquires a nonzero expectation value [33]. Note, however, that these interactions couple next-nearest-neighbor chains. If we start from a lattice model that contains only weak couplings J' and J_χ between nearest-neighbor chains, the couplings g_n and g_ε can only be generated at second order in J' or J_χ, or even be pushed to higher order if we consider other lattices than the triangular one [12]. As a result, even if the marginal current-current interactions (13.13) are less relevant than the interactions in (13.14), either g_+ or g_- can reach strong coupling before the other perturbations depending on the initial conditions of the renormalization group flow. When this happens, the low-energy physics is governed by the interaction that gaps out pairs of chiral modes in different chains, leaving gapless modes on the edge of the two-dimensional array. There is no magnetic order at such fixed point since the spin operators have zero expectation value. The picture is analogous to Fig. 13.1b if we replace the wires by Heisenberg chains and, instead of a chiral fermion, the edge state correspond to a chiral $SU(2)_1$ Wess-Zumino-Novikov-Witten model. This is precisely the edge state of the Kalmeyer–Laughlin chiral spin liquid [16].

We can use the coupled-chain construction to verify other properties of the Kalmeyer–Laughlin chiral spin liquid. For instance, the fractional spin of the bulk quasiparticles follow from a semiclassical analysis of the dominant sine-Gordon potential in the strong coupling limit. Bosonizing the transverse part of the current-current interaction (13.13) generalized for N chains, we find

$$\delta H_{RL} \sim g_+ \sum_l \int \mathrm{d}x \, \cos[\sqrt{4\pi}(\varphi_{l,L} - \varphi_{l+1,R})]. \tag{13.15}$$

In the gapped phase, the bosonic fields are pinned to the minima of the potential, given by $\sqrt{4\pi}(\varphi_{l,L} - \varphi_{l+1,R}) = 2\pi n + \pi$, $n \in \mathbb{N}$. A quasiparticle localized in the bulk corresponds to an excitation in which the value of the fields jumps by $\pm 2\pi$ over some finite region in space. The change in the total spin associated with this kink is

$$\begin{aligned} \Delta S^z &= \sum_l \int \mathrm{d}x \, (J^z_{l,L} + J^z_{l,R}) \\ &= \frac{1}{\sqrt{4\pi}} \sum_l \int \mathrm{d}x \, \partial_x(\varphi_{l,L} - \varphi_{l+1,R}) \\ &= \pm\frac{1}{2}, \end{aligned} \tag{13.16}$$

as expected for spinons. Similarly, we can demonstrate the spinon's anyonic statistics by constructing operators that transport quasiparticles around one another. We can also consider the operators that transport quasiparticles around the entire system in the noncontractible directions of a torus (on a system with periodic boundary conditions), given by [12]

$$U_x = e^{i\sqrt{\pi}\int dx \partial_x \varphi_{l_0,L}(x)}, \qquad U_y = \prod_l e^{i\sqrt{\pi}[\varphi_{l,L}(x_0) - \varphi_{l,R}(x_0)]}, \tag{13.17}$$

where l_0 labels an arbitrary fixed chain and x_0 is an arbitrary point along the chains. It is easy to verify that these operators obey the algebra $U_x U_y = e^{i2\theta} U_y U_x$ with $\theta = \pi/2$, which can be identified with the statistical phase of semions [30].

The coupled-chain construction has been generalized to obtain non-Abelian chiral spin liquids [21, 23] and even a time-reversal-invariant $\mathbb{Z}_2$ spin liquid [28].

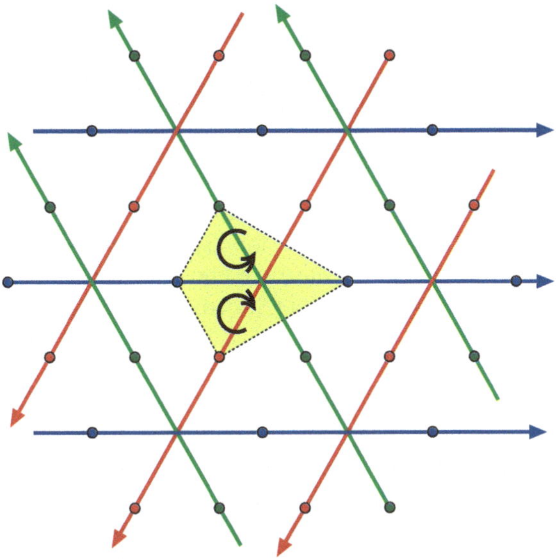

Fig. 13.3 Crossed spin chains that serve as the starting point for the construction of a gapless chiral spin liquid. There are three sets of parallel chains labeled by $q = 1, 2, 3$ (colored in blue, green, and red, respectively). Pairs of chains are coupled at their crossing by three-spin interactions at the triangles where two sites belong to one chain and the third site to the other chain

13.4 Coupling Crossed Chains with Staggered Chirality: Gapless Chiral Spin Liquid

The coupled-chain construction described in the previous section leads to gapped chiral spin liquids, where gapless chiral modes are confined to the edges of an array with open boundary conditions. We will now discuss an alternative construction of a gapless chiral spin liquid [8], which can be viewed as sliding Luttinger liquid phase [25] with gapless chiral modes in the bulk. This construction is inspired by the proposal of a gapless chiral spin in kapellasite, a material described by an extended Heisenberg model on the kagome lattice [8].

As our starting point, we consider three sets of decoupled parallel chains as shown in Fig. 13.3. Each set is labeled by an index $q = 1, 2, 3$, such that a chain in set $q \pm 1$ forms an angle of $\pm 2\pi/3$ measured from a chain in set q. It is convenient to use a coordinate system where, for fixed q, the different parallel chains are distinguished by an index $l \in \mathbb{Z}$ that increases in the direction set by $q + 1$ chains [11]. The spin operator at site j of the lth chain along the q direction is denoted by $\mathbf{S}_q(j, l)$. In the low-energy limit, the effective Hamiltonian for decoupled chains is given by the sum of (13.10) for each chain, with chiral currents denoted as $\mathbf{J}_{q\nu}(x, l)$.

We can then couple pairs of spin chains with local interactions at their crossings. Most natural is a Heisenberg exchange coupling between spins on sites closest to the crossing point. In the geometry of Fig. 13.3, this corresponds to nearest-neighbor exchange J_1 on the underlying kagome lattice. For comparison, the dominant exchange coupling J along the chain directions lies on the diagonals of the hexagons of this kagome lattice [8]. In order to stabilize a chiral spin liquid phase, we also add to the model the three-spin interaction given by

$$\delta H_\chi = J_\chi \sum_{q,j,l} [\mathbf{S}_q(-j-1,l) \times \mathbf{S}_q(-j,l)] \cdot [\mathbf{S}_{q+1}(j+l,j) + \mathbf{S}_{q+1}(j+l-1,j) \\ + \mathbf{S}_{q+2}(-l-1,-j-l) + \mathbf{S}_{q+2}(-l,-j-l)] \,. \tag{13.18}$$

This interaction is defined on triangles where two spins belong to the same chain, labeled by q, l, whereas the third spin belongs to another chain that crosses the link between the first two sites; see Fig. 13.3. The sign the coupling for each triangle is chosen so as to preserve a C_3 rotational symmetry as well as a reflection symmetry with respect to chain directions. This reflection symmetry is such that the chirality is staggered between two triangles that share two sites on the same chain. This symmetry, absent in the Kalmeyer–Laughlin state, plays an important role in the projective symmetry group classification of chiral spin liquids on the triangular and kagome lattices [7].

When taking the continuum limit in the interchain couplings, it becomes important to pay special attention to the transverse coordinate l in the spin operators $\mathbf{S}_q(j,l)$. First, chains with even or odd values of l transform differently, either as site parity or link parity, under reflections by the planes perpendicular to the chain directions (which is not a symmetry of the model with finite J_χ, but can be combined with time reversal to yield an actual symmetry). Thus, we must include in the mode expansion of the low-energy fields both a uniform and a staggered component for the dependence on the transverse coordinate l. Second, we must take into account that, over a finite length Δx much larger than the lattice spacing, each spin chain crosses a number of chains proportional to Δx. For this reason, the continuum limit in the longitudinal direction also entails a coarse-graining of the chain index in the transverse direction. The one-dimensionality of the system in the weak coupling regime is preserved by imposing the usual power-law decay of correlation functions along the chain direction, but short-range correlations in the transverse direction with a length scale set by the ultraviolet cutoff of the theory.

Within this field theory approach, we can analyze the effects of the perturbations allowed by symmetry. The leading perturbations to the decoupled-chain fixed point are relevant operators that involve the staggered magnetization and the dimerization fields:

$$\delta H_{rel} = v\Lambda \sum_q \int \mathrm{d}x\mathrm{d}y \left[\kappa_n \mathbf{n}_{qo}(-x,y) \cdot \mathbf{n}_{q+1,e}(x+y,x) \right. \\ \left. + \kappa_\varepsilon \varepsilon_{qo}(-x,y)\, \varepsilon_{q+1,e}(x+y,x)\right], \tag{13.19}$$

where κ_n and κ_ε are dimensionless coupling constants. The lower indices e, o refer to the uniform and staggered parts of the fields with respect to the transverse coordinate, respectively, which are even or odd under translation by one chain spacing in the transverse direction. As in the coupled-chain construction of the Kalmeyer–Laughlin state, the coupling constants of the relevant operators are generated at second order in the bare couplings of the original lattice model. When κ_n reaches strong coupling before all other perturbations, the two-dimensional system orders magnetically in one of the so-called cuboc phases [11]. When κ_ε prevails, it gives rise to a valence bond crystal. These are the two phases expected in the model of crossed chains weakly coupled by time-reversal-invariant exchange interactions [11]. However,the three-spin interaction in (13.18) generates in the continuum limit various marginal current-current couplings, such as

$$\delta H_{RR} = 2\pi v\lambda \sum_q \int \mathrm{d}x\mathrm{d}y\, \mathbf{J}_{qRe}(-x, y) \cdot \mathbf{J}_{q+1,Re}(x + y, x). \tag{13.20}$$

There is a similar term that couples left-moving currents, but similarly to the case of g_+ and g_- in (13.13), the asymmetry between the corresponding couplings is of first order in J_χ. As a result, an interesting regime becomes possible where the chiral interaction favors strong coupling between right-moving spin currents in all chains, while the left-moving currents remain essentially free. A renormalization group analysis [29] shows that such a regime can be reached for sufficiently large bare values of J_χ, such that λ in (13.20) reaches strong coupling before the relevant operators (13.19).

The simple picture for the ground state in this regime is a chiral spin liquid where each chain harbors only one chiral mode, moving either to the "right" or to the "left", where the directions of "right" and "left" depend on the orientation of the chain and varies with q as illustrated in Fig. 13.3. Note that, due to the C_3 rotational symmetry, which imposes the angles of $\pm 2\pi/3$ between chains with different q values, there is no preferred direction for the propagation of excitations in the bulk. Despite being gapless, the corresponding Hamiltonian for this chiral spin liquid is a stable fixed point of the renormalization group. In fact, gapping out one chiral sector of the theory (say, the right-moving bosons $\varphi_{qR}(x, y)$) implies that all excitations involving the staggered magnetization or the dimerization are gapped, since these fields couple to both chiral sectors; cf. (13.4) and (13.5). Thus, the relevant perturbations (13.19) are ruled out from the low-energy theory and a sliding Luttinger liquid phase becomes stable.

We can calculate physical properties of the gapless chiral spin liquid using the operators in the coupled-chain construction [29]. For instance, the large-distance decay of the spin-spin correlation is dominated by the gapless chiral modes propagating along the chain directions:

$$\langle \mathbf{S}_q(j, l) \cdot \mathbf{S}_{q'}(j', l') \rangle \sim \langle \mathbf{J}_{qL}(x, y) \cdot \mathbf{J}_{q'L}(x', y') \rangle \sim -\frac{\Delta(y - y', \alpha^*)}{(x - x')^2} \delta_{qq'}, \tag{13.21}$$

where $\Delta(y, \alpha^*)$ is a non-universal function that describes short-range correlations in the transverse direction with characteristic length scale α^*. The latter is set by the scale where the marginally relevant operator (13.20) reaches strong coupling and is related to the gap in the right-moving chiral sector. Note that the correlation does not oscillate as a function of the distance in the chain direction and decays with a universal exponent governed by the scaling dimension of the chiral currents in the one-dimensional theory. By contrast, correlation functions for the staggered magnetization and dimerization operators must decay exponentially.

While the gapless chiral spin liquid is not topologically ordered in the same sense of the Kalmeyer–Laughlin state (namely, nontrivial ground state degeneracy on the torus), it can be described as a highly entangled phase of matter. Consider a finite subsystem A with characteristic length L contained within the two-dimensional system. The von Neumann entanglement entropy S of A with the rest of the system can be estimated by

$$S \sim N_{\text{chains}} \times S_{\text{chain}} \sim \frac{c_L}{6} L \ln L, \tag{13.22}$$

where N_{chains} is the number of chains that cross the boundary of A and $S_{\text{chain}} \sim \frac{c_L}{6} \ln L$ is the entanglement entropy associated with each chiral mode with chiral central charge c_L. Therefore, the gapless chiral spin liquid is highly entangled in the sense that it exhibits a logarithmic violation to the area law, a behavior also observed for free fermions with a Fermi surface and strongly correlated systems whose spectrum can be mapped to a large number of one-dimensional modes [34].

It is also interesting to note that this gapless chiral spin liquid is stable against weak disorder [29]. Intuitively, due to the energy gap in one chiral sector, backscattering is suppressed and only forward scattering processes (where "forward" may mean scattering a right mover in a q chain to a right mover in a different $q \pm 1$ direction) are allowed at low energies (Fig. 13.4).

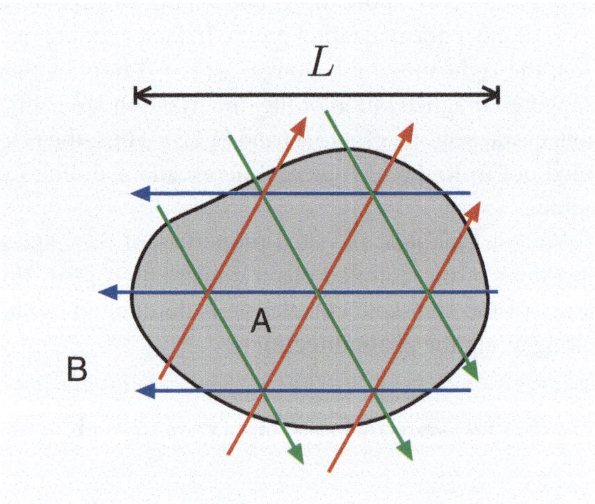

Fig. 13.4 Bipartition of the two-dimensional spin system. The entanglement entropy between the finite subsystem **A** with characteristic length L with the rest of the system **B** is calculated by counting the number of chiral one-dimensional modes that cross the boundary of A

13.5 Conclusion

Coupled-wire constructions have proven a useful tool to study topological phases of matter. The work discussed in this chapter shows that chiral spin liquids can also be approached this way. For instance, we have constructed the Kalmeyer–Laughlin chiral spin liquid using an array of parallel spin chains coupled by three-spin interactions that explicitly break time reversal and reflection symmetries. This is a gapped topological phase equivalent to a fractional quantum Hall effect of bosons at filling factor $\nu = 1/2$, and its topological properties can be verified within the coupled-chain construction without resorting to Chern–Simons gauge fields.

We have also discussed a construction of a gapless chiral spin liquid that starts from weakly coupled crossed chains. In this case, the low-energy picture that emerges corresponds to gapping out one chiral sector of each chain while the other sector remains gapless. This gapless chiral spin liquid also exhibits some exotic physical properties, such as spatially anisotropic power-law-decaying correlation functions and a logarithmic violation of the entanglement area law.

Some interesting open questions pertain to explicitly deriving a relation between the operators in the coupled-chain construction, which stem from the conformal field theory describing the one-dimensional modes, and the matter and gauge field operators in the usual parton construction of quantum spin liquids. Particularly in the case of the gapless chiral spin liquid, a possible duality between the chiral one-dimensional modes and a spinon Fermi surface state with a line Fermi surface protected by the reflection symmetry has been suggested in [29]. Another interesting route to pursue is an alternative construction of chiral spin liquids based on networks of Y junctions of spin chains tuned to a chiral boundary fixed point [9].

Acknowledgements I am grateful to my collaborators on this topic: Eran Sela, Gregory Gorohovsky, and Samuel Bieri. Research at IIP is supported by the Brazilian ministries MEC and MCTIT. I also thank Reinhold Egger for the hospitality at the Heinrich-Heine Universität (HHU), in Düsseldorf, where this chapter was written. My stay at HHU was sponsored by the Alexander von Humboldt Foundation.

References

1. I. Affleck, in *Fields, Strings and Critical Phenomena*, ed. by E. Brézin, J. Zinn-Justin (Les Houches, Session XLIX, Amsterdam: North-Holland, 1988), p. 563
2. I. Affleck, F.D.M. Haldane, Critical theory of quantum spin chains. Phys. Rev. B **36**, 5291 (1987)
3. P.W. Anderson, Resonating valence bonds: a new kind of insulator? Mater. Res. Bull. **8**, 153–160 (1973)
4. G. Baskaran, Novel local symmetries and chiral-symmetry-broken phases in S=1/2 triangular-lattice Heisenberg model. Phys. Rev. Lett. **63**, 2524 (1989)
5. B. Bauer, L. Cincio, B.P. Keller, M. Dolfi, G. Vidal, S. Trebst, A.W.W. Ludwig, Chiral spin liquid and emergent anyons in a Kagome lattice Mott insulator. Nat. Comm. **5**, 5137 (2014)

6. H. Bethe, On the theory of metals 1. eigenvalues and eigenfunction of a linear chain of atoms. Z. Phys. **71**, 205 (1931)
7. S. Bieri, C. Lhuillier, L. Messio, Projective symmetry group classification of chiral spin liquids. Phys. Rev. B **93**, 094437 (2016)
8. S. Bieri, L. Messio, B. Bernu, C. Lhuillier, Gapless chiral spin liquid in a kagome Heisenberg model. Phys. Rev. B **92**, 060407(R) (2015)
9. F. Buccheri, R. Egger, R.G. Pereira, F.B. Ramos, Quantum spin circulator in Y junctions of Heisenberg chains. Phys. Rev. B **97**, 220402 (2018)
10. L.D. Faddeev, L.A. Takhtajan, What is the spin of a spin wave? Phys. Lett. A **85**, 375–377 (1981)
11. S.-S. Gong, W. Zhu, K. Yang, O.A. Starykh, D.N. Sheng, L. Balents, Emergent quasi-one-dimensionality in a kagome magnet: a simple route to complexity. Phys. Rev. B **94**, 035154 (2016)
12. G. Gorohovsky, R.G. Pereira, E. Sela, Chiral spin liquids in arrays of spin chains. Phys. Rev. B **91**, 245139 (2015)
13. M. Greiter, R. Thomale, Non-Abelian statistics in a quantum antiferromagnet. Phys. Rev. Lett. **102**, 207203 (2009)
14. F.D.M. Haldane, Spontaneous dimerization in the $S = 1/2$ Heisenberg antiferromagnetic chain with competing interactions. Phys. Rev. B **25**, 4925(R) (1982)
15. P.-H. Huang, J.-H. Chen, P.R.S. Gomes, T. Neupert, C. Chamon, C. Mudry, Non-Abelian topological spin liquids from arrays of quantum wires or spin chains. Phys. Rev. B **93**, 205123 (2016)
16. V. Kalmeyer, R.B. Laughlin, Equivalence of the resonating-valence-bond and fractional quantum hall states. Phys. Rev. Lett. **59**, 2095 (1987)
17. C.L. Kane, R. Mukhopadhyay, T.C. Lubensky, Fractional quantum hall effect in an array of quantum wires. Phys. Rev. Lett. **88**, 036401 (2002)
18. Y. Kasahara et al., Majorana quantization and half-integer thermal quantum hall effect in a kitaev spin liquid. Nature **559**, 227–231 (2018)
19. A. Kitaev, Anyons in an exactly solved model and beyond. Ann. Phys. **321**, 2–111 (2006)
20. V.E. Korepin, N.M. Bogoliubov, A.G. Izergin, *Quantum Inverse Scattering Method and Correlation Functions* (Cambridge University Press, Cambridge, 1993)
21. P. Lecheminant, A.M. Tsvelik, Lattice spin models for non-Abelian chiral spin liquids. Phys. Rev. B **95**, 140406(R) (2017)
22. C.K. Majumdar, D.K. Ghosh, On next-nearest-neighbor interaction in linear chain. J. Math. Phys. **10**, 1388 (1969)
23. T. Meng, T. Neupert, M. Greiter, R. Thomale, Coupled-wire construction of chiral spin liquids. Phys. Rev. B **91**, 241106(R) (2015)
24. M. Mourigal, M. Enderle, A. Klöpperpieper, J.-S. Caux, A. Stunault, H.M. Ronnow, Fractional spinon excitations in the quantum Heisenberg antiferromagnetic chain. Nat. Phys. **9**, 435–441 (2013)
25. R. Mukhopadhyay, C.L. Kane, T.C. Lubensky, Sliding luttinger liquid phases. Phys. Rev. B **64**, 045120 (2001)
26. M.L. Néel, Propriétés magnétiques des ferrites; ferrimagnétisme et antiferromagnétisme. Ann. Phys. **12**(3), 137–198 (1948)
27. A.A. Nersesyan, A.O. Gogolin, F.H.L. Essler, Incommensurate spin correlations in spin-1/2 frustrated two-leg Heisenberg ladders. Phys. Rev. Lett. **81**, 910 (1998)
28. A.A. Patel, D. Chowdhury, Two-dimensional spin liquids with $\mathbb{Z}_2$ topological order in an array of quantum wires. Phys. Rev. B **94**, 195130 (2016)
29. R.G. Pereira, S. Bieri, Gapless chiral spin liquid from coupled chains on the kagome lattice. SciPost Phys. **4**, 004 (2018)
30. E. Sagi, Y. Oreg, A. Stern, B.I. Halperin, Imprint of topological degeneracy in quasi-one-dimensional fractional quantum hall states. Phys. Rev. B **91**, 245144 (2015)
31. L. Savary, L. Balents, Quantum spin liquids: a review. Rep. Prog. Phys. **80**, 016502 (2017)
32. S.L. Sondhi, K. Yang, Sliding phases via magnetic fields. Phys. Rev. B **63**, 054430 (2001)

33. O.A. Starykh, L. Balents, Dimerized phase and transitions in a spatially anisotropic square lattice antiferromagnet. Phys. Rev. Lett. **93**, 127202 (2004)
34. B. Swingle, Entanglement entropy and the fermi surface. Phys. Rev. Lett. **105**, 050502 (2010)
35. J.C.Y. Teo, C.L. Kane, From luttinger liquid to non-abelian quantum hall states. Phys. Rev. B **89**, 085101 (2014)
36. X.-G. Wen, Mean-field theory of spin-liquid states with finite energy gap and topological orders. Phys. Rev. B **44**, 2664 (1991)
37. X.-G. Wen, F. Wilczek, A. Zee, Chiral spin states and superconductivity. Phys. Rev. B **39**, 11413 (1989)
38. S.R. White, Density matrix formulation for quantum renormalization groups. Phys. Rev. Lett. **69**, 2863 (1992)

Chapter 14
Majorana Zero-Energy Modes in a Magnetic Field-Free Quantum Wire

Mariana Malard

Abstract We report on a proposal for engineering a one-dimensional topological superconductor hosting Majorana zero-energy modes pinned at the edges of a spin-orbit coupled and electron-electron interacting quantum wire subject to a spatially modulated electric field and in proximity to an s-wave superconductor. The combination of the various interactions in the wire produces an effective spin-flip backscattering which opens a gap at two of the four spin-split Fermi points of the system, leaving a helical Luttinger liquid state at the remaining gapless points. This state enters the target topological phase for a certain regime of the proximity-induced pairing. This all-electric setup does not rely on a magnetic field.

14.1 Introduction

A Majorana zero-energy mode (MZM) is an emergent quasi-particle that follows a non-abelian statistics and has the distinguishing property of being its own antiparticle [1]. It is the solid state counterpart of the fermion hypothesized by Ettore Majorana back in 1937 [2] and whose confirmation as a free fundamental particle remains a long-standing puzzle. In contrast, MZMs are expected to be found pinned to defects or edges of materials and setups such as fractional quantum Hall systems [3], cold atoms [4, 5] and hybrid structures realizing topological superconductivity [6–8], to cite some well-known examples. In these systems, the MZMs appear as in-gap states of zero energy [9]. Besides their relevance from a fundamental standpoint, MZMs hold promise of high impact technological applications, notably in the making of a fault tolerant topological quantum computer [10].

The realization that p-wave pairing (that makes a superconductor topological, but is rare in nature) can be emulated in a two-dimensional topological insulator combined with an ordinary s-wave superconductor [11] has led to an intense search for topological superconductivity in different solid state architectures [9]. As known

M. Malard (✉)
Faculdade UnB Planaltina, University of Brasilia, Planaltina, DF 73300-000, Brazil
e-mail: mariana.malard@physics.gu.se

A. Ferraz et al. (eds.), *Strongly Coupled Field Theories for Condensed Matter and Quantum Information Theory*, Springer Proceedings in Physics 239,
https://doi.org/10.1007/978-3-030-35473-2_14

already from the primal toy model of a topological superconductor—the celebrated Kitaev chain of spinless fermions [6]—the spinless character of the pairing is key and, hence, "getting rid of spin" in real systems is a common feature encompassing most proposals. Restricting to one dimensional (1D) realizations, among the most promising are (i) a spin-orbit coupled quantum wire in a magnetic field and coupled to an s-wave superconductor [12, 13] and (ii) a chain of magnetic impurities laid on top of an s-wave superconductor [14]. In both (i) and (ii), a magnetic field opens a gap in the energy spectrum of the system, pushing one of the spin-split bands away from the Fermi level so that the later crosses a single, effectively spinless, quasiparticle band. The states around the Fermi level form the so-called *helical liquid* in which spin and momentum degrees of freedom are locked to each other. In a regime of the pairing, these effectively spinless states enter a topological superconducting phase.

Like the two proposals highlighted above, most schemes for realizing 1D topological superconductors rely on a magnetic field which, although crucial for the very emergence of a spinless state in those systems, ends up damaging the proximity superconducting pairing by canting the spins in the wire [15]. Even more impacting for applications, especially in quantum computing, a magnetic field would reduce robustness against disorder [16, 17] of auxiliary nontopological states that would have to be integrated with the quantum gates [18, 19]. And a strong enough magnetic field would be hard to apply locally [20] in order to avoid it affecting the entire dispositive. From a more fundamental point of view, one could ask whether MZMs could be obtained in one spatial dimension without breaking time reversal symmetry explicitly as is the case with magnetic field-based setups.

In the search for non-magnetic options, *quasi 1D* schemes have been explored using $d_{x^2-y^2}$ [21] or $s_{\pm}$-wave pairing [22], channel-dependent spin-orbit interactions in double-channel wires [23], noncentrosymmetric superconductivity [24], among others [25–33]. All these systems support *paired* MZMs [34]. Unpaired MZMs in the absence of magnetic field have been predicted in a Floquet topological superconductor, with a *time-dependent* spin-orbit interaction [35]. Here, we instead make use of a *spatially* periodic spin-orbit interaction to build a *truly 1D* magnetic field-free topological superconductor supporting *unpaired* MZMs.

We consider a 1D electron system with Rashba and Dresselhaus types of spin-orbit coupling, as well as electron-electron ($e-e$) interactions. An applied electric field which is periodically modulated in space induces the same modulation on the Rashba component. When the Fermi level is tuned so that the two outermost of the four spin-split Fermi momenta sit at the boarders of the Brillouin zone (BZ), a gap opens at those points, resulting in a helical Luttinger liquid (HLL) [36, 37] at the two remaining gapless Fermi points inside the BZ. Kramers theorem—which prevents a spectral gap from opening at the center or at the boundaries of the BZ in a time reversal symmetric system—is here avoided by a *spontaneous* breaking of time reversal symmetry (TRS) at the outer Fermi points. The HLL at the inner Fermi points, however, preserve TRS. A proximity-pairing may turn this HLL into an effectively spinless topological superconductor falling in class BDI of the Altland–Zirnbauer classification [38], thus hosting one unpaired MZM at each end. The emergence of this phase, however, does not follow straightforwardly from topping the HLL with

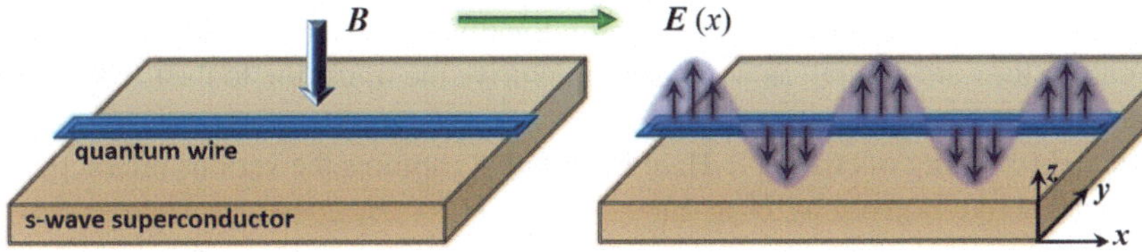

Fig. 14.1 (Color online) A spin-orbit coupled quantum wire immersed in an uniform magnetic field B (left) is replaced by a spin-orbit coupled quantum wire in a spatially modulated electric field $E(\mathrm{x})$ (right). In both setups, a nearby s-wave superconductor induces superconducting pairing in the wire

a superconducting pairing because the later competes with the spin-orbit-induced backscattering yielding the HLL in the first place. Using a renormalization group (RG) approach, we establish that both processes can, in fact, coexist.

In Sect. 14.2 we present a microscopic model capturing the necessary interactions and show schematically how they play out to produce the topological superconductor. As the scheme calls for $e-e$ interactions, a low-energy bosonized version of the model is provided in Sect. 14.3, along with the corresponding RG flow equations yielding the various RG regimes of the model. Combining these regimes we delineate the phase diagram of the model, as presented in Sect. 14.4. Our concluding remarks are given in Sect. 14.5.

14.2 General Picture

Our proposal is to produce a topological superconductor out of the same ingredients used in the setup composed of a spin-orbit coupled quantum wire, an s-wave superconductor and a uniform magnetic field [12, 13], but replacing the later by a spatially modulated electric field, as illustrated in Fig. 14.1. More recently, a similar scheme for realizing topological superconductivity employing a superlattice has been investigated, but in that case the applied periodic field comes from superconducting islands deposited on the sample [39].

In our system, the electrons in the wire experience *Rashba* and *Dresselhaus* spin-orbit interactions which are present in crystals lacking inversion symmetry [40]. The Rashba component can originate also from asymmetries in the setup or material architecture (such as, for example, the inversion asymmetry along the growth direction in semiconductor heterostructures which produces an electric field in that same direction) and from an applied electric field [40]. The latter allows for external control of the Rashba coupling (more than doubling the base value in usual semiconductor quantum wells [41]). For the setup shown in Fig. 14.1, the responsiveness to an electric field means that the Rashba interaction will pick up the same modulation of the applied field. Incidently, also the chemical potential gets locally modulated by the electric field. The electrons in the wire acquire a superconducting pairing due to proximity with an s-wave superconductor. Finally, we take $e-e$ interactions into account for they are crucial, as we shall see, to drive the transition to the topological phase.

The tight-binding microscopic Hamiltonian that captures the various interactions introduced above is given by $H = H_0 + H_{\text{mod}} + H_{sc} + H_{e-e} + h.c.$, where

$$\begin{aligned} H_0 &= \sum_{n=1}^{N}\sum_{\alpha,\alpha'}[-t\sigma^0_{\alpha\alpha'} - i\gamma_D\sigma^x_{\alpha\alpha'} - i\gamma_R\sigma^y_{\alpha\alpha'}]\, c^\dagger_{n,\alpha}c_{n+1,\alpha'} + \\ &+ \sum_{n=1}^{N}\sum_{\alpha}\frac{\mu}{2}\, c^\dagger_{n,\alpha}c_{n,\alpha}, \end{aligned} \tag{14.1}$$

$$\begin{aligned} H_{\text{mod}} &= \sum_{n=1}^{N}\sum_{\alpha,\alpha'}[-i\gamma'_R\cos(2\pi q n)\sigma^y_{\alpha\alpha'}]\, c^\dagger_{n,\alpha}c_{n+1,\alpha'} + \\ &+ \sum_{n=1}^{N}\sum_{\alpha}\frac{\mu'}{2}\cos(2\pi q n)\, c^\dagger_{n,\alpha}c_{n,\alpha}, \end{aligned} \tag{14.2}$$

$$H_{sc} = \sum_{n=1}^{N}\Delta\, c_{n,\uparrow}c_{n,\downarrow}, \tag{14.3}$$

$$H_{e-e} = \sum_{n,n'=1}^{N}\sum_{\alpha,\alpha'} V(n-n')\, c^\dagger_{n,\alpha}c^\dagger_{n',\alpha'}c_{n',\alpha'}c_{n,\alpha}. \tag{14.4}$$

In the above, $c_{n,\alpha}(c^\dagger_{n,\alpha})$ creates (annihilates) an electron at site $n = 1, \ldots, N$ with spin projection $\alpha = \uparrow, \downarrow$ along an arbitrary quantization axis, $\sigma^{(0)}$ is the 2×2 identity matrix and $\sigma^{x(y)}$ is the $x(y)$ Pauli matrix. As for the model parameters, t is a kinetic hopping amplitude, γ_D (γ_R) is the hopping amplitude arising from the uniform Dresselhaus (Rashba) interaction, μ is the uniform chemical potential; γ'_R and μ' are, respectively, the amplitudes of the modulated Rashba hopping and modulated chemical potential, with $2\pi q/a$ the wave number of the modulation (a being the lattice spacing); Δ gives the superconducting pairing energy and $V(n-n')$ the intensity of $e-e$ interactions. The model defined above has time reversal symmetry, i.e. $[H, T] = 0$ with $T = UK$ the time reversal operator composed of an unitary transformation U that flips the electron spin and K the complex conjugation responsible for reversing the momentum.

Defining new operators

$$\begin{aligned} \mathrm{d}_{n,+} &\equiv \frac{1}{\sqrt{2}}\left(-ie^{-i\theta}c_{n,\uparrow} + e^{i\theta}c_{n,\downarrow}\right) \\ \mathrm{d}_{n,-} &\equiv \frac{1}{\sqrt{2}}\left(e^{-i\theta}c_{n,\uparrow} - ie^{i\theta}c_{n,\downarrow}\right) \end{aligned} \tag{14.5}$$

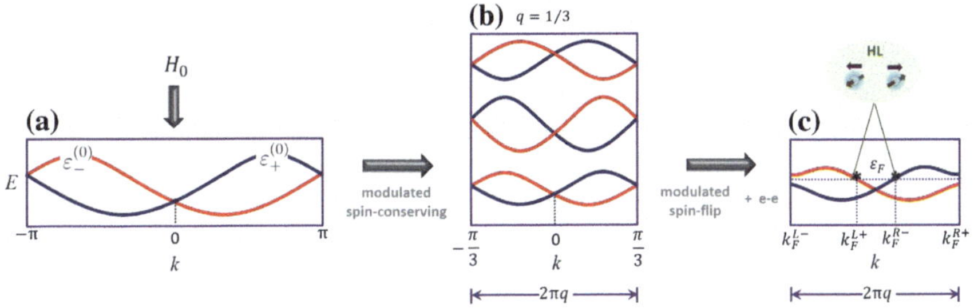

Fig. 14.2 In the presence of a spatial modulation with wave number $2\pi q$ (setting the lattice spacing $a = 1$) with $q = 1/3$, each one of the two spin-split bands in the BZ of a spin-orbit coupled wire **a** splits up in 3 sub-bands folded in a reduced BZ of size $2\pi/3$. Setting the Fermi level so that the outer Fermi points K_F^{R+} and K_F^{L-} match the boundaries of the reduced BZ, an effective spin-flip backscattering lifts the degeneracies of the two lowest sub-bands at the zone boundaries **c**. An HLL emerges in the vicinity of the inner gapless Fermi points K_F^{R-} and K_F^{L+}

where $\tan(2\theta) = \gamma_D/\gamma_R$, and $\tau = \pm$ labels the spin projections along the rotated quantization axis, (14.1)–(14.2) become almost diagonal in spin space:

$$H_0 = \sum_{n=1}^{N}\sum_{\tau}[-t + i\tau\gamma_{\text{eff}}]\, d_{n,\tau}^{\dagger}d_{n+1,\tau} + \frac{\mu}{2}\, d_{n,\tau}^{\dagger}d_{n,\tau}, \tag{14.6}$$

$$H_{\text{mod}} = \sum_{n=1}^{N}\sum_{\tau}\cos(2\pi q n)\left[-\frac{\mu'}{2}\, d_{n,\tau}^{\dagger}d_{n,\tau} + \right.$$
$$\left. + i\tau\gamma_{RR}\, d_{n,\tau}^{\dagger}d_{n+1,\tau} + i\gamma_{RD}\, d_{n,\tau}^{\dagger}d_{n+1,-\tau}\right], \tag{14.7}$$

where $\gamma_{\text{eff}} = \sqrt{\gamma_R^2 + \gamma_D^2}$, $\gamma_{RR} = \gamma_R' \cos(2\theta) = \gamma_R'\gamma_R/\gamma_{\text{eff}}$, $\gamma_{RD} = \gamma_R' \sin(2\theta) = \gamma_R'\gamma_D/\gamma_{\text{eff}}$. All processes in (14.6)–(14.7) conserve spin, except the last one which promotes a spin-flip and results from the interplay (encoded in γ_{RD}) between the modulated Rashba (γ_R') and the uniform Dresselhaus interaction (γ_D).

Diagonalizing H_0 in (14.6) by a Fourier transform results in the well-known spin-split spectrum $\varepsilon_\tau^{(0)}(k) = -2\tilde{t}\cos(ka - \tau q_0 a) + \mu$, with $\tilde{t} = \sqrt{t^2 + \gamma_{\text{eff}}^2}$ and $q_0 a = \arctan(\gamma_{\text{eff}}/t)$. These two bands are depicted inside the Brillouin zone (BZ) in Fig. 14.2a with $a = 1$. Adding the modulated spin-conserving hoppings of H_{mod} in (14.7), the BZ gets reduced to the wave number $2\pi q$ (setting $a = 1$) of the modulation and each band of H_0 splits up in $r = 1/q$ sub-bands gapped at the boundaries of the new reduced BZ. This is illustrated in Fig. 14.2b for $q = 1/3$.

The modulated spin-flip hopping in (14.7) would like to lift the degeneracies at the center and at the boundaries of the reduced BZ through hybridization of the degenerate spin up and spin down states. However, this is forbidden by Kramers' theorem which dictates that the band crossings at the time reversal invariant momenta (TRIM)—$k = 0, \pm\pi/r$ in our reduced BZ—cannot be removed [42]. Mathematically, the inability of the spin-flip term—call it H_{sf}—to open a gap at the TRIM can be seen from the vanishing of its matrix element between time reversal-symmetric single-particle states. Indeed, let $|\alpha\rangle$ and $|T\alpha\rangle$ be, respectively, a single-particle state and its time reversal symmetric counterpart (reversed momentum and spin). Using that $[H_{sf}, T] = 0$ and the properties of the time reversal operator $T^\dagger = T$, $T^2|\alpha\rangle = -|\alpha\rangle$, $\langle T\alpha|T\beta\rangle = \langle\alpha|\beta\rangle$ for any $|\alpha\rangle$ and $|\beta\rangle$, it follows that: $\langle T\alpha|H_{sf}|\alpha\rangle = \langle\alpha|T^\dagger H_{sf}|\alpha\rangle = \langle\alpha|H_{sf}|T\alpha\rangle = \langle T\alpha|H_{sf}|T^2\alpha\rangle = -\langle T\alpha|H_{sf}|\alpha\rangle$, and thus $\langle T\alpha|H_{sf}|\alpha\rangle = 0$. The vanishing of this matrix element implies that a hypothetic single-particle spin-flip backscattering between TRIM states (which would open a gap at those points) does not happen.

Now, if $|\alpha\rangle$ would be a two-electron state, then $T^2|\alpha\rangle = |\alpha\rangle$, meaning that a two-particle spin-flip backscattering is actually allowed. However, the probability of a simultaneous backscattering of two electrons in response to H_{sf} is negligible; unless the electrons interact. In fact, $e - e$ interactions can open a channel for a correlated backscattering with spin-flip if the many-body state is "commensurate" with the lattice modulation. More specifically, for a generic many-body state, there will be four Fermi points $k_F^{x\tau} = \pm k_F + \tau q_0$, with $x = R(L)$ if the sign in front of k_F is positive (negative), $\tau = \pm$ as before, and $k_F = \pi N_e/2Na$, where $N_e(N)$ is the number of electrons (lattice sites). As we shall see in the next section, when the Fermi level is adjusted so that the outer Fermi points k_F^{R+} and k_F^{L-} fall close to the boundaries of the BZ, Coulomb interaction (14.4) and H_{sf} produce an effective sine-Gordon-like backscattering which is quadratic (linear) in the spin-orbit ($e - e$) parameter. This process becomes strongly relevant if the participating couplings fall in a certain regime, breaking time reversal symmetry *spontaneously* at the outer Fermi points via the formation of a spin-density wave (SDW). This enables the detachment of the bands at the boundaries of the BZ, as illustrated in Fig. 14.2c. This "selective" gap opening—that is, occurring only around the Fermi points that are separated exactly by the wave number of the modulation—leading to a partially gapped band structure differs from the usual situation in which an interaction opens a gap throughout the Fermi level.

The inner Fermi points k_F^{R-} and k_F^{L+}, which are not apart by $2\pi q$, are insensitive to the effective backscattering brought about by the modulation and remain gapless, thus supporting a HLL of interacting electrons with spin-momentum locking. Such a HLL is quite particular in that it is neither "holographic", thus differing from the well-known edge states of a quantum spin Hall insulator [42]; nor is it "quasi-helical" [43], i.e. it is also different from helical states achieved in one dimension when a magnetic field breaks time reversal symmetry explicitly.

Finally, the effective "spinless" states at the inner Fermi points may be turned into a p-wave superconductor by adding superconducting pairing, H_{sc} in (14.3). If the chain is finite with open boundaries, degenerate MZMs appear at the left and right ends [6, 44] for, by construction, the Fermi energy is already tuned to the Majoranas' degeneracy level.

Having conceptually outlined our scheme, we next present the low-energy bosonized (effective) theory describing the chain and the RG approach used to obtain the parameter regimes of the theory.

14.3 Low-Energy Bosonized Theory and Renormalization Group Treatment

To properly address $e-e$ interactions, we recast the hamiltonian $H = H_0 + H_{\text{mod}} + H_{sc} + H_{e-e} + h.c.$ discussed in the previous section in a low-energy bosonized form. This is done by first linearizing the spectrum around the system's four Fermi points and rewriting the lattice operators in terms of fermionic right- and left-moving fields in the continuum. A bosonization procedure transforms the fermionic fields into canonically conjugated bosonic fields. The effective theory thus becomes $H = \int \mathrm{d}x\, (\mathcal{H}_{\text{outer}} + \mathcal{H}_{\text{inner}} + \mathcal{H}_{\text{mix}})$ with

$$\begin{aligned}\mathcal{H}_{\text{outer}} &= u[(\partial_x\theta_1)^2 + (\partial_x\phi_1)^2] - \frac{\Delta}{\pi a}\sin\left(\sqrt{\frac{4\pi}{K}}\theta_1\right)\\ &+ \frac{\lambda}{\sqrt{\pi K}a}\cos(\sqrt{4\pi K}\phi_1)\partial_x\theta_1,\end{aligned} \tag{14.8}$$

$$\mathcal{H}_{\text{inner}} = u[(\partial_x\theta_2)^2 + (\partial_x\phi_2)^2] - \frac{\Delta}{\pi a}\sin\left(\sqrt{\frac{4\pi}{K}}\theta_2\right), \tag{14.9}$$

$$\mathcal{H}_{\text{mix}} = \frac{gK}{\pi}\partial_x\phi_1\partial_x\phi_2, \tag{14.10}$$

where $K = (1 + g/(\pi v_F))^{-1/2}$ is the Luttinger parameter determined by the bare Fermi velocity $v_F = 2a\tilde{t}\sin(k_F a)$ and the Coulomb forward scattering of strength g (Here we do not include Coulomb backscattering which is known to be irrelevant in a Luttinger liquid [45] and also in the presence of spin-orbit interactions and superconducting pairing [46]); and $u = v_F/2K$ is the Fermi velocity dressed by $e-e$ interactions. The superconducting pairing Δ appears as the amplitude of a sine-Gordon potential and $\lambda = a\gamma_R'\gamma_D/\gamma_{\text{eff}}$ is the amplitude of a sine-Gordon-like potential encoding an effective spin-flip backscattering induced by the combination of spin-orbit interactions. The $\Delta-$ and $\lambda-$potentials are both renormalized by K. The fields $\phi_i(x)$ and $\theta_i(x)$, $i = 1, 2$, are dual bosonic fields satisfying $v_F\partial_x\theta_i = \mp\partial_t\phi_i$, with $i = 1(2)$ referring to excitations around the outer (inner) Fermi points. The bosonized theory is thus composed of two branches given by $\mathcal{H}_{\text{outer}}$ and $\mathcal{H}_{\text{inner}}$ acting at the corresponding pair of Fermi points and coupled by $\mathcal{H}_{\text{mix}}$. In the non-interacting limit of $g = 0 \rightarrow K = 1$, $\mathcal{H}_{\text{mix}} = 0$ and the theory is described by decoupled and non-renormalized branches $\mathcal{H}_{\text{outer}}$ and $\mathcal{H}_{\text{inner}}$.

Disregarding, for the moment, the coupling between the two branches, the RG solution of the sine-Gordon model in (14.9) yields a strongly relevant superconducting pairing if $K > 1/2$ (weak $e-e$ repulsion), while if $K \leq 1/2$ (strong $e-e$ repulsion) the perturbation is marginally relevant provided its strength Δ is enough to survive the $e-e$ repulsion, satisfying $a\Delta/v_F > (1/K - 2)$. In both cases, a superconducting gap opens up in the inner branch. If, on the other hand, $K \leq 1/2$ with $a\Delta/v_F \leq (1/K - 2)$, the superconducting pairing gets suppressed by strong $e-e$ repulsion, becoming irrelevant.

Coming to (14.8), we see that it describes a competition between the spin-flip backscattering ($\propto \lambda$) and the superconducting pairing ($\propto \Delta$) around the outer Fermi points. The fate of the topological phase is primarily determined by the outcome of this competition. To solve it, initially disregarding the coupling between the two branches, we pass to a Lagrangian formalism in which (14.8) is transformed into the effective action:

$$S_{\text{outer}} = \int \mathrm{d}x\mathrm{d}\tau\Big[\frac{2u}{2}\Big((\partial_x\phi_1)^2 + \frac{1}{(2u)^2}(\partial_\tau\phi_1)^2\Big) - \frac{v_F g_{so}}{\pi a^2}\cos(\sqrt{16\pi K}\phi_1) - \frac{v_F g_{sc}}{\pi a^2}\cos\left(\sqrt{\frac{4\pi}{K}}\theta_1\right)\Big], \tag{14.11}$$

where $\tau = it$ is the imaginary time and where $g_{sc} = a\Delta/v_F$ and $g_{so} = \lambda^2/(4v_F^2)$ are dimensionless coupling constants. Equation (14.11) describes a generalized sine-Gordon theory containing the usual cosine of the ϕ_1-field, as well as the cosine of the dual θ_1-field [47–49]. When $g_{so} = g_{sc}$, (14.11) is invariant under the duality transformation $\phi_1 \leftrightarrow \theta_1$ and $2K \leftrightarrow 1/(2K)$, i.e. is a self-dual sine-Gordon model [50].

Note that although the bare masses g_{so} and g_{sc} of the two cosines originate from completely independent processes (spin-orbit interactions and superconducting pairing, respectively), their respective scaling dimensionalities $\Delta_{so} = 4K$ and $\Delta_{sc} = 1/K$ are controlled by the same parameter K (i.e. by $e-e$ interaction) in a way that prevents the two potentials from being simultaneously minimized.

The RG solution of the model given by (14.11) can be obtained in a one-loop expansion around the Gaussian fixed point $K = 1/2$ as in [50]. Defining $K = 1/2 - g_{ee} + O(g_{ee}^2)$, the RG flow equations for g_{ee}, g_{so} and g_{sc} are:

$$\frac{\mathrm{d}g_{ee}}{\mathrm{d}l} = g_{so}^2 - g_{sc}^2\,, \tag{14.12}$$

$$\frac{\mathrm{d}g_{so}}{\mathrm{d}l} = 4g_{so}g_{ee}\,, \tag{14.13}$$

$$\frac{\mathrm{d}g_{sc}}{\mathrm{d}l} = -4g_{sc}g_{ee}\,, \tag{14.14}$$

where $l = \ln s$, with s a scale factor.

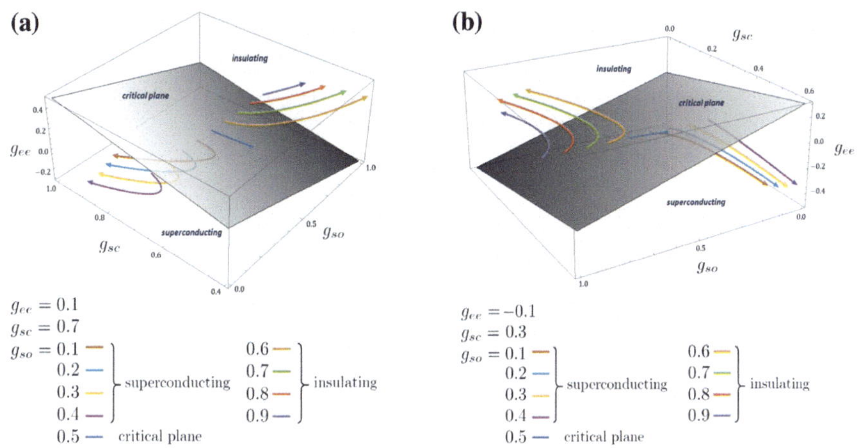

Fig. 14.3 RG flow diagram of the outer branch obtained by numerical solution of the flow equations (14.12)–(14.14). The critical plane equation is: $g_{so} - g_{sc} + 2g_{ee} = 0$. **a** Initial ($l = 0$) $g_{ee} > 0$, **b** Initial ($l = 0$) $g_{ee} < 0$

Figure 14.3 displays the flows obtained from numerical solution of (14.12)–(14.14) for different choices of g_{ee}, g_{so} and g_{sc} initial ($l = 0$) values. For ease of visualization, the flows are separated in two panels—Figs. 14.3a, b—according to the sign of the initial g_{ee}. There are two regimes separated by the critical plane $g_{so} - g_{sc} + 2g_{ee} = 0$ (locus of the theory's fixed points). Below the critical plane, the spin-orbit term is irrelevant whereas the superconducting pairing is *marginally* relevant if $g_{ee} \geq 0$ (Fig. 14.3a) and *strongly* relevant if $g_{ee} < 0$ (Fig. 14.3b). It follows that a gap ($\propto g_{sc}$) opens below the plane, leading to a superconducting phase in the outer branch. Above the critical plane, superconducting pairing goes irrelevant and spin-orbit becomes *strongly* relevant if $g_{ee} > 0$ (Fig. 14.3a) and *marginally* relevant if $g_{ee} \leq 0$ (Fig. 14.3b). As a result, a gap ($\propto g_{so}$) opens up above the plane, sustaining an insulating phase in the outer branch.

14.4 Phase Diagram

To obtain the phase diagram of the full theory the coupling between the outer and inner branches must be addressed. Since the branch-mixing hamiltonian, $\mathcal{H}_{\text{mix}}$ in (14.10), contains the gradient of the ϕ_1-field, the effectiveness of the branch coupling depends on the fate of ϕ_1 in the outer branch. For a strongly relevant g_{so}-perturbation, ϕ_1 gets pinned at one of the minima of the corresponding cosine in (14.11). This pinning occurs at a length scale which is shorter than the length at which the marginal (scaling dimension equal to 2) $\mathcal{H}_{\text{mix}}$ would start to affect the RG flow in the outer branch. As a consequence of the short-length pinning of ϕ_1, $\mathcal{H}_{\text{mix}}$

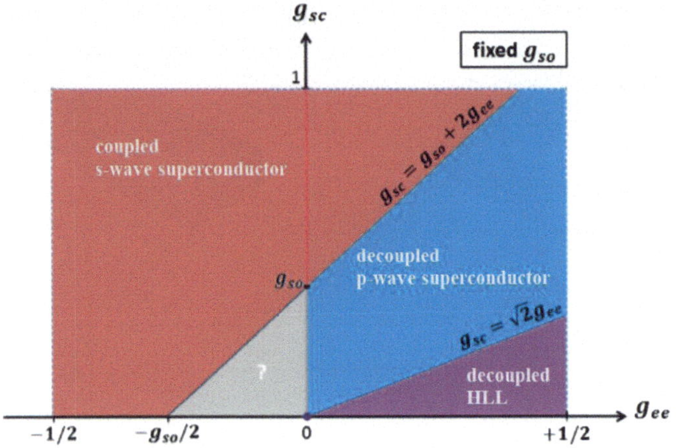

Fig. 14.4 A fixed-g_{so} cut through the 3D phase diagram of the system. The region labeled by a question mark cannot be assessed with the present RG approach

is strongly suppressed and the branches effectively decouple in the regime above the critical plane in Fig. 14.3a (for there $g_{ee} > 0$ means a strong $e-e$ interaction backing a strongly relevant g_{so}-perturbation). But above the critical plane in Fig. 14.3b (where $g_{ee} < 0$ implies a weaker $e-e$ interaction yielding an only marginally relevant g_{so}-perturbation), the pinning of ϕ_1 may be interrupted by $\mathcal{H}_{\text{mix}}$, with the result that the branches may remain coupled throughout the RG scaling. Or $\mathcal{H}_{\text{mix}}$ is too weak to stop the pinning of ϕ_1 above the critical plane even when $g_{ee} < 0$, in which case the branches would decouple also in this regime. These are, however, speculative scenarios; to determine the actual effect of $\mathcal{H}_{\text{mix}}$ in the regime above the critical plane in Fig. 14.3b is beyond the present perturbative RG approach. Differently, below the critical plane the branches definitely remain coupled regardless of the value of g_{ee}. As shown in Figs. 14.3a, b, below the critical plane the g_{sc}-perturbation is the strongly or marginally relevant one, with a pinning effect on θ_1 and, therefore, no suppression upon $\mathcal{H}_{\text{mix}}$.

Figure 14.4 shows a fixed-g_{so} cut through the 3D phase diagram which is obtained by combining the parameter regimes found in Sect. 14.3 for both the outer branch and the inner branch (the later rewritten in terms of g_{ee} and g_{sc}) and taking into account the above scaling analysis of the branch-mixing.

In the "decoupled p-wave superconductor" a strongly relevant spin-orbit perturbation opens a gap at the outer branch and causes the branches to decouple; meanwhile, a marginally relevant superconducting pairing in the inner branch drives the HLL to the desired topological phase. In the "decoupled HLL" the outer branch is the same as in the "decoupled p-wave superconductor", but an irrelevant superconducting pairing is washed out from the inner branch, which thus remains in the HLL state. The "coupled s-wave superconductor" has an irrelevant spin-orbit term in the outer branch so that the theory reduces to two identical branches supporting a strongly or marginally relevant superconducting pairing, and coupled by the branch-mixing term. The result is a conventional superconductor with pairing amplitude renormalized by the $e-e$ forward scattering. At last, the phase signaled with an interrogation mark is the one for which the outer branch cannot be resolved within the present formalism. This parameter regime could, conceivably, be analyzed with a density matrix renormalization group (DMRG).

The "decoupled p-wave superconductor" stretching between the two critical lines in Fig. 14.4 shows that the onset of this phase depends on a compromise between the parameters: for a given g_{so} and $g_{ee} > 0$, g_{sc} must fit within a certain range so that it is neither too strong (that it opens an s-wave superconducting gap throughout the Fermi level), nor too weak (that it cannot change the primitive HLL state). To demonstrate that this regime indeed corresponds to a topological superconducting chain hosting MZMs at its ends, we consider the Hamiltonian which describes a 1D spinless p-wave superconductor:

$$H_{pw}=\int \mathrm{d}x \left[\psi\left(-\frac{\partial_x^2}{2m}-\mu\right)\psi-\Delta_p\psi\left(\frac{i\partial_x}{k_F}\right)\psi\right]+\text{H.c.}, \tag{14.15}$$

where ψ is a spinless fermion field, m is an effective mass, μ is the chemical potential, and Δ_p is the p-wave pairing. By linearizing the spectrum of (14.15) for $\mu > 0$ and writing $\psi(x) = e^{ik_F x}R_+(x) + e^{-ik_F x}L_-(x)$, H_{pw} in (14.15) gets mapped (after dropping rapidly fluctuating terms that average to zero upon integration) onto $\int \mathrm{d}x\, \mathcal{H}_{\text{inner}}^F$, with $\mathcal{H}_{\text{inner}}^F$ given by the fermionic (pre-bosonization) version of (14.9) without $e-e$ interactions ($K = 1$), and with $\Delta \approx 2\Delta_p$.

The Hamiltonian H_{pw}, and therefore $\mathcal{H}_{\text{inner}}^F$, falls in the BDI symmetry class of the Altland–Zirbauer classification, with winding number $W = 1$ for $\mu > 0$ and $\Delta_p \in \mathbb{R}$ [51], implying a single unpaired MZM at each end of the chain. It follows from the above mapping that the same result is valid for the interacting ($0 < K < 1$) model in (14.9), provided $e-e$ interactions do not disrupt the superconducting gap enclosing the MZMs [15, 52–57]. It is then established that the proposed scheme produces a one-dimensional topological superconductor with an unpaired MZM at each end of the wire.

14.5 Summary

We propose a scheme for realizing a magnetic-field-free one-dimensional topological superconductor hosting unpaired MZMs. The system consists of a quantum wire with modulated Rashba and uniform Dresselhaus spin-orbit interactions, as well as $e-e$ interactions, and proximity-coupled to an s-wave superconductor. The microscopic Hamiltonian describing this systems is converted to a low-energy bosonized theory amenable to a RG treatment. The phase diagram is obtained by combining the resulting RG regimes, with the target topological phase emerging from a compromise between the various interactions. The realization of an all-electric device hosting robust MZMs would be fascinating, with relevant implications in topological quantum computing.

References

1. C. Nayak, S. H. Simon, A. Stern, M. Freedman, S. das Sarma, Rev. Mod. Phys. **80**, 1083 (2008)
2. For a review, see F. Wilczek, Nat. Phys. **5**, 614 (2009)
3. G. Moore, N. Read, Nucl. Phys. B **360**, 362 (1991)
4. M. Sato, Y. Takahashi, S. Fujimoto, Phys. Rev. Lett. **103**, 020401 (2009)
5. L. Jiang, T. Kitagawa, J. Alicea, A.R. Akhmerov, D. Pekker, G. Refael, J.I. Cirac, E. Demler, M.D. Lukin, P. Zoller, Phys. Rev. Lett. **106**, 220402 (2011)
6. A.Y. Kitaev, Phys. Usp. **44**, 131 (2001)
7. N. Read, D. Green, Phys. Rev. B **61**, 10267 (2000)
8. D.A. Ivanov, Phys. Rev. Lett. **86**, 268 (2001)
9. For a review, see J. Alicea, Rep. Prog. Phys. **75**, 076501 (2012)
10. For the experimental status up to January 2015, see S. Das Sarma, M. Freedman, C. Nayak, NPJ Quantum Inf. **1**, 15001 (2015)
11. L. Fu, C.L. Kane, Phys. Rev. Lett. **100**, 096407 (2008)
12. R.M. Lutchyn, J.D. Sau, S. DasSarma, Phys. Rev. Lett. **105**, 077001 (2010)
13. Y. Oreg, G. Refael, F. von Oppen, Phys. Rev. Lett. **105**, 177002 (2010)
14. S. Nadj-Perge, I.K. Drozdov, B.A. Bernevig, A. Yazdani, Phys. Rev. B **88**, 020407(R) (2013)
15. E.M. Stoudenmire, J. Alicea, O.A. Starykh, M.P.A. Fisher, Phys. Rev. B **84**, 014503 (2011)
16. A.C. Potter, P.A. Lee, Phys. Rev. B **83**, 184520 (2011)
17. J.D. Sau, S. Tewari, S. Das Sarma, Phys. Rev. B **85**, 064512 (2012)
18. S. Bravyi, A. Kitaev, Phys. Rev. A **71**, 022316 (2005)
19. S. Bravyi, Phys. Rev. A **73**, 042313 (2006)
20. D. Awschalom, N. Samarth, Physics **2**, 50 (2009)
21. C.L.M. Wong, K.T. Law, Phys. Rev. B **86**, 184516 (2012)
22. F. Zhang, C.L. Kane, E.J. Mele, Phys. Rev. Lett. **111**, 056402 (2013)
23. E. Gaidamauskas, J. Paaske, K. Flensberg, Phys. Rev. Lett. **112**, 126402 (2014)
24. S. Nakosai, J.C. Budich, Y. Tanaka, B. Trauzettel, N. Nagaosa, Phys. Rev. Lett. **110**, 117002 (2013)
25. S.B. Chung, J. Horowitz, X.-L. Qi, Phys. Rev. B **88**, 214514 (2013)
26. S. Deng, L. Viola, G. Ortiz, Phys. Rev. Lett. **108**, 036803 (2012)
27. A. Keselman, L. Fu, A. Stern, E. Berg, Phys. Rev. Lett. **111**, 116402 (2013)
28. D. Sticlet, C. Bena, P. Simon, Phys. Rev. B **87**, 104509 (2013)
29. X.-J. Liu, C.L.M. Wong, K.T. Law, Phys. Rev. X **4**, 021018 (2014)
30. E. Dumitrescu, J.D. Sau, S. Tewari, Phys. Rev. B **90**, 245438 (2014)
31. A. Haim, A. Keselman, E. Berg, Y. Oreg, Phys. Rev. B **89**, 220504 (2014)
32. J. Klinovaja, D. Loss, Phys. Rev. B **90**, 045118 (2014)
33. P. Kotetes, Phys. Rev. B **92**, 014514 (2015)
34. For a review, see P. Kotetes, New J. Phys. **15**, 105027 (2013)
35. A.A. Reynoso, D. Frustaglia, Phys. Rev. B **87**, 115420 (2013)
36. C. Wu, B.A. Bernevig, S.-C. Zhang, Phys. Rev. Lett. **96**, 106401 (2006)
37. C. Xu, J.E. Moore, Phys. Rev. B **73**, 045322 (2006)
38. A. Altland, M.R. Zirnbauer, Phys. Rev. B **55**, 1142 (1997)
39. Y. Levine, A. Haim, Y. Oreg, Phys. Rev. B **96**, 165147 (2017)
40. M. Malard, I. Grusha, G.I. Japaridze, H. Johannesson, Phys. Rev. B **84**, 075466 (2011)
41. D. Grundler, Phys. Rev. Lett. **84**, 6074 (2000)
42. For a review, see M. König, H. Buhmann, L.W. Molenkamp, T. Hughes, C.-X. Liu, X.-L. Qi, S.-C. Zhang, J. Phys. Soc. Jpn. **77**, 031007 (2008)

43. B. Braunecker, A. Ström, G.I. Japaridze, Phys. Rev. B **87**, 075151 (2013)
44. L. Fu, Phys. Rev. Lett. **104**, 056402 (2010)
45. T. Giamarchi, *Quantum Physics in One Dimension* (Oxford University Press, Oxford, 2004)
46. A. Schulz, A. De Martino, P. Ingenhoven, R. Egger, Phys. Rev. B **79**, 205432 (2009)
47. J.V. Jose, L.P. Kadanoff, S. Kirkpatrick, D.R. Nelson, Phys. Rev. B **16**, 1217 (1977)
48. D. Boyanovsky, J. Phys. A: Math. Gen. **22**, 2601 (1989)
49. P. Lecheminant, A.O. Gogolin, A.A. Nersesyan, Nucl. Phys. B **639**, 502 (2002)
50. A.O. Gogolin, A.A. Nersesyan, A.M. Tsvelik, *Bosonization and Strongly Correlated Systems* (Cambridge University Press, Cambridge, 1998)
51. S. Tewari, J.D. Sau, Phys. Rev. Lett. **109**, 150408 (2012)
52. E. Sela, A. Altland, A. Rosch, Phys. Rev. B **84**, 085114 (2011)
53. S. Gangadharaiah, B. Braunecker, P. Simon, D. Loss, Phys. Rev. Lett. **107**, 036801 (2011)
54. A. Manolescu, D.C. Marinescu, T.D. Stanescu, J. Phys.: Condens. Matter **26**, 172203 (2014)
55. H. Katsura, D. Schuricht, M. Takahashi, Phys. Rev. B **92**, 115137 (2015)
56. A. Ghazaryan, T. Chakraborty, Phys. Rev. B **92**, 115138 (2015)
57. N.M. Gergs, L. Fritz, D. Schuricht, Phys. Rev. B **93**, 075129 (2016)

Chapter 15
From Graphene to Quantum Computation: An Expedition to the Dirac Sea

Eduardo C. Marino

Abstract Dirac fermions were originally proposed by Dirac as the solutions of his relativistic wave-function equation. From then on they played a crucial role in the relativistic description of quarks and leptons. In the recent years it has been realized that several condensed matter systems contain electrons that, due to the presence of the lattice, behave precisely like Dirac electrons. We explore here the consequences of this fact in materials such as Graphene, Transition Metal Dichalcogenides and in Topological Quantum Computation.

15.1 Introduction

The concept of a Dirac Sea was proposed by Dirac, in order to circumvent the problem of negative energy solutions which appeared out of his relativistic wave-function equation. It was soon realized that the Dirac sea did not exist as such, but, nevertheless, its conceptual framework could be used for the prediction of the existence of antimatter, which was soon experimentally observed.

The Dirac sea, however, has made an spectacular reappearance in the realm of condensed matter systems, most noticeably in polyacetylene in 1+1D, graphene and transition metal dichalcogenides, in 2+1D and Weyl semimetals in 3+1D.

In the case of polyacetylene, we have one carbon atom per site of a linear lattice with lattice parameter a, in an sp^2 hybridization state, which contains one active electron per site in the p_z-orbital. The tight-binding energy band

$$E_0(k) = -2t \cos ka \tag{15.1}$$

therefore, is half-filled, with the Fermi level at $E_F = 0$. Expanding this energy around the points where $E_0(k) = E_F = 0$, namely, $k = \pm\pi/2a$ one readily obtains the linear

E. C. Marino (✉)
Instituto de Física, Universidade Federal do Rio de Janeiro, C.P.68528,
Rio de Janeiro, RJ 21941-972, Brazil
e-mail: marino@if.ufrj.br

A. Ferraz et al. (eds.), *Strongly Coupled Field Theories for Condensed Matter and Quantum Information Theory*, Springer Proceedings in Physics 239,
https://doi.org/10.1007/978-3-030-35473-2_15

dispersion relation, which befits a Dirac fermion [1]. Hence we conclude that the low-energy excitations about the Fermi level have kinematics described by the Dirac, rather than by the Schrödinger equation, being, therefore Dirac fermion excitations. That is how Dirac fermions appear in condensed matter systems.

15.2 Dirac Fermions in Two-Dimensional Systems

15.2.1 *Massless Dirac Fermions in Graphene*

Graphene is a pure carbon structure formed by sp^2-hybridized carbon atoms sitting on the vertices of two interpenetrating triangular lattices, labeled A and B. As in the case of polyacetylene there is a p_z orbital per site, occupied by one electron.

We will apply now the tight-binding approach to graphene, which describes the hopping of these electrons between nearest neighbors, which always belong to the opposite sublattice.

Given a point in a sublattice, it will have three nearest neighbors in the opposite sublattice, located at the points corresponding to the three vectors,

$$\mathbf{d}_1 = \frac{a}{\sqrt{3}}\hat{y} \quad \mathbf{d}_2 = -\frac{a}{\sqrt{3}}\left(\frac{\sqrt{3}}{2}\hat{x} + \frac{1}{2}\hat{y}\right) \quad \mathbf{d}_3 = \frac{a}{\sqrt{3}}\left(\frac{\sqrt{3}}{2}\hat{x} - \frac{1}{2}\hat{y}\right) \tag{15.2}$$

The corresponding tight-binding hamiltonian will be

$$H_{TB} = -t \sum_{\mathbf{R}, i=1,2,3, \sigma=\uparrow,\downarrow} \left[\psi_B^\dagger(\mathbf{R} + \mathbf{d}_i, \sigma)\psi_A(\mathbf{R}, \sigma) + H.C.\right] \tag{15.3}$$

where $\psi_{A,B}^\dagger(\mathbf{r}, \sigma)$is the creation operator of an electron, with spin σ, in the orbital p_z of the carbon atom located at the position $\mathbf{r}$ in sublattice: A, B.

Introducing

$$\psi_{A,B}(\mathbf{r}, \sigma) = \frac{1}{\sqrt{N}} \sum_{\mathbf{k}} e^{i\mathbf{k}\cdot\mathbf{r}} \psi_{A,B}(\mathbf{k}, \sigma), \tag{15.4}$$

we may re-write the hamiltonian as

$$H_{TB} = \sum_{\mathbf{k},\sigma} \psi_A^\dagger(\mathbf{k}, \sigma)\psi_B(\mathbf{k}, \sigma) \left[-t \sum_{i=1,2,3} e^{i\mathbf{k}\cdot\mathbf{d}_i}\right] + H.C. \tag{15.5}$$

Introducing $\Psi^\dagger(\mathbf{k}, \sigma) = (\psi_A^\dagger(\mathbf{k}, \sigma)\psi_B^\dagger(\mathbf{k}, \sigma))$, we may express this as

$$H_{TB} = \sum_{\mathbf{k},\sigma} \Psi^{\dagger}(\mathbf{k},\sigma) \begin{pmatrix} 0 & \phi \\ \phi^{*} & 0 \end{pmatrix} \Psi(\mathbf{k},\sigma) \tag{15.6}$$

where

$$\phi(\mathbf{k}) = -t \sum_{i=1,2,3} e^{i\mathbf{k}\cdot\mathbf{d}_i} \tag{15.7}$$

It follows, immediately, that the energy eigenvalues are

$$E(\mathbf{k}) = \pm|\phi(\mathbf{k})| = \pm t \sqrt{\sum_{i,j=1,2,3} e^{i\mathbf{k}\cdot(\mathbf{d}_i-\mathbf{d}_j)}}. \tag{15.8}$$

This expression, first obtained by Wallace in 1947 [2], describes the two bands of graphene. These touch at the points $\mathbf{K}$, where $\phi(\mathbf{K}) = 0$. It is not difficult to see that for

$$\mathbf{K} = \frac{4\pi}{3a}\hat{x} \quad ; \quad \mathbf{K}' = -\frac{4\pi}{3a}\hat{x} \tag{15.9}$$

we have

$$\phi(\mathbf{K}) = \phi(\mathbf{K}') = 1 + e^{i\frac{2\pi}{3}} + e^{-i\frac{2\pi}{3}} = 0 \tag{15.10}$$

The points $\mathbf{K}$ and $\mathbf{K}'$ are vertices of the hexagonal First Brillouin Zone of graphene and are called "valleys".

Later on, Di Vincenzo and Mele and Semenoff [3], independently, demonstrated that the low-energy excitations of graphene, which occur precisely around the valleys $\mathbf{K}$ and $\mathbf{K}'$, possess a dispersion relation, which corresponds to a relativistic particle and, hence, are governed by the Dirac equation.

Indeed, expanding now the above hamiltonian around each of the valleys $\mathbf{K}$ and $\mathbf{K}'$, we get

$$h_K = -\frac{ta}{\sqrt{3}} \left[-\sum_i \mathbf{p}\cdot\mathbf{d}_i \sin(\mathbf{K}\cdot\mathbf{d}_i)\sigma_x + \sum_i \mathbf{p}\cdot\mathbf{d}_i \cos(\mathbf{K}\cdot\mathbf{d}_i)\sigma_y \right] \tag{15.11}$$

and

$$h_{K'} = -\frac{ta}{\sqrt{3}} \left[-\sum_i \mathbf{p}\cdot\mathbf{d}_i \sin(\mathbf{K}'\cdot\mathbf{d}_i)\sigma_x + \sum_i \mathbf{p}\cdot\mathbf{d}_i \cos(\mathbf{K}'\cdot\mathbf{d}_i)\sigma_y \right]. \tag{15.12}$$

or, equivalently

$$h_K = \frac{\sqrt{3}ta}{2}\left[p_x\sigma_x + p_y\sigma_y\right]$$

$$h_{K'} = \frac{\sqrt{3}ta}{2}\left[-p_x\sigma_x + p_y\sigma_y\right] \tag{15.13}$$

We now identify h_K as the Dirac hamiltonian of a massless particle with velocity $v_F = \frac{\sqrt{3}ta}{2}$, instead of c. Actually $v_F \simeq 10^6$ m/s $\simeq c/300$ and $t \simeq 3$ eV.

As for $h_{K'}$, we can put it on the standard Dirac form by performing a canonical transformation

$$\Psi_{K'}(\mathbf{k},\sigma) \longrightarrow \exp\left\{i\frac{\pi}{2}\sigma_z\right\}\sigma_x\Psi_{K'}(\mathbf{k},\sigma)$$

$$\Psi^\dagger_{K'}(\mathbf{k},\sigma) \longrightarrow \Psi^\dagger_{K'}(\mathbf{k},\sigma)\sigma_x \exp\left\{-i\frac{\pi}{2}\sigma_z\right\}, \tag{15.14}$$

whereupon

$$H_K = \sum_{\mathbf{k},\sigma}\Psi^\dagger_K(\mathbf{k},\sigma)h_K\Psi_K(\mathbf{k},\sigma)$$

$$H_{K'} = \sum_{\mathbf{k},\sigma}\Psi^\dagger_{K'}(\mathbf{k},\sigma)h_{K'}\Psi_{K'}(\mathbf{k},\sigma). \tag{15.15}$$

Here

$$h_K = h_{K'} = v_F\left[p_x\sigma_x + p_y\sigma_y\right] \tag{15.16}$$

The vector Dirac matrices are, then, $\alpha_i = \sigma_i, \quad i = x, y$ and $\beta = \sigma_z$, hence the covariant Dirac matrices are given by

$$\gamma^0 = \sigma_z \ ; \ \gamma^1 = i\sigma_y \ ; \ \gamma^2 = -i\sigma_x \tag{15.17}$$

The energy eigenvalues can now be easily found:

$$E(\mathbf{p}) = \pm v_F|\mathbf{p}| \ ; \quad v_F = \frac{\sqrt{3}ta}{2} \tag{15.18}$$

We see that the energy eigenvalues are such that two opposite cones are formed at each point K, K', where the valence and conduction bands touch. Each set of two opposite Dirac cones corresponds to a "valley". Notice the fact that $E(\mathbf{p})$ is equal for the two valleys, thus expressing the invariance of the system under the time-reversal symmetry, which connects the two valleys.

15.2.2 Massive Dirac Fermions in Transition Metal Dichalcogenides

Transition metal dichalcogenides (TMD), such as WSe_2, WS_2, $MoSe_2$ and MoS_2 are a class of extremely interesting materials, which present a crystal structure similar to graphene, however, with the sites belonging to sublattices A and B being occupied, respectively, by transition metal and chalcogen atoms. The net effect is, again a breakdown of the symmetry between the A and B sublattices of the honeycomb structure, which existed in graphene. This generates a staggered local energy that, ultimately, produces a gap in the energy spectrum. TMD's present peculiar physical properties such as strong electron-hole interaction, leading to exciton states with a binding energy much larger than that of conventional semiconductors.

Transition metal dichalcogenides, form a class of materials that can be described by the massive Dirac equation.

We can model TMD's, by adding to (15.6) a staggered sublattice dependent M term, as follows

$$H_{TB} = \sum_{\mathbf{k},\sigma} \Psi^{\dagger}(\mathbf{k},\sigma) \begin{pmatrix} M & \phi \\ \phi^* & -M \end{pmatrix} \Psi(\mathbf{k},\sigma) \tag{15.19}$$

where $\phi(\mathbf{k})$ is given by (15.7). Notice that the M-term, being proportional to a σ_z matrix in the (A, B) space, describes the staggered energy, provenient from the local atomic asymmetry between the two sublattices in TMD's.

Diagonalizing the hamiltonian, we readily find the energy eigenvalues are now

$$E(\mathbf{k}) = \pm\sqrt{|\phi(\mathbf{k})|^2 + M^2} = \pm\sqrt{t^2 \sum_{i,j=1,2,3} e^{i\mathbf{k}\cdot(\mathbf{d}_i - \mathbf{d}_j)} + M^2}, \tag{15.20}$$

which clearly show the presence of a gap $2M$.

We may re-write the hamiltonian as

$$H_{TB,M} = \sum_{\mathbf{k},\sigma} \Psi^{\dagger}(\mathbf{k},\sigma) h_{TB,M} \Psi(\mathbf{k},\sigma)$$

$$h_{TB,M} = -t\left[\sum_i \cos(\mathbf{k}\cdot\mathbf{d}_i)\sigma_x + \sum_i \sin(\mathbf{k}\cdot\mathbf{d}_i)\sigma_y\right] + M\sigma_z \tag{15.21}$$

where the σ's are Pauli matrices with entries in the (A,B) space.

Now, expanding around the points K and K', we obtain, respectively

$$h_{K,M} = \frac{\sqrt{3}ta}{2}\left[p_x\sigma_x + p_y\sigma_y + M\sigma_z\right]$$

$$h_{K',M} = \frac{\sqrt{3}ta}{2}\left[-p_x\sigma_x + p_y\sigma_y + M\sigma_z\right] \tag{15.22}$$

Then, performing the same canonical transformation (15.14), in $\Psi_{K',\sigma}(\mathbf{k})$, we see that the mass term transforms as

$$M\Psi^{\dagger}_{K',\sigma}\sigma_z\Psi_{K',\sigma} \longrightarrow$$
$$M\Psi^{\dagger}_{K',\sigma}\sigma_x \exp\left\{-i\frac{\pi}{2}\sigma_z\right\}\sigma_z \exp\left\{i\frac{\pi}{2}\sigma_z\right\}\sigma_x\Psi_{K',\sigma} = -M\Psi^{\dagger}_{K',\sigma}\sigma_z\Psi_{K',\sigma} \tag{15.23}$$

The terms linear in the momentum transform as before, hence we obtain, in the massive case [1]

$$H_K = \sum_{\mathbf{k},\sigma}\Psi^{\dagger}_K(\mathbf{k},\sigma)h_K\Psi_K(\mathbf{k},\sigma)$$

$$H_{K'} = \sum_{\mathbf{k},\sigma}\Psi^{\dagger}_{K'}(\mathbf{k},\sigma)h_{K'}\Psi_{K'}(\mathbf{k},\sigma), \tag{15.24}$$

where

$$h_K = v_F\left[p_x\sigma_x + p_y\sigma_y + M\sigma_z\right] \tag{15.25}$$

and

$$h_{K'} = v_F\left[p_x\sigma_x + p_y\sigma_y - M\sigma_z\right] \tag{15.26}$$

Notice that the mass at valley K' has an opposite sign to that at valley K.

A primary mechanism for the existence of a mass term in the systems being examined here, is any asymmetry between the two sublattices A and B. In the case of TMD's, the fact that different atoms occupy each of the two sublattices, naturally generates a σ_z term in the hamiltonian, which expresses such asymmetry. Supposing the local difference in chemical potential between the atoms in the two sublattices is 2Δ, then $M = \Delta$.

15.2.3 Electromagnetic Interaction of Two-Dimensional Dirac Fermions: DC Conductivity in Graphene

The electromagnetic interaction of charged particles constrained to move on a plane, as in the case of electrons in graphene and TMD's can be described by minimally coupling the Dirac field to the vector gauge field A_μ of Pseudo Quantum Electrodynamics [4].

DC-conductivity is given by Kubo's formula,

$$\sigma^{ij} = \lim_{\omega \to 0} \frac{i}{\omega} \langle j^i j^j \rangle (\omega, \mathbf{p} = 0) \tag{15.27}$$

The current correlator is given by the vacuum polarization tensor, namely

$$\langle j^i j^j \rangle (\omega, \mathbf{p}) = \Pi^{ij}(\omega, \mathbf{p}). \tag{15.28}$$

Evaluation of such tensor up to two-loops, gives [1]

$$\sigma^{ij} = \frac{1}{16} \frac{e^2}{\hbar} \left[1 + \frac{92 - 9\pi^2}{18\pi} \alpha_0 \right] \delta^{ij} \pm \frac{1}{2\pi} \left(n + \frac{1}{2} \right) \frac{e^2}{\hbar} \epsilon^{ij}, \tag{15.29}$$

where the two signs correspond to the valleys K and K', respectively, and we use physical units of $\frac{e^2}{\hbar}$.

Summing the contributions of the four flavors in the set: $K\uparrow, K\downarrow, K'\uparrow, K'\downarrow$, we get the total conductivity [5]

$$\begin{aligned} \sigma_T^{ij} &= 4 \times \frac{\pi}{8} \frac{e^2}{h} \left[1 + \frac{92 - 9\pi^2}{18\pi} \alpha_0 \right] \delta^{ij} \\ \sigma_T^{ij} &= \frac{\pi}{2} [1 + 0.056\, \alpha_0] \frac{e^2}{h} \delta^{ij} \simeq 1.76 \frac{e^2}{h} \delta^{ij} \end{aligned} \tag{15.30}$$

Observe that the electrons of each valley give opposite contributions to the transverse conductivity and, consequently, this vanishes. We have, in summary,

$$\begin{aligned} \sigma_T^{xx} &= \frac{\pi}{2} [1 + 0.056\, \alpha_0] \frac{e^2}{h} \simeq 1.76 \frac{e^2}{h} \\ \sigma_T^{xy} &= 0 \end{aligned} \tag{15.31}$$

Even though this is still not so close to the experimental result [6]

$$\sigma_{T,exp}^{ij} \simeq 2.16 \frac{e^2}{h} \delta^{ij},$$

yet, it represents an improvement with respect to the non-interacting minimal conductivity

$$\sigma_{0,T}^{ij} = \frac{\pi}{2}\frac{e^2}{h}\delta^{ij} \simeq 1.57\frac{e^2}{h}\delta^{ij}$$

and is the closest to the experimental value [7].

15.2.4 The Quantum Valley Hall Effect in Graphene

We have just seen that graphene exhibits a longitudinal conductivity, which receives corrections from the electromagnetic interactions. No transverse conductivity, however was obtained in the absence of an external magnetic field.

We now introduce the concept of a “Valley Conductivity”, which is naturally defined in terms of the difference between the conductivities of each valley, namely

$$\sigma_V^{ij} = \sum_{\sigma=\uparrow,\downarrow}\left[\sigma_{K,\sigma}^{ij} - \sigma_{K',\sigma}^{ij}\right] \tag{15.32}$$

The valley conductivity, therefore, relates an applied electric field to the *relative* current between the two valleys.

From (15.29), we conclude that

$$\sigma_V^{ij} = 4\left(n + \frac{1}{2}\right)\frac{e^2}{h}\epsilon^{ij}, \tag{15.33}$$

or

$$\begin{aligned}\sigma_V^{xx} &= 0\\ \sigma_V^{xy} &= 4\left(n + \frac{1}{2}\right)\frac{e^2}{h}\end{aligned} \tag{15.34}$$

The longitudinal valley conductivity vanishes whereas the transverse one is nonzero. Notice that it is identical to the one obtained for the Quantum Hall effect in graphene in the presence of an applied external magnetic field. This is an exact result, by virtue of the Coleman-Hill theorem that rules out the correction of the transverse part of the vacuum polarization tensor by higher order terms.

The fact that these two conductivities are identical can be understood from the fact that both conductivities can be expressed in terms of topological invariants. The emergence of the QVHE in graphene is the consequence of the electromagnetic interactions among the p-electrons of carbon, described by PQED [5]. For this, however, we need to use the full dynamical interaction instead of just the static Coulomb interaction.

15.3 Dirac Fermions in One-Dimensional Systems: Non-Abelian Statistics

15.3.1 Interacting Massless Dirac Fermions

Consider a system of Dirac quasi-particles in one spatial dimension, described by the massless lagrangian density [8]

$$\mathcal{L} = i\overline{\psi}\ \partial\!\!\!/\psi - \mathcal{V}(\psi, \bar{\psi}) \tag{15.35}$$

where $\psi = \begin{pmatrix} \psi_1 \\ \psi_2 \end{pmatrix}$, is a two-component Dirac spinor and $\mathcal{V}$ is an arbitrary potential. Our convention for the γ-matrices is $\gamma^0 = \sigma_x, \gamma^1 = i\sigma_y, \gamma^5 = \gamma^0\gamma^1 = -\sigma_z$.

We also assume invariance under global U(1) and chiral U(1) symmetries, namely

$$\begin{aligned} \psi &\to e^{i\theta}\psi \\ \psi &\to e^{i\theta\gamma^5}\psi \end{aligned} \tag{15.36}$$

that leads to the conservation of charge and chirality.

A Lorentz boost will act on the space-time coordinates x_μ under the vector representation of the Lorentz group, namely, $x^\mu \to \Lambda^\mu{}_\nu x^\nu$, where

$$\Lambda^\mu{}_\nu = \begin{pmatrix} \cosh\omega & -\sinh\omega \\ -\sinh\omega & \cosh\omega \end{pmatrix}, \tag{15.37}$$

in such a way that $\tanh\omega = \mathrm{v}$, where v is the relative velocity between the two reference frames connected by the boost.

The Dirac field, conversely, will transform under the spinor representation of such group, namely

$$\psi \to \left(e^{-s\omega\gamma^5}\right)\psi = \begin{pmatrix} e^{s\omega} & 0 \\ 0 & e^{-s\omega} \end{pmatrix}\psi \tag{15.38}$$

where s is a real parameter, known as the Lorentz spin of the Dirac field. In what follows, we demonstrate that the parameter s determines how a many-particle state-vector behaves under the interexchange of identical particles. It, consequently characterizes the particle statistics, which may be either bosonic ($2s$ = even), fermionic ($2s$ = odd) or anyonic ($2s \neq$ integer).

From (15.37), it follows that the light-cone coordinates $u = x^0 + x^1$ and $v = x^0 - x^1$ transform as

$$u \to e^{-\omega}u \qquad v \to e^{\omega}v \tag{15.39}$$

We want to determine the two-point correlation function of the Dirac field. On the basis of the symmetries of the system, one can write

$$\langle 0|\psi_i(x)\psi_j^\dagger(0)|0\rangle = \begin{pmatrix} f(x) & 0 \\ 0 & g(x) \end{pmatrix}. \tag{15.40}$$

Using (15.38), we may infer that under a Lorentz boost this shall transform as

$$\langle 0|\psi(x)\psi^\dagger(0)|0\rangle \to \begin{pmatrix} e^{s\omega} & 0 \\ 0 & e^{-s\omega} \end{pmatrix}\begin{pmatrix} f(x) & 0 \\ 0 & g(x) \end{pmatrix}\begin{pmatrix} e^{s\omega} & 0 \\ 0 & e^{-s\omega} \end{pmatrix}$$
$$\langle 0|\psi(x)\psi^\dagger(0)|0\rangle \to \begin{pmatrix} e^{2s\omega} f & 0 \\ 0 & e^{-2s\omega} g \end{pmatrix} \tag{15.41}$$

Consequently, the Dirac field correlator may be written in the form

$$\langle 0|\psi(x)\psi^\dagger(0)|0\rangle =$$
$$\begin{pmatrix} \tilde{F}(-x^2)v^{2s} & 0 \\ 0 & \tilde{G}(-x^2)u^{2s} \end{pmatrix} \equiv \begin{pmatrix} F(-x^2)\frac{v^{2s}}{(-x^2)^s} & 0 \\ 0 & G(-x^2)\frac{u^{2s}}{(-x^2)^s} \end{pmatrix} \tag{15.42}$$

where the functions F, G depend on the specific form of the interaction potential $\mathcal{V}$.

We now take the Euclidean limit of the Dirac field correlation functions. Starting from (15.42), we make the analytic continuation $x_0 \to -ix_2^E$. Introducing the complex variable $z = x_1 + ix_2^E$, we have $-x^2 \to x_E^2 \equiv |z|^2$ and we can see that the Wightman correlators (15.42) are mapped into the (Euclidean) functions of the complex variable z,

$$\langle\psi_1(x)\psi_1^\dagger(0)\rangle_S = F(x_E^2)\frac{(-z)^{2s}}{|z|^{2s}}$$
$$\langle\psi_2(x)\psi_2^\dagger(0)\rangle_S = G(x_E^2)\frac{(z^*)^{2s}}{|z|^{2s}}$$
$$\langle\psi_1(x)\psi_2^\dagger(0)\rangle_S = 0 \;\; ; \;\; \langle\psi_2(x)\psi_1^\dagger(0)\rangle_S = 0 \tag{15.43}$$

Introducing the polar representation $z = |z|e^{i\mathrm{Arg}(z)}$, we can re-write those functions as

$$\langle\psi_1(x)\psi_1^\dagger(0)\rangle_S = F(|z|^2)e^{i2s\mathrm{Arg}(-z)} \;\; ; \;\; \langle\psi_2(x)\psi_2^\dagger(0)\rangle_S = G(|z|^2)e^{-i2s\mathrm{Arg}(z)} \tag{15.44}$$

where we chose the cuts of the Arg functions as $-\pi \le \mathrm{Arg(z)} < \pi$ and $0 \le \mathrm{Arg(-z)} < 2\pi$, in such a way that we may write $\mathrm{Arg(-z)} = \mathrm{Arg(z)} + \pi$.

The Euclidean functions, corresponding to field correlators containing the same Dirac fields as above but in a reversed order can be obtained in similar way and are given by [9, 10]

$$\langle \psi_1^\dagger(0)\psi_1(x)\rangle_S = F(|z|^2)e^{i2s\mathrm{Arg}(z)} \quad ; \quad \langle \psi_2^\dagger(0)\psi_2(x)\rangle_S = G(|z|^2)e^{-i2s\mathrm{Arg}(-z)}. \tag{15.45}$$

It follows that

$$\langle \psi_i^\dagger(0)\psi_i(x)\rangle_S = e^{-i2\pi s}\langle \psi_i(x)\psi_i^\dagger(0)\rangle_S. \tag{15.46}$$

The "braiding" or particle exchange operation in this one-dimensional system is a well-defined operation in the framework of the Euclidean functions: functions of opposite arguments will correspond to different ordering of field operators [1].

15.3.2 *Majorana Spinors with Non-Abelian Statistics*

We now introduce the Majorana and anti-Majorana spinor fields

$$\varphi_\pm = \begin{pmatrix} \varphi_{1\pm} \\ \varphi_{2\pm} \end{pmatrix},$$

where

$$\varphi_{i+} = \frac{1}{2}\left(\psi_i + \psi_i^\dagger\right) \quad ; \quad \varphi_{i-} = \frac{1}{2}\left(\psi_i - \psi_i^\dagger\right) \tag{15.47}$$

$$\psi_i = \varphi_{i+} + \varphi_{i-} \quad ; \quad \psi_i^\dagger = \varphi_{i+} - \varphi_{i-} \tag{15.48}$$

Let us analize now the two-point functions of the Majorana and anti-Majorana fields built out of the Dirac field.

$$\langle 0|\varphi_{i+}(x)\varphi_{i+}(y)|0\rangle_W = -\langle 0|\varphi_{i-}(x)\varphi_{i-}(y)|0\rangle_W = \frac{1}{4}\left(\langle 0|\psi_i(x)\psi_i^\dagger(y)|0\rangle_W + \langle 0|\psi_i^\dagger(x)\psi_i(y)|0\rangle_W\right) \tag{15.49}$$

and

$$\langle 0|\varphi_{i-}(x)\varphi_{i+}(y)|0\rangle_W = -\langle 0|\varphi_{i+}(x)\varphi_{i-}(y)|0\rangle_W = \frac{1}{4}\left(\langle 0|\psi_i(x)\psi_i^\dagger(y)|0\rangle_W - \langle 0|\psi_i^\dagger(x)\psi_i(y)|0\rangle_W\right), \tag{15.50}$$

where $i = 1, 2$, represent the two chiralities, namely rightmovers and leftmovers.

Notice that for $i \neq j$, $\langle 0|\varphi_{i\pm}(x)\varphi_{j\pm}(y)|0\rangle_W$ has a trivial analytic structure in terms of the complex variables, implying that nontrivial braiding will only occur involving states for which $i = j$, that is states with the same chirality.

Using (15.44), we obtain

$$\langle\varphi_{1+}(x)\varphi_{1+}(y)\rangle_S = -\langle\varphi_{1-}(x)\varphi_{1-}(y)\rangle_S = \\ \frac{1}{4}F(|x-y|^2)\left[e^{i2s\,\mathrm{Arg}(y-x)} + e^{i2s\,\mathrm{Arg}(x-y)}\right] \tag{15.51}$$

and

$$\langle\varphi_{1-}(x)\varphi_{1+}(y)\rangle_S = -\langle 0|\varphi_{1+}(x)\varphi_{1-}(y)|0\rangle_S = \\ \frac{1}{4}F(|x-y|^2)\left[e^{i2s\,\mathrm{Arg}(y-x)} - e^{i2s\,\mathrm{Arg}(x-y)}\right] \tag{15.52}$$

with similar expressions for the 2-components.

We now investigate the braiding properties of the above functions. For this purpose we use the relation $\mathrm{Arg}(y-x) = \mathrm{Arg}(x-y) + \pi$ and obtain

$$\left[e^{i2s\,\mathrm{Arg}(x-y)} \pm e^{i2s\,\mathrm{Arg}(y-x)}\right] \overrightarrow{x \leftrightarrow y} \\ \left[e^{i2s\,\mathrm{Arg}(y-x)}e^{-i2\pi s} \pm e^{i2s\,\mathrm{Arg}(x-y)}e^{i2\pi s}\right] \tag{15.53}$$

with analogous relations for the corresponding expressions in the φ_2 functions.

When the Dirac field, ψ is either bosonic or fermionic, the two complex phases generated above by the braiding operation are equal, namely $e^{i2\pi s} = e^{-i2\pi s} = \pm 1$ and we conclude that

$$\langle\varphi_\pm(y)\varphi_\pm(x)\rangle_S = e^{i2\pi s}\langle\varphi_\pm(x)\varphi_\pm(y)\rangle_S \tag{15.54}$$

implying that $\varphi_\pm$ will also be bosonic or fermionic.

Conversely, when the Dirac field is an anyon, $e^{i2\pi s} \neq e^{-i2\pi s}$ and we obtain

$$\langle\varphi_{1+}(y)\varphi_{1+}(x)\rangle_S = \cos\delta\langle\varphi_{1+}(x)\varphi_{1+}(y)\rangle_S - i\sin\delta\langle\varphi_{1+}(x)\varphi_{1-}(y)\rangle_S \tag{15.55}$$

$$\langle\varphi_{1-}(y)\varphi_{1+}(x)\rangle_S = -i\sin\delta\langle\varphi_{1+}(x)\varphi_{1+}(y)\rangle_S + \cos\delta\langle\varphi_{1+}(x)\varphi_{1-}(y)\rangle_S \tag{15.56}$$

where in the above expressions $\delta = 2\pi s$.

We conclude that when the Dirac field is an anyon, the $\varphi_\pm$ fields will have non-abelian braiding given by

$$\begin{pmatrix} \langle\varphi_{1+}(y)\varphi_{1+}(x)\rangle_S \\ \langle\varphi_{1-}(y)\varphi_{1+}(x)\rangle_S \end{pmatrix} = \begin{pmatrix} \cos\delta & -i\sin\delta \\ -i\sin\delta & \cos\delta \end{pmatrix} \begin{pmatrix} \langle\varphi_{1+}(x)\varphi_{1+}(y)\rangle_S \\ \langle\varphi_{1+}(x)\varphi_{1-}(y)\rangle_S \end{pmatrix} \tag{15.57}$$

with similar expressions for the φ_2 functions.

15.3.3 Majorana Qubits

We have seen that the interchange of identical particles exhibiting non-Abelian statistics, in 2-particle states, produces matrices of dimension 2, known as the monodromy, or braiding, matrices. These act on the set of degenerate 2-particle states with non-Abelian statistics, which form the qubits, where the information is stored in a process of quantum computation.

The particles with non-Abelian statistics are, therefore the basic building blocks of the qubits. They are, by their turn, obtained from abelian anyons in the form of self-adjoint or anti-self-adjoint combinations, which characterizes them as Majorana and Anti-Majorana quantum states.

The monodromy matrices process the Majorana (Anti-Majorana) qubits, producing a certain output state for a given input state. Now, for the system considered here, whenever the spin/statistics of the basic abelian particles used to build the Majoranas is $s = 1/4$, the monodromy matrices become one of the basic logic gates employed in a quantum computation algorithm, namely, the NOT gate, as we can see in (15.57):

$$\mathcal{M} = -iX, \qquad \text{in which} \qquad X = \begin{pmatrix} 0 & 1 \\ 1 & 0 \end{pmatrix} \tag{15.58}$$

The fact that qubits are made out of quantum states of Majorana quasi-particles has profound implications, because of the peculiar features of such states.

Degeneracy at zero energy is a first property. Suppose

$$\varphi(t, \mathbf{x}) = \int d\mathbf{p}\, \varphi(E(\mathbf{p}), \mathbf{p}) e^{-iE(\mathbf{p})t} e^{i\mathbf{p}\cdot\mathbf{x}},$$

Imposing the Majorana condition in coordinate space

$$\varphi^\dagger(t, \mathbf{x}) = \varphi(t, \mathbf{x}) \Longrightarrow \varphi^\dagger(E, \mathbf{p}) = \varphi(-E, -\mathbf{p}). \tag{15.59}$$

Imposing now the Majorana condition on the energy-momentum space, namely

$$\varphi^\dagger(E, \mathbf{p}) = \varphi(E, \mathbf{p}), \tag{15.60}$$

and using it in (15.59), we get

$$\varphi(E, \mathbf{p}) = \varphi(-E, -\mathbf{p}).$$

Assuming that the Majorana field operator creates an energy eigenstate with eigenvalue $E(\mathbf{p})$, it follows that

$$\begin{aligned} &H|\varphi(E, \mathbf{p})\rangle = E|\varphi(E, \mathbf{p})\rangle \\ &= H|\varphi(-E, -\mathbf{p})\rangle = -E|\varphi(-E, -\mathbf{p})\rangle = -E|\varphi(E, \mathbf{p})\rangle, \\ &H|\varphi(E, \mathbf{p})\rangle = -E|\varphi(E, \mathbf{p})\rangle, \end{aligned} \tag{15.61}$$

The first and last lines together imply that, if a Majorana state $|\varphi(E, \mathbf{p})\rangle$ is an energy eigenstate, then the energy eigenvalue must vanish: $E = 0$. It follows that all the Majorana states are degenerate. The Majorana modes, are also gapless.

15.3.4 Superselecting Sectors and Coherence Protection

We have seen that Majorana and Anti-Majorana states can be written as linear combinations of states carrying opposite charges. Surprisingly, such states cannot be physical [11]. The reason is charged states belong to superselecting sectors of the Hilbert space, which never mix in a physical state. Charge is an observable that, despite its quantum-mechanical nature, does not obey the uncertainty principle. Consequently, a coherent combination of an electron field and its hermitean adjoint, each of which creates states with opposite charge, just cannot be physical [11]. This seldom mentioned fact, is indeed remarkable. It is precisely the reason underlying the coherence robustness of the Majorana qubits. What happens is that the above property precludes the occurrence of isolated Majorana states in the bulk, and therefore they hide in the edges of the system. Nevertheless, pairs of Majorana modes can manifest in the bulk as charged states. It follows that by expressing a Dirac state as a combination of Majoranas, we can use the fact that these cannot be in the bulk to place each one of them far away from the other on the edges of the sample. This construction would naturally make the Majorana pair, or the qubit, immune to the local environmental perturbations, which are responsible for decoherence.

References

1. E.C. Marino, *Quantum Field Theory Approach to Condensed Matter Physics* (Cambridge University Press, UK, 2017)
2. P.R. Wallace, Phys. Rev. **71**, 622 (1947)
3. G.R. Semenoff, Phys. Rev. Lett **53**, 2449 (1984); D.P. Di Vincenzo, E.J. Mele, Phys. Rev. B **29**, 1685 (1984)
4. E.C. Marino, Nucl. Phys. B **408**, 551 (1993)
5. E.C. Marino, L.O. Nascimento, V. Alves, C. Morais Smith, Phys. Rev. X **5**, 011040 (2015)
6. X. Du, I. Skachko, F. Duerr, A. Luican, E.Y. Andrei, Nature **462**, 192 (2009)

7. K. Ziegler, Phys. Rev. B **75**, 233407 (2007)
8. E.C. Marino, J.B. Brozeguini, J. Stat. Mech.: Theory Exper. **P03011**, (2015)
9. E.C. Marino, J.A. Swieca, Nucl. Phys. B170 [FS1], 175 (1980)
10. E.C. Marino, B. Schroer, J.A. Swieca, Nucl. Phys. B200 [FS4], 499 (1982)
11. R.F. Streater, A.S. Wightman, *PCT, Spin and Statistics and All That* (Benjamin/Cummings, Reading, 1964)

Chapter 16
Quantum Dynamics from a Domain Wall Initial State, in Real and Imaginary Time

Jacopo Viti

Abstract I review some recent results regarding the time evolution of 1 + 1 inhomogeneous quantum systems. The main focus is on the so-called Domain Wall initial state, a state such that average particle density or spin magnetization can have a maximal discontinuity.

16.1 The Domain Wall Initial State: Free Fermions

The Domain Wall (DW) initial state [1] is perhaps at the root of one the simplest quantum non-equilibrium protocol in 1 + 1 dimensions. Consider fermions hopping on a one-dimensional chain with a free Hamiltonian

$$H_{\text{free}} = -\frac{1}{2}\sum_{x\in\mathbb{Z}}[c_x^{\dagger}c_{x+1} + c_{x+1}^{\dagger}c_x]. \tag{16.1}$$

In momentum space the Hamiltonian is diagonal and reads

$$H_{\text{free}} = \int_{-\pi}^{\pi} \mathrm{d}k\ \varepsilon(k)c^{\dagger}(k)c(k), \quad \{c^{\dagger}(k), c(k')\} = \delta(k, k') \tag{16.2}$$

where $c(k) = \frac{1}{\sqrt{2\pi}}\sum_{x\in\mathbb{Z}} e^{ikx}c_x$ and $\varepsilon(k) = -\cos(k)$. Suppose moreover that at time $t = 0$ all the particles are on the left half of the chain, i.e. the initial quantum state is a DW initial state

$$|DW\rangle = \prod_{x<0} c_x^{\dagger}|0\rangle, \tag{16.3}$$

J. Viti (✉)
International Institute of Physics, UFRN, Campos Universitário,
Lagoa Nova, Natal 59078-970, Brazil
e-mail: viti.jacopo@gmail.com

ECT, UFRN, Campos Universitário, Lagoa Nova, Natal 59078-970, Brazil

A. Ferraz et al. (eds.), *Strongly Coupled Field Theories for Condensed Matter and Quantum Information Theory*, Springer Proceedings in Physics 239,
https://doi.org/10.1007/978-3-030-35473-2_16

being $|0\rangle$ the Fock vacuum. Intuitively, after time passes, particles will flow from left to right, building a non-trivial density profile along the chain

$$\rho_x(t) \equiv \langle DW(t)|c_x^\dagger c_x|DW(t)\rangle, \tag{16.4}$$

together with a position dependent particle current $j_x(t)$. It is indeed clear that at fixed t for x sufficiently large and negative (positive), $\rho_x(t)$ will be one (zero) and the current vanishing. The density profile in (16.4) and the current can be determined exactly for large space-time separations $x, t \gg 1$ by stationary phase analysis [2]. We will use the shorthand notation $\lim_{\text{hd}}$ (standing for *hydrodynamical*) for the combined limits $\lim_{t\to\infty}\lim_{x\to\infty}$ with $x/t = \xi \in \mathbb{R}$. In a field theory context, the limit is actually called a *continuum limit*. It is instructive to comment the outcome of the stationary phase analysis more in detail. We can introduce the function $n(k, x/t)$ representing the average occupancy of fermionic modes of momentum k, or *fillings*, along a space-time ray ξ. For large space-time separations along a given ray, the density and current are obtained as in thermodynamical equilibrium through the following integrals

$$\lim_{\text{hd}} \rho_x(t) = \int_{-\pi}^{\pi} \frac{\mathrm{d}k}{2\pi}\, n(k,\xi), \quad \lim_{\text{hd}} j_x(t) = \int_{-\pi}^{\pi} \frac{\mathrm{d}k}{2\pi} v(k)\, n(k,\xi), \tag{16.5}$$

$v(k) = \frac{d\varepsilon(k)}{dk}$ being the group velocity of the fermions. The stationary phase analysis shows that the ray dependent fillings satisfy the continuity equation $\partial_t n(k,\xi) + v(k)\partial_x n(k,\xi) = 0$ which can be recast into

$$(\xi - v(k))\partial_\xi n(k,\xi) = 0. \tag{16.6}$$

Provided the boundary conditions $n^{\text{left}}(k) \equiv n(k,-\infty)$ and $n^{\text{right}}(k) \equiv n(k,\infty)$, the solution of (16.6) is a step function

$$n(k,\xi) = n^{\text{left}}(k)\Theta(v(k)-\xi) + n^{\text{right}}(k)\Theta(\xi - v(k)), \tag{16.7}$$

with a discontinuity at $k = v^{-1}(\xi)$. Actually (16.7) can be used [3] to study the time evolution of any free fermionic system that was initially prepared in state of the form $|\Psi\rangle = |\Psi^{\text{left}}\rangle \otimes |\Psi^{\text{right}}\rangle$. The fillings $n^{\text{left}}(k)$ (reps. $n^{\text{right}}(k)$) specify the distribution of momenta in the initial state $|\Psi^{\text{left}}\rangle$ (resp. $|\Psi^{\text{right}}\rangle$). For the DW initial state: $n^{\text{left}} = 1$, $n^{\text{right}} = 0$ and $v(k) = \sin(k)$, giving the exact result [1]

$$\lim_{\text{hd}} \rho_x(t) = \frac{1}{\pi}\arccos(x/t) \tag{16.8}$$

$$\lim_{\text{hd}} j_x(t) = \frac{1}{\pi\sqrt{1-(x/t)^2}}, \tag{16.9}$$

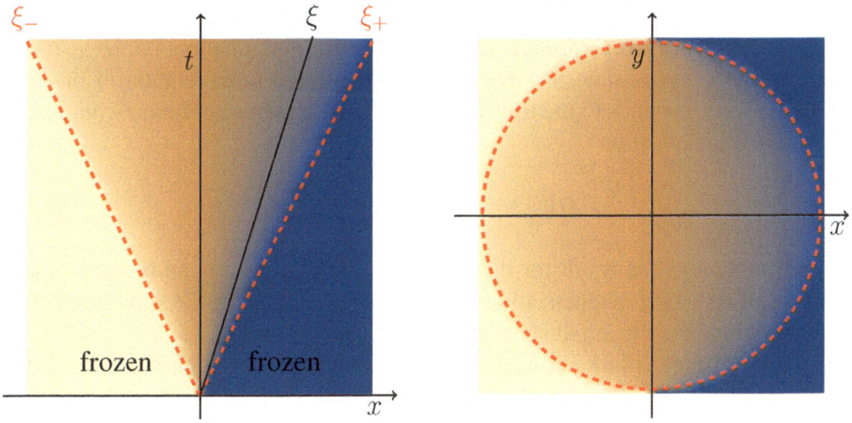

Fig. 16.1 **Left.** `DensityPlot` (with `Mathematica`) of $\lim_{\text{hd}} \rho_x(t)$ given in (16.8). At time $t = 0$, the quantum system is the initial state $|DW\rangle$ with spacial density 1 (corresponding to the colour yellow) on the left and 0 on the right (corresponding to the colour blue). Time-evolving such an initial state with the Hamiltonian in (16.1) results for large space-time separations in the elementary function in (16.8), whose density plot is depicted above. The quantum system is inhomogeneous inside the dashed red lines corresponding to $\xi_\pm = \pm\max_k v(k)$; this is the so-called *light-cone*. Along a given ray ξ (black line), the quantum state is described as in equilibrium with fillings in (16.7). **Right.** `DensityPlot` (with `Mathematica`) of $\lim_{\text{hd}} \rho(x, y; R)$ given in (16.16). Now the inhomogeneous region is contained inside a red dashed circle that is dubbed an *arctic curve* in the jargon of statistical mechanics. Taking $y = it$ and sending $R \to 0$, the statistical mechanics arctic curve is transformed into the quantum mechanical light-cone

depicted in Fig. 16.1 on the left. It is evident that an *inhomogeneous* region is built inside a triangular shape, that is the so-called light-cone. Outside the light-cone, the quantum system appears to be *frozen* in its initial configuration, where correlation functions are trivial. Inside the light-cone correlation functions show short-distance power-law singularities which are position independent. This a an example of two-quantum phases (a trivial outside the light-one and a critical inhomogeneous inside) separated sharply in space-time.

16.2 A Quick Look at Generalized Hydrodynamics

The results of the previous section can be reformulated in a spin-chain framework. The model of interest is now the so-called XXZ spin chain in the gapless regime, with Hamiltonian (L even)

$$H_{XXZ}(\Delta) = -\sum_{x=-\frac{L}{2}}^{\frac{L}{2}-1}\left[\mathbf{s}_x^x\mathbf{s}_{x+1}^x + \mathbf{s}_x^y\mathbf{s}_{x+1}^y + \Delta\left(\mathbf{s}_x^z\mathbf{s}_{x+1}^z - \frac{1}{4}\right)\right]; \tag{16.10}$$

$\mathbf{s}_x^\alpha$ are spin $1/2$ matrices and $\Delta = \cos(\gamma)$. For $\Delta = 0$, the Hamiltonian in (16.10) can be transformed into (16.1) through a Jordan-Wigner transformation, i.e. $H_{XXZ}(0) = H_{\text{free}}$. The DW initial state in (16.3) has then a simple interpretation in the spin language: it corresponds to the following step initial condition for the z-component of the spin magnetization

$$|DW\rangle = |\uparrow\uparrow\ldots\uparrow\downarrow\ldots\downarrow\downarrow\rangle. \tag{16.11}$$

Mutatis mudandis, the result for the fermion density in (16.8) can be recast in an exact result for the z-magnetization profile at $\Delta = 0$ in the XXZ spin chain after time-evolving from the initial state in (16.11).

At $\Delta \neq 0$, time-evolution from the DW initial state was studied numerically in [4] but an exact solution has been possible only recently, exploiting a remarkable generalization of (16.7) to Bethe Ansatz solvable models. In a Thermodynamic Bethe Ansatz formalism [5], the fermionic fillings $n(k) \in [0, 1]$ characterizing the many-body quantum state at equilibrium are replaced by the filling fractions $\vartheta_j(\lambda) \in [0, 1]$ of any species $j = 1, \ldots, n_B$ of bound-states present in the spectrum. The variable $\lambda \in \mathbb{R}$ is called the rapidity and parametrizes both the bare momenta $k_j^{\text{bare}}(\lambda)$ and energy $\varepsilon_j^{\text{bare}}(\lambda)$ of the excitations. The use of the word *bare* will be clarified soon. If a model is solvable by Bethe Ansatz, the bare energy and momenta of a bound state are, in particular, exactly known. It is then also possible to define a bare bound-state group velocity as

$$v_j^{\text{bare}}(\lambda) = \frac{\partial_\lambda \varepsilon_j^{\text{bare}}}{\partial_\lambda k_j^{\text{bare}}}. \tag{16.12}$$

The fillings $\vartheta_j(\lambda)$ at equilibrium have been worked out in same relevant examples, and in particular for generalized Gibbs Ensembles in the XXZ spin chain [5]. The crucial difference between a free fermionic model and an interacting Bethe Ansatz solvable one, is that the group velocity of the quasi-particles is different from the bare one [6], since it actually depends on all the filling fractions $\vartheta_j(\lambda)$, $j = 1, \ldots, n_B$. The effective velocity of propagation of the excitations is termed the *dressed* velocity $v_j^{\text{dressed}}[\vartheta(\lambda)]$, where the notation is chosen to emphasize its dependence on the whole thermodynamical state of the system. Given $v_j^{\text{bare}}(\lambda)$ and all the fillings $\vartheta_j(\lambda)$ ($j = 1, \ldots, n_B$) there is, for the XXZ spin chain [6] and other models, a well defined procedure to determine $v_j^{\text{dressed}}[\vartheta(\lambda)]$. In presence of Bethe Ansatz solvable interactions, (16.7) has been then conjectured to generalize to [7]

$$\vartheta_j(\lambda, \xi) = \vartheta_j^{\text{left}}(\lambda)\Theta(v_j^{\text{dressed}}[\vartheta(\lambda)] - \xi) + \vartheta_j^{\text{right}}(\lambda)\Theta(\xi - v_j^{\text{dressed}}[\vartheta(\lambda)]), \quad j = 1, \ldots, n_B; \tag{16.13}$$

This set of equations together with the dressing transformation of the bare velocity form the content of the so-called *Generalized Hydrodynamics*. In practice, $\vartheta_j^{\text{left}}(\lambda)$ and $\vartheta_j^{\text{right}}(\lambda)$ are known a priori. At a given ray ξ, one first assumes $v_j^{\text{dressed}}(\lambda) = v_j^{\text{bare}}(\lambda)$ and then uses (16.13) to determine the new fillings, after the velocity is dressed and the procedure iterated until convergence is reached. Few iterations for

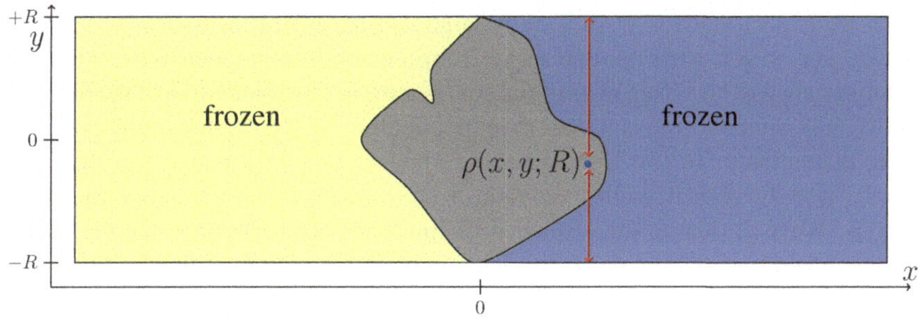

Fig. 16.2 Slab of width $2R$ for time evolution in imaginary time y form a DW initial state $|DW\rangle$, given in (16.3). Outside the inhomogeneous gray region the quantum state is frozen in its initial configuration. The fermion density is $\rho(x, y; R) = 1$ (resp. $\rho(x, y; R) = 0$) for x sufficiently large and negative (resp. positive)

a numerical solution suffice. For the DW initial state in the XXZ spin chain, the Generalized Hydrodynamics equations can be solved exactly. Parametrizing $\gamma = \pi Q/P$ and $Q/P \in \mathbb{Q}$ with P, Q coprimes, it is possible to derive the following result for the magnetization profile [8]

$$\lim_{\mathrm{hd}} \langle DW(t)|\mathbf{s}^z_x|DW(t)\rangle = -\frac{1}{2\pi/P} \arcsin\left(\frac{\xi}{\xi_0}\right), \qquad (16.14)$$

being $\xi_0 = \sin(\gamma)/\sin(P)$. The remarkable feature of the magnetization profile in presence of interactions is its the dependence on the denominator of the fraction γ/π. This indicates a no-where continuous behaviour of (16.14) as a function of the anisotropy parameter Δ. It can be verified however that the continuation of (16.14) to irrational γ is well defined.

16.3 Imaginary Time Evolution

Finally we mention a remarkable connection between the quantum mechanical evolution from a DW initial state and a peculiar phenomena of phase separation in statistical mechanics known as *arctic curves*.

Consider the time evolution from the DW initial state in imaginary time with the Hamiltonian in (16.1) on a slab of width $2R$. Now, see Fig. 16.2, the fermion density on the slab is given by

$$\rho(x, y; R) \equiv \frac{\langle DW|e^{-H_{\mathrm{free}}(R-y)}c^\dagger_x c_x e^{-H_{\mathrm{free}}(R+y)}|DW\rangle}{\langle DW|e^{-2RH_{\mathrm{free}}}|DW\rangle}. \qquad (16.15)$$

Again we expect that for x sufficiently large and negative (positive) $\rho(x, y; R) = 1$ (resp. $\rho(x, y; R) = 0$); i.e. that away from a central inhomogeneous region the quantum state will be frozen in its initial configuration. The calculation of the fermion density in imaginary time is harder than in real time, cf. (16.4), due to the presence of the denominator in (16.15). It can nevertheless be performed [9] in the limit $x \gg 1,\ y \gg 1,\ R \gg 1$, with their ratio fixed, through a similar stationary phase analysis. We will denote the sequence of limits above the hydrodynamical limit $\lim_{\text{hd}}$, in analogy with Sect. 16.1. In such a limit the fermion density inside the strip is

$$\lim_{\text{hd}} \rho(x, y; R) = \frac{1}{\pi} \arccos\left(\frac{x}{\sqrt{R^2 - y^2}}\right), \tag{16.16}$$

which is plotted in Fig. 16.1 on the right. In the imaginary time problem the inhomogeneous region, within which the the fermion density differs from the initial one (i.e. is neither zero or one), is confined into a circle. Such a circle is an example of an arctic curve. In two-dimensional statistical mechanics, arctic curves separate in the thermodynamic limit two phases of a lattice model, one of which is frozen and the other disordered and inhomogeneous. In the frozen phase, in particular, lattice variables cannot fluctuate. A celebrate example of this phenomena is the so-called arctic circle discovered in dimer coverings of the Aztec diamond [10], or equivalently in the six-vertex-model with DW boundary conditions [11]. As discussed in [9], the quantum mechanical curve represented in Fig. 16.1 is obtained from the Hamiltonian limit of the statistical mechanics arctic circle.

Acknowledgements I am grateful to P. Sodano for collecting these proceedings and to J. Dubail and J.-M. Stéphan for many of the ideas and results that are briefly review in this note. I finally thank A. De Luca and M. Collura for joined work in [8].

References

1. T. Antal, Z. Racz, A. Rakos, G.M. Schutz, Phys. Rev. E 59, 4912 (1999), arXiv:cond-mat/9812237; T. Antal, P. L. Krapivsky, A. Rakos, Phys. Rev. E 78, 061115 (2008), arXiv:0808.3514
2. J. Viti, J.-M. Stéphan, J. Dubail, M. Haque, EPL **115**, 40011 (2016)
3. E. Bettelheim, L. Glazman, Phys. Rev. Lett. **109**, 260602 (2012)
4. D. Gobert, C. Kollath, U. Schollwock, G. Schutz, Phys. Rev. E **71**, 36102 (2005)
5. M. Takahashi, *Thermodynamics of One-dimensional Solvable Models* (Cambridge University Press, Cambridge, 1999)
6. L. Bonnes, F.H.L. Essler, A.M. Lauchli, Phys. Rev. Lett. **113**, 187203 (2014)
7. O.A. Castro-Alvaredo, B. Doyon, T. Yoshimura, Phys. Rev. X **6**, 41065 (2016); B. Bertini, M. Collura, J. De Nardis, M. Fagotti, Phys. Rev. Lett. **117**, 207201 (2016)
8. M. Collura, A. De Luca, J. Viti, Phys. Rev. B **97**, 081111 (2018)
9. N. Allegra, J. Dubail, J.-M. Stephan, J. Viti, J. Stat. Mech. 053108 (2016)
10. W. Jockusch, J. Propp, P. Shor, Random domino tilings and the arctic circle theorem, arXiv:math.CO/9801068
11. F. Colomo and A.G. Pronko, J. Stat. Phys. **138**, 662–700 (2010); F. Colomo, A.G. Pronko, P. Zinn-Justin, J. Stat. Mech.: Theor. Exp. L03002 (2010)

Chapter 17
Quantum Thermodynamics at Impurity Quantum Phase Transitions

Abolfazl Bayat, Gabriele De Chiara, Tony J. G. Apollaro, Simone Paganelli, Henrik Johannesson, Pasquale Sodano and Sougato Bose

Abstract The study of quantum thermodynamics, i.e. equilibrium and non-equilibrium thermodynamics of quantum systems, has been applied to various many-body problems, including quantum phase transitions. An important question is whether out-of-equilibrium quantities from this emerging field, such as fluctuations of work, exhibit scaling after a sudden quench. In particular, it is very interesting to explore this problem in impurity models where the lack of an obvious symmetry breaking at criticality makes it very challenging to characterize. Here, by considering a spin emulation of the two impurity Kondo model and performing density matrix

A. Bayat (✉)
Institute of Fundamental and Frontier Sciences, University of Electronic Science and Technology of China, Chengdu 610051, China
e-mail: abolfazl.bayat@uestc.edu.cn

A. Bayat · S. Bose
Department of Physics and Astronomy, University College London, Gower Street, London WC1E 6BT, UK

G. De Chiara
School of Mathematics and Physics, Centre for Theoretical Atomic, Molecular, and Optical Physics, Queen's University Belfast, BT7 1NN London, UK

T. J. G. Apollaro
Department of Physics, University of Malta, MSD 2080, Msida, Malta

S. Paganelli
Dipartimento di Scienze Fisiche e Chimiche, Università dell'Aquila, via Vetoio, 67010 Coppito-LAquila, Italy

H. Johannesson
Department of Physics, University of Gothenburg, 412 96 Gothenburg, Sweden

Beijing Computational Science Research Center, Beijing 100094, China

P. Sodano
International Institute of Physics, Universidade Federal do Rio Grande do Norte, 59078-400 Natal, RN, Brazil
e-mail: pasquale.sodano01@gmail.com

INFN, Sezione di Perugia, Via A. Pascoli, 06123 Perugia, Italy

A. Ferraz et al. (eds.), *Strongly Coupled Field Theories for Condensed Matter and Quantum Information Theory*, Springer Proceedings in Physics 239,
https://doi.org/10.1007/978-3-030-35473-2_17

renormalization group computations, we establish that the irreversible work produced in a quench exhibits finite-size scaling at quantum criticality. Our approach predicts the equilibrium critical exponents for the crossover length and the order parameter of the model, and, moreover, implies a new exponent for the rescaled irreversible work.

17.1 Introduction

Impurities in the bulk of a material are the heart of solid state technologies which is exemplified by the computing revolution driven by the invention of transistors. In fact, even the addition of a single impurity can change the properties of matter. The theory of quantum impurities underpins much of the current understanding of correlated electrons. A case in point is the two-impurity Kondo model (TIKM) [1], with bearing on heavy fermion physics [2], correlation effects in nanostructures [3], spin-based quantum computing [4, 5], and more. The model describes two localized spin-1/2 impurities in an electron gas, coupled by the Ruderman–Kittel–Kasuya–Yosida (RKKY) interaction via their spin exchange with the electrons. In addition to the RKKY coupling, the model exhibits a second energy scale, the Kondo temperature T_K, below which the electrons may screen the impurity spins. For the sake of simplicity, in numerical computations, a spin chain emulation of the TIKM has been introduced which faithfully reproduces its physics [6]. Universal quantum quenches [7] and entanglement properties [8] of the TIKM model have been investigated.

Out-of-equilibrium thermodynamics of closed many-body systems subject to a variation of a Hamiltonian parameter has received considerable attention in the past few years, both experimentally and theoretically [9]. The increasing level of control over few-particle quantum systems has allowed to demonstrate experimentally the information-to-energy conversion and the Jarzynski equality [10–14]. On the one hand, the increasing level of control of simple systems consisting of a few quantum particles has led to the experimental possibility both of building the first quantum engines [15–17] and of investigating nonequilibrium theoretical predictions [11]. On the other hand, studies of the interplay between quantum thermodynamics, many-body physics, and quantum information, have shed light on fundamental aspects of thermalisation of closed quantum systems [19], fluctuation theorems [20], and prospects for quantum coherent thermal machines [21, 22].

A central issue is how the presence of a quantum phase transition (QPT) manifests itself in the out-of-equilibrium thermodynamics after a sudden quench of a Hamiltonian parameter [9, 19, 23–32]. In the sudden quench approach, the thermodynamic properties of a quantum system, initially at thermal equilibrium and experiencing a sudden variation of some global hamiltonian parameter, are investigated. It is now well established [24] that a second-order QPT is signaled by a discontinuity in the derivative of the *irreversible entropy production* (with the derivative taken with respect to the QPT driving parameter which is being quenched), as well as of the *variance of the work* [20]. This is to be contrasted with a first-order QPT, where the derivative of the *average work* (i.e. the first moment of the probability distribution function of the work) exhibits a discontinuity at the transition [28] (with a peak in the irreversible entropy production when the QPT is induced by a local quench [32]). The obvious parallels to the diverging behavior of response functions at a second-order equilibrium QPT prompts the question whether out-of-equilibrium quantities, like the *irreversible work* [20] (which is a measure of the nonadiabaticity of a quantum quench), may also exhibit scaling at criticality. Here, via a novel inroad—studying the quantum thermodynamics for a sudden quench across an impurity quantum critical point—we are able to provide an affirmative answer. Recent studies have also demonstrated a similar conclusion for first order transitions [33].

While for an ordinary bulk QPT the behavior of thermodynamic quantities after a sudden quench reflects the discontinuity of a corresponding equilibrium average value of a global observable [25], the same is not so obvious in an impurity quantum phase transition (iQPT) [34]. Local quenches in many-body quantum systems displaying iQPTs have not been investigated adequately. The lack of such works can be related both to the fact that iQPTs are a relatively new concept compared to the more well-established theory of QPTs classified according to the Ehrenfest-Landau scheme, and, most importantly but related, the identification of an order parameter exhibiting scaling properties according to some critical exponents for the iQPT has been only recently tackled [35, 36]. In ordinary QPTs the fact that the out-of-equilibrium thermodynamics of a global sudden quantum quench highlights the QPT point in the moments of the work probability distribution function (PDF) is due, in the final analysis, to the discontinuity of a corresponding equilibrium average value of a global observable, the latter being the order parameter. This holds, for instance, in spin models where the magnetization and the susceptibility show, respectively, discontinuities for 1st- and 2nd-order QPTs, thus reflecting in nonequilibrium quantum thermodynamics variables, which, as a consequence, inherit also the corresponding universality class scaling behaviour [37]. Moreover, as the irreversible entropy production can be related to the relative entropy of pre- and post-quenched thermal states, the abrupt change induced by the QPT of the latter (at low temperatures) is responsible for the its divergence at the critical point [25]. Based on these, one may ask whether, after a local quench of the impurity coupling, the behaviour of nonequilibrium quantum thermodynamic variables can reveal the iQPT?

In this paper, we elaborate on our results in [38] for the two-impurity Kondo model (TIKM) [39], one of the best studied models supporting an iQPT [2, 40–48]. Here, two spin-1/2 quantum impurities are coupled to each other by a Ruderman–Kittel–Kasuya–Yosida (RKKY) interaction, and, in the simplest variant of the theory [45], to separate bulk reservoirs of conduction electrons by Kondo interactions. When the RKKY interaction dominates, the two impurities form a local spin-singlet state (RKKY phase), while in the opposite limit each of the impurities form a spatially extended singlet state with the electrons in the reservoir to which it is coupled (Kondo-screened phase). We shall show that the iQPT between these two phases is signaled both by the irreversible work production and the variance of work following a sudden quench. Remarkably, the irreversible work shows clear scaling with well-defined critical exponents, related to known equilibrium critical exponents by scaling laws. Moreover, by means of a small quench approximation for the irreversible work production, we are able to link the latter to the two-impurity spin correlation function, which is amenable to experimental determination. While our findings have broad ramifications, they are particularly timely considering recent breakthroughs in designing and performing measurements on tunable nanoscale realizations of the TIKM [3, 49, 50].

17.2 Two Impurity Kondo Model

For the purpose of exploring quantum critical properties of the TIKM, it is sufficient to focus on the spin sector of the model by considering its spin chain emulation which is sufficient to reproduce the underlying physics [6]. This can be emulated by the Kondo spin-chain Hamiltonian $H(K) = \sum_{m=L,R} H_m + H_I$ [6], where

$$\begin{aligned} H_m &= J'\left(J_1 \boldsymbol{\sigma}_1^m \cdot \boldsymbol{\sigma}_2^m + J_2 \boldsymbol{\sigma}_1^m \cdot \boldsymbol{\sigma}_3^m\right) + \\ &\quad + J_1 \sum_{i=2}^{N_m - 1} \boldsymbol{\sigma}_i^m \cdot \boldsymbol{\sigma}_{i+1}^m + J_2 \sum_{i=2}^{N_m - 2} \boldsymbol{\sigma}_i^m \cdot \boldsymbol{\sigma}_{i+2}^m, \\ H_I &= J_1 K \boldsymbol{\sigma}_1^L \cdot \boldsymbol{\sigma}_1^R. \end{aligned} \tag{17.1}$$

Here $m = L, R$ labels the left and right chains with $\boldsymbol{\sigma}_i^m$ the vector of Pauli matrices at site i in chain m, and with J_1 (J_2) nearest- (next-nearest-) neighbor couplings (see Fig. 17.1). In the following we set $J_1 = 1$ as our energy unit. The parameter $J' > 0$ plays the role of antiferromagnetic Kondo coupling and K represents the dimensionless RKKY coupling between the impurity spins $\boldsymbol{\sigma}_1^L$ and $\boldsymbol{\sigma}_1^R$. The total size of the system is thus $N = N_L + N_R$. By fine tuning J_2/J_1 to the critical point $(J_2/J_1)_c = 0.2412$ of the spin chain dimerization transition [51, 52] all logarithmic scaling corrections vanish, allowing for an unambiguous fit of numerical data using the Density Matrix Renormalization Group (DMRG) [53–55]. Indeed, a DMRG study reveals that the Hamiltonian (17.1) faithfully reproduces the features of the iQPT in the TIKM [6].

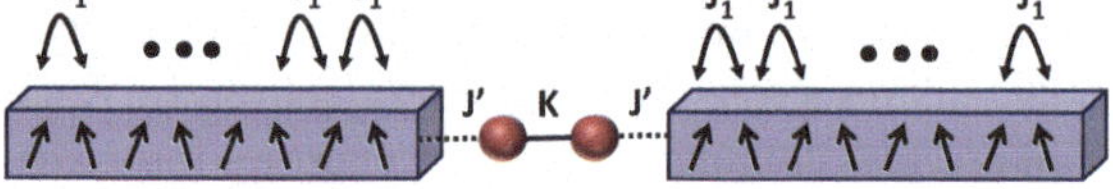

Fig. 17.1 Schematic of the two impurity Kondo mode. The two-impurity Kondo spin chain model consists of two spin-1/2 impurities, each interacting with an array of spin-1/2 particles via a Kondo coupling J'. The two impurity interacts with each other via an inter-impurity RKKY coupling K which serves as the control parameter. By varying K system exhibits a second order quantum phase transition at a critical value $K = K_c$ which depends on the impurity coupling J'

17.3 Thermodynamic Properties: Work Distribution

We assume that the impurity coupling is initially K. The system is at zero temperature in its ground state $|E_0(K)\rangle$ with energy $E_0(K)$. The impurity coupling is then quenched infinitesimally from K to $K + \Delta K$, with the Hamiltonian $H(K)$ suddenly changed to $H(K + \Delta K)$. The work performed on the system becomes a stochastic variable W described by the probability distribution function (PDF) [20]

$$p(W) = \sum_m \left|\langle E'_m | E_0(K)\rangle\right|^2 \delta\left[W - (E'_m - E_0(K))\right], \tag{17.2}$$

where $\{E'_m\}$ and $\{|E'_m\rangle\}$ are the eigenenergies and eigenvectors of $H(K + \Delta K)$, respectively. Notice that the work PDF is an experimentally accessible quantity [56, 57] and that from its knowledge all the statistical moments can be derived as

$$\langle W^n \rangle = \int W^n p(W) dW. \tag{17.3}$$

Due to the nature of the sudden quench in the Hamiltonian, the system is driven out of equilibrium and, by means of the Jarzynski fluctuation relation [10], it is possible to define the so-called *irreversible work*:

$$W_{irr} = \langle W \rangle - \Delta F \geq 0\,, \tag{17.4}$$

where ΔF is the difference between the free energies after and before the quench. Since we assume zero temperature, ΔF is simply the difference of the post- and pre-quench ground state energies. The irreversible work has a simple physical explanation as the amount of energy which has to be taken out from the quenched system in order to bring it to its new equilibrium state which, for our case, is the ground state of $H(K + \Delta K)$ [27]. For the instantaneous quantum quench we have

$$W_{irr} = \langle E_0(K)|\, H(K + \Delta K)\, |E_0(K)\rangle - E'_0(K + \Delta K)\,, \tag{17.5}$$

i.e., the irreversible work is given by the difference between the expectation value of the post-quenched Hamiltonian evaluated on the pre-quenched ground state and the post-quench ground state energy. It is worth emphasizing that the (17.2) and (17.5) are truly out-of-equilibrium quantum thermodynamic quantities, although evaluated at equilibrium due to the sudden quench approximation. In fact, for quasi-static processes, the work PDF would be a delta function peaked at the energy difference between the pre- and post-quenched ground states, whereas the irreversible work would result identically null. Instead, in the sudden quench case, which approximates the case where the quench is performed at a rate much faster than the typical time evolution scale of the pre-quenched ground state, both quantities give a measure of the irreversibility by performing the quench [10, 27].

17.4 Scaling of the Irreversible Work

In order to capture the iQPT between the Kondo regime and the RKKY phase, we introduce the rescaled quantity

$$\widetilde{W}_{irr} = \frac{W_{irr}}{\Delta K^2}, \tag{17.6}$$

and study the variation of $\widetilde{W}_{irr}$ when the coupling K is varied. In this paper we only consider infinitesimal quantum quenches, $\Delta K \ll 1$. In Fig. 17.2a, b we plot the irreversible work $\widetilde{W}_{irr}$ for two impurity couplings $J' = 0.4$ and $J' = 0.5$ respectively. It is clear from the plots that $\widetilde{W}_{irr}$ shows a sharp peak which becomes even more pronounced by increasing the system size N (apart from slightly shifting towards lower values of K's). This signifies that $\widetilde{W}_{irr}$ exhibits non-analytic behaviour at the critical point in the thermodynamic limit. In finite-size systems, such as the ones considered here, the position of the peak determines the critical point K_c which slowly moves towards the left by increasing N.

By considering the specific value of the RKKY coupling K at which $\widetilde{W}_{irr}$ diverges, one can determine numerically the critical point K_c, which then shows a particular

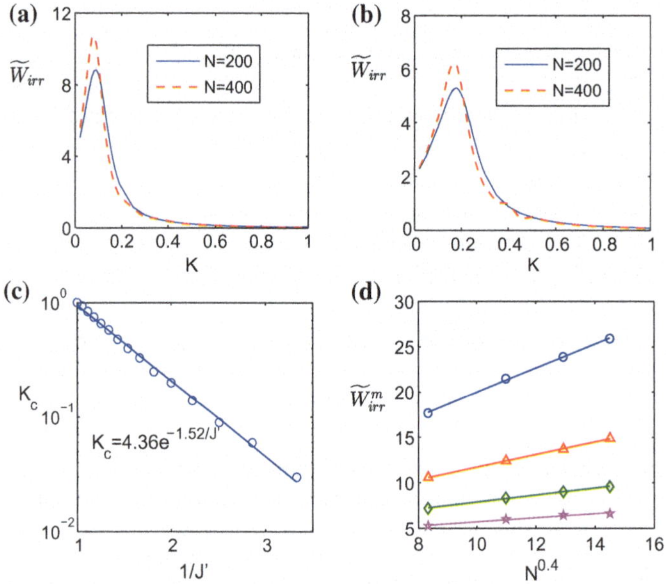

Fig. 17.2 The rescaled irreversible work near criticality. The irreversible work $\widetilde{W}_{irr}$ in terms of K in a chain with **a** $J' = 0.4$; **b** $J' = 0.5$. **c** The critical coupling K_c (blue circles) versus $1/J'$ in a semi-logarithmic scale and its exponential fit (blue line). **d** The maximum of the irreversible work W^m_{irr} versus $N^{0.4}$ and the linear fits. From top to bottom: $J' = 0.4$; $J' = 0.5$; $J' = 0.6$ and $J' = 0.7$. From [38]

dependence on $1/J'$, just as the Kondo temperature T_K (which sets the energy scale for the weak-to-strong of the renormalized Kondo coupling [39]). This can be seen in Fig. 17.2c in which the critical coupling K_c is plotted as a function of $1/J'$. The manifest linear trend in a semi-logarithmic scale confirms that

$$K_c \sim e^{-a/J'} \sim T_K \tag{17.7}$$

for some constant a, in agreement with other studies of the two-impurity Kondo spin chain [6, 35].

In the finite-size systems studied here, the divergence of $\widetilde{W}_{irr}$ at $K = K_c$ appears as a finite peak becoming more prominent for increasing system size, as shown in Fig. 17.2a, b. We define the maximum of the irreversible work as $\widetilde{W}^m_{irr} = \widetilde{W}_{irr}(K = K_c)$. Since $\widetilde{W}^m_{irr}$ increases by increasing the system size N one can try to fit it by an algebraic map of the form

$$\widetilde{W}^m_{irr} \sim N^{\lambda}, \tag{17.8}$$

where λ is a positive exponent. In fact, a perfect match is found for various impurity couplings J' by choosing $\lambda = 0.4$ as depicted in Fig. 17.2d. Note that the exponent λ governs the scaling of a purely non-equilibrium quantity with system size. Note that, whereas for a *global* quench the irreversible work is expected to have a functional dependence on the system size because in (17.4) both the work and the free energy become extensive quantities, it is far from trivial that the same holds for a *local* quench. Nevertheless, in the TIKM here considered, this behavior of $\widetilde{W}^m_{irr}$ is determined by the distinctive nature of the iQPT, where a local change in the RKKY coupling induces a global rearrangement of the ground state of the total system at criticality.

The above analysis for $\widetilde{W}_{irr}$ suggests the Ansatz:

$$\widetilde{W}_{irr} = \frac{A}{|K - K_c|^{\kappa} + BN^{-\lambda}}, \tag{17.9}$$

where A and B are two constants that may vary with J'. This Ansatz is based on the fact that $\widetilde{W}_{irr}$ diverges in the thermodynamic limit as $\widetilde{W}_{irr} \sim |K - K_c|^{-\kappa}$, while for finite-size systems at $K = K_c$ it increases algebraically with the system size as in (17.8). In order to deal with the divergence more conveniently at the critical point we define a normalized function as

$$W_{nor} = (\widetilde{W}^m_{irr} - \widetilde{W}_{irr})/\widetilde{W}^m_{irr}. \tag{17.10}$$

Using the Ansatz of (17.9) one can show that

$$W_{nor} = g(N^{\lambda/\kappa}|K - K_c|), \tag{17.11}$$

where $g(x)$ is a scaling function which can be determined numerically. In order to evaluate the exponent κ we search for the value of κ such that the values of W_{nor} as

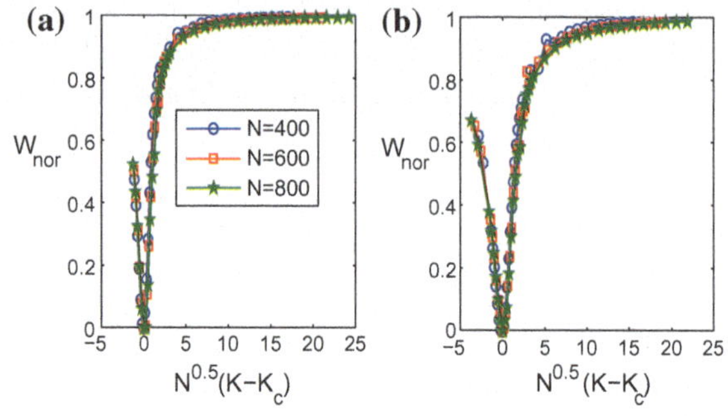

Fig. 17.3 Finite-size scaling. The finite-size scaling of W_{nor} according to the Ansatz of (17.11) for: **a** $J' = 0.4$; **b** $J' = 0.5$. From [38]

a function of $N^{\lambda/\kappa}|K - K_c|$, for various system sizes N, collapse on each other, as shown in Fig. 17.3a, b for two different impurity couplings $J' = 0.4$ and $J' = 0.5$ respectively. As is evident from the figure, using the predetermined exponent $\lambda = 0.4$, one finds that $\kappa = 0.8$.

The irreversible work has been measured in quantum mechanical setups using various methods [13, 14]. Here we follow a different route, showing that, for the present system and for small quenches, one can rely on measuring only the two-impurity correlation function $\langle \boldsymbol{\sigma}_1^L \cdot \boldsymbol{\sigma}_1^R \rangle$ with respect to the ground state. The SU(2) symmetry of the Hamiltonian (17.1) implies that the reduced density matrix of the two impurities is always a Werner state

$$\rho_{1_L,1_R} = \frac{3 + \langle \boldsymbol{\sigma}_1^L \cdot \boldsymbol{\sigma}_1^R \rangle}{12} I_4 - \frac{\langle \boldsymbol{\sigma}_1^L \cdot \boldsymbol{\sigma}_1^R \rangle}{3} |\psi^-\rangle\langle\psi^-|, \tag{17.12}$$

where $|\psi^-\rangle$ is the singlet state, I_4 represents the 4×4 identity matrix and $\langle \boldsymbol{\sigma}_1^L \cdot \boldsymbol{\sigma}_1^R \rangle$ is the two-point correlation function of the impurity spins with respect to the ground state. It is immediate to see that the two-point correlation functions determine all the properties of the two impurities [58], including their entanglement (measured by concurrence [59]) which becomes

$$C = \max\left\{ -\frac{1 + \langle \boldsymbol{\sigma}_1^L \cdot \boldsymbol{\sigma}_1^R \rangle}{2}, 0 \right\}. \tag{17.13}$$

By expanding (17.5) for small ΔK we obtain

$$\widetilde{W}_{irr} = -\frac{1}{2} \frac{\partial \langle \boldsymbol{\sigma}_1^L \cdot \boldsymbol{\sigma}_1^R \rangle}{\partial K}. \tag{17.14}$$

The divergence of $\widetilde{W}_{irr}$ at the critical point and (17.14) suggest that the two-point impurity correlator $\langle \boldsymbol{\sigma}_1^L \cdot \boldsymbol{\sigma}_1^R \rangle$ mimics the behavior of an order parameter, capturing the quantum criticality and showing scaling behavior near the transition. In Fig. 17.4a, b we plot the spin correlator $\langle \boldsymbol{\sigma}_1^L \cdot \boldsymbol{\sigma}_1^R \rangle$ versus the coupling K for two impurity couplings

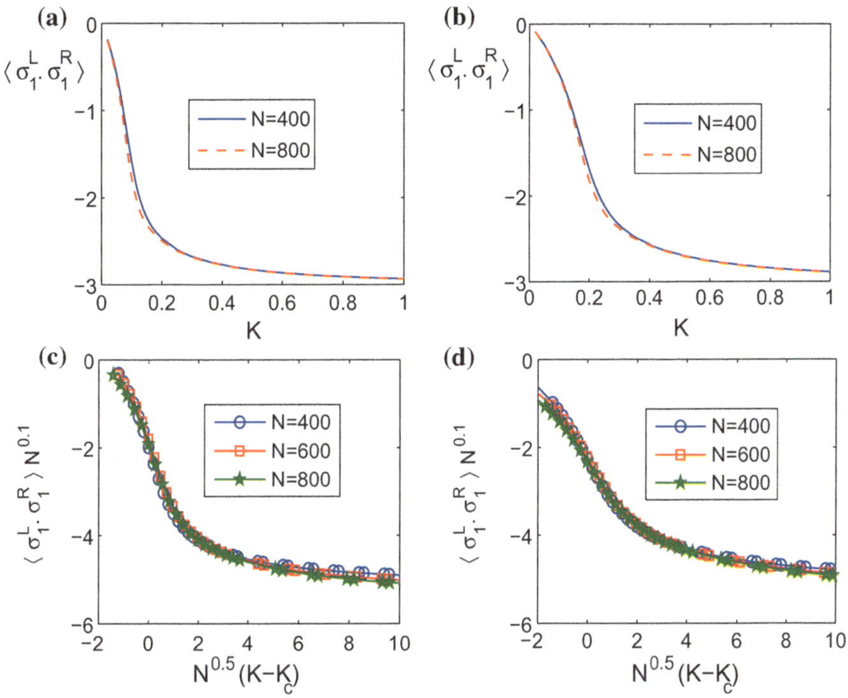

Fig. 17.4 Two-point impurity correlation function. Correlation function $\langle \sigma_1^L \cdot \sigma_1^R \rangle$ of the two impurities versus RKKY coupling K in a chain with **a** $J' = 0.4$; **b** $J' = 0.5$. The finite-size scaling for $\langle \sigma_1^L \cdot \sigma_1^R \rangle$ with **c** $J' = 0.4$; **d** $J' = 0.5$ [38]

$J' = 0.4$ and $J' = 0.5$ respectively. The correlator varies from 0 (for $K = 0$) in the Kondo regime to $\langle \boldsymbol{\sigma}_1^L \cdot \boldsymbol{\sigma}_1^R \rangle = -3$ (for very large K) deep in the RKKY phase. To extract its scaling properties, we make the finite-size-scaling Ansatz

$$\langle \boldsymbol{\sigma}_1^L \cdot \boldsymbol{\sigma}_1^R \rangle = N^{-\beta/\nu} f(N^{1/\nu}|K - K_c|), \tag{17.15}$$

where, in the limit $N \to \infty$, β characterizes scaling of the correlator near criticality, $\langle \boldsymbol{\sigma}_1^L \cdot \boldsymbol{\sigma}_1^R \rangle \sim |K - K_c|^{\beta}$, ν is the exponent governing the divergence of the crossover scale $\xi \sim |K - K_c|^{-\nu}$ [42, 46], and $f(x)$ is a scaling function. In order to determine these critical exponents we identify the values of β and ν such that the plots of $\langle \boldsymbol{\sigma}_1^L \cdot \boldsymbol{\sigma}_1^R \rangle N^{\beta/\nu}$ as a function of $N^{1/\nu}|K - K_c|$ collapse to a single curve for arbitrary system sizes, as shown in Fig. 17.4c, d. The best data collapse is achieved by choosing $\beta = 0.2$ and $\nu = 2$, which are in excellent agreement with the ones found from the analysis of the Schmidt gap [35].

Furthermore, as an alternative way of computing the scaling of the irreversible work $\widetilde{W}_{irr}$, one may directly differentiate both sides of (17.15) with respect to the RKKY coupling K to get

$$\widetilde{W}_{irr} \sim \partial_K \langle \sigma_1^L \cdot \sigma_1^R \rangle \sim N^{(1-\beta)/\nu} f'(N^{1/\nu}|K - K_c|), \tag{17.16}$$

where $f'(x) = \mathrm{d}f/\mathrm{d}x$. The finite-size scaling of (17.16) implies that $\widetilde{W}_{irr} \sim |K - K_c|^{\beta-1}$, which then leads to

$$\kappa = 1 - \beta. \tag{17.17}$$

Moreover, comparing (17.16) and (17.11), we obtain another constraint between the exponents as

$$\kappa = \lambda\nu. \tag{17.18}$$

Equation (17.17) and (17.18) are indeed satisfied for the values found in our numerical analysis as $\lambda = 0.4$, $\nu = 2$, $\beta = 0.2$ and $\kappa = 0.8$, confirming our scaling Ansätze.

It is worth emphasizing that in our local quench problem the energy change, for every finite quench, is always finite and, for an infinitesimal quench ΔK, the irreversible work can be approximated by $W_{irr} \simeq -\Delta K \Delta \langle \sigma_1^L \cdot \sigma_1^R \rangle/2$. Since $\langle \sigma_1^L \cdot \sigma_1^R \rangle$ varies between 0 and 3, then $W_{irr} \leq -3\Delta K/2$, which vanishes for $\Delta K \to 0$. As a consequence, the un-rescaled irreversible work W_{irr} shows no divergences even as $N \to \infty$.

17.5 Variance Analysis of Work

The variance of work is another important non-equilibrium quantity which is defined as

$$\Delta W^2 = \langle W^2 \rangle - \langle W \rangle^2 . \tag{17.19}$$

For convenience, we also rescale the variance as $\Delta\widetilde{W}^2 = \Delta W^2/\Delta K^2$. For a sudden quench one can show that

$$\Delta\widetilde{W}^2 = 3 - 2\langle \sigma_1^L \cdot \sigma_1^R \rangle - \langle \sigma_1^L \cdot \sigma_1^R \rangle^2 . \tag{17.20}$$

The derivative of the rescaled variance with respect to K becomes

$$\partial_K(\Delta\widetilde{W}^2) = 4\left(1 + \langle \sigma_1^L \cdot \sigma_1^R \rangle\right) \widetilde{W}_{irr}. \tag{17.21}$$

Since the correlation function $\langle \sigma_1^L \cdot \sigma_1^R \rangle$ is always finite, both $\widetilde{W}_{irr}$ and $\partial_K(\Delta\widetilde{W}^2)$ diverge at the critical point in the thermodynamic limit. Moreover, $\Delta\widetilde{W}^2$ takes its maximum for values of K slightly smaller than K_c where $\langle \sigma_1^L \cdot \sigma_1^R \rangle = -1$, i.e. the minimum value of K at which the two impurities are entangled [6].

17.6 Out-of-Equilibrium Features

We should emphasise the fact that both the irreversible work and the variance of work are truly out-of-equilibrium quantum thermodynamic quantities, although evaluated at equilibrium due to the sudden small-quench approximation. In fact, for quasi-static processes, the irreversible work would result identically null as the work equals the free energy difference, and the work PDF would be a delta function peaked at the energy difference between the pre- and post-quenched ground states, resulting in zero variance. In the sudden quench case, however, both quantities give certain measures of irreversibility [10, 27]. Moreover, as another feature of non-equilibrium, we should point out that whereas a temperature can be associated to the initial state (which is $T = 0$ in our analysis), the same does not hold after the quench has been performed.

17.7 Summary

In this paper, we have numerically shown that both irreversible work and work variance, as non-equilibrium quantities, signal the impurity quantum phase transition between the Kondo and RKKY regimes in the TIKM. Both quantities exhibit scaling at the quantum critical point, and allow for known equilibrium critical exponents to be extracted. In addition, a new critical exponent κ, governing the behavior of the rescaled irreversible work at the phase transition, is brought to light. Importantly, all out-of-equilibrium quantities considered are amenable to experimental observation in solid-state nanostructures or ultra cold atoms, since ultimately it is sufficient to measure a two-point spin correlation function.

References

1. C. Jayprakash, H.R. Krishna-murthy, J.W. Wilkins, Two-impurity kondo problem. Phys. Rev. Lett. **47**, 737 (1981)
2. B.A. Jones, C.M. Varma, J.W. Wilkins, Low- temperature properties of the two-impurity Kondo Hamiltonian. Phys. Rev. Lett. **61**, 125 (1988)
3. J. Bork, Y.-H. Zhang, L. Diekhöner, Lázló Borda, P. Simon, J. Kroha, P. Wahl, K. Kern, A tunable two-impurity Kondo system in an atomic point contact. Nat. Phys. **7**, 901 (2011)
4. J. Mravlje, A. Ramsak, T. Rejec, Conductance of a molecule with a center of mass motion. Phys. Rev. B **74**, 205320 (2006)
5. S.Y. Cho, R.H. McKenzie, Quantum entanglement in the two-impurity Kondo model. Phys. Rev. A **73**, 012109 (2006)
6. A. Bayat, S. Bose, P. Sodano, H. Johannesson, Entanglement probe of two-impurity Kondo physics in a spin chain. Phys. Rev. Lett. **109**, 066403 (2012)
7. A. Bayat, S. Bose, H. Johannesson, P. Sodano, Universal single-frequency oscillations in a quantum impurity system after a local quench. Phys. Rev. B **92**, 155141 (2015)
8. A. Bayat, Scaling of Tripartite Entanglement at Impurity Quantum Phase Transitions. Phys. Rev. Lett. **118**, 036102 (2017)
9. A. Polkovnikov, K. Sengupta, A. Silva, M. Vengalattore, Colloquium: Nonequilibrium dynamics of closed interacting quantum systems. Rev. Mod. Phys. **83**, 863 (2011)

10. C. Jarzynski, Nonequilibrium equality for free energy differences. Phys. Rev. Lett. **78**, 2690 (1997)
11. S. Toyabe, T. Sagawa, M. Ueda, E. Muneyuki, M. Sano, Experimental Demonstrations of Information-to-Energy Conversion and Validation of the Generalized Jarzynski Equality. Nat. Phys. **6**, 988 (2010)
12. J.V. Koski, V.F. Maisi, J.P. Pekola, D.V. Averin, Experimental realization of a Szilard engine with a single electron. Proc. Natl. Acad. Sci. U.S.A. **111**, 13786 (2014)
13. T.S. Batalhao, A.M. Souza, L. Mazzola, R. Auccaise, R.S. Sarthour, I.S. Oliveira, J. Goold, G. De Chiara, M. Paternostro, R.M. Serra, Experimental reconstruction of work distribution and study of fluctuation relations in a closed quantum system. Phys. Rev. Lett. **113**, 140601 (2014)
14. S. An, J.-N. Zhang, M. Um, D. Lv, Y. Lu, J. Zhang, Z. Yin, H.T. Quan, K. Kim, Experimental test of quantum Jarzynski equality with a trapped ion system. Nat. Phys. **11**, 193 (2015)
15. O. Fialko, D.W. Hallwood, Isolated quantum heat engine. Phys. Rev. Lett. **108**, 085303 (2012)
16. O. Abah, J. Rossnagel, G. Jacob, S. Deffner, F. Schmidt-Kaler, K. Singer, E. Lutz, Single-ion heat engine at maximum power. Phys. Rev. Lett. **109**, 203006 (2012)
17. J. Roßnagel, S.T. Dawkins, K.N. Tolazzi, O. Abah, E. Lutz, F. Schmidt-Kaler, K. Singer, A single-atom heat engine. Science **352**, 325 (2016)
18. G. Maslennikov, S. Ding, R. Habltzel, J. Gan, A. Roulet, S. Nimmrichter, D. Matsukevich, Quantum absorption refrigerator with trapped ions. Nat. Commun. **10**, 202 (2019)
19. J. Eisert, M. Friesdorf, C. Gogolin, Quantum many-body systems out of equilibrium. Nat. Phys. **11**, 124 (2015)
20. M. Campisi, P. Hänggi, P. Talkner, Colloquium: quantum fluctuation relations: foundations and applications. Rev. Mod. Phys. **83**, 771 (2011)
21. J. Goold, M. Huber, A. Riera, L. del Rio, P. Skrzypczyk, The role of quantum information in thermodynamics—a topical review, J. Phys. A: Math. Theor. **49**, 143001 (2016)
22. M.T. Mitchison, *Quantum thermal absorption machines: refrigerators, engines and clocks*, arxiv:1902.02672
23. P. Talkner, E. Lutz, P. Hänggi, Fluctuation theorems: work is not an observable. Phys. Rev. E **75**, 050102(R) (2007)
24. A. Silva, Statistics of the work done on a quantum critical system by quenching a control parameter. Phys. Rev. Lett. **101**, 120603 (2008)
25. R. Dorner, J. Goold, C. Cormick, M. Paternostro, V. Vedral, Emergent thermodynamics in a quenched quantum many-body system. Phys. Rev. Lett. **109**, 160601 (2012)
26. L. Fusco, S. Pigeon, T.J.G. Apollaro, A. Xuereb, L. Mazzola, M. Campisi, A. Ferraro, M. Paternostro, G. De Chiara, Assessing the nonequilibrium thermodynamics in a quenched quantum many-body system via single projective measurements. Phys. Rev. X **4**, 031029 (2014)
27. F. Plastina, A. Alecce, T. J. G. Apollaro, G. Falcone, G. Francica, F. Galve, N. Lo Gullo, R. Zambrini, Irreversible work and inner friction in quantum thermodynamic processes. Phys. Rev. Lett. **113**, 260601 (2014)
28. E. Mascarenhas, H. Bragança, R. Dorner, M. França, Santos, V. Vedral, K. Modi, J. Goold, Work and quantum phase transitions: quantum latency. Phys. Rev. E **89**, 250602 (2014)
29. A. Sindona, J. Goold, N. Lo Gullo, F. Plastina, Statistics of the work distribution for a quenched fermi gas. New J. Phys. **16**, 045013 (2014)
30. S. Paganelli, T.J.G. Apollaro, Irreversible work versus fidelity susceptibility for infinitesimal quenches. Int. J. Mod. Phys. B **31**, 1750065 (2017)
31. F. Cosco, M. Borrelli, P. Silvi, S. Maniscalco, G. De Chiara, Non-equilibrium quantum thermodynamics in Coulomb crystals. Phys. Rev. A **95**, 063615 (2017)
32. T.J.G. Apollaro, G. Francica, M. Paternostro, M. Campisi, *Work Statistics, Irreversible Heat and Correlations Build-up in Joining Two Spin Chains* (2014), arXiv:1406.0648
33. D. Nigro, D. Rossini, E. Vicari, *Scaling properties of work fluctuations after quenches at quantum transitions*, arxiv:1810.04614
34. M. Vojta, Impurity quantum phase transitions. Philos. Mag. **86**, 1807 (2006)
35. A. Bayat, H. Johannesson, S. Bose, P. Sodano, An order parameter for impurity systems at quantum criticality. Nat. Commun. **5**, 3784 (2014)

36. L. Wang, H. Shinaoka, M. Troyer, *Fate of the Kondo Effect and Impurity Quantum Phase Transitions Through the Lens of Fidelity Susceptibility* (2015), arXiv:1507.06991
37. S. Lorenzo, J. Marino, F. Plastina, G. M. Palma, T.J.G. Apollaro, Quantum critical scaling under periodic driving. Sci. Rep. **7**, 5672 (2017)
38. A. Bayat, T.J.G. Apollaro, S. Paganelli, G. De Chiara, H. Johannesson, S. Bose, P. Sodano, Nonequilibrium critical scaling in quantum thermodynamics. Phys. Rev. B **93**, 201106(R) (2016)
39. C. Jayaprakash, H.-R. Krishnamurthy, J. Wilkins, Two-impurity Kondo problem. Phys. Rev. Lett. **47**, 737 (1981)
40. B.A. Jones, C.M. Varma, Critical point in the solution of the two magnetic impurity problem. Phys. Rev. B **40**, 324 (1989)
41. I. Affleck, A.W.W. Ludwig, Exact critical theory of the two-impurity Kondo model. Phys. Rev. Lett. **68**, 1046 (1992)
42. I. Affleck, A.W.W. Ludwig, B.A. Jones, Conformal-field-theory approach to the two-impurity Kondo problem: comparison with numerical renormalization group results. Phys. Rev. B **52**, 9528 (1995)
43. C. Sire, C.M. Varma, H.R. Krishnamurthy, Theory of the non-Fermi-liquid transition point in the two-impurity Kondo model. Phys. Rev. B **48**, 13833 (1993)
44. J. Gan, Mapping the critical point of the two-impurity Kondo model to a two-channel problem. Phys. Rev. Lett. **74**, 2583 (1995)
45. G. Zaránd, C.-H. Chung, P. Simon, M. Vojta, Quantum criticality in a double quantum-dot system, Phys. Rev. Lett. **97**, 166802 (2006)
46. E. Sela, A.K. Mitchell, L. Fritz, Exact crossover Green function in the two-channel and two-impurity Kondo models. Phys. Rev. Lett. **106**, 147202 (2011)
47. A.K. Mitchell, E. Sela, D.E. Logan, Two-channel Kondo physics in two-impurity Kondo models. Phys. Rev. Lett. **108**, 086405 (2012)
48. R.-Q. He, J. Dai, Z.-Y. Lu, Natural orbitals renormalization group approach to the two-impurity Kondo critical point. Phys. Rev. B **91**, 155140 (2015)
49. S.J. Chorley, M.R. Galpin, F.W. Jayatilaka, C.G. Smith, D.E. Logan, M.R. Buitelaar, Tunable Kondo physics in a carbon nanotube double quantum dot. Phys. Rev. Lett. **109**, 156804 (2012)
50. A. Spinelli, M. Gerrits, R. Toskovic, B. Bryant, M. Ternes, A. F. Otte, *Full experimental realisation of the two-impurity Kondo problem*, arXiv:1411.4415v2
51. K. Okamoto, K. Nomura, Fluid-dimer critical point in $S = 1/2$ antiferromagnetic Heisenberg chain with next nearest neighbor interactions. Phys. Lett. A **169**, 433 (1992)
52. S. Eggert, Numerical evidence for multiplicative logarithmic corrections from marginal operators. Phys. Rev. B **54**, 9612 (1996)
53. S.R. White, Density matrix formulation for quantum renormalization groups. Phys. Rev. Lett. **69**, 2863 (1992)
54. U. Schollwöck, The density-matrix renormalization group. Rev. Mod. Phys. **77**, 259 (2005)
55. G. De Chiara, M. Rizzi, D. Rossini, S. Montangero, Density matrix renormalization group for dummies. J. Comput. Theor. Nanosci. **5**, 1277 (2008)
56. R. Dorner, S.R. Clark, L. Heaney, R. Fazio, J. Goold, V. Vedral, Extracting quantum work statistics and fluctuation theorems by single-qubit interferometry. Phys. Rev. Lett. **110**, 230601 (2013); L. Mazzola, G. De Chiara, M. Paternostro, Measuring the characteristic function of the work distribution. Phys. Rev. Lett. **110**, 230602 (2013)
57. A.J. Roncaglia, F. Cerisola, J.P. Paz, Work measurement as a generalized quantum measurement. Phys. Rev. Lett. **113**, 250601 (2014); G. De Chiara, A.J. Roncaglia, J.P. Paz, Measuring work and heat in ultracold quantum gases. New J. Phys. **17**, 035004 (2015)
58. S.Y. Cho, R.H. McKenzie, Quantum Entanglement in the two impurity kondo model. Phys. Rev. A **73**, 012109 (2006)
59. W.K. Wootters, Entanglement of formation of an arbitrary state of two qubits. Phys. Rev. Lett. **80**, 2245 (1998)

Chapter 18
Information Delocalization in Many Body Systems: From MBL Phases to Black Holes

Javier Martinez Magan and Simone Paganelli

Abstract This is a short review of several articles of the authors on the issue of information delocalization in many-body physics, to appear in the proceedings of 'Strongly Coupled Field Theories for Condensed Matter and Quantum Information Theory', held in Natal 2015. Motivated by problems in black hole physics, expander graphs, and MBL phases, we developed an approach to information delocalization based on Mutual Information (MI). Intuitively, given a subsystem A, we looks at the support of the minimal subsystem B which is maximally entangled with A. This support can be computed analytically for random states, setting the intuition for non-equilibrium scenarios. We describe its behavior in a wide range of models, such as chaotic spin chains, Many-Body-Localized (MBL) phases, and systems displaying large-N factorization. For the last class of models, and as expected from the AdS/CFT correspondence, we find similar results when studying entropy evolution in black hole collapse scenarios.

18.1 Introduction

The problem of quantum thermalization has a been a subject of intense debate over more than two decades. The generic paradigm was laid down by Deutsch and Srednicki [11, 42] in the early '90s, and since then it has been analyzed in great detail in many different contexts (see the reviews [8, 13]). The problem of thermalization can be briefly stated as the following question: given microscopic unitarity, how do Gibbs (or generalized) ensembles emerge? Indeed, given an initial pure state $|\psi_0\rangle$, unitary evolution would lead to

---

J. M. Magan
Instituto Balseiro, Río Negro, R8402AGP San Carlos de Bariloche, Argentina

S. Paganelli (✉)
Dipartimento di Scienze Fisiche e Chimiche, Università dell'Aquila, via Vetoio, 67010 Coppito-L'Aquila, Italy
e-mail: simone.paganelli@univaq.it

A. Ferraz et al. (eds.), *Strongly Coupled Field Theories for Condensed Matter and Quantum Information Theory*, Springer Proceedings in Physics 239,
https://doi.org/10.1007/978-3-030-35473-2_18

$$U(t)\,|\psi_0\rangle\langle\psi_0|\,U^\dagger(t) \neq \rho^{\text{Gibbs}}\,. \tag{18.1}$$

Although exact thermality is out of reach, approximate thermality for certain observables $\mathcal{O}$ might appear at sufficiently long times,

$$\langle\psi(t)|\,\mathcal{O}\,|\psi(t)\rangle = \text{Tr}(\rho^{\text{Gibbs}}\,\mathcal{O}) \pm \text{error}\,, \tag{18.2}$$

where the error should vanish in the thermodynamic limit. An interesting well known interpretation from an information theory perspective is: if (18.2) holds for all operators in certain subalgebra A of the full operator algebra of the theory $A \otimes \bar{A}$, then the associated reduced density matrix $\rho_A = \text{Tr}_{\bar{A}}|\psi_t\rangle\langle\psi_t|$ has to be very close to the reduced density matrix of the thermal ensemble $\rho_A^\beta = \text{Tr}_{\bar{A}}\rho_\beta$. In terms of Von Neumann entropies, defined by $S(\rho) = \text{Tr}\rho\log\rho$, at sufficiently long times we expect

$$S_A(t) = S_A^\beta \pm \text{error}\,, \tag{18.3}$$

where the error should again vanish in the thermodynamic limit. On these time scales, entanglement entropy scales linearly with the number of degrees of freedom N_A, and becomes an extensive property of the system: $S_A \propto N_A$. The reason to expect (18.2) and (18.3) was nicely explained in two seminal papers by Lloyd and Page [23, 35], where the properties of typical states in the Hilbert space were studied. One concludes that, if the unitary evolution of the system is sufficiently chaotic so as to drive us from the realm of simple states to the realm of random (or almost random) states, then we expect both relations to be valid at long times.

This extensivity of entanglement entropy has become a proxy for thermal behavior in out-of-equilibrium scenarios, and it has been studied in a wide variety of systems, ranging from spin chains to CFT's and black holes. In the context of black holes, Sekino and Susskind, inspired by [18], used such entanglement extensivity to provide the first proposed definition of what is known as the scrambling time [40]. Below, we will call such entanglement extensivity notion as "Page-scrambling", to distinguish it from the modern approach/definition of information scrambling based on out-of-time order correlation functions [32].

As it turned out, entanglement extensivity can be misleading for the following simple reason. We might have certain classes of non-typical and far from thermal states also showing extensive scaling. A transparent and flagrant example appears when considering local systems on expander graphs, as described in [4]. Expanders graphs are those graphs for which the number of boundary vertices in a certain region scales as the total number of vertices in such region (see [24] for a complete review). Intuitively, these are graphs for which Area $\propto$ Volume. Since vacuum states of local quantum systems typically show area law entanglement, [4] concluded that for quantum systems defined on expander graphs, vacuum states show an entanglement entropy scaling extensively with the system size. Less shocking examples, but still significant for us, are non-local systems, such as the ones described in [14] or the more modern Sachdev-Ye-Kitaev models [20], as studied in [25–27].

Entanglement extensivity as a proxy of thermal behavior also brings certain shortcomings when considering MBL phases. In MBL phases, vacuum states certainly do not show extensive scaling, but quite unexpectedly, in out-of-equilibrium scenarios, such an extensive regime is reached. This regime is expected to be associated with a much different correlation pattern than that of random states, see [30], and this has been a question of much debate.

Therefore, in certain situations, we might need finer grained notions of thermalization in many-body quantum systems than just entanglement extensivity. Motivated by these observations, and inspired by the approach to the information paradox described by Hayden and Preskill in [18], in [29] we proposed a deeper approach based on Mutual Information [29], which we then applied to several models [25, 27, 30], ranging from MBL phases to chaotic spin chains and black holes, as we review below.

18.2 Information Delocalization Through Mutual Information

In this section, we illustrate how both thermalization and localization processes can be characterized by the behavior of quantum information quantities and, specifically, the Mutual Information (MI) between two parties of the whole system under study. Here we summarize the main ideas and give some examples, more details can be found in [29, 30].

Consider a many-body system with a Hilbert space having a tensor-product structure of n single-particle Hilbert spaces

$$\mathscr{H} = \otimes_i \mathscr{H}_i \qquad i = 1, \ldots, n\,, \tag{18.4}$$

and divide it into three mutually non-overlapping subsystems A, B and C, as shown in Fig. 18.1. To each of these we may associate an Hermitian operator algebra (respectively $\mathscr{A}, \mathscr{B}, \mathscr{C}$).

Treating subsystem C as an "environment", the amount of information shared by A and B in a global pure state can be quantified through a distance to state factorization, $||\rho_{AB} - \rho_A \otimes \rho_B||$. To this aim, a suitable quantity is the quantum relative entropy $S(\rho_A \| \rho_B) = \mathrm{Tr}\rho(\log \rho_A - \log \rho_B)$, because it bounds other common definitions [1, 45]. Since in our case A and B are disjoint, the relative entropy coincides with the Mutual Information between A and B, defined from the von Neuman entropy of the reduced density matrices

$$I(A, B) = S(\rho_{AB} \| \rho_A \otimes \rho_B) = S_E(\rho_A) + S_E(\rho_B) - S_E(\rho_{AB}). \tag{18.5}$$

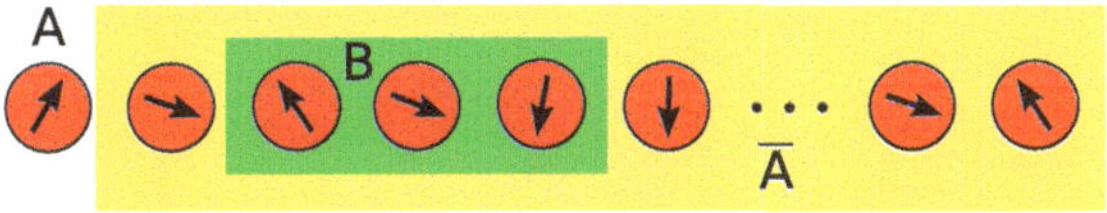

Fig. 18.1 Pictorial scheme of the systems considered. Here $\bar{A} = B \cup C$

The MI gives the total amount of quantum correlations between two subsystems [15] and it is a measure of how much we can learn of A by studying B and viceversa.

Once the subsystem A is fixed, for every B the subadditivity of MI leads to an upper bound for (18.5) $I(A, B) \leq I(A, BC) = I(A, \bar{A}) = I_{\max}$ meaning that, strictly speaking, to gain full information about A, one has to consider the whole complementary subsystem $\bar{A}$. Nevertheless, introducing a tolerance parameter ε, one can ask which is the smallest subsystem B_c s.t. $I_{\max} - I(A, B_c) < \varepsilon$. The number of spins of this minimum B_c

$$\Omega_A^{\varepsilon} = \log_2 \dim \mathscr{H}_{B_c} \tag{18.6}$$

defines what we called *Codification Volume* (CV) of the system A [29]. It quantifies how many spins of the system most part of information of A spreads on. Just as an example, CV is obviously zero if the system A is disentangled from the rest of the universe. For systems with short-range connections, where area law behavior of the entanglement entropy is observed, the CV is expected to be of the order of the number of nearest neighbors. More in general, the way this quantity scales with the size of the system A provides a benchmark for processes as localization or thermalization, as we show hereafter.

18.2.1 Random States

As a case study, we consider a system where a scrambling mechanism produces a state which is random with respect to the Haar measure. Here the CV is given by the average of Ω_A^{ε} over all the possible pure states of $\mathscr{H}$ equally weighted. For a random state of n qubits (with $\dim\mathscr{H} = 2^n$) the entanglement entropy of a subsystem A of a qubits is [35, 41]

$$S_{a,n} = \sum_{i=2^{n-a}+1}^{k=2^n} \frac{1}{i} - \frac{2^a - 1}{2^{n-a+1}}, \tag{18.7}$$

From this result, one can compute the average MI between A and another subsystem B of b qubits $I(A, B)$ obtaining

$$I(A, B) \simeq \begin{cases} \frac{2^{a+b} - 2^{a-b}}{2^{n-a-b+1}} & \text{for } b < \frac{n}{2}, \text{ and } a + b < \frac{n}{2} \\ (2(a+b) - n)\log 2 - \frac{2^{3b+a} - 2^{2n-a-b}}{2^{n+a+b+1}} & \text{for } b < \frac{n}{2}, \text{ and } a + b > \frac{n}{2} \\ 2a \log 2 + \frac{2^{2n-a-b} - 2^{2n+a-b}}{2^{n+a+b+1}} & \text{for } b > \frac{n}{2}. \end{cases} \tag{18.8}$$

This structure is summarized in Fig. 18.2.

To find the CV, we fix a to be independent of n and compute the difference $I(\mathbf{A}, \bar{\mathbf{A}}) - I(\mathbf{A}, \mathbf{B}) \simeq g(a)(2^{-2b+n} - 2^{-n+a})$, where $g(a) = \frac{2^a - 2^{-a}}{2^{a+1}} \sim \mathscr{O}(1)$. Setting this quantity equal to a finite ε, independent of n, we obtain

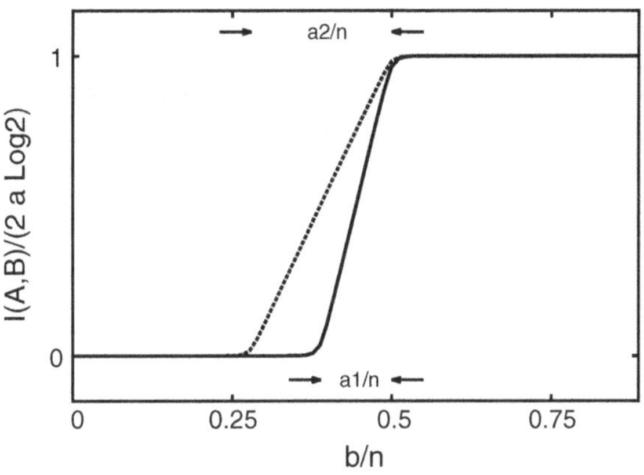

Fig. 18.2 The Mutual Information of a random state between different subsets of the Hilbert space is exponentially suppressed until $b + 2a = \frac{n}{2}$. Then it grows linearly until saturating to its maximum value given by $I(A, B) = 2a$. $a_1/n = 1/9$ (continuous line) $a_2/n = 2/9$ (dotted line)

$$[\Omega_A^\varepsilon]_{\text{average}} \xrightarrow{n\to\infty} \frac{\log(\frac{g(a)}{\varepsilon})}{2\log 2} + \frac{n}{2} \simeq \frac{n}{2}. \tag{18.9}$$

So, for large n, in order to get a full information about a one needs to address half of the system. This is what one also expect to find in a thermalization process.

18.2.2 Single Particle States

Another simple scenario is the single-particle limit, characteristic of the low energy regime of certain many-body systems (e.g. fermionic quasiparticles of Landau-Fermi liquid, bosonic collective modes, spin-waves, etc). In this limit, it is instructive to study analytically two possible states produced by the introduction of some type of disorder: localized states and delocalized random states.

To such end, notice that a generic single-particle can be written as

$$|\psi\rangle = \sum_{r=1}^{n} \psi_r |r\rangle., \tag{18.10}$$

where $|r\rangle$ denotes local excitation basis $|\downarrow\rangle_1 \cdots |\downarrow\rangle_{r-1}|\uparrow\rangle_r|\downarrow\rangle_{r+1}\cdots|\downarrow\rangle_n = |r\rangle$. Simple algebra shows that the entanglement entropy of any desired set of spins A is given by

$$S(\rho_A) = -p_A \log p_A - (1 - p_A)\log(1 - p_A)\ , \tag{18.11}$$

where $p_A = \sum_{r\in A} |\psi_r|^2$.

Let's start by analyzing 1-D localized wavefunctions which are typically characterized by a peak located at a random site j_0 and a localization length ξ. Seting $j_0 = 0$ without loss of generality we have

$$\psi_r \propto e^{-|j|/\xi} \; . \tag{18.12}$$

To calculate the CV, we consider the *peak-to-volume* MI between the central site $j_0 = 0$ (which plays the role of system A) and the set B of sites with $|j| \leq b$. Setting a suitable threshold ε, B_c can be calculated from (18.11) (see also [30]). In the thermodynamic limit one gets

$$I(A, \bar{A}) - I(A, B) = I(1, C) = \varepsilon \simeq \frac{2B_c}{\xi} e^{-B_c/\xi} \; . \tag{18.13}$$

In other words, for a localized single-particle state the CV of the peak-site is of the order of magnitude of $\xi \sim \mathcal{O}(1)$. More importantly, as a byproduct, notice that the scaling behavior of $I(1, C)$ provides another definition of the correlation length. These considerations are expected to remain valid in the many particle scenario.

Now, at the opposite end of the localized/delocalized phase transition, we can study the physical properties of typical states in the single particle subsector. Operationally, up to subleading corrections, this is achieved by assuming that the complex amplitudes in (18.10) are gaussian random variables

$$[\psi_i \psi_j^*] = \frac{1}{n} \delta_{ij} \; , \tag{18.14}$$

where [.] denotes the average of the random variable.

Following [26], we can compute again the MI between the first site and subsystem B, composed by m sites. For $m \gg 1$

$$I(1, B) \simeq \frac{m}{N} \frac{m}{1 - \frac{m}{N}} \tag{18.15}$$

while, for $m \sim \mathcal{O}(1)$, MI vanishes in the thermodynamic limit faster than entanglement entropy itself, a hallmarck of the delocalized phase. For any desired tolerance ε, the size of the minimum B_c containing the required information about A scales as N in the thermodynamic limit.

18.2.3 *Irreversible Growth of the CV in Chaotic Models*

In a chaotic system, we expect that an initial state $|\psi_{\text{in}}\rangle$ evolves towards the sea of random states [23]. We thus expect and irreversible growth of the CV. In order to study such a mechanism, we choose A to be the single spin on the edge site 1 of a 1-D chaotic spin chain, the subsystem B_m to be the set of sites $2, \ldots, m$, and compute the evolution of $I\,(A, B_m)\,(t)$.

The specific form of $\Omega_1^{\varepsilon}(|\psi(t)\rangle)$ depends on the interaction structure of the theory. In particular, Lieb-Robinson's causality bounds [22] limit the allowed growth.

These bounds force information to propagate inside an effective light cone, up to exponentially suppressed corrections. For a sufficiently short amount of time, the information associated with the first spin is expected to have reached only the second spin and we expect:

$$I(A, B_2)(t) \simeq I(A, B_3)(t) \simeq \cdots \simeq I(A, \bar{A})(t) = 2S_A(t). \tag{18.16}$$

Then, at some $t_{1,2}$ the information passes to the third spin. This moment $t_{1,2}$ is signaled by the sudden decrease of $I(A, B_2)(t)$. By now the information should be found inside $I(A, B_3)(t)$, and adding more spins should not increase the MI, again due to Lieb-Robinson bounds. This process should continue until saturation. The set of characteristic time scales $t_{1,m}$ (when $I(A, B_m)(t)$ begins to decrease) is clearly related to the ability of the evolution to *hide* the information associated to A in bigger and bigger subsystems.

As an example, consider the following Hamiltonian:

$$H = -J \sum_{i=1}^{n-1} \sigma_i^z \sigma_{i+1}^z + \frac{J}{2} \sum_{i=1}^{n} \left(3\sigma_i^x - \sigma_i^z\right). \tag{18.17}$$

The Hamiltonian couplings are chosen so as to be far from integrability regions [2]. We review the case of an anti-ferromagnetic initial state (for other initial states see [29]):

$$|\psi_{\rm af}\rangle = |\uparrow\rangle_1 \otimes |\downarrow\rangle_2 \otimes |\uparrow\rangle \otimes \cdots |\downarrow\rangle_n \,, \tag{18.18}$$

and set $n = 10$. The evolution of the MIs between the first site and different blocks of spins is shown in Fig. 18.3, where the expectations previously described appear in a transparent way.

The growth of Ω_1^ε as a function of time is depicted in Fig. 18.4, for a precision $\varepsilon = 0.0001$. The linear growth suggests ballistic propagation of information in the

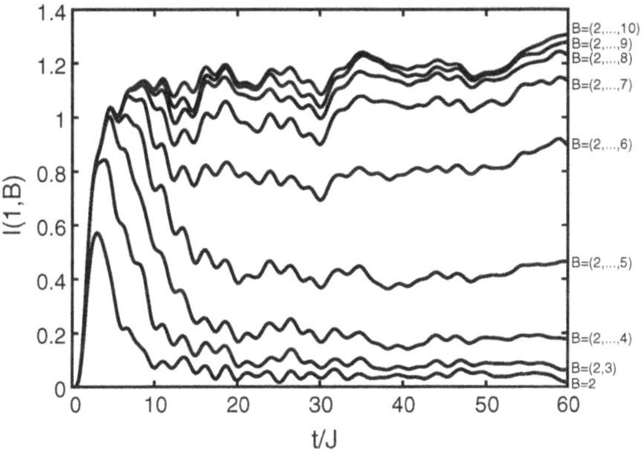

Fig. 18.3 Temporal evolution of the Mutual Information between the site 1 and other possible subsystems of the spin chain described by (18.17) and starting from the initial state $|\psi_{\rm af}\rangle$

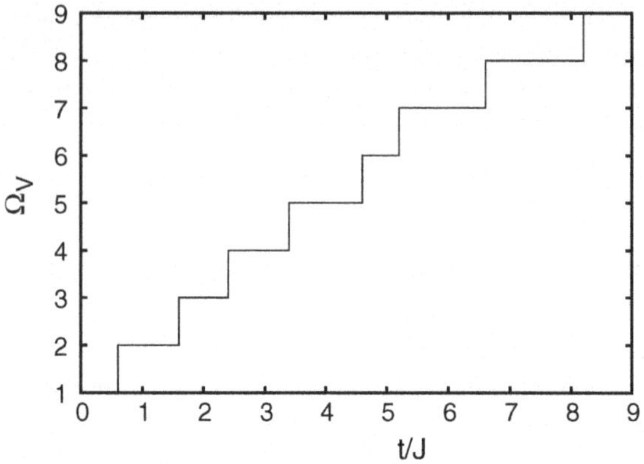

Fig. 18.4 Temporal evolution of $\Omega_1^{\varepsilon}(\rho)$, for $\varepsilon = 0.0001$ and initial state $|\psi_{af}\rangle$(top panel), and $|\psi_{Y+}\rangle$ (bottom panel)

system, with an associated emergent velocity. This emergent velocity is difficult to compute in general, but can be computed systems admitting a gravity dual [33], by means of the Ryu-Takayanagi formula [38].

Finally, it can be checked, see [29], that the process tends to the random state structure described earlier. In particular, the Mutual Information at stationarity shows the same pattern as in (18.2).

18.2.4 Many-Body-Localized Phases

The onset of localization thwarts strong volume-like entanglement, characteristic of systems satisfying the Eigenstate Thermalization Hypothesis. Such chaotic/MBL phase transitions in the eigenstate structure can be studied with entanglement entropy alone, see [5], where area type law was found for the entanglement entropy in the MBL phase. Nevertheless, product states generically evolve into volume-entangled linear combinations of weakly entangled eigenstates, and in these scenarios, our approach might help to unravel the precise structure of information (de)localization.

To this end consider the Hamiltonian studied in [34, 46]:

$$H = \sum_{i=1}^{n-1} J_\perp(\sigma_i^x \sigma_{i+1}^x + \sigma_i^y \sigma_{i+1}^y) + J_z \sigma_i^z \sigma_{i+1}^z + \sum_{i=1}^{n} h_i \sigma_i^z \,, \tag{18.19}$$

where $J_z/J_\perp = 0.2$ and h_i is a random variable with uniform probability distribution in the range $[-\eta, \eta]$. In the numerical simulations we diagonalize the Hamiltonian exactly, with $n = 8$ spins, and average over 1.000 repetitions. When studying the deep MBL phase we will set $\eta = 6$ [34].

The structure of the MI $I(1, B)$ for eigenstates is depicted in Fig. 18.5 (green line). Typically, the information shared by spin 1 is mostly contained in spin 2. QI is

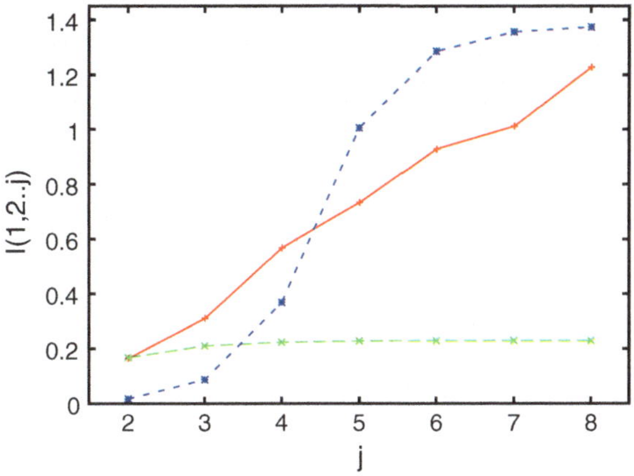

Fig. 18.5 The structure of Mutual Information between the first spin and different size adjacent blocks of spins associated to: the average over the eigenstates of the MBL Hamiltonian (18.19) (green); a random state in the Hilbert space (blue); the average over the eigenstates of a non-integrable Hamiltonian (red)

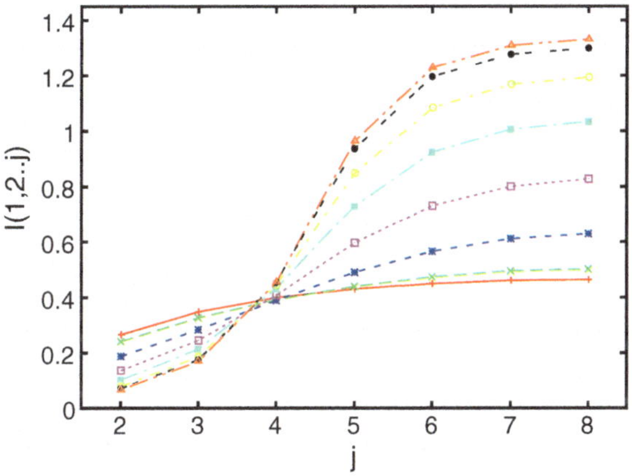

Fig. 18.6 Peak-to-volume MI at stationarity. Each curve correspond to one of the initial states parametrized in (18.21), with $\theta_k = \frac{k\pi}{14}$ with $k = 0, \ldots, 7$. The plot of the MI structure of a random state (red color) is shown for comparison. All curves collapse to the random expectation in the delocalized phase

stored *locally* within nearest neighbors. This was expected from the area law found in [5].

More interesting is to analyze the time evolution of certain product states realizable experimentally [34]. Defining:

$$|\psi\rangle_\theta = \cos\left(\frac{\theta}{2}\right) |\downarrow\rangle + i\, \sin\left(\frac{\theta}{2}\right) |\uparrow\rangle\,, \tag{18.20}$$

we study the following family of initial states parametrized by θ (Fig. 18.6):

$$\begin{aligned}|\Psi_\theta\rangle = |\psi\rangle_\theta \otimes |\psi\rangle_{\pi-\theta} \otimes |\psi\rangle_\theta \otimes |\psi\rangle_{\pi-\theta} \otimes \\ |\psi\rangle_\theta \otimes |\psi\rangle_{\pi-\theta} \otimes |\psi\rangle_\theta \otimes |\psi\rangle_{\pi-\theta}\end{aligned}\quad, \tag{18.21}$$

with $0 \leq \theta \leq \frac{\pi}{2}$. The interesting aspect of this family is that it smoothly interpolates between the "antiferromagnetic" state, for $\theta = 0$, and a product of all spins pointing in the positive y direction for $\theta = \frac{\pi}{2}$. The evolution of the multipartite entangle-

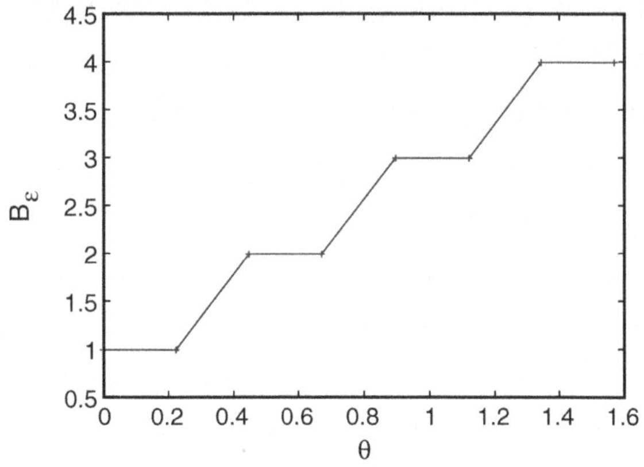

Fig. 18.7 Codification Volume, for $\varepsilon = 0.3$, as a function of θ at stationarity for a many-body-localized Hamiltonian

ment structure of these two states was studied in [29] for the case of a quantum chaotic Hamiltonian, and in both cases information fully delocalizes at long times. On the other hand, for the MBL phase, the MI at stationarity for each initial state $|\Psi_\theta\rangle$ shows an interesting pattern (18.6). The antiferromagnetic $\theta = 0$ case is the horizontal lower line. As we increase θ the structure of MI keeps approaching the structure of the random state [29], which appears also in the figure as the higher line. In a chaotic/delocalized phase, all initial states would collapse to the random expectation. Finally, the delocalization properties can be also studied by establishing certain tolerance ε and finding the codification volume B_c. For $\varepsilon = 0.3$ the result is given in Fig. 18.7.

From (18.6) and (18.7), it is transparent that the amount of delocalization of quantum information in the MBL phase is directly controlled by θ and thus strongly depends on the initial state. This is an interesting peculiar phenomenon of MBL phases which could be studied experimentally as well.

18.3 Large-N Models and Black Holes

One of the main motivations to develop the previous approach to quantum thermalization was to discuss finer grained aspects of information dynamics in black hole physics. In most contexts, black holes are described by distinct classes of large-N theories, such as matrix models [3], gauge theories [12, 16, 31] or the recent SYK [20, 39]. As emphasized in [40], such models share a common aspect: their Hamiltonians couple every degree of freedom with every other. We expect this observation to have generic consequences for their associated information dynamics.

Studying Mutual Information and entanglement dynamics in such systems seems quite challenging at first sight. It would be of great help to have a simple example in which computations can be done explicitly, but that still contains the generic features governing the physics in all other models. In [26, 27] one of us proposed that a system of N fermions with gaussian random interactions would suffice for these purposes. It was later shown in [28] that such expectations turn out to be true in more realistic and strongly interacting models, such as SYK. Finally, some of these aspects of information dynamics were studied in the context of black holes in general relativity in [19], by using the Bekenstein-Hawking formula, where a quite striking match was found. This section describes these advances from a unified perspective.

18.3.1 Random Free Fermions

A system of random free fermions is described by the following (dimensionless) Hamiltonian:

$$H = \sum_{i=1}^{N} c_i^\dagger \, c_i + \eta \sum_{i,j=1}^{N} c_i^\dagger \, V_{ij} \, c_j \;, \qquad (18.22)$$

where $c_i^\dagger$ and c_i create/annihilate spinless fermions and η controls the relative size of interactions. The matrix V is a random matrix taken from the GOE ensemble with zero mean and unit variance (see [17, 43, 44] for different accounts on random matrices). This model can be solved by exact diagonalization of V. If ψ^a are the eigenvectors of V with eigenvalues ε_a, and we define $d_a \equiv \sum_{i=1}^{N} \psi_i^a \, c_i$, then the Hamiltonian can be written as:

$$H = \sum_{a=1}^{N} (1 + \eta \, \varepsilon_a) \, d_a^\dagger \, d_a = \sum_{a=1}^{N} E_a \, d_a^\dagger \, d_a \;. \qquad (18.23)$$

Therefore, any given set $\mathscr{A}$ with cardinality N_p of particles produces an eigenstate:

$$|\Psi^{N_p}\rangle = \prod_{a \in \mathscr{A}} d_a^\dagger \, |0\rangle \;, \qquad (18.24)$$

and all eigenstates arise in this way.

It is known from the theory of random matrices [43, 44] that the eigenvectors are distributed according to the Haar measure on the orthogonal group $O(N)$. For our purposes, this just mean that eigenvectors are normalized random vectors. Mathematically:

$$[\psi_i^a] = 0 \quad [\psi_i^a \, \psi_j^b] = \frac{1}{N} \, \delta_{ab} \, \delta_{ij} \;, \qquad (18.25)$$

where $[p]$ denotes the average of the random variable p over the random matrix ensemble associated to V. The average two point function in an eigenstate with N_p particles is therefore:

$$[C_{ij}^{\Psi}] = [\langle \Psi^{N_p} | c_i^\dagger c_j | \Psi^{N_p} \rangle] = \left[\sum_{a,b=1}^{N} \psi_i^a \psi_j^b \langle \Psi^{N_p} | d_a^\dagger d_b | \Psi^{N_p} \rangle \right] = \sum_{a \in \mathscr{A}} [\psi_i^a \psi_j^a] = \frac{N_p}{N} \delta_{ij} , \tag{18.26}$$

where the last sum runs over the subset $\mathscr{A}$ of particles chosen for $|\Psi^{N_p}\rangle$. Since the system is gaussian, the EE and MI can be directly obtained from the two point function [36]. Given a subsystem A with m degrees of freedom, and the covariance matrix $C_{ij}^{\Psi} = \langle \Psi | c_i^\dagger c_j | \Psi \rangle$, where $i, j \in A$, the entanglement entropy is given by:

$$S_A = - \sum_{i=1}^{m} (\lambda_i \log \lambda_i + (1 - \lambda_i) \log(1 - \lambda_i)) , \tag{18.27}$$

where the λ's are the eigenvalues of C^{Ψ}. Such spectrum is not always possible to find analytically. We can compute it two standard limiting cases: the thermal regime $N_p \gg 1$, and the single particle sector $N_p = 1$, which was described above. For the explicit computations we refer to [26]. For $N_p \gg 1$, and $m \lesssim N_p$, which is the interesting regime in relation to black holes, it can be shown that:

$$[S_m^{\Psi^{N_p}}] = m\, S_1^{N_p} - \frac{m^2}{2(N - N_p)} , \tag{18.28}$$

where:

$$S_1^{N_p} = -\frac{N_p}{N} \log \frac{N_p}{N} - \left(1 - \frac{N_p}{N}\right) \log \left(1 - \frac{N_p}{N}\right) . \tag{18.29}$$

This last expression can just be seen as a thermal entropy per degree of freedom, since N_p (the total energy) can be related to an effective emergent temperature [26]. The second term in (18.28) can be seen as a small deviation from thermal behavior. This behavior was already discussed in section (18.2), when considering full-fledge random states in the Hilbert space. Only the size of the deviations from thermality changed, and the MI behaves qualitatively as in (18.8).

We now discuss out-of-equilibrium processes. Beginning with an initial state with $\bar{N}$ excited oscillators:

$$|\Psi_{\text{in}}\rangle = c_1^\dagger c_2^\dagger \dots c_{\bar{N}}^\dagger |0\rangle , \tag{18.30}$$

and writing the unitary evolution in the decoupled oscillator basis $d_a^\dagger$:

$$|\Psi(t)\rangle = \sum_{j,k,\dots,l} \psi_1^j \psi_2^k \cdots \psi_{\bar{N}}^l \, e^{i\,t\,(E_j + E_k + \cdots + E_l)} d_j^\dagger d_k^\dagger \cdots d_l^\dagger |0\rangle . \tag{18.31}$$

allows to write the covariance matrix as:

$$
\begin{aligned}
\mathscr{C}_{ij}(t) &= \langle \Psi(t)| \, c_i^\dagger c_j \, |\Psi(t)\rangle \\
&= \sum_{j,l;k,m;n,p} \psi_1^j \cdots \psi_{\bar{N}}^l \psi_1^k \cdots \psi_{\bar{N}}^m \psi_i^n \psi_j^p \, e^{-it(E_j+\cdots+E_l-E_k-\cdots-E_m)} \times \\
&\times \langle 0| \, d_j \ldots d_l \, d_n^\dagger \, d_p \, d_k^\dagger \ldots d_m^\dagger \, |0\rangle \, .
\end{aligned} \tag{18.32}
$$

But since eigenvectors of random matrices are almost random vectors, it is simple to conclude that on average:

$$
i \neq j \longrightarrow [\, \mathscr{C}_{ij}(t) \,] = 0 \, . \tag{18.33}
$$

The only surviving entries are the diagonal ones, the expectation values of all number operators. The effective permutation symmetry of the Hamiltonian (18.22) implies that the relaxation of initially excited particles is the same for all of them $[\langle c_\uparrow^\dagger c_\uparrow \rangle](t) = n_\uparrow(t)$, and the same holds for the initially non-excited ones $[\langle c_\downarrow^\dagger c_\downarrow \rangle(t)] = n_\downarrow(t)$. Up to small deviations, entanglement entropy simply evolves as:

$$
[S_A] = -\sum_{i=1}^{M} [n_i(t) \log n_i(t) + (1 - n_i(t)) \log(1 - n_i(t))] \, . \tag{18.34}
$$

Entanglement entropy is extensive all times, and so the MI and codification volume behave in the same way as for random state.

Given formulas (18.26), (18.28), (18.33), (18.34), we can now summarize the features we expect to remain valid in realistic models, like SYK or large-N gauge theories:

- Entanglement entropy is extensive at all times and its functional form is that of (18.34), in terms of occupation numbers.
- Information is delocalized instantaneously at all times, as defined by MI in section (18.2).
- The reason is that the mean field approximation becomes exact in the thermodynamic limit.

In the next section, we review a more abstract (but generic) method that was used in [28] to prove these aspects in realistic, strongly interacting black hole models.

18.3.2 K-Body Ensembles, SYK and de Finetti Theorems

The easiest generalization of the random free fermion models is to the so-called random k-body ensembles [6, 21], defined by the following Hamiltonian:

$$
H = \sum_{\substack{1 \le i_1 < \cdots < i_k \le N \\ 1 \le j_1 < \cdots < j_k \le N}} J_{i_1 \cdots i_k; j_1 \cdots j_k} c_{i_1}^\dagger \cdots c_{i_k}^\dagger c_{j_1} \cdots c_{j_k} \, . \tag{18.35}
$$

Each term of the Hamiltonian contains k annihilation/creation operators, and $J_{i_1\cdots i_k;j_1\cdots j_k}$ are random numbers with the only constraint of ensuring a hermitian Hamiltonian. A cousin model is the SYK model [20, 39]:

$$H = i^{k/2} \sum_{1\le i_1<\cdots<i_k\le N} J_{i_1\cdots i_k}\chi_{i_1}\cdots\chi_{i_k}\,, \tag{18.36}$$

containing N Majorana fermions interacting through random k-body interactions. Strikingly, this is arguably the best toy model of black hole physics, given that it saturates the bound on chaos [20, 32].

The crucial observation for our concerns is that both models show an emergent $O(N)$ symmetry on average, which contains the permutation group as a subgroup. Such feature turns out to severely constrain the state structure, as dictated by the so-called de-Finetti theorems [10]. The application of this theorem to large-N theories and gravity was described by one of us in [28]. We briefly review the theorem and main results here.

The classic version of the theorem is very old [10]. It considers probability distributions $p_N(x_1, x_2, \ldots, x_N)$ invariant under permutations. Such distributions are "M-exchangeable" if there is another permutation symmetric distribution $p_M(x_1, x_2, \ldots, x_M)$ of $M > N$ variables satisfying $p_N(x_1, x_2, \ldots, x_N) = \mathrm{tr}_{M-N}$ $p_M(x_1, x_2, \ldots, x_M)$. If there exists such extended distribution for any M, we say p_N to be infinitely exchangeable. The theorem asserts that infinitely exchangeable distributions are convex combinations of identically distributed uncorrelated ones:

$$p_N(x_1, x_2, \ldots, x_N) = \int d\mu[p(x)]\, p(x_1)p(x_2)\cdots p(x_N)\,. \tag{18.37}$$

In the previous formula $\mu[p(x)]$ is a measure on the space of probability distributions of one variable. If p_N is only M-exchangeable, such convex combination is just an approximation with an error depending on the ratio N/M. Theorem 18.37 is a paradigmatic result, showing how global symmetries can constrain the state structure.

We apply this theorem to SYK type models here, but as described in [28] other large-N models can be considered as well. Consider again the state (18.30) in which the first N fermions has been excited. Unitarily evolving with (18.36) or (18.35) and averaging, it can be shown that [25]:

$$[\rho(t)] = [U(t)|\psi_{\mathrm{in}}\rangle\langle\psi_{\mathrm{in}}|U^\dagger(t)] = \sum_{a_1a_2\cdots a_N} p_{a_1a_2\cdots a_N}(t)|a_1a_2\cdots a_N\rangle\langle a_1a_2\cdots a_N|\,, \tag{18.38}$$

On average, only the diagonal entries $p_{a_1a_2\cdots a_N}(t) \equiv [|\langle a_1a_2\cdots a_N|U(t)|\psi_{\mathrm{in}}\rangle|^2]$ survive. Decoherence work at all times at a microscopic level. Therefore, the reduced density matrix of the initially excited fermions reads:

$$[\rho^A(t)] = \sum_{a_1a_2\cdots a_N} p_{a_1a_2\cdots a_N}(t)|a_1a_2\cdots a_N\rangle\langle a_1a_2\cdots a_N|\,, \tag{18.39}$$

Since $\rho^A_{\text{in}} = |1_1 \cdots 1_N\rangle\langle 1_1 \cdots 1_N|$ and the evolution display permutation symmetry, the diagonal probability distribution $p_{a_1 a_2 \cdots a_N}(t)$ is permutation invariant, and the theorem (18.37) applies:

$$[\rho^A] \simeq \sum_{a_1 a_2 \cdots a_N} \int d\mu[p,t]\, p(a_1)p(a_2)\cdots p(a_N)|a_1 a_2 \cdots a_N\rangle\langle a_1 a_2 \cdots a_N| \,, \tag{18.40}$$

with $\mu[p,t]$ is a time dependent measure on the space of binary distributions. Binary distributions are just specified by a real number $p \in [0,1]$. Setting $\rho(p) = p|0\rangle\langle 0| + (1-p)|1\rangle\langle 1|$ we find:

$$[\rho^A(t)] \simeq \int d\mu[p,t]\rho(p)^{\otimes m} \,. \tag{18.41}$$

The first conclusion is that SYK states are close to separable (classical mixtures) at all times. There is very little quantum entanglement in the large-N limit. In regards to entropy evolution we find:

$$m\int d\mu[p,t]S(\rho(p)) \leq S(\rho^A(t)) \leq m\int d\mu[p,t]S(\rho(p)) + S(\mu[p,t]) \,. \tag{18.42}$$

Since $S(\mu[p,t])$ does not increase with m, entropy evolution is extensive at large m.[1] We conclude that large and small subsystems share the same characteristic time scales, those of $\mu[p,t]$. In particular, two point functions are controlled by $\mu[p,t]$, and therefore they decay time scale controls the relaxation time scale of the global system as well. This is intimately connected with the fact that information spreads instantaneously over all degrees of freedom. In the next section we will see how these aspects show in real black hole models.

18.3.3 Black Holes

The main conclusion of the previous approach to information dynamics when applied to a black hole like system (large-N theories), is that information delocalizes instantaneously, entanglement entropy evolves extensively and the process relaxes in a time scale controlled by that of a single degree of freedom. In collaboration with Aron Jansen, in [19] we checked these aspects in exact black hole collapse scenarios in general relativity. We accomplished this by numerically computing the Bekenstein-Hawking entropy formula, which in units of the Planck length is just:

$$S_{\text{BH}} = \frac{A}{4} \,, \tag{18.43}$$

[1] Such relation can be used to bound the quantum distance to the maximally mixed state, see [28].

More concretely we considered GR coupled to a scalar field, and choose an initial state containing a black hole with entropy $S_{\rm i} = A_{\rm i}/4$ and a scalar field profile containing enough energy to backreact on the geometry. As time evolves the scalar field collapses towards the black hole, increasing its entropy to $S_{\rm f} = A_{\rm f}/4$. Here we will quote the main results.

First, it is remarkable that extensivity in the entropy evolution follows directly from (18.43), since given any two disconnected horizon patches $\mathscr{A}$ and $\mathscr{B}$:

$$S_{\mathscr{A}\cup\mathscr{B}}(t) = S_{\mathscr{A}}(t) + S_{\mathscr{B}}(t)\,. \qquad (18.44)$$

Mutual information between different horizon patches thus vanishes at all times, and the structure of information delocalization is akin to that found for SYK type models, including the statement that the characteristic time scale for global relaxation does not depend on the size of the chosen patch. To find such characteristic time scale we need the specific law governing the evolution of geometric quantities near equilibrium. In this regime the scalar field is described by so-called quasinormal modes, see [7] for a review, which behave as damped harmonic oscillators,

$$\phi(r,t) = Ae^{-\omega_{\rm I} t}\,(\cos(\omega_{\rm R} t+\delta)\phi_I(r)+\sin(\omega_{\rm R} t+\delta)\phi_R(r))\,. \qquad (18.45)$$

Although there is a whole spectrum of quasinormal frequencies, at late times only the lowest mode $\omega = \omega_{\rm R} + i\,\omega_{\rm I}$ survives. For different reasons, it turns out that the correct ansatz for the black hole entropy near equilibrium is:

$$\delta S(t) \equiv S_{\rm f} - S(t) = A\,e^{-2\omega_{\rm I} t}\,(\cos(2\omega_{\rm R} t+\delta) + B\,)\;, \qquad (18.46)$$

where ω is the lowest quasinormal mode, and A and δ parametrise the initial amplitude and phase. The damping shift B, on the contrary, does not depend on the initial conditions and its physics is still to be understood. Quite strikingly, for apparent horizons it is such that oscillations saturate hawking area theorems [19], and it might be a good smoking gun for microscopic models of black holes. Luckily, on a first approach, it is not needed for connection the information/entropy dynamics of black holes and microscopic models.

The previous ansatz was numerically confirmed in different space-time dimensions and for different scalar field masses in [19]. It matches the one expected from the results of the previous section. Formulas (18.34) and (18.28) relate the evolution of the entropy to the evolution of the number operator. This happens for gauge theories as well [27]. Since number operators arise by squaring fields, they decay with twice the lowest quasinormal frequency, as in (18.46).

18.4 Summary and Discussion

In this article, we have reviewed an approach to quantum thermalization developed in [29], and its subsequent application to different examples, such as local spin systems, MBL phases, large-N theories and black holes [26–28, 30].

The approach is based on Mutual Information. Intuitively, given a subsystem A, it asks for the support of minimal size subsystem B maximally entangled with A. There were various motivations to study this object. First, the approach to the information paradox developed by Hayden and Preskill in [18], based on the structure of entanglement and quantum information in random subsystems (or systems undergoing random Hamiltonian evolution) naturally leads to these considerations. Second, it complements the study of entanglement entropy, which has been extensively explored in the context of quantum thermalization. In such a context, and at least for sufficiently chaotic systems, such entropy is expected to saturate to the thermal one, and therefore become extensive in the subsystem size. On the other hand, this extensivity does not grant the system to be "thermalized", and more proxies are needed. The pattern of Mutual Information contributes to a deeper understanding of the state structure, at least for certain systems.

This approach can be analytically considered for random states (18.2) and sets the intuition for more realistic ones. The conclusion, as expected, is that the support is of the order of the entropy of the system. Such observation sets the expectations for chaotic systems at late times, which has been confirmed above (18.2.3). From a different perspective, such pattern of Mutual Information can be used to define notions of correlation lengths in local systems and it is useful for the chaotic/MBL phase transition, as discussed above.

Finally, we went out and study these quantities, including entanglement entropy, in microscopic models of black holes, such as SYK and gauge theories. We also studied them in black holes themselves, by using the Bekenstein-Hawking formula. The key was to find a simple system, that of random free fermions, in which computations are transparent enough, but that still contains the generic aspects to be extracted. Unluckily, the results, in this case, do not really allow to distinguish the approach based on entanglement entropy to that based on Mutual Information. This is because Mutual information vanishes at all times, implying that information instantaneously delocalizes over the system. To unravel further patterns of information dynamics in black hole physics one needs even finer quantities, such as out-of-time ordered correlation functions [32]. These quantities do not really tell us about how quantum information delocalizes over the different degrees of freedom in the system, but how it gets hidden in increasingly growing operators, as described in [37].

Acknowledgements Javier Magan work is supported by an It From Qubit grant by the Simons foundation.

References

1. K.M.R. Audenaert, J. Eisert, Continuity bounds on the quantum relative entropy. J. Math. Phys. **46**, 102104 (2005)
2. M.C. Bañuls, J.I. Cirac, M.B. Hastings, Strong and Weak Thermalization of Infinite Nonintegrable Quantum Systems. Phys. Rev. Lett. **106**, 050405 (2001)
3. T. Banks, W. Fischler, S.H. Shenker, L. Susskind, M-theory as a mtrix model: A conjecture. Phys. Rev. D. **55**, 5112 (1997)
4. J.L.F. Barbon, J.M. Magán, Fast Scramblers. Horizons and Expander Graphs. JHEP **1208**, 016 (2012)
5. B. Bauer, N. Nayak, Area laws in a many-body localized state and its implications for topological order. J. Stat. Mech. **2013**, P09005 (2013)
6. L. Benet, H.A. Weidenmueller, Review of the k-body embedded ensebmles of gaussian random matrices. J. Phys. A **36**, 3569 (2003)
7. E. Berti, V. Cardoso, A.O. Starinets, Class. Quant. Grav **26**, 163001 (2009)
8. F. Borgonovi, F.M. Izrailev, L.F. Santos, V.G. Zelevinsky, Quantum chaos and thermalization in isolated systems of interacting particles. Phys. Rep. **626**, 1 (2016)
9. C.M. Caves, C.A. Fuchs, R. Schack, Unkown quantum states: the quantum de Finetti representation. J. Math. Phys. **43**, 4537 (2002)
10. de Finetti, B.: Le prevision: ses lois logiques, ses sources subjectives. Annales de l' I.H.P. , tome 7, (1937) 1
11. J.M. Deutsch, Quantum statistical mechanics in a closed system. Phys. Rev. A **43**, 2046 (1991)
12. W. Edward, Anti de Sitter space and holography. Adv. Theor. Math. Phys. **2**, 253 (1998)
13. C. Gogolin, J. Eisert, Equilibration, thermalisation, and the emergence of statistical mechanics in closed quantum systems. Rep. Prog. Phys. **79**, 5 (2016)
14. G. Gori, S. Paganelli, A. Sharma, P. Sodano, A. Trombettoni, Explicit Hamiltonians Inducing Volume Law for Entanglement Entropy in Fermionic Lattices. Phys. Rev. B. **91**, 245138 (2015)
15. B. Groisman, S. Popescu, W. Andreas, Quantum, classical, and total amount of correlations in a quantum state. Phys. Rev. A **72**, 032317 (2005)
16. S.S. Gubser, I.R. Klebanov, A.M. Polyakov, Gauge theory correlators from noncritical stringc-theory. Phys. Lett. **428**, 105 (1998)
17. Haake, F.: Quantum Signatures of Chaos. Springer-Verlag Berlin Heidelberg, **54** (2010)
18. P. Hayden, J. Preskill, Black holes as mirrors: Quantum information in random subsystems. J. High Energy Phys. **0709**, 120 (2007)
19. A. Jansen, J.M. Magan, Black hole collapse and democratic models. Phys. Rev. D **94**, 104007 (2016)
20. A. Kitaev, A simple model of quantum holography. Talks at KITP, April 7, 2015 and May 27, 2015
21. V.K. Kota, N.D. Chavda, Embedded random matrix ensembles from nuclear structures and their recent applications. Int. J. Mod. Phys. E **27**, 1830001 (2018)
22. E.H. Lieb, D.W. Robinson, The finite group velocity of quantum spin systems. Commun. Math. Phys **28**, 251 (1972)
23. S. Lloyd, H. Pagels, Complexity as thermodynamic depth. Ann. Phys. **188**, 186 (1988)
24. A. Lubotzky, Expander Graphs in Pure and Applied Mathematics. Bull. Amer. Math. Soc. **49**, (2012)
25. J.M. Magan, K-local microscopic diffusion at the Sachdev-Ye-Kitaev model. arXiv:1612.06765 [hep-th]
26. J.M. Magan, Random free fermions: An analytical example of eigenstate thermalization. Phys. Rev. Lett **116**, 030401 (2016)
27. J.M. Magan, Black holes as random particles: entanglement evolution in infinite range and matrix models. JHEP **1608**, 081 (2016)
28. J.M. Magan, De Finetti theorems and entanglement in large-N theories and gravity. Phys. Rev. D **96**, 086002 (2017)

29. J.M. Magan, S. Paganelli, Codification volume of an operator algebra and its irreversible growth through thermal processes. Phys. Rev. A **90**, 032103 (2014)
30. J.M. Magan, S. Paganelli, V. Oganesyan, Multipoint entanglement in disordered systems. Phys. Lett. A. **381**, 535 (2017)
31. J.M. Maldacena, The Large N limit of superconformal field theories and supergravity. Adv. Theor. Math. Phys. **2**, (1998)
32. J. Maldacena, S.H. Shenker, D.A. Huse, A bound on chaos. J. High Energy Phys. **08**, 106 (2016)
33. M. Mezei, D.J. Stanford, On entanglement spreading in chaotic systems. High Energ. Phys. **65**, 2017 (2017)
34. A. Nanduri, H. Kim, D. Stanford, Entanglement spreading in a many-body localized system. Phys. Rev. B **90**, 064201 (2014)
35. D.N. Page, Average entropy of a subsystem. Phys. Rev. Lett. **71**, 1291 (1993)
36. I. Peschel, Entanglement in solvable many-particle models. Braz. J. Phys **42**, 267 (2012)
37. D. Roberts, D. Stanford, A. Streicher, Operator growth in the SYK model. arXiv:1802.02633 [hep-th]
38. S. Ryu, T. Takayanagi, Holographic Derivation of Entanglement Entropy from the anti-de Sitter Space/Conformal Field Theory Correspondence. Phys. Rev. Lett. **96**, 181602 (2006)
39. S. Sachdev, J.W. Ye, Gapless spin fluid ground state in a random, quantum Heisenberg ferromagnet. Phys. Rev. Lett. **70**, 3339 (1993)
40. Y. Sekino, L. Susskind, Fast scramblers. J. High Energy Phys. **10**, 065 (2008)
41. S. Sen, Average entropy of a quantum subsystem. Phys. Rev. Lett **77**, 1 (1996)
42. M. Srednicki, Chaos and quantum thermalization. Phys. Rev. E **50**, 888 (1994)
43. T. Tao, *Topics in Random Matrix Theory*. Graduate Studies in Mathematics, vol. 32 (2011)
44. T. Tao, V. Vu, Random matrices: Universal properties of eigenvectors. Random Matrices: Theory Appl. **01** (2012)
45. M.M. Wolf, F. Verstraete, M.B. Hastings, J.I. Cirac, Area Laws in Quantum Systems: Mutual Information and Correlations. Phys. Rev. Lett. **100**, 070502 (2008)
46. M. Žnidarič, T. Prosen, P. Prelovšek, Many-body localization in the Heisenberg XXZ magnet in a random field. Phys. Rev. B **77**, 064426 (2008)

Chapter 19
Breaking the Area Law: The Rainbow State

Giovanni Ramírez, Javier Rodríguez-Laguna and Germán Sierra

Abstract An exponential deformation of a 1D critical Hamiltonian, with couplings falling on a length scale h^{-1}, gives rise to ground states whose entanglement entropy follows a volume law, i.e. the area law is violated maximally. The ground state is now in the so-called *rainbow phase*, where valence bonds connect sites on the left half with their symmetric counterparts on the right. Here we discuss some of the most relevant features of this rainbow phase, focusing on the XX and Heisenberg models. Moreover, we show that the rainbow state can be understood either as a thermo-field double of a conformal field theory with a temperature proportional to h or as a massless Dirac fermion in a curved spacetime with constant negative curvature proportional to h. Finally, we introduce a study of the time-evolution of the rainbow state after a quench to a homogeneous Hamiltonian.

19.1 Introduction

Quantum many-body systems are models which allow us to illustrate important notions about macroscopic physics, e.g. magnetic behaviour, in terms of microscopic elementary interactions between the constituents of that system. In addition to their physical interest, the development of new methods for their study has given

G. Ramírez (✉)
Instituto de Investigación en Ciencias Físicas y Matemáticas (ICFM-USAC), Universidad de San Carlos de Guatemala, Guatemala City, Guatemala
e-mail: ramirez@ecfm.usac.edu.gt

J. Rodríguez-Laguna
Departamento de Física Fundamental, Universidad Nacional de Educacion a Distancia (UNED), Madrid, Spain
e-mail: jvrlag@gmail.com

G. Sierra
Instituto de Física Teórica (UAM-CSIC), Madrid, Spain
e-mail: german.sierra@uam.es

A. Ferraz et al. (eds.), *Strongly Coupled Field Theories for Condensed Matter and Quantum Information Theory*, Springer Proceedings in Physics 239,
https://doi.org/10.1007/978-3-030-35473-2_19

an impulse to other fields such as quantum integrability [1], quantum groups [2], quantum computation and information [3] or quantum simulators [4].

Early studies of many-body quantum mechanics used to make the assumption that each particle moves under the effective field created by all the others, i.e. Hartree-Fock or mean-field type methods [5]. These techniques are very successful to explain many properties of electrons in solids, through the use of the Fermi liquid approximation or Density Functional Theory [6, 7]. Nonetheless, they are unable to take completely into account the effect of strong correlations, which are a key in many magnetic properties of materials, superconductivity [8, 9], quantum Hall effect [10] or topological insulators [11].

Furthermore, the advent of new technologies such as cold atoms in optical lattices or trapped ions [12, 13], allows to engineer quantum systems in which strong correlations are not avoided, but looked for. The reasons can be to mimic other quantum systems or to harness the specific effects of quantum correlations to profit from them, building better computation and communication technologies.

Quantum entanglement is defined as the property of those pure states which do not allow a description as product states, i.e. *non-factorizable* states. For a composite system divided in two parts A and B with Hilbert space $\mathcal{H} = \mathcal{H}_A \otimes \mathcal{H}_B$, a factorizable state can be written as $|\psi\rangle = |\psi_A\rangle \otimes |\psi_B\rangle$, where states $|\psi_A\rangle$ and $|\psi_B\rangle$ describe A and B respectively. Factorizability of states can be determined using the Schmidt decomposition [14], all states in $\mathcal{H}$ can be expressed as

$$|\psi\rangle = \sum_{i=1}^{\chi} \lambda_i \, |a_i\rangle \otimes |b_i\rangle \ ,$$

where $|a_i\rangle$ and $|b_i\rangle$ are orthonormal states of $\mathcal{H}_A$ and $\mathcal{H}_B$ respectively, and the Schmidt coefficients $\lambda_i \in \mathbf{R}^+$ and $\sum_i \lambda_i = 1$. The Schmidt number (or Schmidt rank) χ is bounded by the dimension of Hilbert spaces of A and B, i.e. $\chi \leq \min\{\dim\{\mathcal{H}_A\}, \dim\{\mathcal{H}_B\}\}$. A state is factorizable if $\chi = 1$, and if $\chi > 1$ the state is entangled.

Factorizability defines absence of entanglement. In order to quantify entanglement it is convenient to consider an observer which is only allowed to access subsystem A. Even when the global state is pure, $|\psi\rangle$, the subsystem accessible to A can be *mixed*. Thus, its quantum-mechanical description is performed via a reduced density matrix $\rho_A = \mathrm{Tr}_{\bar{A}} \, |\psi\rangle\langle\psi|$, where $\bar{A}$ is the complementary of subsystem A. We define the von Neumann's entropy of this reduced density matrix

$$S_A = -\mathrm{tr}\,\{\rho_A \log(\rho_A)\} \ .$$

Let us remark that for a pure state, $S_A = S_B$. The von Neumann's entropy satisfies $S_A \geq 0$, with $S_A = 0$ only for factorizable states. Von Neumann's entropy is also called the entanglement entropy (EE).

Why is S_A called an entropy? There is a deep relation with the concept of entropy in statistical mechanics and information theory. Indeed, von Neumann's entropy is a quantum analogue of Gibbs entropy. But in contrast, it is not related to thermal fluctuations. The EE can be argued to measure quantum correlations between subsystems. In classical information theory, the action of sending a message is also viewed as the action of correlating the sender and the receiver. The average amount of information contained in that message is measured with Shannon's entropy [15]. Von Neumann's entropy is just Shannon's entropy of the eigenvalues of the reduced density matrix, which can be regarded as a probability distribution. Quantum information theory builds upon this deep relation between entanglement and information.

The goal of this paper is to present some details of the rainbow model and then to summarise some results previously obtained. This work is organised as follows. First we present the rainbow model and some details of the concentric valence bond states. After that, we present a summary of the results previously obtained, we include the references where more details were discussed.

19.2 Concentric Valence Bond States

Consider a spinless fermion chain of L sites, whose dynamics is described by the Hamiltonian

$$H = -\sum_{i=1}^{L} t_i \; c_i^\dagger c_{i+1} + \text{h.c.} \tag{19.1}$$

where t_i represents the hopping amplitudes between sites i and $i+1$, $c_i^\dagger$ and c_i are, respectively, the fermionic creation and annihilation operators on site i, which satisfy anticommutation relations: $\{c_j, c_k^\dagger\} = \delta_{jk}$, $\{c_j, c_k\} = \{c_j^\dagger, c_k^\dagger\} = 0$.

Hamiltonian (19.1), which is quadratic in fermionic operators, is also called free fermion Hamiltonian. Moreover, free fermion Hamiltonians are solvable in terms of single-body states that are occupied by particles which move independently of each other. Diagonalising the hopping matrix, $T_{ij} = t_i(\delta_{j,i+1} + \delta_{j,i-1})$, $Tv_k = \varepsilon_k v_k$, allows to obtain the single-body energy levels, ε_k, and the single-body modes, $v_{k,i}$, which determine a canonical transformation

$$b_k^\dagger = \sum_i v_{k,i} c_i^\dagger \ , \tag{19.2}$$

where $v_{k,i}$ is an unitary matrix, thus the new operators $b_k^\dagger$ also follow fermionic commutation relations, i.e. b_k are also fermionic operators.

All eigenstates of the Hamiltonian (19.1) have the form

$$|\psi\rangle = \prod_{k\in\Omega} b_k^\dagger \, |0\rangle \ , \tag{19.3}$$

where Ω is a subset of $\{1, \cdots, L\}$ denoting the occupied modes and $|0\rangle$ is the Fock vacuum, which is annihilated by the operators c_i. The energy for the state (19.3) is $E = \sum_\Omega \varepsilon_k$. Therefore, the ground state (GS) is given by filling up all modes with lowest energy, i.e. $\Omega = \{k \mid \varepsilon_k < 0\}$. In absence of diagonal terms in the hopping matrix T, particle-hole symmetry forces $\varepsilon_k = -\varepsilon_{L+1-k}$, so the number of particles of the ground states is $|\Omega| = n_F = L/2$, i.e. we will operate at half-filling.

The correlation matrix, C, has elements defined by $C_{ij} \equiv \langle GS| c_i^\dagger c_j |GS\rangle$, which, in terms of the single-body modes is

$$C_{ij} = \sum_{k=1}^{n_F} \bar{v}_{k,i} v_{k,j} \ . \tag{19.4}$$

Let us now describe the family of local Hamiltonians whose GS approaches asymptotically the rainbow state . For bookkeeping convenience, let us number the sites as half-integers from $-(L-1)/2$ to $(L-1)/2$, and the corresponding links as integers, see Fig. 19.1. The rainbow Hamiltonian then reads

$$H \equiv -\frac{J}{2} c^\dagger_{\frac{1}{2}} c_{-\frac{1}{2}} - \frac{J}{2} \sum_{i=\frac{1}{2}}^{L-\frac{3}{2}} e^{-hi} \left[c_i^\dagger c_{i+1} + c^\dagger_{-i} c_{-(i+1)} \right] + \text{h.c.} \tag{19.5}$$

where $h \in \mathbf{R}^+$ is the inhomogeneity parameter, and we may also define $\alpha \equiv \exp(-h/2)$, as it is done in Fig. 19.1. Via the Jordan-Wigner transformation, this Hamiltonian is equivalent to the XX model for a spin-1/2 chain. For $h = 0$ we recover the standard uniform 1D Hamiltonian of a spinless fermion model with open boundary conditions (OBC). Its low energy properties are captured by a conformal field theory (CFT) with central charge $c = 1$: the massless Dirac fermion theory, or equivalently (upon bosonization) a Luttinger liquid with Luttinger parameter $K = 1$.

For $h \gg 1$ we obtain the Hamiltonian used to illustrate a violation of the area law for local Hamiltonians [16]. On the other hand, for $h < 0$ and truncating the chain to the sites $i > 0$, one obtains a Hamiltonian which has the scale-free structure of Wilson's approach to the Kondo impurity problem [17]. Models where t_i is a hyperbolic function of the site index i were considered in order to enhance the energy gap [18].

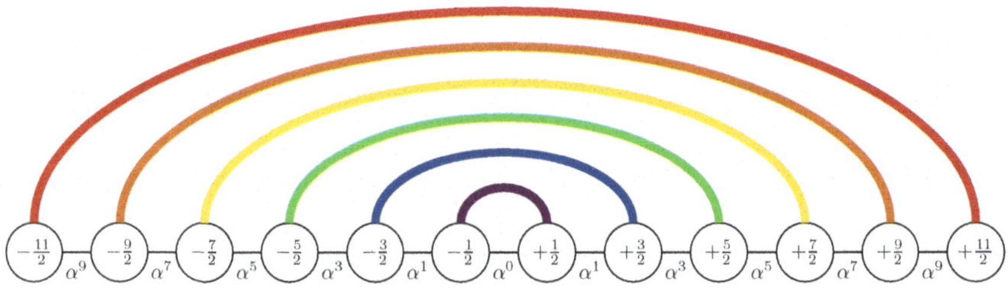

Fig. 19.1 Rainbow state representation, showing the $(-k, +k)$ valence bonds above the central link. Each bond contributes as $\log(2)$ to the entanglement entropy, thus the entanglement entropy of the left (or right) half of the chain is $L \log 2$. The hopping amplitudes are given in terms of $\alpha = \exp(-h/2)$

It is worth to notice the striking similarity between our system and the Kondo chain [19]. Indeed, let us divide our inhomogeneous chain into three parts: central link, left sites and right sites. The left and right sites correspond, in our analogy, to the spin up and down chains used in Wilson's chain representation of the Kondo problem. In both cases, they form a system of free fermions, with exponentially decaying couplings. In the Kondo chain, notwithstanding, the central link becomes a magnetic impurity, which renders the full system non-gaussian.

19.3 Results

In a first paper [20] we analysed the deformation of the critical local 1D Hamiltonian to explore a smooth crossover between a log law and a volume law for the EE. We presented the details of the Heisenberg XX model. The value $h = 0$ corresponds to the uniform model, which is described by the CFT, and whose entanglement entropy could also be obtained by means of the Fisher-Hartwig conjecture [21]. In the $h \to \infty$ limit the GS becomes the rainbow state.

We used a graphical representation for the correlation matrix which operates as follows. For any matrix element, C_{ij}, we draw a line inside the unit circle with a colour which marks the strength of the correlation between those points. A finitely-correlated state will be characterised by a correlation matrix whose representation is given by short lines which do not go deep inside the unit circle. A conformal state, with infinite correlation length, is characterised by a certain self-similar structure in the geodesic pattern. Realisations of the rainbow state correlations are also very easy to spot. Figure 19.2 shows the structure of the correlation matrix for a 1D system with periodic boundary conditions (PBC) for a different systems. The intensity of each line connecting two sites is related to $\left|C_{ij}\right|$.

We studied numerically the EE of blocks containing ℓ sites starting from the left extreme of the chain within the GS of Hamiltonian (19.5). For large values of h we observe a characteristic *tent shape* in the EE, i.e. an approximately linear growth up to $\ell = L$ followed by a symmetric linear decrease, giving the volumetric behaviour. Figure 19.3 shows the EE for different values of h for a system of $N = 32$ sites. As the value of h decreases, the slope decreases and ripples start to appear and for $h = 0$ they recover the parity oscillations characteristics of the von Neumann entropy with OBC. For positive h, the slope was empirically shown to be given by $h/6$.

The analysis of the entanglement spectrum showed very interesting connections between conformal growth $S \sim \log(L)$ and volumetric growth $S \sim L$. Indeed, the spectrum is approximately equally spaced, with an entanglement spacing ΔL that decays with the system size as $1/\log(L)$ at the conformal point and as $1/L$ for rainbow system. We have also found that the EE is approximately proportional to the inverse of the entanglement spacing, in wide regions of the parameter space, which generalises previous known results for critical and massive systems.

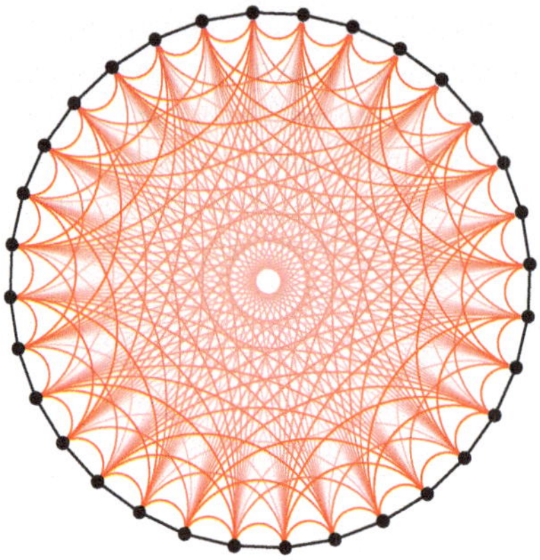

Fig. 19.2 Structure of the correlation matrix defined by the elements C_{ij} for a system of $N = 32$ with periodic boundary conditions. **Left:** for a homogeneous system. **Right:** for a rainbow system

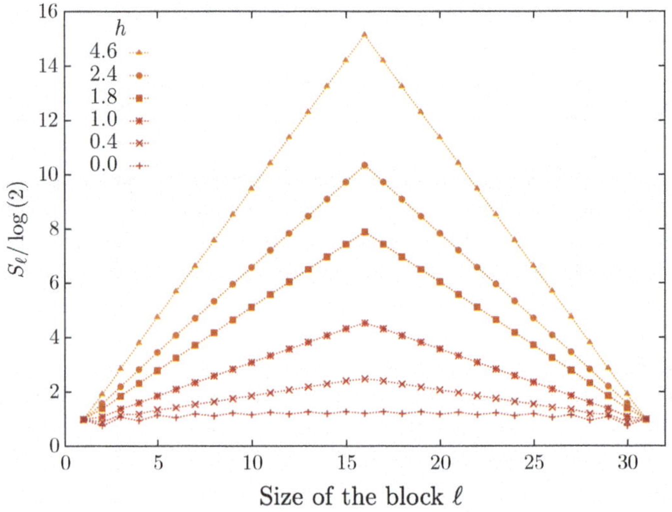

Fig. 19.3 Block entropy S_ℓ, for a system of size $L = 16$ (32 sites). Notice the *tent shape* appear for small inhomogeneities in the system, denoting volumetric growth of the entanglement entropy

In a second paper we have extended the analysis using field-theory methods [22]. We showed that the system can be described as a conformal deformation of the homogeneous case $h = 0$, given by the following transformation

$$\tilde{x} = \text{sign}(x)\frac{e^{h|x|} - 1}{h}\ , \tag{19.6}$$

which maps the interval $x \in [-L, L]$ to $\tilde{x} \in [-\tilde{L}, \tilde{L}]$, where $\tilde{L} = (e^{hL} - 1)/h$.

If we expand the local operators c_n into slow modes, $\psi_R(x)$ and $\psi_L(x)$ around the Fermi points $\pm k_F$

$$\frac{c_n}{\sqrt{a}} \simeq e^{ik_F x}\psi_L(x) + e^{-ik_F x}\psi_R(x) , \tag{19.7}$$

located at the position $x = an \in (-\mathscr{L}, \mathscr{L})$, where a is the lattice spacing and $\mathscr{L} = aL$ and, at half-filling $k_F = \pi/(2a)$ is the Fermi momentum. With this expansion we were able to obtain the Hamiltonian

$$H \simeq i \int_{-\tilde{L}}^{\tilde{L}} \mathrm{d}\tilde{x} \left[\tilde{\psi}_R^\dagger \partial_{\tilde{x}} \tilde{\psi}_R - \tilde{\psi}_L^\dagger \partial_{\tilde{x}} \tilde{\psi}_L \right] , \tag{19.8}$$

which represents the free fermion Hamiltonian for a chain of length $2\tilde{L}$ under transformation (19.6) with the the fermion fields given by

$$\tilde{\psi}_{R,L}(\tilde{x}) = \left(\frac{\mathrm{d}\tilde{x}}{\mathrm{d}x}\right)^{-1/2} \psi_{R,L}(x) . \tag{19.9}$$

Thus, using conformal invariance and substituting L by $\tilde{L}$, we were able to transform the EE for a block of one-half of the critical system [23]

$$S_L^{CFT} = c\log(L)/6 + c' , \tag{19.10}$$

into the EE of the rainbow system

$$S_L = \frac{c}{6}\log\left(\frac{e^{hL} - 1}{h}\right) + c' , \tag{19.11}$$

which is shown in Fig. 19.4 as a function of the half-chain size to check the theoretical prediction for different values of h.

We also showed that the corresponding conformal transformations suggests the definition of a temperature $T = 1/\beta = h/(2\pi)$. Based on the entanglement spectrum, which is given the eigenvalues of the entanglement Hamiltonian

$$H_E = \sum_p \varepsilon_p b_p^\dagger b_p + f_0 , \tag{19.12}$$

where b_p and $b_p^\dagger$ are fermion operators and the entanglement energies are given approximately by

$$\varepsilon_p \simeq \Delta_L p, \quad \text{for } p = \begin{cases} \pm\frac{1}{2}, \pm\frac{3}{2}, \ldots, \pm\frac{L-1}{2}, & L : \text{even} , \\ 0, \pm 1, \pm 2, \ldots, \pm(L-2), & L : \text{odd} , \end{cases} \tag{19.13}$$

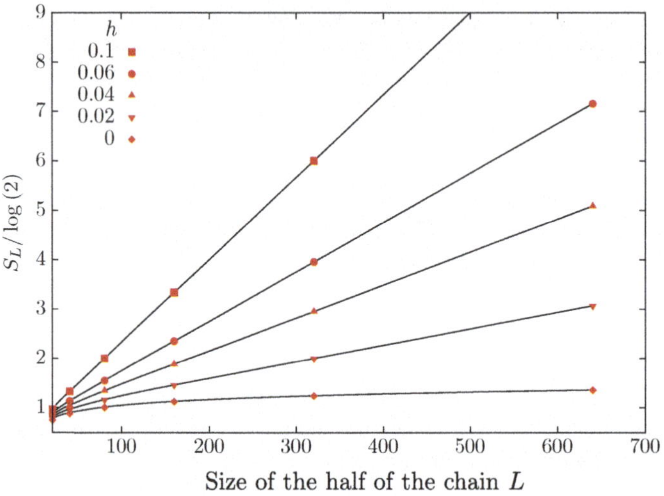

Fig. 19.4 Comparing numerical results of the half-chain entropy of the rainbow system for different values of h with the theoretical prediction

where the level spacing Δ_L is related to the half-chain entropy. Moreover, the single-body entanglement energies fulfil

$$\varepsilon_p \simeq \beta \varepsilon_p^{CFT} = \left(\frac{2\pi}{h}\right)\left(\frac{\pi}{L}p\right) = \frac{2\pi^2}{z}p \,. \tag{19.14}$$

Thus, the appearance of a volume law entropy is linked to the existence of an effective temperature for the GS of the rainbow model that was finally identified with a thermo-field state

$$|\psi\rangle = \sum_n e^{-\beta E_n/2} |n\rangle_L |n\rangle_R \,, \tag{19.15}$$

where $|n\rangle_R$ and $|n\rangle_L$ correspond to the homogeneous GS for the right and left parts of the chain with $h = 0$. This striking result points towards an unexpected connection with the theory of black holes and the emergence of space-time from entanglement. These intriguing connections were further explored within the framework of CFT [24]. Furthermore, we also showed how the deformation on the system accounts for the change in the dispersion relation, the single-particle wave functions in the vicinity of the Fermi point and the half-chain von Neumann and Rényi entropies. Finally, we show how to extend the rainbow Hamiltonian to more dimensions in a natural way and we to checked that the EE of the 2D analogue grows as the area of the block.

In a third paper [25] we applied methods of 2D quantum field theory in curved space-time to determine the entanglement structure of the rainbow phase. We showed that the rainbow system can be described by a massless Dirac fermion on a Riemannian manifold with constant negative curvature everywhere except at the centre, equivalent to a Poincaré metric with a strip removed. We used this identification to apply CFT for inhomogeneous 1D quantum systems [26] to provide accurate predictions for the smooth part of the n-order Rényi EE for blocks of different types such as the block at the chain's edge given by the bipartition $A = [-L, x]$, $B = [x, L]$

$$S_n^{CFT}(x) = \frac{n+1}{12n} \ln Y(x) , \tag{19.16}$$

where

$$Y(x) = 8e^{-h|x|} \frac{e^{hL}-1}{\pi L} \cos\left(\frac{\pi}{2} \frac{e^{h|x|}-1}{e^{hL}-1}\right) , \tag{19.17}$$

or for a block at an arbitrary position given by the bipartition $A = [x_1, x_2]$, $B = [-L, x_1] \cup [x_2, L]$

$$S_n^{CFT}(x_1, x_2) = \frac{n+1}{12n} \ln 4Y(x_1, x_2) + E_n , \tag{19.18}$$

where E_n is a non-universal constant, and with $e^{\sigma} = e^{-h|x|}$ we have

$$Y(x_1, x_2) = \frac{e^{\sigma(x_1)+\sigma(x_2)}\, 16\tilde{L}^2}{\pi^2 \cos\left(\frac{\pi(\tilde{x}_1+\tilde{x}_2)}{4\tilde{L}}\right)} \sin\left(\frac{\pi(\tilde{x}_1-\tilde{x}_2)}{4\tilde{L}}\right)^2 \cos\left(\frac{\pi\tilde{x}_1}{2\tilde{L}}\right) \cos\left(\frac{\pi\tilde{x}_2}{2\tilde{L}}\right) . \tag{19.19}$$

Furthermore, we also showed that for a physical temperature $T = J(x_0)$, i.e. a temperature in the range of the energies spanned by the values of the hopping amplitude for a point $x_0 \in [-L, L]$, the system splits into three regions: $x < -x_0$, $x \in [-x_0, x_0]$ and $x > x_0$ (cf. Fig. 19.5). The central region still behaves as if it were at $T = 0$ while the two extremes as if they were at $T \to \infty$. Thus, the entropy of a block is obtained by adding the contributions on each region

$$S(x) \sim \begin{cases} (L - |x|) \ln 2, & x \in (-L, -x_0) , \\ (L - x_0) \ln 2 + (x_0 - |x|)h/6, & x \in (-x_0, x_0) , \\ (L - 2x_0 + x) \ln 2, & x \in (x_0, L) . \end{cases} \tag{19.20}$$

Fig. 19.5 Illustration for the finite temperature behaviour of the rainbow chain. Let us find x_0 such that $T \approx J(x_0) = J(-x_0)$. Then, for $|x| > |x_0|$, $J(x) \ll T$ and the system behaves as if it were at infinite temperature. On the other hand, for $|x| < |x_0|$ the system behaves as if at zero temperature

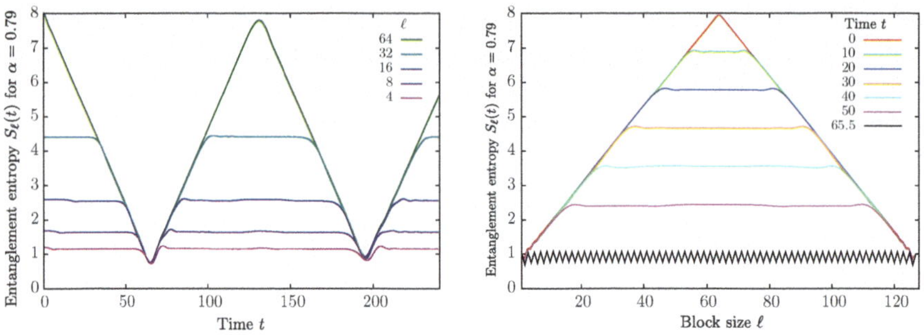

Fig. 19.6 Time evolution of the entanglement entropy. **Left:** for blocks of size ℓ starting from the leftmost site. Notice the time periodicity and the strict order: the entropy of a larger block is always strictly larger than the entropy of a smaller one. **Right:** for the tent shape (cf. Fig. 19.3) of the entropy

In a fourth paper [27], we study the time-evolution of the rainbow state and the dimer state, after a quench to a homogeneous Hamiltonian in 1D. The subsequent evolution of the EE presents very intriguing features. First, the EE of the half-chain of the rainbow decreases linearly with time and, after it reaches a minimal value, it increases again, eventually reaching (approximately) the initial state. Blocks of smaller sizes only decrease after a certain transient time, which can not be explained directly via a Lieb-Robinson bound [28], since the quench is global (cf. Fig. 19.6). There has been experimental proposals resulting in maximal entanglement with different technologies [29] and simulated quench with integrated photonic chips [30].

The dimer state, on the other hand presents an opposite behaviour: the EE grows linearly for all blocks, reaching a maximally entangled state which resembles the rainbow state. Afterwards, the EE decreases again, cyclically. We also focus on the correlation between pairs of sites which suggests the motion of quasiparticles.- Thus, we will attempt a theoretical explanation in terms of an extension of the quasiparticle picture by Calabrese and Cardy [31].

Acknowledgements We would like to thank Pasquale Sodano for organising the conferences hosted at the International Institute of Physics (Natal) and for his efforts to collect the contributions for the Proceedings. We would also like to thank Silvia N. Santalla for useful conversations. We acknowledge financial support from the grants FIS2015-69167-C2-1-P, QUITEMAD+ S2013/ICE-2801 and SEV-2016-0597.

References

1. R.J. Baxter, Phys. A. **106**(1–2), 18 (1981)
2. C. Gómez, M. Ruiz-Altaba, G. Sierra, *Quantum Groups in Two-Dimensional Physics*. Cambridge Lecture Notes in Physics (Cambridge University Press, Cambridge, 1996)
3. M.A. Nielsen, I.L. Chuang, *Quantum Computation and Quantum Information: 10th Anniversary Edition*, 10th edn. (Cambridge University Press, Cambridge, 2010)

4. M. Lewenstein, A. Sanpera, V. Ahufinger, B. Damski, A. Sen(De), U. Sen, Adv. Phys. **56**(2), 243 (2007)
5. N.W. Ashcroft, N.D. Mermin, *Solid State Physics* (Harcourt College Publishers, San Diego, 1976)
6. P. Hohenberg, W. Kohn, Phys. Rev. **136**, B864 (1964)
7. W. Kohn, L.J. Sham, Phys. Rev. **140**, A1133 (1965)
8. J. Bardeen, L.N. Cooper, J.R. Schrieffer, Phys. Rev. **106**, 162 (1957)
9. J. Bardeen, L.N. Cooper, J.R. Schrieffer, Phys. Rev. **108**, 1175 (1957)
10. R.B. Laughlin, Phys. Rev. B **23**, 5632 (1981)
11. L. Fu, C.L. Kane, Phys. Rev. B **76**, 045302 (2007)
12. I. Bloch, J. Dalibard, W. Zwerger, Rev. Mod. Phys. **80**, 885 (2008)
13. M. Lewenstein, A. Sanpera, V. Ahufinger, *Ultracold Atoms in Optical Lattices: Simulating Quantum Many-Body Systems* (Oxford University Press, Oxford, 2012)
14. E. Schmidt, Math. Ann. **63**(4), 433 (1907)
15. C. Shannon, Bell Syst. Tech. J. **27**(3), 379 (1948)
16. G. Vitagliano, A. Riera, J.I. Latorre, New J. Phys. **12**(11), 113049 (2010)
17. K. Okunishi, T. Nishino, Phys. Rev. B **82**, 144409 (2010)
18. H. Ueda, H. Nakano, K. Kusakabe, T. Nishino, Progr. Theor. Phys. **124**(3), 389 (2010)
19. K.G. Wilson, Rev. Mod. Phys. **47**, 773 (1975)
20. G. Ramírez, J. Rodríguez-Laguna, G. Sierra, J. Stat. Mech. **2014**(10), P10004 (2014)
21. B. Jin, V. Korepin, J. Stat. Phys. **116**, 79 (2004)
22. G. Ramírez, J. Rodríguez-Laguna, G. Sierra, J. Stat. Mech. **2015**(6), P06002 (2015)
23. P. Calabrese, J. Cardy, J. Stat. Mech. **2004**(06), P06002 (2004)
24. E. Tonni, J. Rodríguez-Laguna, G. Sierra, J. Stat. Mech. **2018**(4), 043105 (2018)
25. J. Rodríguez-Laguna, J. Dubail, G. Ramírez, P. Calabrese, G. Sierra, J. Stat. Mech. **50**(16), 164001 (2017)
26. J. Dubail, J.M. Stéphan, J. Viti, P. Calabrese, SciPost Phys. **2**, 002 (2017)
27. G. Ramírez, J. Rodríguez-Laguna, G. Sierra, Quenched dynamics of valence bond states. Work in progress
28. E.H. Lieb, D.W. Robinson, Commun. Math. Phys. **28**(3), 251 (1972)
29. B. Alkurtass, L. Banchi, S. Bose, Phys. Rev. A **90**, 042304 (2014)
30. I. Pitsios, L. Banchi, A.S. Rab, M. Bentivegna, D. Caprara, A. Crespi, N. Spagnolo, S. Bose, P. Mataloni, R. Osellame, F. Sciarrino, Nat. Commun. **8**, 1569 (2017)
31. P. Calabrese, J. Cardy, J. Stat. Mech. **2005**(04), P04010 (2005)

Index

A. Ferraz et al. (eds.), *Strongly Coupled Field Theories for Condensed Matter and Quantum Information Theory*, Springer Proceedings in Physics 239,
https://doi.org/10.1007/978-3-030-35473-2

The manufacturer's authorised representative in the EU is Springer Nature Customer Service Centre GmbH, Europaplatz 3, 69115 Heidelberg, Germany. If you have any concerns regarding our products, please contact ProductSafety@springernature.com

Printed and bound by CPI Group (UK) Ltd, Croydon, CR0 4YY
30/09/2026
02225591-0017